# Lecture Notes in Mathematics

## Volume 2381

**Editors-in-Chief**

Jean-Michel Morel, City University of Hong Kong, Kowloon Tong, China

Bernard Teissier, IMJ-PRG, Paris, France

**Series Editors**

Karin Baur, University of Leeds, Leeds, UK

Michel Brion, UGA, Grenoble, France

Rupert Frank, LMU, Munich, Germany

Annette Huber, Albert Ludwig University, Freiburg, Germany

Davar Khoshnevisan, The University of Utah, Salt Lake City, UT, USA

Ioannis Kontoyiannis, University of Cambridge, Cambridge, UK

Angela Kunoth, University of Cologne, Cologne, Germany

Ariane Mézard, IMJ-PRG, Paris, France

Mark Podolskij, University of Luxembourg, Esch-sur-Alzette, Luxembourg

Mark Policott, Mathematics Institute, University of Warwick, Coventry, UK

László Székelyhidi, MPI for Mathematics in the Sciences, Leipzig, Germany

Gabriele Vezzosi, UniFI, Florence, Italy

Anna Wienhard, MPI for Mathematics in the Sciences, Leipzig, Germany

This series reports on new developments in all areas of mathematics and their applications - quickly, informally and at a high level. Mathematical texts analysing new developments in modelling and numerical simulation are welcome. The type of material considered for publication includes:

1. Research monographs
2. Lectures on a new field or presentations of a new angle in a classical field
3. Summer schools and intensive courses on topics of current research.

Texts which are out of print but still in demand may also be considered if they fall within these categories. The timeliness of a manuscript is sometimes more important than its form, which may be preliminary or tentative. Please visit the LNM Editorial Policy (https://drive.google.com/file/d/1MOg4TbwOSokRnFJ3ZR3ciEeK s9hOnNX_/view?usp=sharing)

*Titles from this series are indexed by Scopus, Web of Science, Mathematical Reviews, and zbMATH.*

Yves Martinez-Maure

# Hedgehog Theory

 Springer

Yves Martinez-Maure
Institut de Mathématiques de Jussieu - Paris
Rive Gauche
Sorbonne University
Paris, France

ISSN 0075-8434         ISSN 1617-9692   (electronic)
Lecture Notes in Mathematics
ISBN 978-3-032-02807-5         ISBN 978-3-032-04298-9   (eBook)
https://doi.org/10.1007/978-3-032-04298-9



This Springer imprint is published by the registered company Springer Nature Switzerland AG
The registered company address is: Gewerbestrasse 11, 6330 Cham, Switzerland

# Declarations

**Competing Interests** The author has no competing interests to declare that are relevant to the content of this manuscript.

# Contents

# Chapter 1
# Introduction

Although known since Antiquity, it was not until the twentieth century that the elementary and natural notion of a convex body began to reveal the full extent and richness of its applications in many mathematical fields as varied as number theory, differential or integral geometry, discrete and combinatorial geometry, optimization, functional analysis, probability, or stochastic geometry. The Brunn-Minkowski theory, initiated in the seminal works of H. Brunn and H. Minkowski on the turn of the twentieth century, became the classical heart of convex geometry. It can be regarded in the Euclidean vector space $\mathbb{R}^{n+1}$ as the fruit of the union of only two elementary notions: Minkowski addition and volume [167]. In particular, the mixed volumes, which satisfy the famous Alexandrov-Fenchel inequality which subsumes many classical geometric inequalities as special cases, are born from this union. The reader is referred to Rolf Schneider's book [167] for a comprehensive introduction to convex geometry, and more particularly to its Chap. 5 for the basic properties of the mixed volumes.

## 1.1 Aims of the Monograph

As we will see in Chap. 2, hedgehogs are the geometrical objects that describe the Minkowski differences (i.e., formal differences) of arbitrary convex bodies in $\mathbb{R}^{n+1}$. This monograph aims to summarize the core of hedgehog theory and its main applications to other mathematical fields.

**The first aim** of the present monograph is to give a comprehensive and self-contained introduction to hedgehog theory which emerges naturally when one tries to associate a geometric object with any formal difference of convex bodies (in other words, when we attempt to take into account the inverse operation of the Minkowski addition by considering a wider general class of geometric objects than that of convex bodies alone).

© The Author(s), under exclusive license to Springer Nature Switzerland AG 2026
Y. Martinez-Maure, *Hedgehog Theory*, Lecture Notes in Mathematics 2381,
https://doi.org/10.1007/978-3-032-04298-9_1

**Fig. 1.1** $C^2$-hedgehogs as differences of $C_+^2$ convex bodies

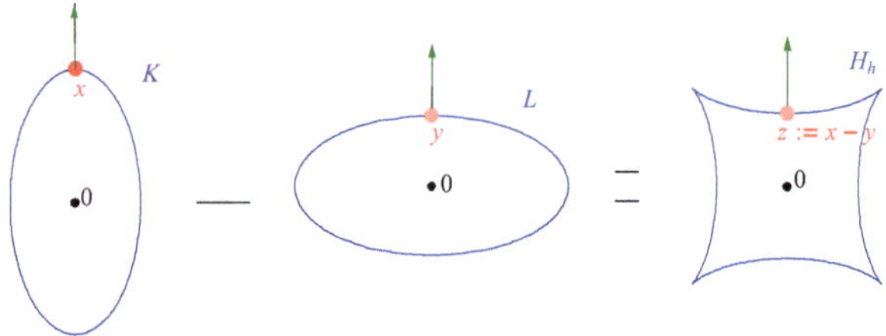


Classical real hedgehogs can be regarded as the geometrical realizations of formal differences of arbitrary convex bodies in $\mathbb{R}^{n+1}$. For instance, hedgehogs with a $C^2$-support function on the unit sphere $\mathbb{S}^n$ of $\mathbb{R}^{n+1}$ can be constructed geometrically in $\mathbb{R}^{n+1}$ as 'Minkowski differences' of convex bodies of class $C_+^2$ (i.e., the boundaries of which are $C^2$-hypersurfaces with positive Gaussian curvature): as shown in Fig. 1.1, we can construct the difference of two such convex bodies $K$ and $L$ by subtracting, for every $u \in \mathbb{S}^n$, the points $x$ and $y$ of their respective boundaries $\partial K$ and $\partial L$ that correspond to the outer unit normal $u$. When this is done for every $u \in \mathbb{S}^n$, we obtain a hedgehog whose $C^2$-support function $h$ is the difference $k - l$ of the respective support functions, $k$ and $l$, of the two convex bodies.

The idea of considering Minkowski differences of convex bodies may be traced back to papers by A.D. Alexandrov and H. Geppert [53] in the 1930s. Many classical notions for convex bodies extend to hedgehogs and quite a number of classical results find their counterparts. Of course, a few adaptations can be necessary. In particular, volumes have to be replaced by their algebraic versions.

The attention of specialists in convex geometry is drawn to the fact that this monograph never consider the so-called Minkowski difference $K \div L$ of $K$ and $L$ as defined in [167, p. 146]. Moreover, in hedgehog theory, the notation $K - L$ denotes a formal difference of convex bodies or its geometrical realization in the form of a hedgehog (such as the hedgehog $\mathcal{H}_h$ on the right side of Fig. 1.1 is the geometrical realization of the formal difference $K - L$ of convex bodies bordered by ellipses shown on the left): we then say that $\mathcal{H}_h$ is the Minkowski difference $K - L$. In particular, $K - L$ will never denote the set $\{x - y \,|\, x \in K, y \in L\}$ as is often the case in convex geometry.

**The second aim** of this monograph is to show the diversity and extent of applications of hedgehogs (together with their generalizations) in various fields. The notion of a hedgehog has, in particular, proved useful in the study of convex bodies and to geometrize certain analytical problems by considering functions as support functions. The usefulness of hedgehogs in the study of convex bodies is mainly due to the fact that they may offer the possibility to provide appropriate decompositions of the convex bodies under consideration into sums of hedgehogs. One of the first achievements of the theory is for instance the construction of counterexamples to an old conjectured characterization of the 2-sphere [95, 102, 141], which has important consequences in terms of Monge-Ampère PDE's on the 2-sphere. The conjecture raised in the mid 1930s by A.D. Alexandrov was that if $S$ in $\mathbb{R}^3$ is a closed convex surface of class $C_+^2$, whose principal curvatures $k_1$ and $k_2$ satisfy the following inequality

$$(k_1 - c)(k_2 - c) \leq 0,$$

with some constant $c$, then $S$ must be a sphere of radius $r = 1/c$. In the case $S$ is real analytic, this conjecture was proved by Alexandrov himself [1, 2], an independent proof was given by Münzner [131] who also proved the conjecture in the case where $S$ is a surface of revolution [130]. Koutroufiotis strengthened this last result, proving that it suffices to assume that $S$ has an enveloping circular cylinder [78]. However, in the general case the conjecture remained open for almost 70 years. From the hedgehog point of view, the natural approach was of course to split $S$ into this sum $S\left(0_{\mathbb{R}^3}; r\right) + \left(S - S\left(0_{\mathbb{R}^3}; r\right)\right)$, and to study only the hedgehog term $\left(S - S\left(0_{\mathbb{R}^3}; r\right)\right)$ by adapting classical techniques for convex bodies to hedgehogs, the question then being whether or not this hedgehog term is necessarily reduced to a single point. This approach made it possible to obtain a series of counterexamples to Alexandrov's conjecture. It also permitted us to prove the conjecture for convex surfaces of constant with, and to give a new proof for analytic surfaces [95]. It is worth noticing that the question is still under scrutiny for other classes of differentiability given its importance in PDE's theory [52, 61].

## 1.2  Outline of the Remaining Sections

The set $\mathcal{K}^{n+1}$ of convex bodies of $\mathbb{R}^{n+1}$, equipped with Minkowski addition and multiplication by nonnegative real numbers, forms a commutative semigroup, having the cancellation property, with scalar operator. Of course, it does not constitute a vector space since there is no subtraction in $\mathcal{K}^{n+1}$. Now formal differences of convex bodies of $\mathbb{R}^{n+1}$ form a vector space $\mathcal{H}^{n+1}$ in which $\mathcal{K}^{n+1}$ is a cone that spans the entire space. It is thus natural to consider the multilinear extension of the mixed volume $v : \left(\mathcal{K}^{n+1}\right)^{n+1} \to \mathbb{R}$ to a symmetric $(n+1)-$linear form on $\mathcal{H}^{n+1}$. We still denote this extension by $v$. We enter in 'hedgehog theory' when we seek to associate a geometric realization with each element of $\mathcal{H}^{n+1}$. We will see in Chap. 2 that there are different ways of proceeding depending on the class of convex bodies which we choose to work with. We may regard hedgehogs as envelopes parametrized by their Gauss map (Sect. 2.2) if the support functions are assumed to be $C^2$ (we can also define the notion of a hedgehog with a $C^1$ support function as an envelope but such a hedgehog does not necessarily represent a difference of two convex bodies and can be a highly singular object: Sect. 2.5); we can also define hedgehogs as Legendrian fronts (Sect. 2.6) if the support functions are assumed to be $C^\infty$; another approach is to make use of Euler Calculus (Sect. 2.4) if we consider polytopes or convex bodies with an analytic support function; finally, in the most general case, when the support functions are differences of support functions of arbitrary convex bodies, we can define hedgehogs proceeding by induction with respect to the dimension, replacing support sets by 'support hedgehogs' (Sect. 2.3).

We will of course give first examples, particular cases (such as 'projective hedgehogs') properties and tools regarding these different variants of the notion of a hedgehog. In particular, we will introduce different types of indexes of a point with respect to a hedgehog, and present first applications, some of which are related to orthogonal projection techniques, to the study of hedgehogs (Sects. 2.4, 2.8 and 2.9). The Kronecker index will, in particular, be used to introduce the algebraic $(n + 1)$-dimensional volume of hedgehogs in $\mathbb{R}^{n+1}$. We will also say a few words on projective and polarity dualities applied to hedgehogs (Sect. 2.7). We will use our presentation of $C^{\infty}$-hedgehogs as Legendrian fronts to describe their generic singularities, and we will present a first open problem, raised by R. Langevin, G. Levitt and H. Rosenberg (Problem 2.6.1): Does there exist a generic projective hedgehog without any swallowtail in $\mathbb{R}^3$? Partial results will be given in Sect. 10.2.

As already mentioned in the introduction, the classical Brunn-Minkowski theory is based on the Minkowski addition of convex bodies, combined with the notion of volume. Having already glimpsed in the first two sections some ideas for extending this theory to hedgehogs, we will undertake a first more in-depth and more systematic study of this extension to $C^2$-hedgehogs in Chap. 3. After briefly recalling and completing the necessary basic tools and their geometric interpretations, we will present a partial extension of the Alexandrov-Fenchel inequality to hedgehogs. We will then consider the extension to hedgehogs of particular cases of the Alexandrov-Fenchel inequality (isoperimetric inequalities, quadratic Minkowskian inequalities, Brunn-Minkowski type inequalities, etc), as well as some of their geometric consequences. In some cases, the classical inequalities for convex bodies will extend without any modification to hedgehogs by simply replacing the geometric quantities involved by their algebraic versions (this will be for instance the case for the isoperimetric inequality in the plane). But, of course, in most cases, an additional condition will be necessary. We will continue Chap. 3 by giving a stability estimate for the Alexandrov-Fenchel inequality. To end Chap. 3, we will mention an application of hedgehogs to the study of the Blaschke diagram. Here we have to recall that a part of the boundary of the Blaschke diagram must correspond to an unknown sharp inequality of the form $V \geq f(S, M)$, where $V$, $S$ and $M$ respectively denote the volume, surface area and integral of mean curvature of a convex body in $\mathbb{R}^3$.

In Chap. 4, we will consider a series of special convex bodies, hedgehogs, and multihedgehogs, which are also called $N$-hedgehogs, ($N \in \mathbb{N}^*$): an envelope of a family of cooriented planes of $\mathbb{R}^{n+1}$ will be called an $N$-**hedgehog** if, for an open dense set of $u \in \mathbb{S}^n$, it has exactly $N$ cooriented support planes with normal vector $u$. Thus, ordinary $C^2$-hedgehogs are merely 1-hedgehogs. We will extend to hedgehogs a series of classical notions for convex bodies. For instance, we will start by extending the notion of width to hedgehogs. As an application of our study, we will give an example of a noncircular algebraic curve of constant width whose equation is relatively simple, which answers a problem raised by S. Rabinowitz (Sect. 4.1.2). In passing, we will study various concepts related to convex bodies.

In particular, we will study the relationship between planar projective hedgehogs (which are the planar hedgehogs of constant width 0) and Zindler curves (which are the planar closed curves of which all chords that divide the curve perimeter - or area - in a half, have the same length) in Sect. 4.2.2. We will then rely on a notion of symplectic area to introduce and study Zindler-type surfaces in $\mathbb{R}^4$. Section 4.5 will aim to motivate the development of a Brunn-Minkowski theory for minimal surfaces by continuing the pioneering works by R. Langevin, G. Levitt, H. Rosenberg and E. Toubiana [81, 82, 159].

This Chap. 4 will also be an opportunity to discover a first series of applications of hedgehog theory to analysis. In Sect. 4.3, we will consider the cosine transform, which associates to any continuous function $f : \mathbb{S}^n \to \mathbb{R}$ the map $T_f : \mathbb{R}^{n+1} \to \mathbb{R}$ defined by

$$T_f(x) := \int_{\mathbb{S}^n} |\langle x, v \rangle|\, f(v)\, d\sigma(v),$$

where $\langle ., . \rangle$ is the standard inner product and $\sigma$ the spherical Lebesgue measure. We will prove that the cosine transform, which often appears in convex geometry, is a bounded linear operator from $C(\mathbb{S}^n; \mathbb{R})$ to $C^2(\mathbb{S}^n; \mathbb{R})$. It follows that the boundaries of zonoids (resp. generalized zonoids) whose generating measure have a continuous density with respect to $\sigma$ can be considered as $C^2$-hedgehogs. We will study such hedgehogs. Recall that zonotopes are the Minkowski sums of line segments, and that zonoids are (necessarily centrally symmetric) convex bodies that are the limit, in the sense of the Hausdorff metric, of a sequence of zonotopes. Zonoids play an important role in various areas such as the theory of vector measures, Banach space theory or stochastic geometry. We will obtain a local property of zonoids whose generating measure have a continuous density with respect to $\sigma$. We then define projection hedgehogs (resp. mixed projection hedgehogs) and interpret their support functions in terms of $n$-dimensional volume (resp. mixed volume). Finally, this study will lead us to consider the extension of the Minkowski problem (in differential geometry, the one of the existence, uniqueness and regularity of a closed convex hypersurface with preassigned curvature function) to hedgehogs. The classical Minkowski problem played an important role in the development of the theory of elliptic Monge-Ampère equations. The study of its extension to hedgehogs will be the subject of Chap. 5. In Sect. 4.4, we will study the existence of a nontrivial $C^2$-hedgehog in $\mathbb{R}^3$ that is hyperbolic (i.e., with an everywhere nonpositive curvature function), in order to determine the validity of the characterization of the 2-sphere conjectured by A.D. Alexandrov. This question amounts to studying a partial differential inequation. We will prove this Alexandrov conjecture in some particular cases, such as the case when the surface is assumed to be of constant width, and give a counterexample in the general case. In passing, we will consider the discrete version of hyperbolic hedgehogs. After a brief presentation of hedgehog polytopes (also called polyhedral hedgehogs) in $\mathbb{R}^3$, we will introduce two notions of hyperbolicity (weak and strong hyperbolicity) for hedgehog polytopes of $\mathbb{R}^3$ and give examples. Our example of a strongly hyperbolic polytope is obtained by a discretization of our counterexample to Alexandrov's conjecture. In Sect. 4.6, we will give a geometric proof of the Sturm-Hurwitz theorem in the framework of planar mutihedgehogs. We will take the opportunity to present a series of geometric consequences and inequalities. We will end Chap. 4 by a detailed study of planar general hedgehogs (i.e., Minkowski differences of arbitrary convex bodies of $\mathbb{R}^2$). Our way of introducing general hedgehogs (proceeding by induction on $n$ and replacing support sets by 'support hedgehogs') makes clear that a perfect understanding of planar hedgehogs is a prerequisite to a study of general hedgehogs of $\mathbb{R}^{n+1}$. In particular, we will: $(i)$ study their length measures and solve the extension of the Christoffel-Minkowski Problem to plane hedgehogs; $(ii)$ characterize support functions of plane convex bodies among support functions of plane hedgehogs and support functions of plane hedgehogs among continuous functions; $(iii)$ study the mixed area of hedgehogs in $\mathbb{R}^2$ and give an extension of the classical Minkowski inequality (and thus of the isoperimetric inequality) to hedgehogs.

Chapter 5 is entirely devoted to the extension of the Minkowski problem to hedgehogs. We already encounter it in Sects. 4.3 and 4.4 (the existence of hyperbolic hedgehogs is naturally a subproblem of it). The classical Minkowski problem is a fundamental question in the Brunn-Minkowski theory. It asks for necessary and sufficient conditions on a nonnegative Borel measure on the unit sphere $\mathbb{S}^n$ of $\mathbb{R}^{n+1}$ to be the surface area measure of some convex body $K$ in $\mathbb{R}^{n+1}$, unique up to translation. When restricting to the class of convex bodies of $\mathbb{R}^{n+1}$ whose surface area measures have a density with respect to spherical Lebesgue measure on $\mathbb{S}^n$, the classical Minkowski problem can be formulated as that of the existence, uniqueness and regularity of a closed convex hypersurface with preassigned curvature function. As already mentioned, this problem, which played an important role in the development of the theory of elliptic Monge-Ampère equations, has a natural extension to hedgehogs. But for non convex hedgehogs, the problem becomes much more difficult, even for $n = 2$ and for $C^\infty$-hedgehogs, since it essentially boils down to the question of solutions of Monge-Ampère equations of mixed type on the unit sphere $\mathbb{S}^n$. In Sect. 5.1, we will mainly formulate the uniqueness question and give first partial results. In Sect. 5.2, we will consider Gauss rigidity and Gauss infinitesimal rigidity for hedgehogs of $\mathbb{R}^3$ (regarded as Minkowski differences of closed convex surfaces of $\mathbb{R}^3$ with positive Gaussian curvature). As noticed by I. Izmestiev [72, 73], Gauss rigidity (Gauss infinitesimal rigidity) can be interpreted as uniqueness (resp. « infinitesimal » uniqueness) in the Minkowski problem, that is in the problem of prescribing the $n$th surface area measure of a polytope $P$ of $\mathbb{R}^{n+1}$ on the unit sphere $\mathbb{S}^n$ (resp. the Gaussian curvature of smooth strictly convex closed hypersurface of $\mathbb{R}^{n+1}$ as a function of the outer unit normal). The uniqueness part of the Minkowski problem extended to hedgehogs will already have been studied in the previous subsection. In particular, we will have already seen different ways of constructing pairs of non-congruent hedgehogs that share the same curvature function (i.e., inverse of the Gaussian curvature). If we consider a 1-parameter family of $C^2$-hedgehogs $\left(\mathcal{H}_{h_t}\right)_{t\in[0,1]}$ all having the same curvature function, we do not know whether these hedgehogs are congruent in $\mathbb{R}^3$. However, we will prove a theorem of volume preservation under curvature-preserving deformations: Under an appropriate differentiability condition of the family with respect to the parameter, we will prove that all the hedgehogs of the considered family have the same algebraic volume!

Like convex bodies, hedgehogs are completely determined by (and can be identified with) their support functions. Adopting a projective viewpoint, we will prove in Chap. 6 that any holomorphic function $h : \mathbb{C}^n \to \mathbb{C}$ can be regarded as the 'support function' of a 'complex hedgehog' $\mathcal{H}_h$, which is defined by a holomorphic parametrization $x_h : \mathbb{C}^n \to \mathbb{C}^{n+1}$ in the complex Euclidean space $\mathbb{C}^{n+1}$. In the same vein, we will introduce the notion of evolute of such a hedgehog $\mathcal{H}_h$ in $\mathbb{C}^2$, and a natural (but apparently hitherto unknown) notion of complex curvature, which will allow us to interpret this evolute as the locus of the centers of complex curvature. We notice that the complex linear space of holomorphic functions defined up to a similitude on the unit disc $\mathbb{D}$ of $\mathbb{C}$ can be endowed with a scalar product which can be interpreted as a mixed symplectic area, and we give a sharp estimation of the (symplectic) area of $x_h (\mathbb{D})$ using the energy of the loop $x_h : \mathbb{S}^1 = \mathbb{R}/2\pi\mathbb{Z} \to \mathbb{C}^2$, $\theta \mapsto x_h \left(e^{i\theta}\right)$, in the case where $h : \mathbb{D} \to \mathbb{C}$ is the sum of a power series $\sum h_n z^n$ with radius of convergence $R > 1$. Our hope is to spark further research giving elements of a 'theory of mixed volumes for complex hedgehogs' (replacing Euclidean volumes by symplectic ones). We will next return to real hedgehogs, but in $\mathbb{R}^{2n}$ endowed with a linear complex structure $J$. We will introduce and study the notion of evolute of a hedgehog with a smooth support function in $\left(\mathbb{R}^{2n}, J\right)$. We will particularly focus on $\mathbb{R}^4$ endowed with a linear Kähler structure determined by the datum of a pure unit quaternion. In parallel, we will study the symplectic area of the images of the oriented Hopf circles under hedgehog parametrizations and introduce a quaternionic curvature function for such an image. Finally, we will consider hedgehogs in $\mathbb{R}^{4n}$ regarded as a hyperkähler vector space.

Of course, the classical hedgehog theory is not restricted to Euclidean spaces. Chapter 7 will be devoted to a short introduction and study of hedgehog theory in non-Euclidean spaces. In [44], F. Fillastre introduced and studied '$\Gamma$-convex bodies' (or, 'Fuchsian convex bodies'), which are the closed convex sets of the Lorentz-Minkowski space $\mathbb{L}^{n+1}$ that are globally invariant under the action of some Fuchsian group $\Gamma$. In this paper, F. Fillastre gave a 'reversed Alexandrov–Fenchel inequality' and thus a 'reversed Brunn-Minkowski inequality'. This work permits to introduce 'Fuchsian hedgehogs' whose 'support functions' are differences of support functions of two $\Gamma$-convex bodies (see the remark on page 314 in [44]). In Sect. 7.2, we will give a detailed study of plane Lorentzian and Fuchsian hedgehogs, including a series of Fuchsian analogues of classical geometrical inequalities (which are also reversed as compared to classical ones). For an application to marginally trapped surfaces, a short introduction to hedgehogs of $\mathbb{L}^3$ will be given later in Sect. 8.2.1. This brief introduction of Lorentzian hedgehogs of $\mathbb{L}^3$ can be easily extended to higher dimensions.

On another note, we will see that, like convexity, the notion of a hedgehog is affine. In Sect. 7.3, we will define the notion of a hedgehog of $\mathbb{R}^{n+1}$ regarded as an affine space over itself. Extending the Euclidean or affine space $\mathbb{R}^{n+1}$ by adding points at infinity (which we regard as corresponding to families of parallel lines, and which together make up a hyperplane at infinity), we will see that we can define a hedgehog of the real projective space $\mathbb{P}^{n+1} := \mathbb{R}P^{n+1}$ as a hedgehog of $\mathbb{R}^{n+1}$ regarded as the complement of any projective hyperplane of $\mathbb{P}^{n+1}$, and thus a hedgehog of $\mathbb{R}^{n+1}$ as a hedgehog of $\mathbb{P}^{n+1}$. Thus any hedgehog of $\mathbb{P}^{n+1} = \mathbb{S}^{n+1}/\{-Id, Id\}$ is contained in the complement of a projective hyperplane of $\mathbb{P}^{n+1}$, and can therefore be regarded as a hedgehog of $\mathbb{S}^{n+1}$ that is contained in an open hemisphere, say the open hemisphere with center $p \in \mathbb{S}^{n+1}$ which we will denote by $\mathbb{S}_p^{n+1}$. Using the gnomonic projection from $\mathbb{S}_p^{n+1}$ onto the tangent hyperplane $p + T_p\mathbb{S}^{n+1}$ to $\mathbb{S}^{n+1}$ at $p$, we then retrieve hedgehogs of $\mathbb{R}^{n+1}$ (which we identify with $p + T_p\mathbb{S}^{n+1}$).

Similarly, regarding the hyperbolic space $\mathbb{H}^{n+1}$ as the upper sheet of the hyperboloid with equation $\langle x, x \rangle_L = -1$ in $\mathbb{R}^{n+2}$ endowed with the Lorentzian inner product given by

$$\langle x, y \rangle_L = \sum_{i=1}^{n+1} x_i y_i - x_{n+2} y_{n+2},$$

for $x = (x_1, \ldots, x_{n+2})$ and $y = (y_1, \ldots, y_{n+2})$, that is,

$$\mathbb{H}^{n+1} = \left\{ x \in \mathbb{R}^{n+2} \,|\, \langle x, x \rangle_L = -1, x_{n+2} > 0 \right\},$$

we can make a gnomonic projection (which preserves geodesics) from $\mathbb{H}^{n+1}$ onto the interior $\mathbb{B}^{n+1}$ of the (Euclidean) unit ball of $\mathbb{R}^{n+1}$ (identified with the affine hyperplane $\mathbb{R}^{n+1} \times \{1\}$ of $\mathbb{R}^{n+2}$). This gnomonic projection $g : \mathbb{H}^{n+1} \to \mathbb{B}^{n+1}$ sends $m \in \mathbb{H}^{n+1}$ to the intersection point of the linear line $\mathbb{R}m$ with $\mathbb{R}^{n+1} \simeq \mathbb{R}^{n+1} \times \{1\}$. Considering hedgehogs of $\mathbb{R}^{n+1}$ that are included in $\mathbb{B}^{n+1} \subset \mathbb{R}^{n+1} \times \{1\}$, and taking their images under the radial projection

$$\rho : \mathbb{B}^{n+1} \to \mathbb{H}^{n+1} \subset \mathbb{R}^{n+2}$$
$$x \mapsto x/\sqrt{|\langle x, x \rangle_L|} \quad ,$$

which is the inverse $g^{-1}$ of the gnomonic projection, we can then introduce hedgehogs of $\mathbb{H}^{n+1}$. These hedgehog hypersurfaces, which are envelopes of smooth families of cooriented (totally geodesic) hyperplanes of $\mathbb{H}^{n+1}$, will be called 'g-hedgehogs' of $\mathbb{H}^{n+1}$, where the letter g stands for indicating that these hypersurfaces are 'geodesically hedgehog hypersurfaces'. This change of names ('g-hedgehogs' instead of 'hedgehogs') aims to differentiate these g-hedgehogs from another class of hedgehogs introduced in Sect. 7.4, namely 'h-hedgehogs' of $\mathbb{H}^{n+1}$. In our definition of h-hedgehogs, horospheres will play in $\mathbb{H}^{n+1}$ the role assigned to cooriented hyperplanes in $\mathbb{R}^{n+1}$, and the corresponding ideal points the one of the unit normal vectors. Indeed, as we will recall, the best analogue to Euclidean hyperplanes in $\mathbb{H}^{n+1}$ are not actually the totally geodesic hyperplanes of $\mathbb{H}^{n+1}$, but the horospheres. We will notice that these two notions of hedgehogs in $\mathbb{H}^{n+1}$ (g-hedgehogs and h-hedgehogs) correspond to the two natural notions of convexity in hyperbolic space: geodesical convexity and horospherical convexity, which is stronger. We will define the signed h-width of a h-hedgehog $\mathcal{H}_h$ of $\mathbb{H}^{n+1}$ in direction of an arbitrary ideal point $u \in \mathbb{S}_\infty^n$ (here, $\mathbb{S}_\infty^n$ denotes the ideal boundary sphere at infinity of $\mathbb{H}^{n+1}$), and then give a simple condition for a h-hedgehog of $\mathbb{H}^{n+1}$ to be of constant h-width.

The focus of Chap. 8 is the study of marginally trapped surfaces. We recall that a closed embedded spacelike 2-surface of a 4-dimensional spacetime is said to be trapped if its mean curvature vector is everywhere timelike. Trapped surfaces were introduced by R. Penrose [146] to study singularities of spacetimes. They appeared in a natural way earlier in the work of Blaschke, in the context of conformal and Laguerre geometry [19]. These surfaces play an extremely important role in general relativity where they are of central importance in the study of black holes, those regions of spacetime where everything is trapped, and nothing can escape, even light. The limiting case of marginally trapped surfaces (i.e., surfaces whose mean curvature vector is everywhere lightlike) plays the role of apparent horizons of black holes. Mathematically, marginally trapped surfaces are regarded as spacetime analogues of minimal surfaces in Riemannian geometry. Even though they received considerable attention both from mathematicians and physicists, these surfaces are still not very well understood. In Chap. 8, we will try to argue and to show through fundamental examples that (a very huge class of) singular marginally trapped surfaces arise naturally from a 'lightlike co-contact structure', exactly in the same way as Legendrian fronts arise from a contact one (by projection of a Legendrian submanifold to the base of a Legendrian fibration), and that there is an adjunction relationship between both notions. We especially focus our interest on marginally trapped hedgehogs and study their relationships with Laguerre geometry and Brunn-Minkowski theory.

It is conjectured since long that any convex body in $n$-dimensional Euclidean space $\mathbb{R}^n$ has an interior point lying on normals through $2n$ distinct boundary points. This concurrent normals conjecture has been proved for $n = 2$ and $n = 3$ by E. Heil in [64–66]. J. Pardon put forward a proof for $n = 4$ under a smoothness assumption on the boundary [144]. For $n \geq 5$, it was only known that any convex body in $\mathbb{R}^n$ has an interior point lying on normals through six distinct boundary points. However Zamfirescu has shown that, in the sense of Baire category based on the Hausdorff distance between convex sets, most interior points of most convex bodies lie on infinitely many normals [196]. The main aim of Chap. 9 is to present a new approach and contributions to the study of this conjecture by making use of hedgehog theory.

In most of the papers on concurrent normals to a convex body $K$ with a smooth boundary $\partial K$ in $\mathbb{R}^{n+1}$, the focal of $\partial K$ is regarded as the complement of the set of points $x \in \mathbb{R}^{n+1}$ such that the square of the distance function from $x$ induces a

Morse function on $\partial K$:

$$d_x : \partial K \to \mathbb{R}$$
$$y \mapsto \|x - y\|^2,$$

where $\|.\| : \mathbb{R}^{n+1} \to \mathbb{R}_+$ is the Euclidean norm. In Chap. 9, we will adopt another point of view. For any $x \in \mathbb{R}^{n+1}$, we will consider the support function of $\partial K$ with respect to $x$, that is $h_x : \mathbb{S}^n \to \mathbb{R}$, $u \mapsto h(u) - \langle x, u \rangle$, where $h : \mathbb{S}^n \to \mathbb{R}$ is the support function of $\partial K$, and we will regard the focal $\mathcal{F}_{\nabla h}$ of $\partial K$ as the complement of the set of points $x \in \mathbb{R}^{n+1}$ such that $h_x : \mathbb{S}^n \to \mathbb{R}$ is a Morse function. In other words, we will regard the focal of $\partial K$ as the subset $\mathcal{F}_{\nabla h}$ of $\mathbb{R}^{n+1}$ on which the number and nature of the critical points of $h_x$ change. We will thus begin our study by a detailed study of focal hypersurfaces of hedgehogs in Sect. 9.1. For $n \in \{3, 4\}$, we will then prove that any normal through a boundary point to any convex body $K$ (with a smooth enough support function) in $\mathbb{R}^n$ passes arbitrarily close to the set of interior points of $K \cup L$ lying on normals through at least 6 distinct points of $\partial K$, where $L$ is the body bounded by the smallest convex parallel hypersurface to $\partial K$ whose unit normal points in the opposite direction. Motivated by this work published in 2022 [114], Grebennikov and Panina gave a proof of almost the same fact for any $n \geq 3$ via bifurcation theory [57]. Finally, we will prove that it is not true that for any convex body $K$ of $\mathbb{R}^4$, there are at least 8 normal lines passing through the center of the minimal spherical shell of $K$.

Chapter 10 will be devoted to miscellaneous questions regarding hedgehogs or making use of hedgehog techniques. Section 10.1 will deal with the convolution of hedgehogs. The question that will be studied in Sect. 10.2 will be to understand what are the generic singularities that are inevitable during an eversion of the sphere following a generic path of hedgehogs. We will see that such a path of hedgehogs necessarily includes hedgehogs carrying positive swallowtails. Finally, we will show through an example that hedgehogs are not only a generalization of convex bodies, but also a way of thinking about convex hypersurfaces in conjunction with their spherical images. Our example will consist in giving a set of conditions that is necessary and sufficient for the existence and uniqueness up to translations of a 3-dimensional polytope $P$ in $\mathbb{R}^3$ having $N$ facets with given unit outward normal vectors $n_1, \ldots, n_N$ and corresponding facet perimeters $L_1, \ldots, L_N$.

# Chapter 2
# Background on Classical Real Hedgehogs

## 2.1  Genesis: Sums and Formal Differences of Convex Bodies

In this section, we recall for the convenience of the reader the background on real hedgehogs. The set $\mathcal{K}^{n+1}$ of all convex bodies of $(n+1)$-Euclidean vector space $\mathbb{R}^{n+1}$ is usually equipped with Minkowski addition (see Fig. 2.1) and multiplication by nonnegative real numbers, which are respectively defined by:

$$(i) \quad \forall (K, L) \in \left(\mathcal{K}^{n+1}\right)^2, \, K + L = \{x + y \, | \, x \in K, y \in L\};$$
$$(ii) \quad \forall \lambda \in \mathbb{R}_+, \forall K \in \mathcal{K}^{n+1}, \, \lambda.K = \{\lambda x \, | \, x \in K\}.$$

It does not constitute a vector space since there is no subtraction in $\mathcal{K}^{n+1}$: not for every pair $(K, L) \in \left(\mathcal{K}^{n+1}\right)^2$ does there exist an $X \in \mathcal{K}^{n+1}$ such that $L + X = K$. Now, in the same way as we construct the group $\mathbb{Z}$, of integers from the monoid $\mathbb{N}$ of nonnegative integers, we can construct the vector space $\mathcal{H}^{n+1}$ of formal differences of convex bodies from $\mathcal{K}^{n+1}$. We can then regard $\mathcal{K}^{n+1}$ as a cone of $\mathcal{H}^{n+1}$ that spans the entire space.

As in the convex case, we will characterize these formal differences by their support functions. Recall that every convex body $K \in \mathcal{K}^{n+1}$ is determined by its support function $h_K : \mathbb{R}^{n+1} \longrightarrow \mathbb{R}$, $u \mapsto h_K(u) = \sup\{\langle x, u \rangle \, | \, x \in K\}$, or equivalently by its restriction to $\mathbb{S}^n$ [167, Section 1.7]; note that for all $u \in \mathbb{S}^n$, $h_K(u)$ is simply the signed distance from the origin to the support hyperplane with normal vector $u$. More precisely, for all convex body $K \in \mathcal{K}^{n+1}$, the support function $h_K : \mathbb{R}^{n+1} \longrightarrow \mathbb{R}$ is sublinear (i.e., positively homogeneous and subadditive), and conversely, for any sublinear function $h : \mathbb{R}^{n+1} \longrightarrow \mathbb{R}$, there exists a unique convex body $K \in \mathcal{K}^{n+1}$ with support function $h$ (ib.).

The starting point of hedgehog theory is therefore the desire to work in the vector space $\mathcal{H}^{n+1}$ of formal differences of convex bodies from $\mathcal{K}^{n+1}$, or equivalently in the vector space spanned by the support functions of convex bodies in $C\left(\mathbb{S}^n; \mathbb{R}\right)$.

© The Author(s), under exclusive license to Springer Nature Switzerland AG 2026    11
Y. Martinez-Maure, *Hedgehog Theory*, Lecture Notes in Mathematics 2381,
https://doi.org/10.1007/978-3-032-04298-9_2

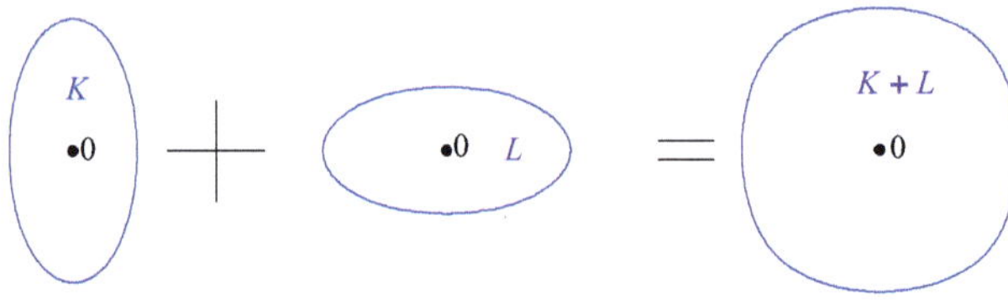

**Fig. 2.1**  For any convex bodies $K, L$ in $\mathbb{R}^{n+1}$, $K + L = \{x + y \mid x \in K, y \in L\}$ is still a convex body

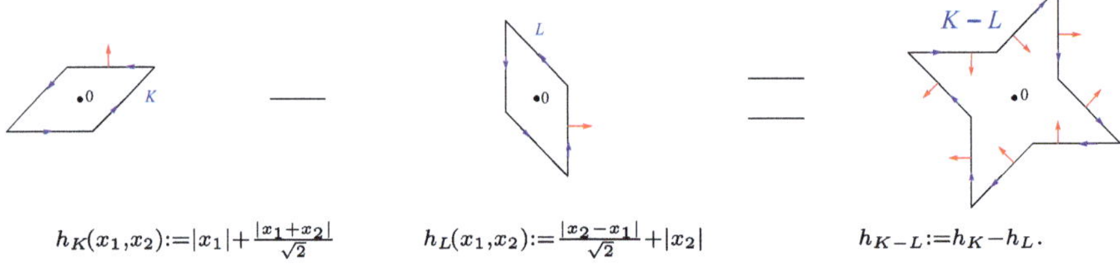

$$h_K(x_1, x_2) := |x_1| + \frac{|x_1 + x_2|}{\sqrt{2}} \qquad h_L(x_1, x_2) := \frac{|x_2 - x_1|}{\sqrt{2}} + |x_2| \qquad h_{K-L} := h_K - h_L.$$

**Fig. 2.2**  The hedgehog that describes the formal differencek K − L

But of course, we would prefer to deal with a geometric notion of difference of convex bodies rather than with purely formal differences. The object of the next subsections will be precisely to show how such geometric differences of convex bodies can be constructed. In other words, we will see below how to associate a geometric object, called a hedgehog, to any formal difference of two convex bodies $K - L$, $(K, L \in \mathcal{K}^{n+1})$, or equivalently to any difference of two sublinear functions $k, l : \mathbb{R}^{n+1} \to \mathbb{R}$. Here is, for instance, the hedgehog that describes the formal difference $K - L$ of the two plane convex bodies $K, L \subset \mathbb{R}^2$ with respective support functions $h_K(x_1, x_2) := |x_1| + \frac{|x_1 + x_2|}{\sqrt{2}}$ and $h_L(x_1, x_2) := \frac{|x_2 - x_1|}{\sqrt{2}} + |x_2|$, $\left((x_1, x_2) \in \mathbb{R}^2\right)$ (Fig. 2.2):

In summary, hedgehog theory will consist essentially in:

1. considering each formal difference of convex bodies of $\mathbb{R}^{n+1}$ as a geometric object in $\mathbb{R}^{n+1}$, called a **hedgehog**;
2. extending the mixed volume $v : \left(\mathcal{K}^{n+1}\right)^{n+1} \to \mathbb{R}$ to a symmetric $(n + 1)$-linear form on $\mathcal{H}^{n+1}$;
3. extending certain parts of the Brunn-Minkowski theory to $\mathcal{H}^{n+1}$.

For $n \leq 2$, it goes back to a paper by H. Geppert [53] who introduced the notion of a hedgehog under the German names *stützbare Bereiche* $(n = 1)$ and *stützbare Flächen* $(n = 2)$. But unfortunately, this work fell into oblivion for half a century.

## 2.2  Hedgehogs with a $C^2$-support Function

### 2.2.1  Basics on Hedgehogs with a $C^2$-support Function

Here, we will follow more or less [82]. As recalled above, every convex body $K \subset \mathbb{R}^{n+1}$ is determined by its support function $h_K : \mathbb{S}^n \longrightarrow \mathbb{R}$, where $h_K(u)$ is defined by $h_K(u) = \sup\{\langle x, u \rangle \mid x \in K\}$, $(u \in \mathbb{S}^n)$, that is, as the signed distance from the origin to the support hyperplane with normal vector $u$. In particular, every closed convex hypersurface of class $C^2_+$ (i.e., $C^2$-hypersurface with positive Gaussian curvature) is determined by its support function $h$ (which must be of class $C^2$ on $\mathbb{S}^n$ [167, p. 115]) as the envelope $\mathcal{H}_h$ of the family of hyperplanes with equation $\langle x, u \rangle = h(u)$. This envelope $\mathcal{H}_h$ is described analytically by the following system of equations

$$\begin{cases} \langle x, u \rangle = h(u) \\ \langle x, . \rangle = dh_u(.) \end{cases}.$$

The second equation is obtained from the first by performing a partial differentiation with respect to $u$. From the first equation, the orthogonal projection of $x$ onto the line spanned by $u$ is $h(u)u$, and from the second one, the orthogonal projection of $x$ onto $u^\perp$ is the gradient of $h$ at $u$ (see Fig. 2.3). Therefore, for each $u \in \mathbb{S}^n$, $x_h(u) = h(u)u + (\nabla h)(u)$ is the unique solution of this system.

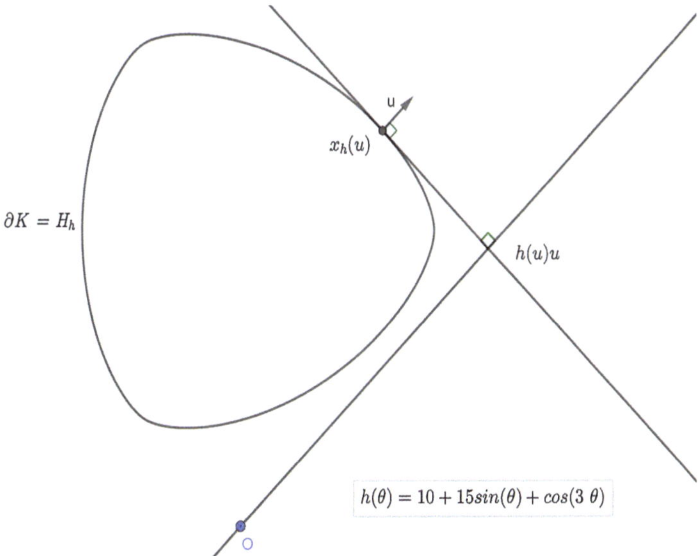


**Fig. 2.3**  Envelope parametrized by its Gauss map

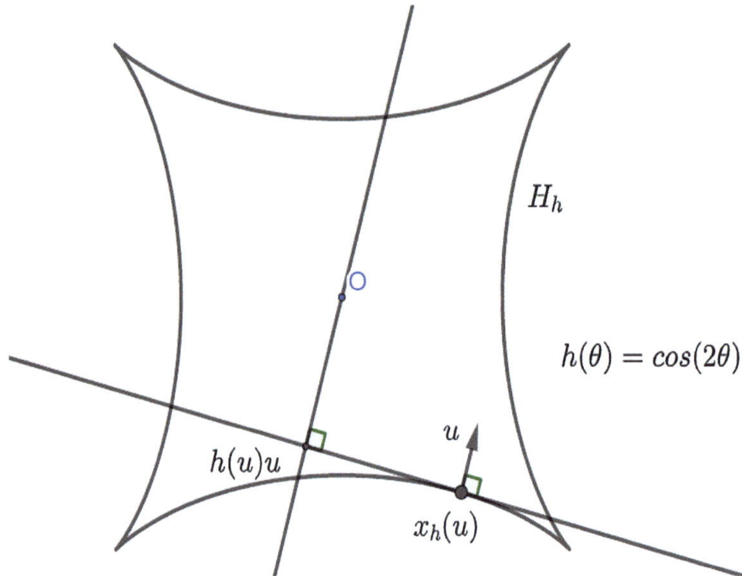


**Fig. 2.4**  Plane hedgehog with $C^2$-support function

Now, for any $C^2$-function $h$ on $\mathbb{S}^n$, the envelope $\mathcal{H}_h$ is in fact well-defined (even if $h$ is not the support function of a convex hypersurface). Its natural parametrization $x_h : \mathbb{S}^n \to \mathcal{H}_h$, $u \mapsto h(u)u + (\nabla h)(u)$ can be interpreted as the inverse of its Gauss map, in the sense that: at each regular point $x_h(u)$ of $\mathcal{H}_h$, $u$ is a normal vector to $\mathcal{H}_h$. We say that $\mathcal{H}_h$ is the hedgehog (or $C^2$-**hedgehog**) with support function $h$ (see Fig. 2.4). Note that $x_h$ depends linearly on $h$. For all $u \in \mathbb{S}^n$, we will consider that the support hyperplane with equation $\langle x, u \rangle = h(u)$ is cooriented (transversally oriented) by its unit normal vector $u$.

Since the parametrization $x_h$ can be regarded as the inverse of the Gauss map, the Gaussian curvature $\kappa_h$ of $\mathcal{H}_h$ at $x_h(u)$ is given by $\kappa_h(u) = 1/\det[T_u x_h]$, where $T_u x_h$ is the tangent map of $x_h$ at $u$. Therefore, singularities are the very points at which the Gaussian curvature is infinite. For every $u \in \mathbb{S}^n$, the tangent map of $x_h$ at the point $u$ is $T_u x_h = h(u)\, Id_{T_u \mathbb{S}^n} + H_h(u)$, where $H_h(u)$ is the symmetric endomorphism associated with the Hessian $\left(\nabla^2 h\right)_u$ of $h$ at $u$. Consequently, if $\lambda_1(u) \le \ldots \le \lambda_n(u)$ are the eigenvalues of the Hessian $\left(\nabla^2 h\right)_u$ of $h$ at $u$ then $R_1(u) := (\lambda_1 + h)(u) \le \ldots \le R_n(u) := (\lambda_2 + h)(u)$ can be interpreted as the principal radii of curvature of $\mathcal{H}_h$ at $x_h(u)$, and the so-called ***curvature function*** $R_h(u) := 1/\kappa_h(u) = \det[T_u x_h]$ is given for all $u \in \mathbb{S}^n$ by

$$\begin{aligned} R_h(u) &= \det\left[h(u)\, Id_{T_u \mathbb{S}^n} + H_h(u)\right] \\ &= \det\left[H_{ij}(u) + h(u)\,\delta_{ij}\right] \\ &= (R_1 \ldots R_n)(u) \end{aligned}$$

where $\delta_{ij}$ are the Kronecker symbols and $\left(H_{ij}\left(u\right)\right)$ the Hessian of $h$ at $u$ with respect to an orthonormal frame on $\mathbb{S}^n$.

Note that curvature function $R_h := 1/\kappa_h$ is thus well-defined and continuous on $\mathbb{S}^n$. In particular, the Minkowski Problem (the problem of prescribing the Gauss curvature) arises naturally for hedgehogs (see Chap. 5).

**Remarks**

1. In computations, it is often more convenient to replace $h$ by its positively $1$−homogeneous extension to $\mathbb{R}^{n+1}\setminus\{0\}$, which is given by

$$\varphi\left(x\right) := \|x\|\,h\left(\frac{x}{\|x\|}\right),$$

   for $x \in \mathbb{R}^{n+1}\setminus\{0\}$, where $\|.\|$ is the Euclidean norm on $\mathbb{R}^{n+1}$. A straightforward computation gives:

   (i)  $x_h$ is the restriction of the Euclidean gradient of $\varphi$ to the unit sphere $\mathbb{S}^n$;
   (ii)  For all $u \in \mathbb{S}^n$, the tangent map $T_u x_h$ identifies with the restriction to $\mathbb{S}^n$ of the symmetric endomorphism associated with the Hessian of $\varphi$ at $u$.

2. **Any hedgehog $\mathcal{H}_h$ with a $C^2$-support function $h : \mathbb{S}^n \to \mathbb{R}$ can be regarded as the Minkowski difference $K - L$ of two convex bodies** (or of two hypersurfaces by considering their boundaries) $K$, $L$ **of class $C_+^2$** in $\mathbb{R}^{n+1}$. Indeed, given any $h \in C^2\left(\mathbb{S}^n; \mathbb{R}\right)$, for all large enough real constants $r$, the functions $h + r$ and $r$ are the support functions of convex hypersurfaces of class $C_+^2$, and such that $h - (h + r) - r$.

**Orientation**

The hedgehog $\mathcal{H}_h$ of $\mathbb{R}^{n+1}$ with support function $h \in C^2\left(\mathbb{S}^n; \mathbb{R}\right)$ will be regarded as the oriented (possibly singular) hypersurface $x_h\left(\mathbb{S}^n\right)$ image of $\mathbb{S}^n$, equipped with its canonical orientation, under the map $x_h : \mathbb{S}^n \to \mathcal{H}_h \subset \mathbb{R}^{n+1}$. Note that for all $u \in \mathbb{S}^n$ such that $\kappa_h\left(u\right) > 0$ (resp. $\kappa_h\left(u\right) < 0$), the orientation of the tangent space $T_u\mathbb{S}^n$ is preserved (resp. reversed) by the tangent map $T_u x_h : T_u\mathbb{S}^n \to T_{x_h(u)}\mathcal{H}_h = T_u\mathbb{S}^n$.

**Volume and Surface Area**

From an analytical point of view, we obtain exactly the same formulas as in the convex case for area, volume and mixed volumes. We will come back to this in more detail in the next subsection, but we already briefly recall how these notions can be introduced and interpreted in the $C^2$-hedgehog case.

Using the curvature function, we can define an ***(algebraic) area measure***:

$$s(h, \Omega) := \int_{\Omega} R_h (u) \, d\sigma (u),$$

($\Omega \subset \mathbb{S}^n$ Borel set), and thus an ***(algebraic) area***:

$$s(h) := \int_{\mathbb{S}^n} R_h (u) \, d\sigma (u)$$

which can be interpreted as the difference $s_+ (h) - s_- (h)$, where $s (h)$ (resp. $s_- (h)$) denotes the total area of the regions of $\mathcal{H}_h$ on which the Gauss curvature $\kappa_h$ is positive (resp. negative).

Any $C^2$-hedgehog $\mathcal{H}_h$ of $\mathbb{R}^{n+1}$ is a (possibly singular and self-intersecting) parametrized hypersurface $x_h : \mathbb{S}^n \to \mathcal{H}_h \subset \mathbb{R}^{n+1}$ equipped with a transverse orientation defined as follows: at each regular point $x_h (u)$ of $\mathcal{H}_h$, the ***usual transverse orientation*** of $\mathcal{H}_h$ is given by the normal vector $sgn \, [R_h (u)] \, u$, where $sgn \, (.)$ is the signum function and $R_h$ the curvature function of $\mathcal{H}_h$ (that is, the inverse $\kappa_h^{-1}$ of the Gauss curvature $\kappa_h$ of $\mathcal{H}_h$). The ***Kronecker index*** $i_h (x)$ of a point $x \in \mathbb{R}^{n+1} \backslash \mathcal{H}_h$ with respect to $\mathcal{H}_h$ can be defined as the degree of the map

$$\mathcal{U}_{(h,x)} : \mathbb{S}^n \to \mathbb{S}^n, \ u \longmapsto \frac{x_h(u) - x}{\|x_h(u) - x\|} \, ;$$

$i_h (x)$ may be interpreted as the algebraic intersection number of an oriented half-line with origin $x$ with the hypersurface $\mathcal{H}_h$ equipped with its transverse orientation (number independent of the oriented half-line for an open dense set of directions). The usual transverse orientation and the Kronecker index are thus mutually associated. It is worth noting that if we let $\tilde{h} (u) = -h (-u)$ for all $u \in \mathbb{S}^n$, where $h \in C^2 (\mathbb{S}^n; \mathbb{R})$, then the hedgehogs $\mathcal{H}_h = x_h (\mathbb{S}^n)$ and $\mathcal{H}_{\tilde{h}} = x_{\tilde{h}} (\mathbb{S}^n)$ are identical as hypersurfaces of $\mathbb{R}^{n+1}$ except that they have opposite transverse orientations when $n + 1$ is odd. Indeed

$$x_{\tilde{h}} (-u) = x_h (u) \ \text{ for all } u \in \mathbb{S}^n,$$

but

$$sgn \left[ R_{\tilde{h}} (-u) \right] (-u) = (-1)^{n+1} sgn \, [R_h (u)] \, u,$$

and thus

$$i_{\tilde{h}} (x) = (-1)^{n+1} i_h (x) \ \text{ for all } x \in \mathbb{R}^{n+1} \backslash \mathcal{H}_h.$$

The Kronecker index played an important role in obtaining a counterexample to the old uniqueness conjecture of A.D. Alexandrov that we mentioned in our introduction. For $n + 1 = 2$, the Kronecker index $i_h(x)$ is nothing but the winding number of $\mathcal{H}_h$ around $x$: it counts the total number of times that $\mathcal{H}_h$ winds around $x$. For instance, the index is equal to $-1$ at any interior point of the hedgehog represented on Fig. 2.4, since the curve winds once clockwise around the point. A straightforward computation proves that, for all $x \in \mathbb{R}^{n+1} \setminus \mathcal{H}_h$, we have

$$i_h(x) = \frac{1}{\omega_n} \int_{\mathbb{S}^n} \frac{h(u) - \langle x, u \rangle}{\|x_h(u) - x\|^{n+1}} R_h(u)\, d\sigma(u),$$

where $R_h$ is the curvature function and $\sigma$ the spherical Lebesgue measure on $\mathbb{S}^n$.

The *(algebraic $(n + 1)$-dimensional) volume* of a hedgehog $\mathcal{H}_h \subset \mathbb{R}^{n+1}$ can then be defined by

$$v_{n+1}(h) := \int_{\mathbb{R}^{n+1} \setminus \mathcal{H}_h} i_h(x)\, d\lambda(x),$$

where $\lambda$ denotes the Lebesgue measure on $\mathbb{R}^{n+1}$, and it satisfies

$$v_{n+1}(h) = \frac{1}{n + 1} \int_{\mathbb{S}^n} h(u) R_h(u)\, d\sigma(u),$$

where $R_h$ is the curvature function and $\sigma$ the spherical Lebesgue measure on $\mathbb{S}^n$. For instance, in the example of Fig. 2.4 the 2-dimensional volume (or algebraic area) $v_2(h)$ of $\mathcal{H}_h$ in $\mathbb{R}^2$, also denoted by $a(h)$, is equal to minus the area contained in the curve. As for convex bodies of class $C_+^2$, we introduce the mixed curvature function $R_{(h_1, \ldots, h_n)}$ and the *mixed (algebraic $(n + 1)$-dimensional) volume* $v_{n+1}(h_1, \ldots, h_{n+1})$ of a family $\left( \mathcal{H}_{h_1}, \ldots, \mathcal{H}_{h_{n+1}} \right)$ of $C^2$-hedgehogs of $\mathbb{R}^{n+1}$ in such a way that

$$v(h_1, \ldots, h_{n+1}) = \frac{1}{n + 1} \int_{\mathbb{S}^n} h_1(u) R_{(h_2, \ldots, h_{n+1})}(u)\, d\sigma(u),$$

where $R_{(h_2, \ldots, h_{n+1})}$ is the mixed curvature function of $\left( \mathcal{H}_{h_2}, \ldots, \mathcal{H}_{h_{n+1}} \right)$ (see Proposition 3.1.1).

### 2.2.2 *The Example of Projective Hedgehogs*

Concerning the spherical image of the classical models of the real projective plane in $\mathbb{R}^3$, such as the Boy surface see Fig. 2.5 or the Steiner Roman surface see Fig. 2.6, Hilbert and Cohn-Vossen have written in *Geometry and the imagination*: "Unfortunately, the way in which it is distributed over the unit sphere has not yet been studied" (see e.g., [5] for a study of Boy and Steiner surfaces). A $C^2$-hedgehog $\mathcal{H}_h$ of $\mathbb{R}^{n+1}$ is said to be *projective* if its support function $h : \mathbb{S}^n \to \mathbb{R}$ is odd (or

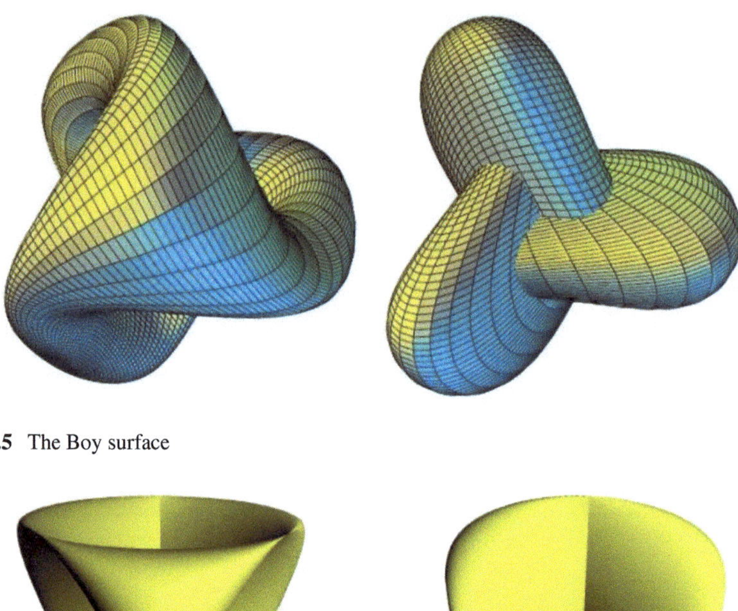

**Fig. 2.5** The Boy surface

**Fig. 2.6** The Steiner Roman surface

antisymmetric), that is such that $h(-u) = -h(u)$ for all $u \in \mathbb{S}^n$. For such a $C^2$-hedgehog $\mathcal{H}_h$, each pair of antipodal points $-u$, $u$ on $\mathbb{S}^n$ correspond to one and the same point $x_h(-u) = x_h(u)$ on $\mathcal{H}_h$. So not too singular projective hedgehogs $\mathcal{H}_h$ of $\mathbb{R}^3$ can be regarded as models of the real projective plane $\mathbb{RP}^2$ whose Gauss map is a bijection from the model onto $\mathbb{RP}^2$. Here is, for instance, a ***hedgehog version of the Steiner Roman surface***: $\mathcal{H}_h$, where $h(x, y, z) = x\left(x^2 - 3y^2\right) + 2z^3$, $(x, y, z) \in \mathbb{S}^2 \subset \mathbb{R}^3$. This model is represented on Fig. 2.7. As ***the Steiner Roman surface*** (see Fig. 2.6), which can be parametrized by $f : \mathbb{S}^2 \to \mathbb{R}^3$, $(x, y, z) \mapsto (yz, zx, xy)$, it has a threefold axis of symmetry and three lines of self-intersection whose end points are singular points of the same topological type as Whitney umbrellas without the handle.

Recall that ***the Boy surface*** is an immersion of the real projective plane in $\mathbb{R}^3$ discovered by W. Boy in 1901. This model of the real projective plane has no other singularity than lines of self-intersection and a single triple point (see Fig. 2.5).

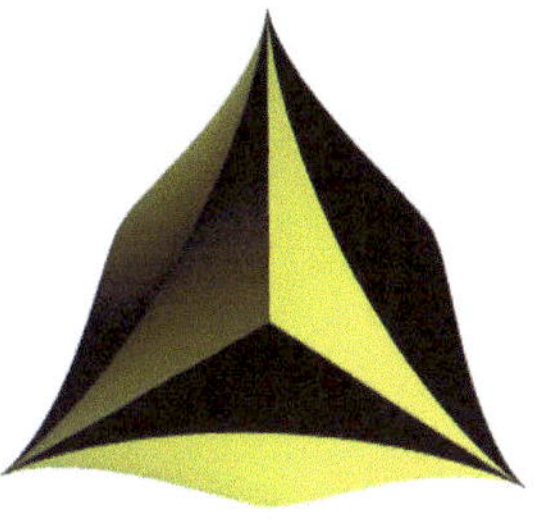

**Fig. 2.7**  A hedgehog version of the Steiner Roman surface

Beware of plane representations of projective hedgehogs of $\mathbb{R}^3$. They may be deceptive regarding singularities. For instance, as we will see later, the apparent contour of the projective hedgehog version $\mathcal{H}_h$ of the Roman surface shown in Fig. 2.7 is entirely composed of singular points of $\mathcal{H}_h$. We will see in Sect. 4.1 that projective $C^2$-hedgehogs of $\mathbb{R}^{n+1}$ are nothing but '$C^2$-hedgehogs of zero constant width' of $\mathbb{R}^{n+1}$.

## 2.3  Construction of General Hedgehogs

Recall that any $C^2$-hedgehog of $\mathbb{R}^{n+1}$ can be regarded as the Minkowski difference $K - L$ of two convex bodies (or of two hypersurfaces by considering their boundaries) $K$, $L$ of class $C_+^2$ in $\mathbb{R}^{n+1}$ (see Sect. 2.2.1). In [100], the author extended hedgehog theory by regarding hedgehogs of $\mathbb{R}^{n+1}$ as all the Minkowski differences $K - L$ of arbitrary convex bodies $K, L \in \mathcal{K}^{n+1}$. The trick is simply to replace 'support sets' by 'support hedgehogs', and to define hedgehogs inductively as collections of lower-dimensional 'support hedgehogs'. More precisely, our construction of general hedgehogs is based on the following three remarks.

*(i)* In $\mathbb{R}$, every convex body $K$ is determined by its support function $h_K$ as the segment $[-h_K(-1), h_K(1)]$, where $-h_K(-1) \leq h_K(1)$, so that the difference $K - L$ of two convex bodies $K, L$ can be defined as an oriented segment of $\mathbb{R}$: $K - L := [-(h_K - h_L)(-1), (h_K - h_L)(1)]$.

*(ii)* If $K$ and $L$ are two convex bodies of $\mathbb{R}^{n+1}$ then for all $u \in \mathbb{S}^n$, their support sets with unit normal $u$, say $K_u$ and $L_u$, can be identified with convex bodies of the linear hyperplane of $\mathbb{R}^{n+1}$ that is orthogonal to $u$, and this linear hyperplane, say $u^\perp$, may be identified with the $n$-dimensional Euclidean vector space $\mathbb{R}^n$. In fact [167, Theorem 1.7.2], the support set $K_u = \{x \in K \mid \langle x, u \rangle = h_k(u)\}$ is given by $K_u = \{h_K(u)u\} + K_{u^\perp}$, where $K_{u^\perp}$ is the convex body of $u^\perp$ with support function $h'_K(u; v) - \lim_{i \downarrow 0} [h_K(u + tv) - h_K(u)]/t$.

*(iii)* Addition of two convex bodies $K, L \subset \mathbb{R}^{n+1}$ corresponds to that of their support sets with same unit normal vector: $(K + L)_u = K_u + L_u$ for all $u \in \mathbb{S}^n$; therefore, the difference $K - L$ of two convex bodies $K, L \subset \mathbb{R}^{n+1}$ must be defined in such a way that $(K - L)_u = K_u - L_u$ for all $u \in \mathbb{S}^n$.

A natural way of defining geometrically general hedgehogs as differences of arbitrary convex bodies is therefore to proceed by induction with respect to the dimension by extending the notion of **support set** *with normal vector u* to a notion of **support hedgehog** *with normal vector u*. Let $\mathcal{H}_h \in \mathcal{H}^{n+1}$ be an arbitrary hedgehog of $\mathbb{R}^{n+1}$ (i.e., a Minkowski difference of two arbitrary convex bodies $K, L \in \mathcal{K}^{n+1}$ such that $h = h_K - h_L$). To any $u \in \mathbb{S}^n$, we associate the hedgehog $H_h^u$ of $u^\perp$ with support function $h'(u; v) = \lim_{t \downarrow 0} (h(u + tv) - h(u)) / t$. The hedgehog $\mathcal{H}_h$ is then given by the datum of all the support hedgehogs $\mathcal{H}_h^u := \{h(u)u\} + H_h^u$, $(u \in \mathbb{S}^n)$.

The above definition makes clear that a perfect understanding of plane hedgehogs is a prerequisite to a study of general hedgehogs of $\mathbb{R}^{n+1}$. We will give a complete study of plane general hedgehogs in Sect. 4.7.

In the polytopal case, hedgehogs are also known under the name of *'virtual polytopes'*. The notion of a virtual polytope was independently introduced by several authors (see, e.g., [124] or [154]). Let us give an example in $\mathbb{R}^2$. Let $K$ and $L$ be the convex bodies of $\mathbb{R}^2$ with support function $h_K(x) = |\langle x, e_1 \rangle| + |\langle x, e_2 \rangle|$ and $h_L(x) = |\langle x, e_3 \rangle| + |\langle x, e_4 \rangle|$, where $\langle ., . \rangle$ is the standard inner product on $\mathbb{R}^2$, $(e_1, e_2)$ the canonical basis of $\mathbb{R}^2$ and $e_3, e_4 \in \mathbb{R}^2$ the unit vectors given by $e_3 = \frac{1}{\sqrt{2}}(e_1 + e_2)$ and $e_4 = \frac{1}{\sqrt{2}}(e_1 - e_2)$. These convex bodies are two squares whose formal difference $K - L$ can be realized geometrically as the hedgehog with support function $h = h_K - h_L$, which is a regular octagram constructed by connecting every third consecutive vertex of a regular octagon (i.e., a regular star polygon with Schläfli symbol {8/3}): see Fig. 2.8.

R. Schneider showed in [169] how in the typical case (in the sense of Baire category) general hedgehogs that are differences of strictly convex bodies are highly singular objects.

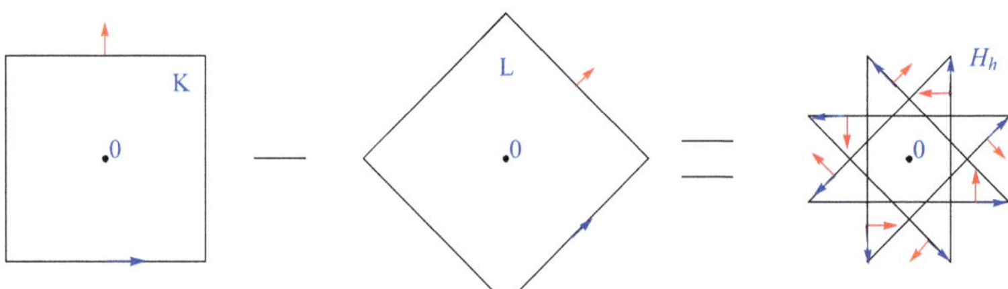


**Fig. 2.8**  Octagram obtained as the difference of two squares

## 2.4  Construction of Hedgehogs via Euler Calculus

There is a close relationship between Minkowski addition of convex bodies and convolution of their characteristic functions with respect to the Euler characteristic [59, 164, 185]: If $A$ and $B$ are compact convex subsets of $\mathbb{R}^{n+1}$, then

$$\mathbf{1}_A * \mathbf{1}_B = \mathbf{1}_{A+B},$$

where $*$ denotes the convolution product with respect to the Euler characteristic and $A + B$ the usual Minkowski sum of $A$ and $B$. Introducing the convolution inverse of the characteristic function for certain classes of convex bodies (such that polytopes or convex bodies with an analytic support function), we are led to a new mode of geometrically conceiving formal differences of convex bodies: the hedgehog $\mathcal{H}_h$ representing the formal difference $K - L$ can then be conceived from its 'Euler index', which is given by

$$\mathbf{1}_h := \mathbf{1}_K * (\mathbf{1}_L)^{-1} = (-1)^{n+1} \left( \mathbf{1}_K * \mathbf{1}_{\overset{o}{-L}} \right),$$

where $-\overset{o}{L}$ is the reflection of the interior $\overset{o}{L}$ through the origin $0_{\mathbb{R}^{n+1}}$ [105]. The relationship between Minkowski addition and convolution of characteristic or index functions can then be extended to certain classes of hedgehogs (ib.)

The present subsubsection is thus devoted to this approach of hedgehogs. We have chosen the framework of hedgehogs with analytic support functions (we will refer to them as *'analytic hedgehogs'* or '$C^{\omega}$-hedgehogs') even if some of our results still hold with a few adaptations under weaker assumptions.

The subsection is organized as follows. For the convenience of the reader Sect. 2.4.1 briefly summarizes basic notions and results from Euler's integral calculus. Section 2.4.2 presents the main results and Sect. 2.4.3 the proofs. Unless explicitly stated otherwise, the results of this subsection are essentially taken from [105].

### *2.4.1  Euler Calculus*

Euler calculus is an integration theory built with the Euler characteristic $\chi$ as a finitely additive measure. Born in the sheaf theory, it has applications to algebraic topology, to stratified Morse theory, for reconstructing objects in integral geometry and for enumeration problems in computational geometry and sensor networks [35]. The short survey papers by P. Schapira [164] and O. Viro [185] played an important role in the development of this theory.

Now we thus briefly summarize very basic notions and results from Euler calculus. We refer the reader to [35] for the proofs and more information on Euler calculus and its applications.

**Tame Sets** In Euler calculus, the measurable sets are the tame sets in some fixed 0-minimal structure. We will not recall here the definition of tame subsets in a fixed 0-minimal structure. It can be found in the classical surveys on Euler calculus, e.g., in [184]. Classical examples include polyconvex sets, semialgebraic sets and subanalytic sets. Here, we will only need to know some basic facts that we will summarize below. In particular, we will need to know that the union and intersection of two tame sets are again tame.

**Euler Characteristic** Fix an 0-minimal structure $\mathcal{O}$ on a topological space $X$. Definable functions between two spaces are those whose graphs are in $\mathcal{O}$. The Euler characteristic $\chi : \mathcal{O} \to \mathbb{Z}$ admits the following combinatorial definition:

Any tame set $A \in \mathcal{O}$ is definably homeomorphic to a finite disjoint union of open simplices $\coprod_i \sigma_i$ and we set:

$$\chi(A) = \sum_i (-1)^{\dim(\sigma_i)}.$$

Algebraic topology asserts that this quantity is well-defined, that is, independent of the decomposition into open simplices. This combinatorial Euler characteristic is a topological invariant. It is also a homotopy invariant for compact finite cell complexes (but not for non-compact spaces).

**Examples**

1. Euler characteristic can be regarded as a generalization of cardinality. For a finite discrete tame set $A$, $\chi(A)$ is the cardinality of $A$:

$$\chi(A) = \#A;$$

2. A closed orientable 2-manifold $S$ has Euler characteristic $2 - 2g$, where $g$ denotes the genus of $S$;
3. If $A$ is a compact contractible tame set, then $\chi(A) = 1$;
4. Any open $n$-ball of $\mathbb{R}^n$ has Euler characteristic $(-1)^n$;
5. The $n$-dimensional sphere $\mathbb{S}^n$ has Euler characteristic $1 + (-1)^n$;
6. The Euler characteristic of any odd-dimensional compact manifold is equal to zero.

**Remarks**

1. Euler calculus relies on the following additivity property:

**Proposition 2.4.1** *For any pair $\{A, B\}$ of tame subsets of $X$, we have:*

$$\chi(A \cup B) = \chi(A) + \chi(B) - \chi(A \cap B).$$

**2.** Euler characteristic is multiplicative under cross products:

**Proposition 2.4.2** *For any pair $\{E, F\}$ of tame sets, we have:*

$$\chi\left(E \times F\right) = \chi\left(E\right).\chi\left(F\right).$$

Note that these additivity and multiplicativity properties generalize the ones of cardinality of sets.

**Euler Integral**  The above additivity property suggests to define a measure over tame sets via:

$$\int_X \mathbf{1}_A\left(x\right) d\chi = \chi\left(A\right)$$

where $\mathbf{1}_A$ is the characteristic function over a tame subset $A$ of $X$. A function $f :$ $X \to \mathbb{Z}$ is said to be constructible if it has finite range and if all its level sets $f^{-1}\left(\{s\}\right)$ are tame subsets of $X$. Let $CF\left(X\right)$ denote the $\mathbb{Z}$-module of all $\mathbb{Z}$-valued constructible functions on $X$. The Euler integral is defined to be the homomorphism $\int_X : CF\left(X\right) \to \mathbb{Z}$ given by:

$$\int_X f d\chi := \sum_{s=-\infty}^{+\infty} s\chi\left[f^{-1}\left(\{s\}\right)\right].$$

Alternately, writing $f \in CF\left(X\right)$ as $f = \sum_i c_i \mathbf{1}_{\sigma_i}$, where $X = \bigsqcup_i \sigma_i$ is a decomposition of $X$ into a finite disjoint union of open cells and where $c_i \in \mathbb{Z}$, we have:

$$\int_X f d\chi := \sum_i c_i \chi\left(\sigma_i\right) = \sum_i c_i \left(-1\right)^{\dim(\sigma_i)}.$$

**Convolution**  On a finite-dimensional real vector space $V$, a convolution operator with respect to Euler characteristic is defined as follows:

$$\forall\left(f, g\right) \in CF\left(V\right)^2, \qquad \left(f * g\right)\left(x\right) = \int_V f\left(y\right) g\left(x - y\right) d\chi\left(y\right).$$

Convolution is a commutative, associative operator providing $CF\left(V\right)$ with the structure of an algebra.

**Proposition 2.4.3**  *$(CF\left(V\right), +, *)$ is a commutative ring with multiplicative identity element $\mathbf{1}_{\{0_V\}}$.*

**Relationship with Minkowski Addition**  There is a well-known relationship between Minkowski addition and convolution with respect to the Euler characteristic [59, 164, 185]:

**Groemer's Theorem** *([59]) Let A and B be two compact convex subsets of $\mathbb{R}^{n+1}$. We have*

$$\mathbf{1}_A * \mathbf{1}_B = \mathbf{1}_{A+B},$$

*where $*$ denotes the convolution product with respect to the Euler characteristic and $A + B$ the usual Minkowski sum of A and B.*

This relationship will be our starting point for constructing hedgehogs via Euler calculus.

### 2.4.2 Minkowski Inversion with Respect to $\chi$ and Euler Index

In this section, given a convex body $K$ in $\mathbb{R}^{n+1}$, we will often need $\overset{o}{K}$ and $\partial K$ to be tame subsets of $\mathbb{R}^{n+1}$. It is why we will restrict ourselves to analytic hedgehogs (resp. convex bodies).

The following result can be regarded as a Minkowski inversion theorem since $\mathbf{1}_{\{0_{\mathbb{R}^{n+1}}\}}$ is the multiplicative identity of $\left(CF\left(\mathbb{R}^{n+1}\right), +, *\right)$:

**Theorem 2.4.1** *Let K in $\mathbb{R}^{n+1}$ be a convex body of class $C_+^{\omega}$. We have*

$$(-1)^{n+1}\left(\mathbf{1}_K * \mathbf{1}_{-\overset{o}{K}}\right) = \mathbf{1}_{\{0_{\mathbb{R}^{n+1}}\}},$$

*where $-\overset{o}{K}$ denotes the reflection of $\overset{o}{K}$ through the origin $0_{\mathbb{R}^{n+1}}$. In other words, the convolution inverse of the characteristic function of K is given by:*

$$(\mathbf{1}_K)^{-1} = (-1)^{n+1}\mathbf{1}_{-\overset{o}{K}}.$$

**Remarks**

1. Of course, if $K$ is a convex body reduced to a point $a$ of $\mathbb{R}^{n+1}$, then the convolution inverse of the characteristic function of $K$ is given by:

$$(\mathbf{1}_K)^{-1} = \mathbf{1}_{\{-a\}}.$$

2. In [154], Pukhlikov and Khovanskii gave a similar Minkowski inversion theorem in the polytopal case: for every convex polytope $K$ in $\mathbb{R}^{n+1}$, we have

$$(-1)^{\dim K}\left(\mathbf{1}_K * \mathbf{1}_{-relint K}\right) = \mathbf{1}_{\{0_{\mathbb{R}^{n+1}}\}},$$

where $relint K$ is the relative interior of $K$, that is, the interior of $K$ in the smallest affine subspace that contains $K$.

**Euler Index**

**Definition 2.4.1** Let $\mathcal{H}_h$ be a $C^\omega$-hedgehog of $\mathbb{R}^{n+1}$, and let $K$, $L$ in $\mathbb{R}^{n+1}$ be convex bodies of class $C_+^\omega$ such that $\mathcal{H}_h$ is representing the formal difference $K - L$. Define the Euler index of $\mathcal{H}_h$ by

$$\mathbf{1}_h := \mathbf{1}_K * (\mathbf{1}_L)^{-1} = (-1)^{n+1} \left( \mathbf{1}_K * \mathbf{1}_{\overset{o}{-L}} \right),$$

where $-\overset{o}{L}$ denotes the reflection of $\overset{o}{L}$ through the origin $0_{\mathbb{R}^{n+1}}$.

**Remarks**

1. Using Groemer's theorem, which we have recalled above, and the fact that the convolution product $*$ is commutative, associative and admits $\mathbf{1}_{\{0_{\mathbb{R}^{n+1}}\}}$ as unity, it is easy to check that $\mathbf{1}_h$ is independent of the choice of the pair $(K, L)$ of convex bodies of class $C_+^\omega$ such that $\mathcal{H}_h$ is representing $K - L$.
2. Given any $C^\omega$-hedgehog $\mathcal{H}_h$ of $\mathbb{R}^{n+1}$, for every large enough $r > 0$, $k := h + r$ and $l := r$ are the respective support functions of two convex bodies $K$ and $L$ such that $\mathcal{H}_h$ is representing the formal difference $K - L$. Indeed, $h = k - l$ and if $r$ is large enough then, for all $u \in \mathbb{S}^n$, the principal radii of curvature of $\mathcal{H}_k$ at $x_k(u)$, which are the eigenvalues of the tangent map $T_u x_k = T_u x_h + r\, Id_{T_u \mathbb{S}^n}$, are all positive.

Furthermore, Groemer's theorem admits the following extension to analytic hedgehogs:

**Theorem 2.4.2** *Let $\mathcal{H}_f$ and $\mathcal{H}_g$ be two analytic hedgehogs of $\mathbb{R}^{n+1}$. We have*

$$\mathbf{1}_f * \mathbf{1}_g = \mathbf{1}_{f+g}.$$

This can easily be deduced from Groemer's theorem by using the above Minkowski inversion theorem. We will leave it to the reader to write down the details.

**Relationship with Kronecker Index**

**Theorem 2.4.3** *Let $\mathcal{H}_h$ be a $C^\omega$-hedgehog of $\mathbb{R}^{n+1}$, and let $K$, $L$ in $\mathbb{R}^{n+1}$ be convex bodies of class $C_+^\omega$ such that $\mathcal{H}_h$ is representing the formal difference $K - L$.*

*For any $x \in \mathbb{R}^{n+1} \setminus \mathcal{H}_h$, the Euler index $\mathbf{1}_h(x) := (-1)^{n+1} \left( \mathbf{1}_K * \mathbf{1}_{\overset{o}{-L}} \right)(x)$ of $\mathcal{H}_h$ at $x$ is equal to $i_h(x)$, that is, to the degree of the map*

$$\mathcal{U}_{(h,r)} : \mathbb{S}^n \to \mathbb{S}^n, \ u \longmapsto \frac{x_h(u) - x}{\|x_h(u) - x\|}.$$

*In other words, the Kronecker index $i_h$ is nothing but the restriction of the Euler index to $\mathbb{R}^{n+1}\setminus\mathcal{H}_h$.*

## Expressions for the Kronecker Index

We then give new expressions for the Kronecker index resorting only to the support functions and the Euler characteristic.

**Theorem 2.4.4** *Let $\mathcal{H}_h \subset \mathbb{R}^{n+1}$ be a $C^\omega$-hedgehog. Fix $x \in \mathbb{R}^{n+1}\setminus\mathcal{H}_h$ and let $h_x : \mathbb{S}^n \to \mathbb{R}$ be the support function of $\mathcal{H}_h$ with respect to $x$:*

$$h_x(u) := \langle x_h(u) - x, u \rangle = h(u) - \langle x, u \rangle.$$

*The Kronecker index $i_h(x)$ is given by*

$$i_h(x) = 1 + (-1)^{n+1}\,\chi_h^-(x) = \chi_h^+(x) + (-1)^{n+1},$$

*where $\chi_h^-(x) := \chi\left[(h_x)^{-1}(]-\infty, 0[)\right]$ and $\chi_h^+(x) := \chi\left[(h_x)^{-1}(]0, +\infty[)\right]$.*

**Corollary 2.4.1** *Under the assumptions of the previous theorem, we have:*

$$\forall x \in \mathbb{R}^{n+1}\setminus\mathcal{H}_h, \quad i_h(x) = \begin{cases} 1 - \frac{1}{2}\chi_h(x) & \text{if } n+1 \text{ is even} \\[2mm] \frac{1}{2}\left(\chi_h^+(x) - \chi_h^-(x)\right) & \text{if } n+1 \text{ is odd,} \end{cases}$$

*where $\chi_h(x) := \chi\left[(h_x)^{-1}(\{0\})\right]$, $\chi_h^-(x) := \chi\left[(h_x)^{-1}(]-\infty, 0[)\right]$ and $\chi_h^+(x) := \chi\left[(h_x)^{-1}(]0, +\infty[)\right]$.*

**Remarks**

1. From these results, if $n+1$ is even then, for any $x \in \mathbb{R}^{n+1}\setminus\mathcal{H}_h$, the knowledge of one of the four integers $\chi_h(x)$, $\chi_h^-(x)$, $\chi_h^+(x)$ and $i_h(x)$ implies that of the three others.
2. For $n+1 = 2$, Corollary 2.4.1 gives

$$i_h(x) = 1 - \frac{1}{2}n_h(x),$$

   where $n_h(x)$ denotes the number of cooriented support lines of $\mathcal{H}_h$ through $x$, that is, the number of zeros of $h_x : \mathbb{S}^1 \to \mathbb{R}$, $u \mapsto h(u) - \langle x, u \rangle$ (remind that the Euler characteristic is a generalization of cardinality). In Sect. 2.8, we will extend this formula to any $C^2$-hedgehog (Theorem 2.8.1).
3. For $n+1 = 3$, we will also prove in Sect. 2.8 that if $\mathcal{H}_h$ is a $C^2$-hedgehog in $\mathbb{R}^3$ then, for every $x \in \mathbb{R}^3\setminus\mathcal{H}_h$, we have (see Theorem 2.8.2):

$$i_h(x) = r_h^+(x) - r_h^-(x),$$

where $r_h^-(x)$ (resp. $r_h^+(x)$) denotes the number of connected components of $\mathbb{S}^2 - h_x^{-1}(\{0\})$ on which $h_x(u) := h(u) - \langle x, u \rangle$ is negative (resp. positive).

### Euler Index at a Point of $\mathcal{H}_h$

We are now going to consider the case when $x$ is a point of $\mathcal{H}_h$. We begin by the case $n + 1 = 2$.

**Theorem 2.4.5** *Let $\mathcal{H}_h$ be a $C^\omega$-hedgehog of $\mathbb{R}^2$. At a simple regular point $x :=$ $x_h(u)$ of $\mathcal{H}_h$, the Euler index $\mathbf{1}_h(x)$ is equal to the value taken by the Kronecker index $i_h$ on the connected component of $\mathbb{R}^2 \backslash \mathcal{H}_h$ towards which the unit normal vector $-u$ is pointing to. At a simple cusp point $c$ of $\mathcal{H}_h$, the Euler index $\mathbf{1}_h(c)$ is equal to the value taken by the Kronecker index $i_h$ on the connected component of $\mathbb{R}^2 \backslash \mathcal{H}_h$ that lies, in a neighborhood $\Omega$ of $c$, on the same side of $\mathcal{H}_h$ as the evolute of $\mathcal{H}_h \cap \Omega$.*

### Remarks

1. Generic singularities of plane $C^\infty$-hedgehogs are cusp points (see Sect. 2.6.3).
2. This result can be extended to hedgehogs $\mathcal{H}_h$ of $\mathbb{R}^2$ that are Minkowski differences $K - L$ of two convex polygons $K$ and $L$. For instance, if we start again with the example of the octagram $\mathcal{H}_h = K - L$ of two squares presented in Fig. 2.8, the Euler index of $\mathcal{H}_h$ is such that $\left( \mathbf{1}_K * \mathbf{1}_{\underset{-L}{o}} \right) = \mathbf{1}_h$.

Figure 2.4 is describing this relation by means of representations in $\mathbb{R}^2$. As can be seen on this figure, where the red arrows are representing unit normal vectors $u$, at a simple non-angular point $x$ of $\mathcal{H}_h$, the Euler index $\mathbf{1}_h(x)$ is equal to the value taken by the Kronecker index $i_h$ on the connected component of $\mathbb{R}^2 \backslash \mathcal{H}_h$ towards which the normal vector $-u$ is pointing to. The blue arrows just indicate the orientation of $\mathcal{H}_h$ (Fig. 2.9).

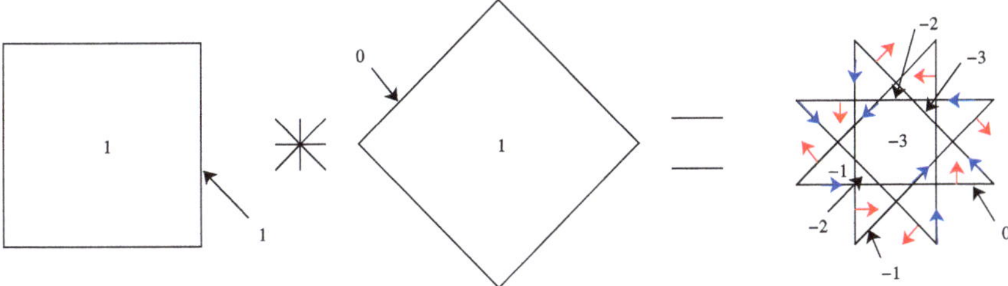


**Fig. 2.9** Euler index of a Minkowski difference of two squares

In higher dimensions, the question is more complicated at the singular points. However, the result remains true at the simple regular points.

**Theorem 2.4.6** *Let $\mathcal{H}_h \subset \mathbb{R}^{n+1}$ be a $C^\omega$-hedgehog. At a simple regular point $x :=$ $x_h(u)$ of $\mathcal{H}_h$, the Euler index $\mathbf{1}_h(x)$ is equal to the value taken by the Kronecker index $i_h$ on the connected component of $\mathbb{R}^{n+1} \setminus \mathcal{H}_h$ towards which the unit normal vector $-u$ is pointing to.*

### 2.4.3  Proof of the Results

***Proof of Theorem 2.4.1***  By the definition of the convolution product, we have

$$\left(\mathbf{1}_K * \mathbf{1}_{\underset{-K}{\overset{o}{}}}\right)(x) := \int_{\mathbb{R}^{n+1}} \mathbf{1}_K(y)\,\mathbf{1}_{\underset{-K}{\overset{o}{}}}(x - y)\,d\chi(y) \quad \text{for } x \in \mathbb{R}^{n+1}.$$

Fix $x \in \mathbb{R}^{n+1}$. The range of $F_x : \mathbb{R}^{n+1} \to \mathbb{R},\ y \longmapsto \mathbf{1}_K(y)\,\mathbf{1}_{\underset{-K}{\overset{o}{}}}(x - y)$ is included in $\{0, 1\}$ and

$$\forall y \in \mathbb{R}^{n+1}, \quad F_x(y) = 1 \Leftrightarrow y \in K \cap \left(\overset{o}{K} + \{x\}\right).$$

By the definition of Euler integral, we thus obtain

$$\left(\mathbf{1}_K * \mathbf{1}_{\underset{-K}{\overset{o}{}}}\right)(x) := \int_{\mathbb{R}^{n+1}} F_x(y)\,d\chi(y) = \chi\left[K \cap \left(\overset{o}{K} + \{x\}\right)\right].$$

If $x = 0_{\mathbb{R}^{n+1}}$ then $K \cap \left(\overset{o}{K} + \{x\}\right) = \overset{o}{K}$ and hence $\left(\mathbf{1}_K * \mathbf{1}_{\underset{-K}{\overset{o}{}}}\right)(x) = (-1)^{n+1}$ since $\overset{o}{K}$ is homeomorphic to an open $(n + 1)$-ball.

Assume $x \neq 0_{\mathbb{R}^{n+1}}$. If $K \cap \left(\overset{o}{K} + \{x\}\right) = \varnothing$ then $\chi\left[K \cap \left(\overset{o}{K} + \{x\}\right)\right] = 0$. Hence, we may assume that $K \cap \left(\overset{o}{K} + \{x\}\right) \neq \varnothing$. In this case, $\overset{o}{K} \cap \left(\overset{o}{K} + \{x\}\right)$ is homeomorphic to an open $(n + 1)$-ball and its boundary is the disjoint union of $\partial K \cap \left(\overset{o}{K} + \{x\}\right)$ and $K \cap \partial(K + \{x\})$, where the boundary of a convex body $L$

is denoted by $\partial L$. Therefore, $K \cap \left( \overset{o}{K} + \{x\} \right)$ is then the disjoint union of $\overset{o}{K} \cap$ $\left( \overset{o}{K} + \{x\} \right)$ and $\partial K \cap \left( \overset{o}{K} + \{x\} \right)$, which is homeomorphic to an open $n$-ball, so that

$$\chi \left[ K \cap \left( \overset{o}{K} + \{x\} \right) \right] = \chi \left[ \overset{o}{K} \cap \left( \overset{o}{K} + \{x\} \right) \right] + \chi \left[ \partial K \cap \left( \overset{o}{K} + \{x\} \right) \right]$$

$$= (-1)^{n+1} + (-1)^n$$

$$= 0,$$

which achieves the proof.    ■

Before we prove Theorems 2.4.3 and 2.4.4, we need to state and prove some intermediate results and properties.

**Proposition 2.4.4** *Under assumptions of Theorem 2.4.3, we have:*

$$\mathbf{1}_h (x) = \begin{cases} i_h (x) = 0 & \text{if } x \notin K + (-L) \\[2ex] (-1)^{n+1} \left( 1 - \chi \left[ (K + \{-x\}) \cap \partial L \right] \right) & \text{if } x \in (K + (-L)) \setminus \mathcal{H}_h. \end{cases}$$

***Proof*** We have

$$\mathbf{1}_K * \mathbf{1}_{-\overset{o}{L}} = \mathbf{1}_K * (\mathbf{1}_{-L} - \mathbf{1}_{-\partial L}) = (\mathbf{1}_K * \mathbf{1}_{-L}) - (\mathbf{1}_K * \mathbf{1}_{-\partial L}),$$

where $-L$ (resp. $-\partial L$) denotes the reflection of $L$ (resp. $\partial L$) through the origin. Now, we have $\mathbf{1}_K * \mathbf{1}_{-L} = \mathbf{1}_{K+(-L)}$ by Groemer's theorem, so that

$$\mathbf{1}_K * \mathbf{1}_{-\overset{o}{L}} = \mathbf{1}_{K+(-L)} - (\mathbf{1}_K * \mathbf{1}_{-\partial L}).$$

Let $x \in \mathbb{R}^{n+1}$. The range of $F_x : \mathbb{R}^{n+1} \to \mathbb{R}$, $y \longmapsto \mathbf{1}_K (y) \mathbf{1}_{-\partial L} (x - y)$ is included in $\{0, 1\}$ and

$$\forall y \in \mathbb{R}^{n+1}, \quad F_x (y) = 1 \Leftrightarrow y \in K \cap (\partial L + \{x\}).$$

By the definition of Euler integral, we thus obtain

$$(\mathbf{1}_K * \mathbf{1}_{-\partial L}) (x) := \int_{\mathbb{R}^{n+1}} F_x (y) \, d\chi (y) = \chi \left[ K \cap (\partial L + \{x\}) \right].$$

Using the translation $y \mapsto y - x$, we deduce that

$$(\mathbf{1}_K * \mathbf{1}_{-\partial L})(x) = \chi\left[(K + \{-x\}) \cap \partial L\right].$$

First assume $x \notin K + (-L)$. Then $\mathbf{1}_{K+(-L)}(x) = 0$ and $(\mathbf{1}_K * \mathbf{1}_{-\partial L})(x) = 0$ since $(K + \{-x\}) \cap \partial L \neq \varnothing$ would imply $x \in K + (-\partial L)$. Consequently

$$\mathbf{1}_h(x) := (-1)^{n+1}\left(\mathbf{1}_K * \mathbf{1}_{-\overset{o}{L}}\right)(x)$$

$$= (-1)^{n+1}\left(\mathbf{1}_{K+(-L)}(x) - (\mathbf{1}_K * \mathbf{1}_{-\partial L})(x)\right)$$

$$= 0.$$

Since $x_h(\mathbb{S}^n) \subset K + (-L)$, we also have $i_h(x) = 0$ and thus $\mathbf{1}_h(x) = i_h(x)$. Now assume $x \in (K + (-L)) \setminus \mathcal{H}_h$. Then we obtain

$$\mathbf{1}_h(x) = (-1)^{n+1}\left(1 - \chi\left[(K + \{-x\}) \cap \partial L\right]\right).$$

$\blacksquare$

Recall that we say that two submanifolds $S_1$ and $S_2$ of a manifold $M$ are transverse, and we write $S_1 \pitchfork S_2$, if $T_m M = T_m S_1 + T_m S_2$ for all $m \in S_1 \cap S_2$.

**Proposition 2.4.5** *Let $\mathcal{H}_h$ be a $C^\omega$-hedgehog of $\mathbb{R}^{n+1}$ and let $K, L$ in $\mathbb{R}^{n+1}$ be two convex bodies of class $C_+^\omega$ such that $\mathcal{H}_h$ is representing the formal difference $K - L$. For every $x \in \mathbb{R}^{n+1}$ such that $(K + \{-x\}) \cap L \neq \varnothing$ and $\partial(K + \{-x\}) \pitchfork \partial L$, the following properties hold:*

*(i)* $(h_x)^{-1}(\{0\}) \approx \partial(K + \{-x\}) \cap \partial L$;
*(ii)* $(h_x)^{-1}(]-\infty, 0]) \approx \partial(K + \{-x\}) \cap L$;
*(iii)* $(h_x)^{-1}([0, +\infty[) \approx (K + \{-x\}) \cap \partial L$;

*where $\approx$ is the homeomorphism relation and $(h_x)(u) := h(u) - \langle x, u \rangle$, $(u \in \mathbb{S}^n)$.*

***Proof*** *(i)* It follows from the assumptions that $(K + \{-x\}) \cap L$ is a strictly convex body with interior points, and thus that its support function

$$f : \mathbb{S}^n \longrightarrow \mathbb{R}$$
$$u \mapsto \sup\{\langle p, u \rangle \,|\, p \in (K + \{-x\}) \cap L\}$$

is continuously differentiable [167]. Denote by $k$ and $l$ the respective support functions of $K$ and $L$ and let $k_x(u) := k(u) - \langle x, u \rangle$ for all $u \in \mathbb{S}^n$. Notice that the zeros of $h_x = k_x - l$ are the points $u \in \mathbb{S}^n$ such that the support hyperplanes with exterior normal vector $u$ of $K + \{-x\}$ and $L$ coincide. Note that such an

$u \in (h_x)^{-1}(\{0\})$ cannot be a regular point of $x_f$. So, we can consider the continuous map

$$\phi : (h_x)^{-1}(\{0\}) \to \partial (K + \{-x\}) \cap \partial L$$
$$u \longmapsto x_f(u) := (\nabla f)(u) + f(u) u$$

To check that it defines a homeomorphism from the compact $(h_x)^{-1}(\{0\})$ to $\partial (K + \{-x\}) \cap \partial L$, it suffices to prove that it is a bijection.

Let $p \in \partial (K + \{-x\}) \cap \partial L$. Since $\partial (K + \{-x\}) \pitchfork \partial L$, there exists a pair of non-antipodal points $v$ and $w$ on $\mathbb{S}^n$, such that

$$p = x_{k_x}(v) = x_l(w) .$$

Let $\gamma$ denote the shortest arc between $v$ and $w$ on $\mathbb{S}^n$. Since we have clearly $h_x(v) < 0$ and $h_x(w) > 0$, there exists some $u \in \gamma$ such that $h_x(u) = 0$. It remains to prove that such an $u \in \gamma$ is unique and such that $\phi(u) = p$. For $\xi \in \mathbb{S}^n$, let $H_{k_x}(\xi)$ and $H_l(\xi)$ (resp. $H_{k_x}^-(\xi)$ and $H_l^-(\xi)$) denote the respective support hyperplanes (resp. halfspaces) with exterior normal vector $\xi$ of $K + \{-x\}$ and $L$. Note that: ($\alpha$) The segment with endpoints $x_{k_x}(u)$ and $x_l(u)$, say $\sigma(u)$, is passing through the complementary of $H_{k_x}^-(v) \cup H_l^-(w)$; ($\beta$) $H_{k_x}(u) = H_l(u) = \left(x_{k_x}(u) x_l(u)\right) + \left(v^\perp \cap w^\perp\right)$, where $\xi^\perp$ is the vector subspace orthogonal to $\xi \in \mathbb{S}^n$ and $\left(x_{k_x}(u) x_l(u)\right)$ the line through $x_{k_x}(u)$ and $x_l(u)$.

Let $u_1, u_2 \in \gamma \cap (h_x)^{-1}(\{0\})$. From ($\alpha$) and ($\beta$) with $u = u_1$ and $u = u_2$, it follows that the support hyperplanes $H_{k_x}(u_1) = H_l(u_1)$ and $H_{k_x}(u_2) = H_l(u_2)$ of the convex hull of $(K + \{-x\}) \cup L$ must coincide (in order that all the endpoints of the segments $\sigma(u_1)$ and $\sigma(u_2)$ lie in each of the support halfspaces $H_l^-(u_1)$ and $H_l^-(u_2)$, see Fig. 2.10). Therefore, there exists a unique $u \in \gamma$ such that $h_x(u) = 0$ and it satisfies $\phi(u) = p$.

To complete the proof it is sufficient to observe that any crossing of $(h_x)^{-1}(\{0\})$ on $\mathbb{S}^n$ from $(h_x)^{-1}(]-\infty, 0])$ to $(h_x)^{-1}([0, +\infty[)$ corresponds to a crossing of $\partial (K + \{-x\}) \cap \partial L$ on $\partial (K + \{-x\})$ (resp. $\partial L$) from $\partial (K + \{-x\}) \cap \overset{0}{L}$ to $\partial (K + \{-x\}) \cap \left(\mathbb{R}^{n+1} \backslash L\right)$ (resp. from. $\left(\mathbb{R}^{n+1} \backslash (K + \{-x\})\right) \cap \partial L$ to $\left(\overset{0}{K + \{-x\}}\right) \cap \partial L$, which results from the proof of (i).  ∎

The following corollary is immediate.

**Corollary 2.4.2** *Under the assumptions of the previous proposition, we have:*

$$\begin{cases} \chi \left[\partial (K + \{-x\}) \cap \partial L\right] = \chi_h(x) \\ \chi \left[\partial (K + \{-x\}) \cap L\right] = \chi_h^-(x) + \chi_h(x) \\ \chi \left[(K + \{-x\}) \cap \partial L\right] = \chi_h(x) + \chi_h^+(x) \end{cases}$$

*where* $\chi_h(x) := \chi \left[(h_x)^{-1}(\{0\})\right]$, $\chi_h^-(x) := \chi \left[(h_x)^{-1}(]-\infty, 0[)\right]$ *and* $\chi_h^+(x) := \chi \left[(h_x)^{-1}(]0, +\infty[)\right]$.

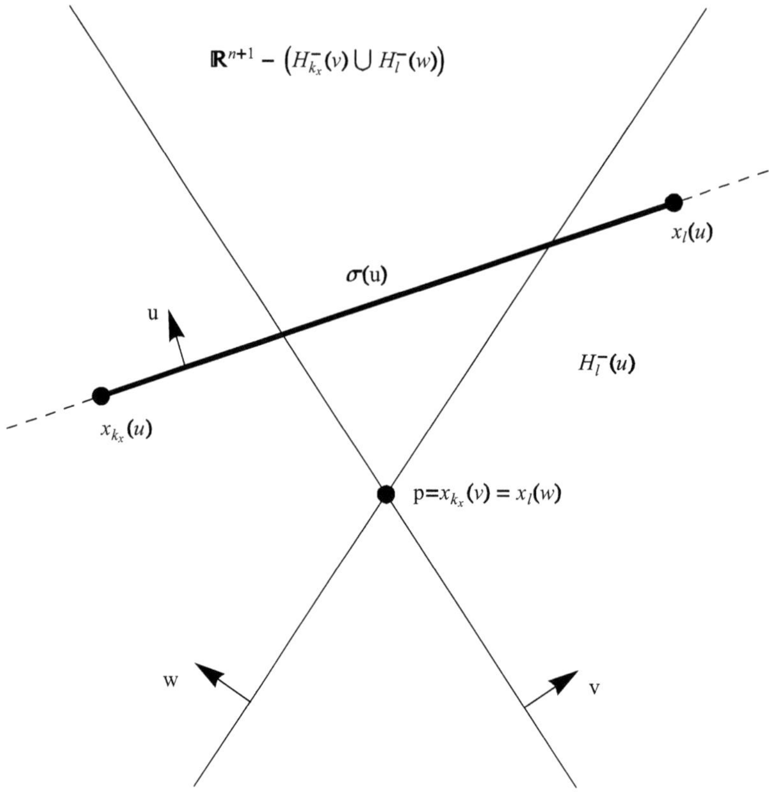


**Fig. 2.10** Projection view onto the plane $(\mathbb{R}v + \mathbb{R}w)^{\perp}$

**Lemma 2.4.1** *Let $\mathcal{H}_h$ be a $C^\omega$-hedgehog of $\mathbb{R}^{n+1}$ and let $K$, $L$ in $\mathbb{R}^{n+1}$ be two convex bodies of class $C_+^\omega$ such that $\mathcal{H}_h$ is representing the formal difference $K - L$. For any $x \in \mathbb{R}^{n+1} \backslash \mathcal{H}_h$, we have:*

$$\mathbf{1}_h\left(x\right) = i_h\left(x\right) = 1 - (-1)^n \chi_h^-\left(x\right) = 0$$

*if $\partial\left(K + \{-x\}\right)$ and $\partial L$ are externally tangent (that is, if they intersect in exactly one point and the intersection of their interior is empty).*

***Proof*** Let $a = b - x$ be the point of tangency of $\partial\left(K + \{-x\}\right)$ and $\partial L$, where $(a, b) \in K \times L$. By Proposition 2.4.4, we have

$$\mathbf{1}_h\left(x\right) = (-1)^{n+1}\left(1 - \chi\left[\left(K + \{-x\}\right) \cap \partial L\right]\right).$$

Since $\left(K + \{-x\}\right) \cap \partial L = \{a\}$, this implies $\mathbf{1}_h\left(x\right) = 0$.

Let $u$ be the point of $\mathbb{S}^n$ such that $a = x_{k_x}(u) = x_l(-u)$. For all $\varepsilon > 0$, $x_\varepsilon := x + \varepsilon u$ is such that $(K + \{-x_\varepsilon\}) \cap L = \varnothing$ and hence $x_\varepsilon \notin K + (-L)$. Therefore, $i_h(x_\varepsilon) = 0$ for all $\varepsilon > 0$ and hence $i_h(x) = 0$.

Finally, by noticing that $\chi_h^-(x)$ is constant on each connected component of $\mathbb{R}^{n+1} - \mathcal{H}_h$ and that $(h_x)^{-1}(]-\infty, 0[)$ is homeomorphic to an open $n$-ball $B_n$ when the Euclidean norm of $x$ is sufficiently large, we see that

$$\chi_h^-(x) = \chi(B_n) = (-1)^n \,,$$

which achieves the proof. $\qquad\blacksquare$

**Lemma 2.4.2** *Let $\mathcal{H}_h \subset \mathbb{R}^{n+1}$ be an analytic hedgehog. For every $x \in \mathbb{R}^{n+1} \backslash \mathcal{H}_h$, the index $1_h(x)$ is given by*

$$\mathbf{1}_h(x) = 1 + (-1)^{n+1}\, \chi_h^-(x) \,,$$

*where $\chi_h^-(x) := \chi\left[(h_x)^{-1}(]-\infty, 0[)\right]$.*

***Proof*** From Proposition 2.4.4 and Lemma 2.4.1, we can assume without loss of generality that $x \in (K + (-L)) \backslash \mathcal{H}_h$ and $\partial(K + \{-x\}) \pitchfork \partial L$. Then, by Proposition 2.4.4 and Corollary 2.4.2, we have:

$$\mathbf{1}_h(x) = (-1)^{n+1}\left(1 - \left(\chi_h(x) + \chi_h^+(x)\right)\right).$$

But

$$\chi_h^-(x) + \chi_h(x) + \chi_h^+(x) = \chi(\mathbb{S}^n) \quad \text{and} \quad \chi(\mathbb{S}^n) = 1 + (-1)^n \,,$$

so that

$$\mathbf{1}_h(x) = 1 + (-1)^{n+1}\, \chi_h^-(x) \,.$$

$\qquad\blacksquare$

***Proof of Theorems 2.4.3 and 2.4.4*** Let $\mathcal{H}_{\widetilde{h}} \subset \mathbb{R}^{n+1}$ be the hedgehog with support function $\widetilde{h}(-u) = -h(u)$, $(u \in \mathbb{S}^n)$. Note that $\mathcal{H}_h$ and $\mathcal{H}_{\widetilde{h}}$ have:

- the same geometric realization since $x_{\widetilde{h}}(-u) = x_h(u)$ for all $u \in \mathbb{S}^n$;
- the same transverse orientation (resp. opposite transverse orientations) at each point $x_{\widetilde{h}}(-u) = x_h(u)$ if $n + 1$ is even (resp. odd).

Therefore $i_{\widetilde{h}} = (-1)^{n+1} i_h$ on $\mathbb{R}^{n+1} \backslash \mathcal{H}_h$. Thus if we prove that, under assumptions of Theorem 2.4.3, $i_h(x) = \chi_h^+(x) + (-1)^{n+1}$ for all $x \in \mathbb{R}^{n+1} \backslash \mathcal{H}_h$, then

$$i_h(x) = (-1)^{n+1} i_{\widetilde{h}}(x)$$

$$= (-1)^{n+1}\left(\chi_{\widetilde{h}}^+(x) + (-1)^{n+1}\right)$$

$$= 1 + (-1)^{n+1} \chi_{\widetilde{h}}^{+} (x)$$

$$= 1 + (-1)^{n+1} \chi_{h}^{-} (x) \quad \text{for all } x \in \mathbb{R}^{n+1} \backslash \mathcal{H}_h,$$

and hence $i_h = \mathbf{1}_h$ on $\mathbb{R}^{n+1} \backslash \mathcal{H}_h$ by Lemma 2.4.2. So it remains only to prove that:

$$\forall x \in \mathbb{R}^{n+1} \backslash \mathcal{H}_h, \quad i_h (x) = \chi_h^{+} (x) + (-1)^{n+1} .$$

Since $i_h (x)$ is equal to 0 and $(h_x)^{-1} (]0, +\infty[)$ homeomorphic to an open $n$-ball when the distance of $x$ from the origin is sufficiently large, it suffices to prove that the map $x \mapsto i_h (x) - \left( \chi_h^{+} (x) + (-1)^{n+1} \right)$ is constant on $\mathbb{R}^{n+1} \backslash \mathcal{H}_h$. Since the maps $x \mapsto i_h (x)$ and $x \mapsto \chi_h^{+} (x)$ are constant on each connected component of $\mathbb{R}^{n+1} - \mathcal{H}_h$, we only need to prove that $i_h (x) - \chi_h^{+} (x)$ remains constant whenever $x$ crosses $\mathcal{H}_h$ transversally at a regular point.

Recall that, at a regular point $x_h (u)$ of $\mathcal{H}_h$, the transverse orientation of $\mathcal{H}_h$ is given by $sgn \, [R_h (u)] \, u$, where $sgn$ is the sign function and $R_h$ the curvature function of $\mathcal{H}_h$. Therefore, the Kronecker index $i_h (x)$ decreases by one unit whenever $x$ crosses $\mathcal{H}_h$ transversally at a simple regular point $x_h (u)$ in the direction of $sgn \, [R_h (u)] \, u$. Thus it is sufficient to prove that $\chi_h^{+} (x)$ also decreases by one unit whenever $x$ crosses $\mathcal{H}_h$ transversally at a simple regular point $x_h (u)$ in the direction of $sgn \, [R_h (u)] \, u$.

Let $x_h (u)$ be a simple regular point of $\mathcal{H}_h$. As the point $x_h (u)$ is regular, the curvature function of $\mathcal{H}_h$ is nonzero at $u$: $R_h (u) \neq 0$. Recall that $R_h (u)$ is the product of the principal radii of curvature $R_h^1 (u)$, ..., $R_h^n (u)$ of $\mathcal{H}_h$ at $u$, which are defined as the eigenvalues of $x_h$ at $u$. Denote by $p$ (resp. $q$) the number of principal radii of curvature of $\mathcal{H}_h$ at $u$ that are positive (resp. negative), $\left( (p, q) \in \mathbb{N}^2 \text{ and } p + q = n \right)$.

Let us consider the variation of $\chi_h^{+} (x)$ when $x$, moving on the normal line to $\mathcal{H}_h$ at $x_h (u)$, crosses $\mathcal{H}_h$ at $x_h (u)$ in the direction of transverse orientation (that is, in the direction of $(-1)^q u$). We first consider the case where the sectional curvature $\sigma_{x_h(u)}$ of $\mathcal{H}_h$ at $x_h (u)$ is positive (i.e., $(p, q) = (n, 0)$ or $(0, n)$). In the sequel of the proof, $B^n$ will denote an open $n$-ball. If $q = 0$, then the effect of the crossing on $\chi_h^{+} (x)$ is to add $\chi (B^n) - \chi (\mathbb{S}^n)$, that is $- 1$, to $\chi_h^{+} (x)$. If $q = n$, then the effect of the crossing on $\chi_h^{+} (x)$ is to add $(-1)^{n+1} \chi (B^n)$, that is $- 1$, to $\chi_h^{+} (x)$. Thus, in both cases, the effect of this crossing in the direction of transverse orientation is that $\chi_h^{+} (x)$ decreases by one unit.

We now turn to the case where $p$ and $q$ are nonzero. If we consider $(h_x)^{-1} (\{0\})$, which is a (not necessarily connected) smooth orientable hypersurface of $\mathbb{S}^n$ for any $x \in \mathbb{R}^{n+1} \backslash \mathcal{H}_h$ (since $\nabla h_x (u) \neq 0$ whenever $h_x (u) = 0$), the effect of the crossing in the direction of transverse orientation can then be viewed as a surgery performed on the hypersurface. If $q$ is even (resp. odd), the "surgery" consists in cutting out a piece of hypersurface homeomorphic to $\mathbb{S}^{q-1} \times D^p$ (resp. $D^q \times \mathbb{S}^{p-1}$) and replacing it by a piece of hypersurface homeomorphic to $D^q \times \mathbb{S}^{p-1}$ (resp. $\mathbb{S}^{q-1} \times D^p$), where $D^{m+1}$ is the closed $m$-ball bounded by $\mathbb{S}^m$, $(m \in \mathbb{N})$. Recall that such a surgery is possible by the fact that $\mathbb{S}^{p-1} \times \mathbb{S}^{q-1}$ can be regarded as the boundary of $\mathbb{S}^{q-1} \times D^p$ or as the boundary of $D^q \times \mathbb{S}^{p-1}$. When we consider $(h_x)^{-1}([0, +\infty[)$, the effect of the "surgery" is to remove (resp. to add) a cell complex that is homeomorphic to $D^p \times B^q$ if $q$ is even (resp. odd). Since Euler characteristic is multiplicative under cross products, the effect of the crossing on $\chi_h^+(x)$ is thus to add $(-1)^{q+1} \chi(B^q)$, that is $-1$.  ∎

**Proof of Corollary 2.4.1**  By Theorem 2.4.4, if $n + 1$ is even, for every $x \in \mathbb{R}^{n+1} \setminus \mathcal{H}_h$, we have $i_h(x) = 1 + \chi_h^-(x) = \chi_h^+(x) + 1$, and hence $i_h(x) = 1 + \frac{1}{2}\left(\chi_h^-(x) + \chi_h^+(x)\right)$. Since $\chi_h^-(x) + \chi_h(x) + \chi_h^+(x) = \chi(\mathbb{S}^n) = 1 + (-1)^n$, it follows that $i_h(x) = 1 - \frac{1}{2}\chi_h(x)$.

Now, if $n + 1$ is odd then, for every $x \in \mathbb{R}^{n+1} \setminus \mathcal{H}_h$, $i_h(x) = 1 - \chi_h^-(x) = \chi_h^+(x) - 1$ and hence $i_h(x) = \frac{1}{2}\left(\chi_h^+(x) - \chi_h^-(x)\right)$.  ∎

**Proof of Theorem 2.4.5**  We will give later a proof valid in any dimension $n + 1$, $(n \in \mathbb{N}^*)$, (cf. proof of Theorem 2.4.6). However, in order to deal with the special case of cusp points, we present here a slightly different proof in the plane.

Let $K, L$ in $\mathbb{R}^2$ be convex bodies of class $C_+^\omega$ such that $\mathcal{H}_h$ is representing the formal difference $K - L$ in $\mathbb{R}^2$. We will denote by $k$ and $l$ their respective support functions. Following the proof of Proposition 2.4.4 for $n + 1 = 2$, we obtain

$$\mathbf{1}_h(x) = 1 - \chi\left[(K + \{-x\}) \cap \partial L\right],$$

since $x := x_h(u) = x_k(u) + (-x_l(u)) \in K + (-L)$.

Note that $\partial(K + \{-x\})$ and $\partial L$ are internally tangent at the point $x_l(u)$ since $x_l(u) = x_{k_x}(u)$, where $k_x(u) := k(u) - \langle x, u \rangle$, $\left(u \in \mathbb{S}^1\right)$. Here 'internally' means that the two convex curves lie in the same side of their common tangent. Since $x := x_h(u)$ is assumed to be a regular point of $\mathcal{H}_h$, we have $R_h(u) \neq 0$ and thus $R_{k_x}(u) \neq R_l(u)$.

If $R_h(u) > 0$, then $R_{k_x}(u) > R_l(u)$, so that, in a neighborhood of the tangent point, $(\partial L) \setminus \{x_l(u)\}$ lies in the interior of $K + \{-x\}$. It follows that

$$\chi\left[(K + \{-x\}) \cap \partial L\right] = \frac{1}{2}\left(\chi\left[\partial(K + \{-x\}) \cap \partial L\right] - 1\right) = \frac{1}{2}n_h'(x),$$

where $n_h'(x) = \chi\left(\left\{v \in \mathbb{S}^1 \setminus \{u\} \mid h_x(v) = 0\right\}\right)$. Thus $\mathbf{1}_h(x)$ is then equal to $1 - \frac{1}{2}n_h'(x)$, which is the value taken by $i_h$ on the connected component of $\mathbb{R}^2 \setminus \mathcal{H}_h$ towards which the unit normal vector $-u$ is pointing to.

If $R_h(u) < 0$, then $R_{k_x}(u) < R_l(u)$, so that, in a neighborhood of the tangent point, $(\partial(K + \{-x\})) \setminus \{x_l(u)\}$ lies in the interior of $L$. It follows that

$$\chi\left[(K + \{-x\}) \cap \partial L\right] = \frac{1}{2}\left(\chi\left[\partial(K + \{-x\}) \cap \partial L\right] + 1\right) = \frac{1}{2}n_h'(x) + 1,$$

where $n'_h(x) = \chi\left(\{v \in \mathbb{S}^1 \setminus \{u\} \,|\, h_x(v) = 0\}\right)$. Thus $\mathbf{1}_h(x)$ is then equal to $-\frac{1}{2}n'_h(x)$, which is the value taken by $i_h$ on the connected component of $\mathbb{R}^2 \setminus \mathcal{H}_h$ towards which the unit normal vector $-u$ is pointing to.

Following the same approach for a simple cusp point $c := x_h(v)$ and noticing that $R_h = R_{k_x} - R_l$ changes sign at $v$, we obtain

$$\mathbf{1}_h(c) = 1 - \frac{1}{2}n'_h(c),$$

where $n'_h(c) = \chi\left(\{v \in \mathbb{S}^1 \setminus \{v\} \,|\, h_x(v) = 0\}\right)$, which is the required value for $\mathbf{1}_h(c)$. $\blacksquare$

***Proof of Theorem 2.4.6*** Let $K, L$ in $\mathbb{R}^{n+1}$ be convex bodies of class $C^\omega_+$ such that $\mathcal{H}_h$ is representing the formal difference $K - L$ in $\mathbb{R}^{n+1}$. Denote by $k$ and $l$ their respective support functions. Following the proof of Proposition 2.4.4, we obtain

$$\mathbf{1}_h(x) = (-1)^{n+1}\left(1 - \chi\left[(K + \{-x\}) \cap \partial L\right]\right),$$

since $x := x_h(u) = x_k(u) + (-x_l(u)) \in K + (-L)$. Note that $\partial(K + \{-x\})$ and $\partial L$ are internally tangent at the point $x_l(u)$ since $x_l(u) = x_{k_x}(u)$, where $k_x(u) := k(u) - \langle x, u \rangle$, $(u \in \mathbb{S}^n)$.

The result is the consequence of the following four observations:

*(i)* The proof of Proposition 2.4.4 can be adapted to obtain $\chi\left[(K + \{-x\}) \cap \partial L\right] = \chi_h(x) + \chi_h^+(x)$ in the present case;

*(ii)* $\chi_h^-(x) + \chi_h(x) + \chi_h^+(x) = \chi(\mathbb{S}^n) = 1 + (-1)^n$;

*(iii)* At $x = x_h(u)$, $\chi_h^- : \mathbb{R}^{n+1} \to \mathbb{Z}$, $p \mapsto \chi\left[(h_p)^{-1}(]-\infty, 0[)\right]$ takes the same value as the one it takes on the connected component of $\mathbb{R}^{n+1} \setminus \mathcal{H}_h$ towards which $-u$ is pointing to;

*(iv)* On this connected component, $i_h(p) = 1 + (-1)^{n+1}\chi_h^-(p)$ by Theorem 2.4.4. $\blacksquare$

### 2.4.4  Further Remarks

**Euler Characteristic of an Analytic Hedgehog** Let $\mathcal{H}_h$ in $\mathbb{R}^{n+1}$ be an analytic hedgehog. Define its Euler characteristic by:

$$\chi(\mathcal{H}_h) := \int_{\mathbb{R}^{n+1}} \mathbf{1}_h(x)\, d\chi(x).$$

**Proposition 2.4.6** *Any analytic hedgehog of $\mathbb{R}^{n+1}$ has Euler characteristic 1.*

***Proof*** Let $\mathcal{H}_h$ be a $C^\omega$-hedgehog of $\mathbb{R}^{n+1}$ and let $K, L \subset \mathbb{R}^{n+1}$ be convex bodies of class $C_+^\omega$ such that $\mathcal{H}_h$ is representing the formal difference $K - L$. By the definitions of $\chi\,(\mathcal{H}_h)$ and $\mathbf{1}_h$, we have:

$$\chi\,(\mathcal{H}_h) := \int_{\mathbb{R}^{n+1}} (-1)^{n+1} \left(\mathbf{1}_K * \mathbf{1}_{\overset{o}{-L}}\right) (x)\, d\chi\,(x).$$

Convolution is a commutative, associative operator providing $CF\left(\mathbb{R}^{n+1}\right)$ with the structure of an algebra and by reversing the order of integration, we get immediately [35, Lemma 19.1, p. 36]:

$$\int_{\mathbb{R}^{n+1}} (f * g)\, d\chi = \left(\int_{\mathbb{R}^{n+1}} f\, d\chi\right)\left(\int_{\mathbb{R}^{n+1}} g\, d\chi\right) \quad \text{for all } f, g \in CF\left(\mathbb{R}^{n+1}\right).$$

Thus

$$\chi\,(\mathcal{H}_h) = (-1)^{n+1} \left(\int_{\mathbb{R}^{n+1}} \mathbf{1}_K\,(x)\, d\chi\,(x)\right)\left(\int_{\mathbb{R}^{n+1}} \mathbf{1}_{\overset{o}{-L}}\,(x)\, d\chi\,(x)\right),$$

that is, $\chi\,(\mathcal{H}_h) = (-1)^{n+1}\, \chi\,(K)\, \chi\left(\overset{o}{-L}\right) = (-1)^{n+1}\, \chi\,(D)\, \chi\left(\overset{o}{D}\right) = 1$, where $D$ is the closed $(n+1)$-ball bounded by $\mathbb{S}^n$ in $\mathbb{R}^{n+1}$, $(n \in \mathbb{N})$.   ∎

**Mixed Volume of Analytic Hedgehogs** As a consequence of Theorems 2.4.2 and 2.4.3, we have:

Given hedgehogs with support functions $h_1, \ldots, h_{n+1} \in C^\omega\,(\mathbb{S}^n; \mathbb{R})$, the real function $P : \mathbb{R}^{n+1} \longrightarrow \mathbb{R}$ given by

$$P\,(\alpha_1, \ldots, \alpha_{n+1}) := v_{n+1}\left(\sum_{k=1}^{n+1} \alpha_k h_k\right) = \int_{\mathbb{R}^{n+1}} \left(\mathbf{1}_{\alpha_1 h_1} * \ldots * \mathbf{1}_{\alpha_{n+1} h_{n+1}}\right)(x)\, d\lambda\,(x),$$

where $\lambda$ denotes the Lebesgue measure on $\mathbb{R}^{n+1}$, is a homogeneous polynomial the coefficients of which are the mixed volumes of $\mathcal{H}_{h_1}, \ldots, \mathcal{H}_{h_{n+1}}$ up to a constant factor.

For some other consequences of the results of this subsection, we refer the reader to [105].

## 2.5   Hedgehogs with a $C^1$-support Function

This subsection can be omitted in a first reading. In Sect. 2.2, we introduced the notion of a hedgehog $\mathcal{H}_h$ with a $C^2$-support function $h : \mathbb{S}^n \to \mathbb{R}$, and we saw that its natural parametrization $x_h : \mathbb{S}^n \to \mathcal{H}_h, u \mapsto h(u)u + (\nabla h)\,(u)$ can be interpreted as the inverse of its Gauss map. It is worth noting that if $h : \mathbb{S}^n \to \mathbb{R}$ is only $C^1$, the envelope $\mathcal{H}_h$ is still well defined and parametrized by $x_h : \mathbb{S}^n \to \mathcal{H}_h$, $u \mapsto h(u)u + (\nabla h)(u)$. Now, in this case, the hedgehog $\mathcal{H}_h$ may not correspond to some Minkowski difference of two convex bodies. For instance, we know that a plane hedgehog that can be regarded as the Minkowski difference of two planar convex bodies is necessarily a rectifiable curve [100] (see Sect. 4.7), which is not the case for any hedgehog with a $C^1$-support function. In fact, as we will see, such a hedgehog can even be a nowhere differentiable fractal curve of infinite length. The results of this subsection are essentially taken from [96].

In 1872, K. Weierstrass astounded the mathematical world by giving an example of a family of real functions that are continuous on the whole real line without being differentiable at any point. ( « Über continuirliche Functionen eines reellen Arguments, die für keinen Werth des letzteren einen bestimmten Differentialquotienten besitzen », gelesen in der Königl. Akademie der Wissenschaften am 18 Juli 1872 », pp. 71–74). In this subsection, we use such a Weierstrass' function to construct an example of a fractal (projective) hedgehog with a $C^1$-support function.

**Theorem 2.5.1 (K. Weierstrass, 1872)** *Let $f$ be a real function of the form*

$$f(x) = \sum_{n=0}^{+\infty} a^n \cos\left(b^n \pi x\right)$$

*where $a \in\, ]0, 1[$, $b$ is an odd natural number and $ab > 1 + \frac{3\pi}{2}$. The function $f$ is continuous everywhere and differentiable nowhere.*

**Corollary 2.5.1** *Let $h$ be a real function of the form*

$$h(\theta) = \sum_{n=1}^{+\infty} \left(1/\alpha^n\right) \sin\left(\beta^n \theta\right),$$

*where $\beta$ is an odd natural number and $\alpha$ is a real number such that $\alpha > \beta$ and $\beta^2 > \alpha\left(1 + \frac{3\pi}{2}\right)$. The function $h$ is of class $C^1$ on $\mathbb{R}$ but its derivative is nowhere differentiable.*

***Proof*** By the Weierstrass $M$-test, the series

$$\sum \left(1/\alpha^n\right) \sin\left(\beta^n \theta\right) \quad \text{and} \quad \sum \left(\beta/\alpha\right)^n \cos\left(\beta^n \theta\right)$$

converge uniformly on $\mathbb{R}$ since $0 < 1/\alpha < 1$ and $0 < \beta/\alpha < 1$. Consequently, as $u_n(\theta) = (1/\alpha)^n \sin\left(\beta^n \theta\right)$ is a function of class $C^1$ on $\mathbb{R}$ the derivative of which is $u'_n(\theta) = (\beta/\alpha)^n \cos\left(\beta^n \theta\right)$, the function $h$ is of class $C^1$ on $\mathbb{R}$ and

$$h'(\theta) = \sum_{n=1}^{+\infty} (\beta/\alpha)^n \cos\left(\beta^n \theta\right) \quad \text{for all } \theta \in \mathbb{R}.$$

Now, given the conditions imposed on $\alpha$ and $\beta$, this derivative $h'$ is nowhere differentiable by Theorem 2.5.1. ∎

**Theorem 2.5.2**   *There exists a fractal projective hedgehog $\mathcal{H}_h$ in $\mathbb{R}^2$. More precisely, if $h : \mathbb{S}^1 = \mathbb{R}/2\pi\mathbb{Z} \to \mathbb{R}, \theta \mapsto h(\theta)$ is a Möbius function of the form*

$$h(\theta) = \sum_{n=1}^{+\infty} \left(1/\alpha^n\right) \sin\left(\beta^n \theta\right),$$

*where $\beta$ is an odd natural number, and $\alpha$ is a real number such that $\alpha > \beta$ and $\beta^2 > \alpha \left(1 + \frac{3\pi}{2}\right)$, then the hedgehog $\mathcal{H}_h$ satisfies the following properties:*

*(i)   the curve $\mathcal{H}_h$ is continuous but nowhere differentiable;*
*(ii)   the curve $\mathcal{H}_h$ has infinite length.*

**Proof**   From the previous corollary, the Möbius function $h$ is of class $C^1$ but its derivative $h'$ is nowhere differentiable. It follows immediately that the natural parametrization of $\mathcal{H}_h$, namely

$$x_h : \quad \begin{aligned} I = [0, 2\pi] &\longrightarrow \mathcal{H}_h \subset \mathbb{R}^2 \\ \theta &\longmapsto \left(x_h^1(\theta), x_h^2(\theta)\right) = h(\theta)u(\theta) + h'(\theta)u'(\theta) \end{aligned},$$

where $u(\theta) = (\cos\theta, \sin\theta)$, is continuous everywhere but nowhere differentiable.
   Now, the length of $\mathcal{H}_h$, namely

$$L(h) = \sup_{\sigma} \sum_{k=1}^{n} \|x_h(\theta_k) - x_h(\theta_{k-1})\|,$$

where the supremum is taken over all partitions $\sigma = (\theta_0, \ldots, \theta_n)$ of $I$, is finite if and only if both components of $x_h = \left(x_h^1, x_h^2\right)$ are of bounded variation on $I$. But in this case, the functions $x_h^1$ and $x_h^2$ are almost everywhere differentiable on $I$, in contradiction with the fact that $x_h = \left(x_h^1, x_h^2\right)$ is nowhere differentiable. ∎

   Moreover, the partial sums of the series $\sum \left(1/\alpha^n\right) \sin\left(\beta^n\theta\right)$, where $n \geq 1$, define a sequence $\left(\mathcal{H}_{h_n}\right)_{n \geq 1}$ of projective hedgehogs the natural parametrizations of which converge uniformly to that of $\mathcal{H}_h$. So, a fair approximation of $\mathcal{H}_h$ is given by $\mathcal{H}_{h_n}$ for a large enough $n$. Taking only $n = 5$, we obtain the representation of $\mathcal{H}_h$ given in Fig. 2.11 for $\alpha = 8$ and $\beta = 7$.
   In a preprint entitled "Fractal solutions to the problem of the two-dimensional floating body", which is a joint work of the author with D. Rochera, it is deduced the existence of fractal Zindler curves of the same Haussdorff dimension as the graph of a Weierstrass function.
   In a sense, this construction looks like that of the Koch curve (see e.g.,[43]): each step introduces new singular points as indicated in Fig. 2.12. But in the present case, there also appear self-intersections.

**Fig. 2.11** A fractal projective hedgehog

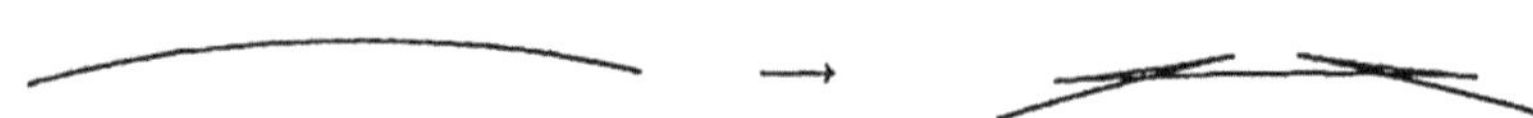

**Fig. 2.12** Introduction of new singular points

## 2.6  $C^\infty$-hedgehogs as Legendrian Fronts

This subsection can be omitted in a first reading. In this subsection, we will see that $C^\infty$-hedgehogs (that is, hedgehogs with a $C^\infty$-smooth support function) are wavefronts in the sense of contact geometry. Therefore, $C^\infty$-hedgehogs only have Legendrian singularities.

### 2.6.1  Contact Manifolds and Metric Contact Manifolds

A **contact structure** on an oriented $(2n + 1)$-dimensional $C^\infty$-manifold $M$ is the datum of a smooth field $V$ of tangent hyperplanes on $M$, called **contact hyperplanes**, satisfying the following condition of maximal non-integrability: any (and hence every) 1-form $\alpha$ defining $V$ (i.e., such that $V = Ker\,(\alpha)$) satisfies $\alpha \wedge (d\alpha)^n \neq 0$ everywhere on $M$. Any 1-form $\alpha$ defining such a maximally non-integrable hyperplane field $V$ on $M$ is called a **contact form** on $M$. Given such a contact structure (or a contact form $\alpha$ defining it), the pair $(M, V)$ (or the pair $(M, \alpha)$

if we want to fix the contact form defining $V$) is then called a ***contact manifold***. On $(M, \alpha)$, the ***Reeb vector field*** $\xi_\alpha$ associated to the contact form $\alpha$ is defined to be the unique smooth vector field satisfying

$$\alpha\left(\xi_\alpha\right) = 1 \qquad \text{and} \qquad \xi_\alpha \in Ker\left(d\alpha\right).$$

A submanifold $L$ of a contact manifold $(M, V)$ is said to be *integral* if $T_m L \subset V_m$ for all $m \in L$. A ***Legendrian submanifold*** of $(M, V)$ is an integral submanifold of $(M, V)$ with maximal dimension $n = \left(\dim M - 1\right)/2$. A fibration of a contact manifold is said to be *Legendrian* if all its fibers are Legendrian submanifolds.

Let $i : L \to E$ be an immersed Legendrian submanifold $L$ in the total space of a Legendrian fibration $\pi : E \to B$. The restriction $x = \pi \circ i : L \to B$ of $\pi$ to $L$ is called a ***Legendrian map***, and its image $x\left(L\right)$ in $B$ is called its ***Legendrian front*** or ***wavefront***.

**Example** Unit tangent bundles of Riemannian manifolds are among the most classical examples of contact manifolds. Let us recall briefly how this is done. Let $(M, g)$ be a Riemannian manifold and let

$$UTM = \{u \in TM \mid g\left(u, u\right) = 1\}$$

be its unit tangent bundle with canonical projection $\pi : UTM \to M$; the metric $g$ induces a contact form $\alpha$ (and thus a contact structure $V$) on $UTM$ as follows: for any $u \in UTM$ and $v \in T_u\left(UTM\right)$, we let

$$\alpha_u\left(v\right) = g\left(u, T_u\pi\left(v\right)\right),$$

where $T_u\pi\left(v\right) = \pi_*\left(v\right)$ is the pushforward along $\pi$ of the vector $v$. Moreover, $\pi : UTM \to M$ is an example of a Legendrian fibration.

In particular, if we let

$$\alpha_{(x,u)} := \langle u, dx \rangle = \sum_{i=0}^{n+1} u_i dx_i$$

for all $(x, u) \in U\mathbb{R}^{n+1} = \mathbb{R}^{n+1} \times \mathbb{S}^n$, where $(x_1, \cdots, x_{n+1}; u_1, \cdots, u_{n+1})$ are the canonical coordinate functions on $U\mathbb{R}^{n+1} = \mathbb{R}^{n+1} \times \mathbb{S}^n \subset \mathbb{R}^{2n+2}$, we obtain a contact manifold $\left(U\mathbb{R}^{n+1}; \alpha\right)$.

A ***contactomorphism*** from a contact manifold $(M_1, V_1)$ to a contact manifold $(M_2, V_2)$ is a diffeomorphism $f : M_1 \to M_2$ that preserves the contact structure, i.e., such that $Tf\left(V_1\right) = V_2$, where $Tf : TM_1 \to TM_2$ denotes the tangent map of $f$. If $V_i = Ker\left(\alpha_i\right)$, $(i = 1, 2)$, this is equivalent to the existence of a nowhere zero function $\lambda : M_1 \to M_2$ such that $f^*\alpha_2 = \lambda\alpha_1$.

**Example** Another example of a contact manifold is defined as follows: on the manifold $T\mathbb{S}^n \times \mathbb{R}$, where the tangent bundle $T\mathbb{S}^n$ is identified with

$$\left\{ (u, p) \in \left(\mathbb{R}^{n+1}\right)^2 \mid \|u\| = 1 \text{ and } \langle u, p \rangle = 0 \right\}$$

($\|.\|$ and $\langle ., . \rangle$ denoting respectively the Euclidean norm and scalar product in $\mathbb{R}^{n+1}$), we define a contact form $\beta$ by putting $\beta_{(u,p,z)} := dz - p\,du$ for all $(u, p, z) \in T\mathbb{S}^n \times \mathbb{R}$. Moreover

$$f : U\mathbb{R}^{n+1} = \mathbb{R}^{n+1} \times \mathbb{S}^n \to T\mathbb{S}^n \times \mathbb{R}$$
$$(x, u) \mapsto (u, x - \langle x, u \rangle u, \langle x, u \rangle)$$

is a diffeomorphism such that $f^*\beta = \alpha$, and hence a contactomorphism from $\left(U\mathbb{R}^{n+1}, \alpha\right)$ to $(T\mathbb{S}^n \times \mathbb{R}, \beta)$.

A ***metric contact manifold*** is defined to be a tuple $(M, g, \alpha, J)$, where $(M, g)$ is a Riemannian manifold, $\alpha$ a smooth 1-form on $M$ and $J$ a section of the endomorphism bundle $End\,(TM)$ which satisfy the following three conditions:

*(i)* $\alpha\,(\xi_\alpha) = 1$, where $\xi_\alpha$ is the metric dual of $\alpha$;
*(ii)* $d\alpha\,(X, Y) = g\,(JX, Y)$ for any vector fields $X, Y$ on $M$;
*(iii)* $J^2 X = -X + \alpha\,(X)\,\xi_\alpha$ for any vector field $X$ on $M$.

Then $(M, Ker\,(\alpha))$ is a contact manifold (i.e., $\alpha \wedge (d\alpha)^n \neq 0$ on $M$), $\xi_\alpha$ is the Reeb vector field associated to $\alpha$, $J\xi_\alpha = 0$ and $g$ is determined by $\alpha$ and $J$ through the equality $g\,(X, Y) = \alpha\,(X)\,\alpha\,(Y) + d\alpha\,(X, JY)$, (see e.g., [176]).

**Example** In the case of hedgehogs of $\mathbb{R}^{n+1}$, we will consider the metric contact manifold $\left(U\mathbb{R}^{n+1}, g, \alpha, J\right)$, where $g$ is the Riemannian product metric on $U\mathbb{R}^{n+1} = \mathbb{R}^{n+1} \times \mathbb{S}^n$ and $J : TU\mathbb{R}^{n+1} \to TU\mathbb{R}^{n+1}$, $(X, Q) \mapsto (Q, \langle X, q \rangle q - X)$.

### 2.6.2  $C^\infty$-hedgehogs as Legendrian Fronts

Let us consider first the case where $(M, g) = \left(\mathbb{R}^{n+1}, g_{can}\right)$, where $g_{can} = \langle ., . \rangle$ is the canonical Euclidean metric. Let $\mathcal{H}_h$ be a hedgehog of $\mathbb{R}^{n+1}$ with support function $h \in C^\infty\,(\mathbb{S}^n; \mathbb{R})$. Let us recall that its natural parametrization $x_h : \mathbb{S}^n \to \mathbb{R}^{n+1}$, $u \mapsto x_h\,(u) = h(u)u + (\nabla h)\,(u)$ can be interpreted as the inverse of its Gauss map. Thus it appears that

$$i_h : \mathbb{S}^n \to U\mathbb{R}^{n+1} = \mathbb{R}^{n+1} \times \mathbb{S}^n$$
$$u \mapsto (x_h\,(u)\,, u)$$

is the immersion of a Legendrian submanifold in $U\mathbb{R}^{n+1}$ of which $\mathcal{H}_h$ is the Legendrian front in $\mathbb{R}^{n+1}$ and $x_h = \pi \circ i_h$ the corresponding Legendrian map. Recall that on $U\mathbb{R}^{n+1}$, the contact form and the associated Reeb vector field are respectively given by

$$\alpha_{(x,u)} := \langle u, dx \rangle = \sum_{i=0}^{n+1} u_i dx_i \qquad \text{and} \qquad \xi(x,u) := \left(u; 0_{T_u\mathbb{S}^n}\right),$$

for all $(x,u) \in U\mathbb{R}^{n+1}$, where $(x_1, \cdots, x_{n+1}; u_1, \cdots, u_{n+1})$ are the canonical coordinate functions on $U\mathbb{R}^{n+1} = \mathbb{R}^{n+1} \times \mathbb{S}^n \subset \mathbb{R}^{2n+2}$.

Thus, hedgehogs of $\mathbb{R}^{n+1}$ are the Legendrian fronts of those Legendrian submanifolds of $\left(U\mathbb{R}^{n+1}, \alpha\right)$ whose Legendrian maps can be interpreted as the inverse of the Gauss map of their image (i.e., of the Legendrian front).

$$\begin{array}{ccc} & i_h & \\ \mathbb{S}^n & \to & i_h\left(\mathbb{S}^n\right) \subset \left(U\mathbb{R}^{n+1}, Ker\left(\alpha\right)\right) \\ & & \\ x_h \searrow & & \downarrow \pi \\ & & \\ & & \mathcal{H}_h \subset \mathbb{R}^{n+1}. \end{array}$$

### 2.6.3   Generic Singularities of Smooth Hedgehogs

As we have just seen, $C^\infty$-hedgehogs of $\mathbb{R}^{n+1}$ can be regarded as Legendrian fronts. Therefore they have only Legendre singularities. Since their natural parametrizations $x_h : \mathbb{S}^n \to \mathbb{R}$ form an open dense set among all Legendrian maps of the Legendrian fibration $\pi : \left(U\mathbb{R}^{n+1}, Ker\left(\alpha\right)\right) \to \mathbb{R}^{n+1}$, $(x,u) \mapsto x$ [90], Arnold's works can be used to classify their generic singularities for $n \leq 5$ [6].

In particular, generic singularities of $C^\infty$-hedgehogs are **cusp points** in $\mathbb{R}^2$ (Fig. 2.13), **cuspidal edges** and **swallowtails** in $\mathbb{R}^3$ (Fig. 2.14). Swallowtails are the cusp points of cuspidal edges. Elliptic and hyperbolic regions, which are defined by the sign of the Gauss curvature $\kappa_h = 1/R_h$, are separated by cuspidal edges on which the curvature function $R_h$ is equal to 0 (or, loosely speaking, on which the Gauss curvature $\kappa_h$ becomes infinite): see Fig. 2.14a. Note that we can distinguish two types of swallowtails (**negative** or **positive**) according to the sign of the Gaussian curvature on the tail: see Fig. 2.14b and c. More precisely, there exists an open dense subset $\mathcal{U}$ of $C^\infty\left(\mathbb{S}^2; \mathbb{R}\right)$ in the $C^4$-topology such that: for all $h \in \mathcal{U}$, the singularities of $\mathcal{H}_h$ are all equivalent to one of the three models of singularities represented in Fig. 2.14.

Moreover, for such a hedgehog, R. Langevin, G. Levitt and H. Rosenberg proved the following counting formula on $\mathbb{S}^2$ [82]:

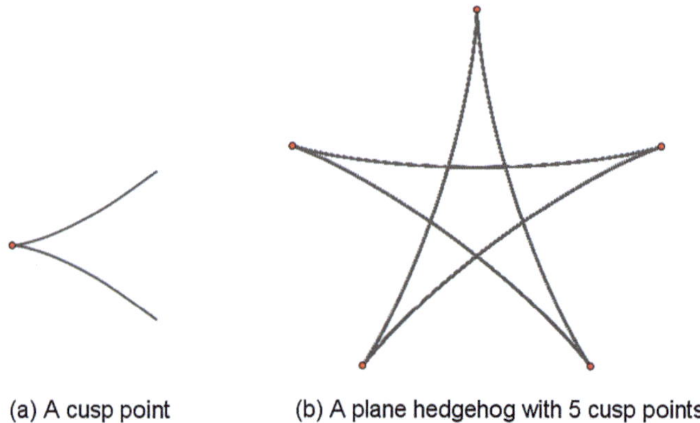

(a) A cusp point                    (b) A plane hedgehog with 5 cusp points

**Fig. 2.13**  Singularities of generic plane hedgehogs are cusp points

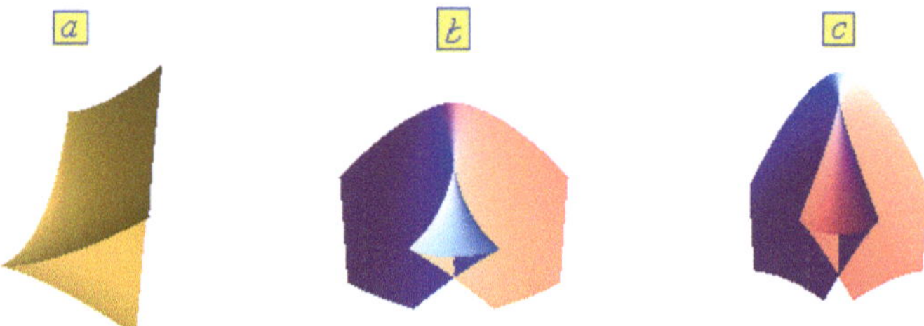

**Fig. 2.14**  Singularities of generic hedgehogs of $\mathbb{R}^3$

**Proposition 2.6.1**  *Given* $\mathcal{H}_h$ *a generic* $C^\infty$*-hedgehog of* $\mathbb{R}^3$*, we have*

$$r^+ - r^- = \frac{q^+ - q^-}{2} + 1, \qquad (2.6.1)$$

*where* $q^-$ *(resp.* $q^+$*) is the number of negative (resp. positive) swallowtails of* $\mathcal{H}_h$*, and* $r^-$ *(resp.* $r^+$*) the number of its hyperbolic (resp. elliptic) regions.*

***Proof of Proposition 2.6.1***  The authors proceeded by 3 steps in order to establish this result:

- First, they considered the following path of hedgehogs

$$\gamma: \quad [0,1] \to H_3$$
$$t \mapsto \mathcal{H}_{h+tr},$$

where $H_3$ is the linear space of $C^\infty$-hedgehogs defined up to a translation in $\mathbb{R}^3$, and $r \in \mathbb{R}^*_+$ is large enough to ensure that $\mathcal{H}_{h+r}$ is the boundary of a convex body of class $C^\infty_+$;

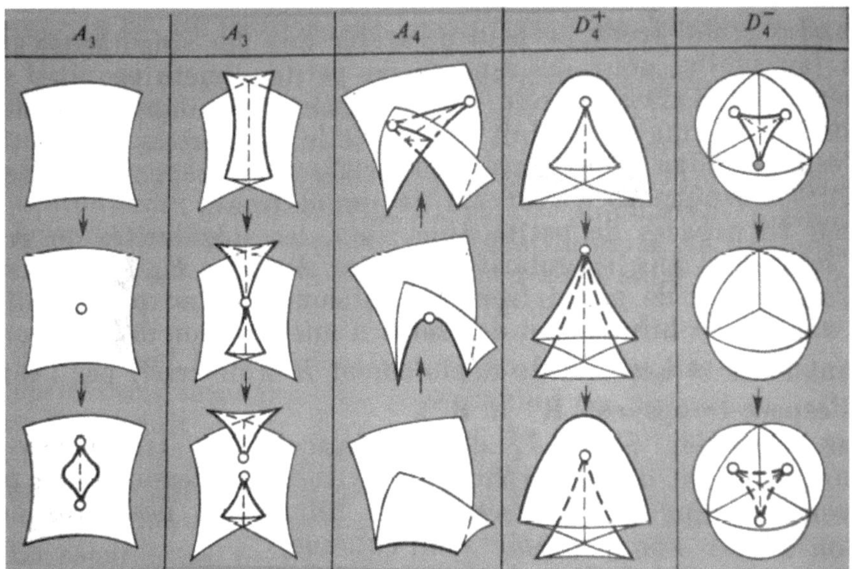

**Fig. 2.15** The 5 types of wavefront metamorphoses occurring in generic 1-parameter families (extracted from [7])

- Then, they noted that this path of hedgehogs can be made generic so that when $t$ varies, the qualitative changes in the singularities of $\mathcal{H}_{h+tr}$ are all of one of the 5 types of generic wave front metamorphoses described by Arnold (see Fig. 2.15);
- Finally, they checked that no one of these metamorphoses changes the quantity $r^+ - r^- - \left(q^+ - q^-\right)/2$.

∎

In order to conclude this subsection, let us mention the following unsolved problem raised by R. Langevin, G. Levitt and H. Rosenberg in [82]:

**Problem 2.6.1** Does there exist a generic projective hedgehog without any swallowtail in $\mathbb{R}^3$?

In [115], G. Panina and the author discussed the discrete (i.e., the piecewise linear) counterpart of the same problem. More precisely, G. Panina and the author defined swallowtails and cuspidal edges for the discrete case, derived an analogous counting formula, and presented a discussion on the open problem of existence of a generic projective hedgehog without swallowtails in this setting of polyhedral hedgehogs. In Sect. 10.2, we will present a partial answer to Problem 2.6.1 by proving that in every generic path of hedgehogs performing the eversion of the sphere in $\mathbb{R}^3$, there exists a hedgehog that have positive swallowtails We will see that it is easy to check that there is also a hedgehog that have negative swallowtails by considering the generic metamorphoses of Fig. 2.15.

## 2.7  A Few Words on Duality

To conclude this background section, let us say a word on duality for hedgehogs, which will be useful in many circumstances. In the setting of convex bodies of $(n+1)$-Euclidean vector space $\mathbb{R}^{n+1}$, we can distinguish two main notions of duality which are closely related: projective duality and polarity duality with respect to the origin $o$. Let us briefly recall these notions and consider how they can be extended and interpreted in the setting of $C^2$-hedgehogs.

**Projective Duality**

Let $\mathbb{P}$ be the projective space $P(E)$, where $E = \mathbb{R}^{n+2}$. Recall that each point $[x]$ of $\mathbb{P}$ is of the form

$$[x] = \left\{ \lambda\,(x_1, \ldots, x_{n+2}) \,\middle|\, \lambda \neq 0 \text{ and } (x_1, \ldots, x_{n+2}) \in E \setminus \{0\} \right\},$$

and that $[x_1, \ldots, x_{n+2}]$ are called homogeneous coordinates. We say that a subset $K$ of $\mathbb{P}$ is a **convex body** *of* $\mathbb{P}$ if, for any hyperplane $H$ in $\mathbb{P} \setminus K$, $K$ is a convex body of $\mathbb{R}^{n+1} = \mathbb{P} \setminus H$. Let $K$ be such a convex body of $\mathbb{P}$. Recall that projective duality yields a bijection between $\mathbb{P}^* = P(E^*)$ and the set $\mathcal{H}(E)$ of hyperplanes of $\mathbb{P}$, by assigning to each point $\{\lambda f \mid \lambda \neq 0\}$ of $\mathbb{P}^*$, $f$ being a nonnull linear form $f \in E^*$, the projective hyperplane $P(Ker f)$. Thus we can identify $\mathbb{P}^*$ with $\mathcal{H}(E)$. The **dual convex body** of $K$, say $K^*$, can thus be defined in $\mathbb{P}^* = \mathcal{H}(E)$ to be the closure of the set of all hyperplanes disjoint from $K$. We can then check that $K^*$ is a convex body, and if $K$ is smooth and strictly convex, then so is $K^*$. The **projective Legendre transform** $\mathcal{L}_K : \partial K \to \partial K^*$ is then the bijection given by $\mathcal{L}_K(x) = T_x(\partial K)$, where $\partial K$ and $\partial K^*$ denote the respective boundaries of $K$ and $K^*$, and $T_x(\partial K)$ the support hyperplane of $\partial K$ at $x$. Since we have a canonical isomorphism between $E$ and its bidual $E^{**}$, which allows us to identify $E$ and $E^{**}$, we can see $K^{**}$ as a convex body of $\mathbb{P}$, and check that $K^{**} = K$.

Let us adapt this definition to $C^2$-hedgehogs of $(n+1)$-Euclidean vector space $\mathbb{R}^{n+1}$. Let $\mathcal{H}_h$ be such a hedgehog. We can see $\mathcal{H}_h$ in $\mathbb{P}$ by adding the hyperplane at infinity to $\mathbb{R}^{n+1}$. More precisely, we imbed $E = \mathbb{R}^{n+1}$ in $\mathbb{P}$ by

$$x = (x_1, \ldots, x_{n+1}) \in E \mapsto [x_1, \ldots, x_{n+1}, 1] \in \mathbb{P},$$

so that $\mathbb{R}^{n+1}$ is identified with the set of $[x] \in \mathbb{P}$ whose last homogeneous coordinate is nonzero.

The family of support hyperplanes $(H_h(u))_{u \in \mathbb{S}^n}$ with equation $\langle x, u \rangle = h(u)$ then defines the dual hypersurface, say $\mathcal{H}_h^*$ in $\mathbb{P}^* = \mathcal{H}(E)$, that is parametrized by the map assigning to each $u \in \mathbb{S}^n$ the hyperplane with homogeneous linear equation $\langle x, u \rangle - h(u)\, x_{n+2} = 0$, where $x = (x_1, \ldots, x_{n+1})$. Equipping $E = \mathbb{R}^{n+2}$ with the standard Euclidean structure, we can then see $\mathcal{H}_h^*$ as the hypersurface of $\mathbb{P}$ that is parametrized by

$$x_h^* : \mathbb{S}^n \subset \mathbb{R}^{n+1} \to \mathbb{P}, \, u = (u_1, \ldots, u_{n+1}) \mapsto [u_1, \ldots, u_{n+1}, -h(u)].$$

In the case where the support function $h$ does not vanish on $\mathbb{S}^n$, we can thus regard $\mathcal{H}_h^*$ as the starlike hypersurface of $\mathbb{R}^{n+1}$ that is parametrized by

$$x_h^* : \mathbb{S}^n \to \mathbb{R}^{n+1}, u \mapsto -\frac{u}{h(u)}.$$

**Polarity Duality with Respect to the Origin**

In order to define a duality for convex bodies in $(n+1)$-Euclidean vector space $\mathbb{R}^{n+1}$ itself, we have to counterbalance the removal of the hyperplane at infinity by the selection of a distinguished point in $\mathbb{R}^{n+1}$, say the origin $o$ of $\mathbb{R}^{n+1}$. Let $\mathcal{K}_o^{n+1}$ denote the set of all convex bodies of $\mathbb{R}^{n+1}$ with the origin $o$ as an interior point. For every $K \in \mathcal{K}_o^{n+1}$, the *polar* (dual) *body* of $K$ (with respect to $o$) can be defined by

$$K^o = \left\{ x^* \in \left(\mathbb{R}^{n+1}\right)^* \,\middle|\, x^*(y) \le 1 \text{ for all } y \in K \right\},$$

or by

$$K^o = \left\{ x \in \mathbb{R}^{n+1} \,\middle|\, \langle x, y \rangle \le 1 \text{ for all } y \in K \right\}$$

by using the isomorphism between $V = \mathbb{R}^{n+1}$ and $V^*$ induced by the scalar product to identify both spaces. One can then easily check that $K^o \in \mathcal{K}_o^{n+1}$, and prove that the polarity correspondence $p_o : \mathcal{K}_o^{n+1} \to \mathcal{K}_o^{n+1}$, $K \mapsto K^o$ is a duality, that is:

$$\forall K \in \mathcal{K}_o^{n+1}, K^{oo} := \left(K^o\right)^o = K.$$

For every $K \in \mathcal{K}_o^{n+1}$, the radial function of the polar body $K^o$ is defined by

$$\rho_{K^o}(u) := \max\left\{ \lambda \in \mathbb{R}_+ \,\middle|\, \lambda u \in K^o \right\} \text{ for all } u \in \mathbb{S}^n,$$

and related to the support function $h_K$ of $K$ by

$$\rho_{K^o} = \frac{1}{h_K}.$$

Now we can extend the notion of polar to $C^2$-hedgehogs whose support function does not vanish on $\mathbb{S}^n$. For such a hedgehog, the polar of $\mathcal{H}_h$ (with respect to $o$) can be defined to be the starlike hypersurface $\mathcal{H}_h^o$ parametrized by

$$x_h^o : \mathbb{S}^n \to \mathbb{R}^{n+1}, u \mapsto \frac{u}{h(u)}.$$

We then notice that polarity (with respect to $o$) coincides up to a sign with projective duality in the case that $K \in \mathcal{K}_o^{n+1}$ (resp. the support function of the hedgehog does not vanish on $\mathbb{S}^n$). For this reason, it will sometimes happen that we confuse both notions and simply speak of the dual hypersurface of $\mathcal{H}_h$ (with $h$ non-vanishing on $\mathbb{S}^n$) to be the starlike hypersurface $\mathcal{H}_h^*$ parametrized by

$$x_h^* : \mathbb{S}^n \to \mathbb{R}^{n+1}, u \mapsto \frac{u}{h(u)}.$$

Note that for all $u \in \mathbb{S}^n$, $x_h^*(u)$ (resp. $x_h^o(u)$) is a normal vector to $\mathcal{H}_h$ at $x_h(u)$, and $x_h(u)$ is a normal vector to $\mathcal{H}_h^*$ (resp. $\mathcal{H}_h^o$) at $x_h^*(u)$ (resp. $x_h^o(u)$). It is also worth to note that the polarity correspondence $\mathcal{H}_h \mapsto \mathcal{H}_h^o$ can be regarded as the composition $\mathcal{I} \circ \mathcal{P}$, where $\mathcal{I}$ denotes the inversion with respect to the unit sphere, defined by $\mathcal{I}(x) = x/\|x\|^2$ for all $x \neq 0$, and $\mathcal{P}$ the map that assigns to each hedgehog $\mathcal{H}_h$ (with non-vanishing support function), its **pedal hypersurface** (with respect to the origin) $\mathcal{P}(\mathcal{H}_h)$ that is obtained by assigning to each $x_h(u) \in \mathcal{H}_h$ the foot $h(u)u$ of the perpendicular from the origin to the support hyperplane of $\mathcal{H}_h$ at $x_h(u)$:

$$\mathcal{H}_h \xrightarrow{\ \mathcal{P}\ } \mathcal{P}(\mathcal{H}_h) \xrightarrow{\ \mathcal{I}\ } \mathcal{H}_h^o = (\mathcal{I} \circ \mathcal{P})(\mathcal{H}_h)$$
$$x_h(u) \longmapsto h(u)u \longmapsto x_h^o(u) = u/h(u)$$

Let us recall that in the planar case, cusp points of $\mathcal{H}_h$ correspond to inflection points of $\mathcal{H}_h^*$, and multiple points of $\mathcal{H}_h$ correspond to multiple tangent lines to $\mathcal{H}_h^*$, that is, to lines that are tangent to $\mathcal{H}_h^*$ at more than one points (Fig. 2.16).

**A Few Words About the Generic Case in $\mathbb{R}^3$**  We have recalled above that there exists an open dense subset $\mathcal{U}$ of $C^\infty(\mathbb{S}^2; \mathbb{R})$ in the $C^4$-topology such that: for all

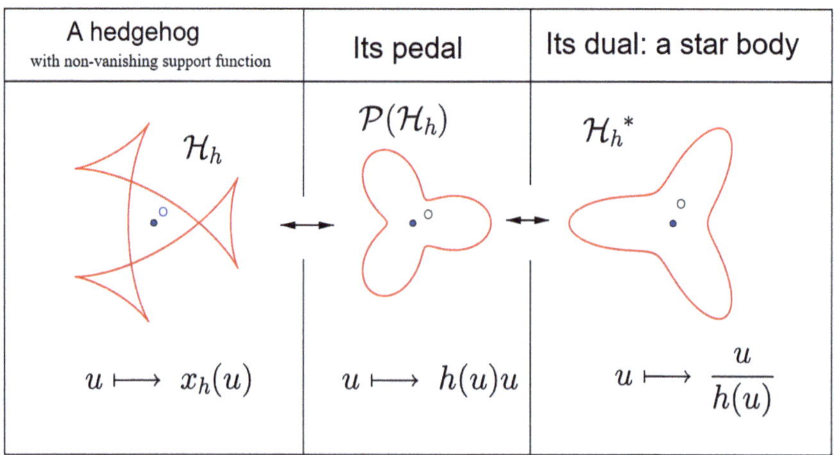


**Fig. 2.16**  A plane $C^2$ hedgehog, its pedal, and its dual

$h \in \mathcal{U}$, the singularities of $\mathcal{H}_h$ are all equivalent to one of the three models of singularities represented in Fig. 2.14, that is to a cuspidal edge or a (negative or positive) swallowtail. The singularities of the Gauss map of a surface dual to such a generic hedgehog $\mathcal{H}_h$ are parabolic curves (i.e., curves along which the Gaussian curvature vanishes), and (elliptic or hyperbolic) cusps of the Gauss map (which are also called **godrons**). Of course, parabolic curves are separating elliptic and hyperbolic regions of $\mathcal{H}_h^*$. Recall that the parabolic curves consist of the points where there is a unique (but double) asymptotic direction, and that godrons are the parabolic points at which the unique asymptotic direction is tangent to the parabolic curve. For a generic hedgehog $\mathcal{H}_h$, parabolic curves of $\mathcal{H}_h^*$ correspond to cuspidal edges of $\mathcal{H}_h$, and (elliptic or hyperbolic) godrons to (negative or positive) swallowtails. Of course, the self-intersection curves of swallowtails of $\mathcal{H}_h$ then correspond to planes doubly tangent to $\mathcal{H}_h^*$. Recall that a godron is said to be elliptic or positive (resp. hyperbolic or negative) if, when we tend towards the godron on the parabolic line, the half-asymptotic curves directed to the hyperbolic region point towards (resp. away from) the godron.

Remind that an asymptotic direction at a point $m$ of a smooth surface $M^2$ in $\mathbb{R}^3$ is a direction in which $M^2$ has zero sectional curvature. An asymptotic curve on $M^2$ is a curve whose direction at every point is an asymptotic direction. The asymptotic curves form a pair of transverse foliations on the hyperbolic regions of $M^2$ and a family of cusps on a parabolic curve, except at godrons, at which the unique asymptotic direction is tangent to the parabolic curve. For a generic immersion $x :$ $M^2 \to \mathbb{R}^3$ of a smooth surface $M^2$ in $\mathbb{R}^3$, there are three topologically distinct types of configurations of the asymptotic curves near a godron (see Fig. 2.17), and godrons of $M^2$ are characterized on parabolic curves of $M^2$ by the property of being in the closure of the set of geodesic inflections of asymptotic curves (which is generically a smooth curve of points, the so-called **flecnodal curve**). One of the three types of configurations of asymptotic curves near a godron (Fig. 2.17a) occurs at hyperbolic cusps, and the other two (Fig. 2.17b and c) at elliptic cusps. Similarly, there exists an open dense subset $\mathcal{U}$ of $C^\infty \left( \mathbb{S}^2; \mathbb{R} \right)$ in the $C^4$-topology such that: for all $h \in \mathcal{U}$, the foliations induced on $\mathbb{S}^2$ by the asymptotic curves of $\mathcal{H}_h$ have the same three topological types of singularities near (the spherical representation of) a swallowtail, and at the source $\mathbb{S}^2$, swallowtails of $\mathcal{H}_h$ are characterized on cuspidal edges of $\mathcal{H}_h$ by the property of being in the closure of the set of inflection points of asymptotic curves. One of the three types of configurations of (the spherical representation of) asymptotic curves near (the spherical representation of) a swallowtail (Fig. 2.17a) occurs for positive swallowtails, and the other two (Fig. 2.17b and c) for negative swallowtails [90].

This duality can for instance be explained by the fact that if we associate to every point $u$ of the northern hemisphere $\mathbb{S}_+^2$ of $\mathbb{S}^2$, the point $(a, b, c)$ of $\mathbb{R}^3$ such that $ax + by + z = c$ be an equation of the plane through $x_h(u)$ that is orthogonal to $u$, then we obtain a smooth surface $M^2$ of which the parabolic points (which are the singularities of the Gauss map of $M^2$) exactly correspond to the singular points of $x_h \left( \mathbb{S}_+^2 \right)$, the godrons to the swallowtails of $x_h \left( \mathbb{S}_+^2 \right)$, and the asymptotic curves

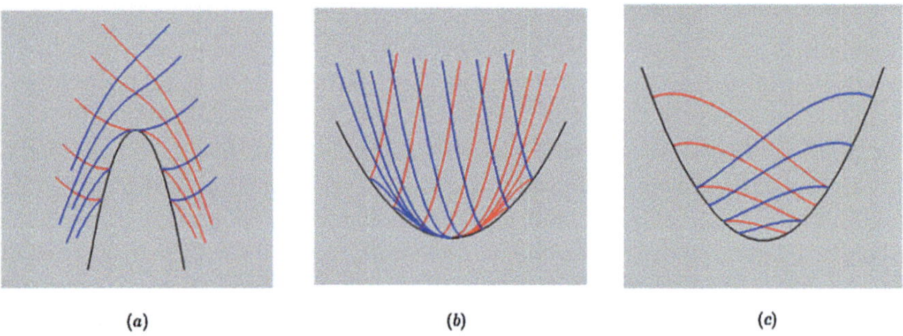

**Fig. 2.17** The three distinct topological types of configurations of asymptotic curves near a Gaussian cusp (extracted from [11])

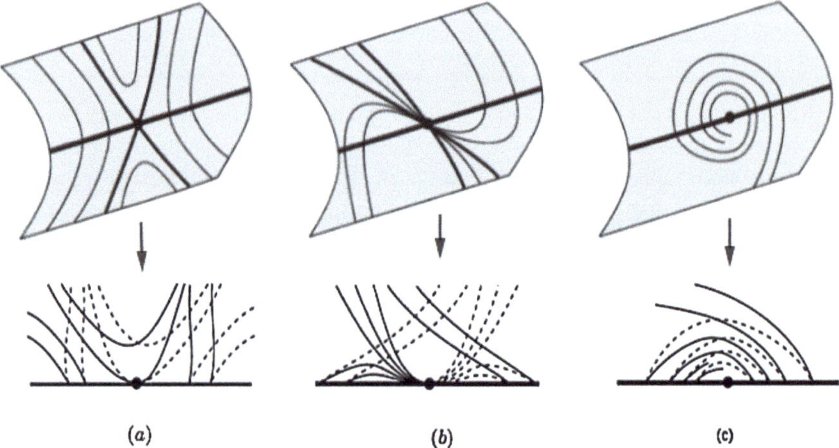

**Fig. 2.18** The three types of configurations of Fig. 2.17 are obtained by projection (figure extracted from [11])

to those of $x_h \left( \mathbb{S}^2_+ \right)$. For more details and proofs see [11, 74, 90] and the references herein. Note that the three types of configurations shown in Fig. 2.17 are obtained by projection as indicated in Fig. 2.18.

Later on, we will make use of these duality notions on different occasions. We will also see how we may adapt the concept of duality to the context of complex hedgehogs or that of a non-Euclidean space.

## 2.8   Kronecker Index: Locus of Zeros and Projections

We will discover in the first subsection how the Kronecker index $i_h(x)$ is closely related to the geometry of the set of zeros of $h_x : \mathbb{S}^n \to \mathbb{R}$, $u \mapsto h(u) - \langle x, u \rangle$, which is the support function of the hedgehog $\mathcal{H}_{h_x} = \mathcal{H}_h - \{x\}$. We will see in the secund subsection that $C^2$-hedgehogs of $\mathbb{R}^{n+1}$ behave well under projections. More precisely, we will see how we can deduce information on a $C^2$-hedgehog by considering its images under orthogonal projections onto hyperplanes.

### 2.8.1   *Kronecker Index and Locus of Zeros of the Support Function*

The Kronecker index $i_h(x)$ of a point $x \in \mathbb{R}^{n+1} \backslash \mathcal{H}_h$ with respect to $\mathcal{H}_h$ in $\mathbb{R}^{n+1}$ is closely related to the locus of zeros of $h_x : \mathbb{S}^n \to \mathbb{R}$, $u \mapsto h(u) - \langle x, u \rangle$, which is the support function of $\mathcal{H}_{h_x} = \mathcal{H}_h - \{x\}$. It is important to recall here that the study of the Kronecker index together with orthogonal projection techniques adapted to hedgehogs (Theorem 2.8.3) will be the main ingredient in the resolution of the uniqueness conjecture of A.D. Alexandrov (see Sect. 4.4). For $n + 1 = 2$, we have the following simple relationship.

**Theorem 2.8.1 ([93])** *Let $\mathcal{H}_h$ be a $C^2$-hedgehog of $\mathbb{R}^2$. For every $x \in \mathbb{R}^2 \backslash \mathcal{H}_h$, the Kronecker index $i_h(x)$ (that is, the winding number of $\mathcal{H}_h$ around $x$) is given by*

$$i_h(x) = 1 - \frac{1}{2} n_h(x),$$

*where $n_h(x)$ denotes the number of cooriented support lines of $\mathcal{H}_h$ through $x$, that is, the number of zeros of $h_x : \mathbb{S}^1 \to \mathbb{R}$, $u \mapsto h(u) - \langle x, u \rangle$.*

*Figure 2.19 illustrates this result using the example of the hedgehog $\mathcal{H}_h$ with support function $h(\theta) := \cos(2\theta)$, $(\theta \in \mathbb{R}/2\pi\mathbb{Z})$.*

***Proof of Theorem 2.8.1*** Let $(x_1, x_2)$ be the standard coordinates in $\mathbb{R}^2$. Since $\mathcal{H}_h$ is obtained from $\mathcal{H}_{h_x}$ by the translation of vector $x$, we may assume that $x$ is the origin $0_{\mathbb{R}^2} = (0, 0)$ of $\mathbb{R}^2 \backslash \mathcal{H}_h$ even if that means replacing $\mathcal{H}_h$ by $\mathcal{H}_{h_x}$. If $0_{\mathbb{R}^2} = (0, 0) \in \mathbb{R}^2$, we obtain

$$i_h(0_{\mathbb{R}^2}) = \frac{1}{2\pi} \int_0^{2\pi} \frac{h(\theta)(h + h'')(\theta)}{(h^2 + h'^2)(\theta)} \, d\theta$$

$$= 1 + \frac{1}{2\pi} \int_0^{2\pi} \frac{h(\theta)h''(\theta)) - h'(\theta)^2}{(h^2 + h'^2)(\theta)} \, d\theta,$$

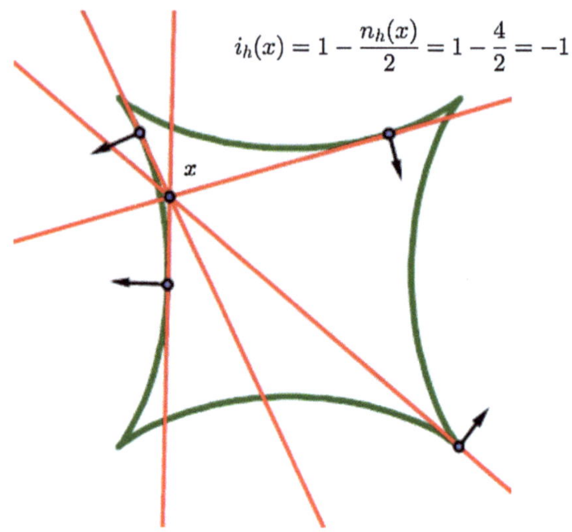

**Fig. 2.19** Evaluating $i_h(x)$ by considering $n_h(x)$



where $h(\theta) = h(\cos\theta, \sin\theta)$. In other words, we have

$$i_h\left(0_{\mathbb{R}^2}\right) = 1 + \frac{1}{2\pi}\int_{\Gamma_h}\omega,$$

where $\Gamma_h$ is the oriented curve of $\mathbb{R}^2$ that is parametrized by

$$\gamma_h : [0, 2\pi] \to \mathbb{R}^2, \ \theta \longmapsto \left(h(\theta), h'(\theta)\right),$$

and $\omega$ the 1-differential form given by

$$\omega(x_1, x_2) = \frac{x_1 dx_2 - x_2 dx_1}{x_1^2 + x_2^2} \quad \text{for all} \quad (x_1, x_2) \neq (0, 0) \ \text{in} \ \mathbb{R}^2.$$

Now the integral of $\omega/2\pi$ around $\Gamma_h$ is the winding number of $\Gamma_h$ around the origin $0_{\mathbb{R}^2}$, which measures the total algebraic number of turns of $\Gamma_h$ around $0_{\mathbb{R}^2}$. (i.e., that $\gamma_h(\theta)$ makes around $0_{\mathbb{R}^2}$ as $\theta$ varies from 0 to $2\pi$). This number is positive if $\Gamma_h$ turns counterclockwise, and negative if $\Gamma_h$ turns clockwise. Since the $x_1$-coordinate of $\gamma_h$ strictly increases (resp. decreases) when the $x_2$-coordinate of $\gamma_h$ is positive (resp. negative), this number must be equal to the opposite of $n_h\left(0_{\mathbb{R}^2}\right)$, and so we have well $i_h\left(0_{\mathbb{R}^2}\right) = 1 - \frac{1}{2}n_h\left(0_{\mathbb{R}^2}\right)$. $\blacksquare$

**Definition 2.8.1** Let $\mathcal{H}_h$ be a $C^2$-hedgehog of $\mathbb{R}^{n+1}$. The $i_h$-interior, or simply **interior**, of $\mathcal{H}_h$ (resp. the $i_h$-exterior, or simply the **exterior**, of $\mathcal{H}_h$) in $\mathbb{R}^{n+1}$ is defined to be the subset of $\mathbb{R}^2$ given by:

$$I_h := \left\{ x \in \mathbb{R}^2 \backslash \mathcal{H}_h \mid i_h(x) \neq 0 \right\}$$

$$\left( \text{resp. } E_i(\mathcal{H}_h) \text{ or } E(\mathcal{H}_h) := \left\{ x \in \mathbb{R}^2 \backslash \mathcal{H}_h \mid i_h(x) = 0 \right\} \right).$$

**Corollary 2.8.1 ([93])** *For every $C^2$-hedgehog $\mathcal{H}_h$ of $\mathbb{R}^2$,*

$$C_h := \left\{ x \in \mathbb{R}^2 \backslash \mathcal{H}_h \mid i_h(x) = 1 \right\},$$

*is a convex part (possibly empty) of $\mathbb{R}^2$. We call $C_h$ the **convex interior** of $\mathcal{H}_h$.*

***Proof of Corollary 2.8.1*** We can assume without loss of generality that the integral of $h$ over $\mathbb{S}^1$ is nonnegative since $C_h = C_{\tilde{h}}$, where $\tilde{h}(u) = -h(-u)$, $\left(u \in \mathbb{S}^1\right)$. From Theorem 2.8.1, $C_h$ is the set of $x \in \mathbb{R}^2$ such that the function $u \mapsto h_x(u) - \langle x, u \rangle$ does not vanish on $\mathbb{S}^1$. It can of course be empty. If $x \in C_h$, then $h_x : \mathbb{S}^1 \to \mathbb{R}$ must remain positive on $\mathbb{S}^1$ by continuity of $h_x$ (the integral of $h_x$ over $\mathbb{S}^1$ is equal to the one of $h$, and thus nonnegative). $C_h$ can thus be written

$$C_h = \bigcap_{u \in \mathbb{S}^1} P_h^-(u),$$

where $P_h^-(u)$ is the open half-plane with inequation $\langle x, u \rangle < h(u)$. Therefore, $C_h$ is a convex of $\mathbb{R}^2$ as an intersection of convex subsets of $\mathbb{R}^2$. ∎

Theorem 2.8.1 teaches us, in particular, that any $C^2$-hedgehog of $\mathbb{R}^2$ with empty convex interior turns its convexity outwards (see the proof of Proposition 1 in [93] for more details).

**Corollary 2.8.2 ([93])** *Let $\mathcal{H}_h$ be a $C^2$-hedgehog of $\mathbb{R}^2$ with empty convex interior: $C_h = \varnothing$. If $u \in \mathbb{S}^1$ is a regular point of $x_h : \mathbb{S}^1 \to \mathbb{R}^2$, then the support line with equation $\langle x, u \rangle = h(u)$ does not meet the exterior of $\mathcal{H}_h$ in the vicinity of $x_h(u)$.*

An immediate consequence of Corollary 2.8.2 is that any such hedgehog $\mathcal{H}_h$ (for instance a projective hedgehog $\mathcal{H}_h$ or a hedgehog $\mathcal{H}_h$ of zero algebraic length or, equivalently, of zero mean width: see Chaps. 3 and 4) is contained in the convex hull of its singularities and such that $x_h : \mathbb{S}^1 \to \mathbb{R}^2$ admits at least 4 singular points. From this last remark, we can deduce by duality the following particular case of the ***tennis ball theorem***, which states that any closed simple smooth curve on $\mathbb{S}^2$ dividing the sphere into two parts of equal area must have at least four inflection points [8].

**Corollary 2.8.3 ([93])**  *Let $C$ be a closed simple smooth curve of $\mathbb{S}^2$ that is everywhere transverse to the meridians. If $C$ has at most 3 inflection points, then $C$ is contained in an open hemisphere of $\mathbb{S}^2$. In particular, if $C$ divides $\mathbb{S}^2$ into two parts of equal area, then $C$ has at least four inflection points.*

Here, an inflection point is simply a zero of the geodesic curvature, that is, a point of at least second order tangency of the curve with a great circle of $\mathbb{S}^2$.

***Proof of Corollary 2.8.3***  By virtue of assumptions, $C$ admits a parametrization of the form

$$
\gamma_h : \quad \mathbb{S}^1 \longrightarrow \mathbb{S}^2 \subset \mathbb{R}^2 \times \mathbb{R}
$$
$$
u \longmapsto \frac{1}{\sqrt{1+h(u)^2}}(u, h(u)),
$$

where $h : \mathbb{S}^1 \to \mathbb{R}$ is the support function of a $C^2$ hedgehog $\mathcal{H}_h$ of $\mathbb{R}^2$. The curves $C$ and $\mathcal{H}_h$ can be regarded as two dual curves: the section of $\mathbb{S}^2$ (resp. $\mathbb{R}^2 \times \{-1\}$) by the linear plane that is orthogonal to $(x_h(u), -1)$ (resp. $\gamma_h(u)$) is the great circle of $\mathbb{S}^2$ that is tangent to $C$ at $\gamma_h(u)$ (resp. is the support line with unit normal vector $(u, -1)$ of the plane hedgehog $\mathcal{H}_h \times \{-1\} \subset \mathbb{R}^2 \times \{-1\}$). Therefore, inflection points of $C$ exactly correspond to singular points of $x_h : \mathbb{S}^1 \to \mathbb{R}$. This can be confirmed by the direct calculation of the geodesic curvature of $C$:

$$
K_g = \left( \frac{1+h^2}{1+\|x_h\|^2} \right)^{\frac{3}{2}} R_h.
$$

Now assume that $C$ has at most 3 inflection points. It then results that $x_h : \mathbb{S}^1 \to \mathbb{R}^2$ has at most 3 singular points, and it follows from Corollary 2.8.2 that $\mathcal{H}_h$ has a nonempty convex interior. Let $x \in C_h$. From Theorem 2.8.1, no support line of $\mathcal{H}_h$ passes through $x$. It follows that the section of $\mathbb{S}^2$ by the linear plane orthogonal to $(x, -1)$ is a great circle that does not meet $C$. Indeed, if this great circle contained a point $\gamma_h(u)$ of $C$ then $(x, -1)$ would belong to the support line of $\mathcal{H}_h$ that is the section of $\mathbb{R}^2 \times \{-1\}$ by the linear plane that is orthogonal to $\gamma_h(u)$. Therefore, $C$ is contained in one of the two hemispheres separated by the linear plane that is orthogonal to $(x, -1)$. ∎

The following corollary is also an immediate consequence of Theorem 2.8.1.

**Corollary 2.8.4 ([93])**  *Let $\mathcal{H}_h$ be a $C^2$-hedgehog of $\mathbb{R}^2$ such that $h(-u) = -h(u)$ for some $u \in \mathbb{S}^1$ (i.e., a plane $C^2$-hedgehog two of the support lines of which are coincident). Then $C_h = \varnothing$, and thus*

$$
v_2(h) := \int_{\mathbb{R}^2 \setminus \mathcal{H}_h} i_h(x)\, d\lambda(x) \leq 0,
$$

*where $\lambda$ denotes the Lebesgue measure on $\mathbb{R}^2$.*

In particular, if $\mathcal{H}_h$ is a $C^2$-hedgehog of $\mathbb{R}^2$ that is projective (i.e., such that $h$ is odd), or such that

$$\int_0^{2\pi} h\,(\theta)\,d\theta = 0 \text{ or, equivalently, } \int_0^{2\pi} (h\,(\theta + \pi) + h\,(\theta))\,d\theta = 0$$

(i.e., of zero algebraic length or, equivalently, of zero mean width: see Chaps. 3 and 4), then $v_2\,(h) \leq 0$.

As a consequence, we can deduce that the map $h \mapsto \sqrt{-v_2\,(h)}$ is a norm associated with a scalar product on the linear space of projective $C^2$-hedgehogs (resp. of $C^2$-hedgehogs of zero mean width) defined up to a translation in $\mathbb{R}^2$. This last result will be generalized in Chap. 3.

**The Dimension 3 Case**

Recall that the Kronecker index $i_h\,(x)$ of a point $x \in \mathbb{R}^3 \backslash \mathcal{H}_h$ with respect to $\mathcal{H}_h$ can be defined as the degree of the map

$$\mathcal{U}_{(h,x)} : \mathbb{S}^2 \to \mathbb{S}^2, \, u \longmapsto \frac{x_h(u) - x}{\|x_h(u) - x\|} \, ;$$

$i_h\,(x)$ may be interpreted as the algebraic intersection number of an oriented half-line with origin $x$ with the hypersurface $\mathcal{H}_h$ equipped with its transverse orientation (number independent of the oriented half-line for an open dense set of directions). Note that for any $C^2$-hedgehog $\mathcal{H}_h$ of $\mathbb{R}^3$, and any $x \in \mathbb{R}^3 \backslash \mathcal{H}_h$, the set $h_x^{-1}\,(\{0\})$ consists of a finite number of disjoint simple smooth closed curves of $\mathbb{S}^2$ on which $h_x$ changes sign cleanly.

**Theorem 2.8.2 ([102])** *Let $\mathcal{H}_h$ be a $C^2$-hedgehog of $\mathbb{R}^3$. For every $x \in \mathbb{R}^3 \backslash \mathcal{H}_h$, the Kronecker index $i_h\,(x)$ is given by*

$$i_h\,(x) = r_h^+\,(x) - r_h^-\,(x),$$

*where $r_h^-\,(x)$ (resp. $r_h^+\,(x)$) denotes the number of connected components of $\mathbb{S}^2 \backslash h_x^{-1}\,(\{0\})$ on which $h_x(u) := h(u) - \langle x, u \rangle$ is negative (resp. positive).*

***Proof of Theorem 2.8.2*** For the convenience of the reader, we recall the main steps of the proof. For all $x \in \mathbb{R}^3 \backslash \mathcal{H}_h$, we have $\nabla\,(h_x)\,(u) \neq 0$ whenever $h_x\,(u) = 0$, $(u \in \mathbb{S}^2)$. Therefore, for all $x \in \mathbb{R}^3 \backslash \mathcal{H}_h$, the set $h_x^{-1}\,(\{0\})$ consists of a finite number, say $c_h\,(x)$, of disjoint simple smooth closed spherical curves on which $h_x$ changes sign cleanly. Note that $c_h\,(x) = r_h^-\,(x) + r_h^+\,(x) - 1$. Then, the proof relies on the following two lemmas.

**Lemma 1** *The map $x \mapsto i_h(x) - \left(r_h^+(x) - r_h^-(x)\right)$ is constant on $\mathbb{R}^3 \backslash \mathcal{H}_h$.*

[**Proof** The first step consists in noticing that $x \mapsto r_h^-(x)$, $x \mapsto r_h^+(x)$ and thus $x \mapsto c_h(x)$ are constant on each connected component of $\mathbb{R}^3 \setminus \mathcal{H}_h$. The second one, consists in proving that $x \mapsto i_h(x) - \left(r_h^+(x) - r_h^-(x)\right)$ remains constant as $x$ transversally crosses an elliptic (resp. hyperbolic) region of $\mathcal{H}_h$. As $x$ transversally crosses a simple elliptic region of $\mathcal{H}_h$ at $x_h(u)$ from locally convex to locally concave side, we must distinguish two cases: $(i)$ If $u$ is pointing towards the locally concave side, then $i_h(x)$ decreases by one unit whereas $r_h^-(x)$ increases by one unit, and $r_h^+(x)$ remains constant; $(ii)$ If $u$ is pointing towards the locally convex side, then $i_h(x)$ and $r_h^+(x)$ increases by one unit whereas $r_h^-(x)$ remains constant. As $x$ transversally crosses a simple hyperbolic region of $\mathcal{H}_h$ at $x_h(u)$ in the direction of $-u$, which is the unit normal at $x_h(u)$ since $x_h(u)$ is hyperbolic, then $i_h(x)$ decreases by one unit and there are exactly two possibilities: $(i)$ If $c_h(x)$ increases by one unit then $r_h^-(x)$ increases by one unit and $r_h^+(x)$ remains constant; $(ii)$ If $c_h(x)$ decreases by one unit then $r_h^+(x)$ decreases by one unit and $r_h^-(x)$ remains constant].

**Lemma 2** *If $\|x\|$ is sufficiently large, then $c_h(x) = 1$.*

[**Proof** This second lemma essentially follows from the fact that $x_h : \mathbb{S}^2 \to \mathcal{H}_h$ can be interpreted as the inverse of the Gauss map].

Lemma 2 implies that $r_h^-(x) = r_h^+(x) = 1$ when $\|x\|$ is sufficiently large, and thus the theorem follows from Lemma 1. $\blacksquare$


### 2.8.2 Orthogonal Projection Techniques

Let $\mathcal{H}_h$ be a $C^2$-hedgehog of $\mathbb{R}^{n+1}$. It is possible to deduce information on $\mathcal{H}_h$ by considering its images under orthogonal projections onto hyperplanes. We proceed as follows. For any $\xi \in \mathbb{S}^n$ we consider the restriction $h_\xi$ of $h$ to the great sphere $\mathbb{S}_\xi^{n-1} = \mathbb{S}^n \cap \xi^\perp$, where $\xi^\perp$ is the linear subspace orthogonal to $\xi$. Note that $h_\xi$ is the support function of the hedgehog $\mathcal{H}_{h_\xi}$ that is the image of $x_h\left(\mathbb{S}_\xi^{n-1}\right)$ under the orthogonal projection onto $\xi^\perp$:

$$\mathcal{H}_{h_\xi} = \pi_\xi \left[x_h \left(\mathbb{S}_\xi^{n-1}\right)\right],$$

where $\pi_\xi$ is the orthogonal projection onto the hyperplane $\xi^\perp$. In order to illustrate our point, consider the case where $n+1 = 2$. Then, the index of a point $x \in \xi^\perp - \mathcal{H}_{h_\xi}$ with respect to $\mathcal{H}_{h_\xi}$ (i.e., the winding number of $\mathcal{H}_{h_\xi}$ around $x$) gives us information on the curvature of $\mathcal{H}_h$ on the line $\{x\} + \mathbb{R}\xi$. For every $\xi \in \mathbb{S}^2$, $P_\xi$ will denote the oriented plane vector of $\mathbb{R}^3$ with unit normal vector $\xi$, and $S_\xi^+$ the half unit sphere given by $\langle u, \xi \rangle \geq 0$, $\left(u \in \mathbb{S}^2\right)$.

**Theorem 2.8.3** *Let $\mathcal{H}_h$ be a $C^2$-hedgehog of $\mathbb{R}^3$, and let $x$ be a regular value of the map $x_h^\xi = \pi_\xi \circ x_h : \mathbb{S}^2 \to P_\xi$. The index of $x \in P_\xi \setminus \mathcal{H}_{h_\xi}$ with respect to $\mathcal{H}_{h_\xi}$ is* *given by*

$$i_{h_\xi}(x) = n_h^\xi(x)^+ - n_h^\xi(x)^-,$$

*where $n_h^\xi(x)^+$ (resp. $n_h^\xi(x)^-$) is the number of $u \in \mathbb{S}_\xi^+$ such that $x_h(u)$ is an elliptic (resp. a hyperbolic) point of $\mathcal{H}_h$ lying on the line $\{x\} + \mathbb{R}\xi$. See Fig. 2.20 for an illustation.*

***Proof of Theorem 2.8.3*** Let $n_h^\xi(x)$ be the algebraic number of intersections of the oriented line passing through $x$ and directed by $\xi$, say $D_x(\xi)$, with the surface $x_h\left(\mathbb{S}_\xi^+\right)$ equipped with its transverse orientation: $n_h^\xi(x)$ is given by

$$n_h^\xi(x) = n_h^\xi(x)^+ - n_h^\xi(x)^-,$$

where $n_h^\xi(x)^+$ (resp. $n_h^\xi(x)^-$) is the number of $u \in \mathbb{S}_\xi^+$ such that $x_h(u)$ is an elliptic (resp. a hyperbolic) point of $\mathcal{H}_h$ lying on the line $\{x\} + \mathbb{R}\xi$. Indeed, the tangent map $T_u x_h : T_u \mathbb{S}^2 \to T_{x_h(u)}\mathbb{S}^2 = T_u \mathbb{S}^2$ retains or reverses the orientation depending on whether $R_h(u) = \det[T_u x_h]$ is $> 0$ or $< 0$, $\left(u \in \mathbb{S}^2\right)$. It therefore suffices to prove that

$$i_{h_\xi}(x) = n_h^\xi(x).$$

Let $(x_1, x_2, x_3)$ be the standard coordinates in $\mathbb{R}^3$. Without loss of generality, we can identify $P_\xi$ to the plane with equation $x_3 = 0$ (and thus with the Euclidean vector plane $\mathbb{R}^2$), and assume that $x$ is its origin $0_{\mathbb{R}^2}$. The index $i_{h_\xi}(x)$ is then the winding number of $\mathcal{H}_{h_\xi}$ around $x \in P_\xi \setminus \mathcal{H}_{h_\xi}$, and

$$i_{h_\xi}(x) = \frac{1}{2\pi} \int_{\mathcal{H}_{h_\xi}} \omega,$$

where $\omega$ the 1-differential form given by

$$\omega(x_1, x_2) = \frac{x_1 dx_2 - x_2 dx_1}{x_1^2 + x_2^2} \quad \text{for all} \quad (x_1, x_2) \neq (0,0) \text{ in } \mathbb{R}^2.$$

This index $i_{h_\xi}(x)$ can also be regarded as the winding number of $x_h\left(\mathbb{S}_\xi^1\right)$ around the oriented line $D_x(\xi)$, passing through $x$ and directed by $\xi$. In other words, $i_{h_\xi}(x)$ is given by

$$i_{h_\xi}(x) = \frac{1}{2\pi} \int_{x_h\left(\mathbb{S}_\xi^1\right)} \omega,$$

which can be checked by an easy calculation. So we have

$$i_{h_\xi}(x) = \frac{1}{2\pi} \int_{\partial S} \omega,$$

where $S$ denotes the surface $x_h \left( \mathbb{S}_\xi^+ \right)$ equipped with its transverse orientation. Since $x$ is a regular value of the map $x_h^\xi = \pi_\xi \circ x_h : \mathbb{S}^2 \to P_\xi$, there exists a small closed disc, say $D$, centered at $x$ in $P_\xi$ whose inverse image under $x_h^\xi$ is empty or admits a partition of the form

$$\left( x_h^\xi \right)^{-1}(D) = \bigcup_{k=1}^{N} D_k,$$

where $x_h^\xi = \pi_\xi \circ x_h$ is a $C^1$-diffeomorphism from $D_k$ onto $D$ for all $k \in \{1, \ldots, N\}$. By the Stokes's formula, we have

$$\frac{1}{2\pi} \int_{\partial S} \omega = \sum_{k=1}^{N} \frac{1}{2\pi} \int_{\partial S_k} \omega,$$

where $S_k$ denotes the surface $x_h(D_k)$ equipped with its transverse orientation. Now the oriented boundary $\partial S_k$ of $S_k$ turns exactly once around the oriented line $D_x(\xi)$ and this turn is counted positively or negatively depending on whether the curvature function $R_h$ is $> 0$ or $< 0$ on $S_k$. Therefore, it comes

$$i_{h_\xi}(x) = \frac{1}{2\pi} \int_{\partial S} \omega = n_h^\xi(x)^+ - n_h^\xi(x)^- .$$

$\blacksquare$

**Remark** The following is a classical question in geometric tomography: *What can we say about an object given some information about its projections?* In Sect. 4.4., we will make use of Theorem 2.8.3 to give a counter-example to an old conjectured characterization of the sphere. S. Myroshnychenko proved that two hedgehogs in dimension $n + 1 \geq 3$ coincide up to a translation and a reflection in the origin, if their projections on any two-dimensional plane are directly congruent and have no rigid motion symmetries (his proof relies on a nice general result on continuous functions on $\mathbb{S}^n$) [132].

**Theorem 2.8.4 ([93])** *Every $C^2$ projective hedgehog of $\mathbb{R}^{n+1}$ lies in the convex hull of its singularities.*

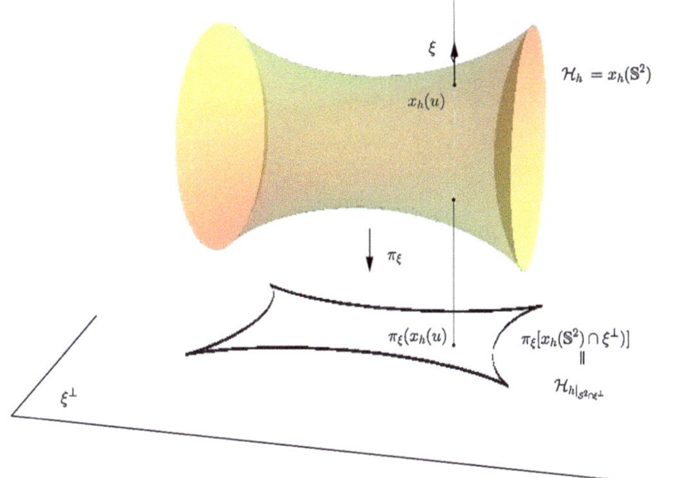


**Fig. 2.20**  Illustration of Theorem 2.8.3

***Proof of Theorem 2.8.4***  The proof proceeds by induction with respect to the dimension $n + 1$. For $n + 1 = 2$, the result is a consequence of Corollary 2.8.2. Make the induction assumption that the assertion is true for every $C^2$ projective hedgehog of $\mathbb{R}^n$ ($n \geq 2$). Suppose there exists some $C^2$ projective hedgehog $\mathcal{H}_h$ in $\mathbb{R}^{n+1}$ that does not lie in the convex hull of its singularities. Then there exists a support hyperplane of this convex hull that does not contain any singularity of $\mathcal{H}_h$. Note that $\mathcal{H}_h$ is strictly convex at any point $x_h\,(u)$ that lies on a such a hyperplane (i.e., all the principal radii of curvature of $\mathcal{H}_h$ are nonzero and have the same sign at $x_h\,(u)$). Now let $H$ be any linear hyperplane that is perpendicular to such a support hyperplane of $\mathcal{H}_h$. The restriction of $h$ to the unit sphere $\mathbb{S}^n \cap H$ of $H$ is the support function of a projective hedgehog of $H$ that is the orthogonal projection of $x_h\,(\mathbb{S}^n \cap H)$ onto $H$. This projective hedgehog of $H$ does not lie in the convex hull of its singularities in contradiction with the induction assumption since $H$ can be identified with $\mathbb{R}^n$.  ∎

## Projected Algebraic Area

Let $\mathcal{H}_h$ be a $C^2$-hedgehog of $\mathbb{R}^3$. For all $\xi \in \mathbb{S}^2$, let

$$v_2^\xi\,(h) := \frac{1}{2} \int_{\mathbb{S}^2} |\langle u, \xi \rangle|\, R_h\,(u)\, d\sigma\,(u)\,.$$

By translation invariance of the algebraic volume of $C^2$-hedgehogs in $\mathbb{R}^3$, we have

$$0 = \int_{\mathbb{S}^2} \langle u, \xi \rangle \, R_h\,(u)\,d\sigma\,(u) = \int_{\mathbb{S}_\xi^+} |\langle u, \xi \rangle|\, R_h\,(u)\,d\sigma\,(u)$$

$$- \int_{\mathbb{S}_\xi^-} |\langle u, \xi \rangle|\, R_h\,(u)\,d\sigma\,(u),$$

where $S_\xi^+$ (resp. $S_\xi^-$) is the half unit sphere given by $\langle u, \xi \rangle \geq 0$ (resp. $\langle u, \xi \rangle \leq 0$), $\left( u \in \mathbb{S}^2 \right)$. Therefore

$$v_2^\xi\,(h) = \int_{\mathbb{S}_\xi^+} |\langle u, \xi \rangle|\, R_h\,(u)\,d\sigma\,(u) = \int_{\mathbb{S}_\xi^-} |\langle u, \xi \rangle|\, R_h\,(u)\,d\sigma\,(u).$$

As we will see in Chap. 3, $R_h\,(u)\,u d\sigma\,(u)$ is the vector algebraic area element around $x_h\,(u) \in \mathcal{H}_h$ corresponding to the spherical area element $d\sigma\,(u)$ around $u \in \mathbb{S}^2$. So, $v_2^\xi\,(h)$ can be interpreted as the projected algebraic area of $\mathcal{H}_h$ onto the plane $P_\xi := \xi^\perp$. This ***projected algebraic area*** can obviously be written

$$v_2^\xi\,(h) = \int_{\Omega_\xi\,(h)} n_h^\xi\,(x)\,d\lambda\,(x)$$

where $\Omega_\xi\,(h)$ is the set of the regular values of the restriction of $\pi_\xi \circ x_h$ to $S_\xi^+$, and $\lambda$ the Lebesgue measure on $P_\xi := \xi^\perp$.

As an immediate consequence of Theorem 2.8.3, we then obtain the following.

**Corollary 2.8.5 ([93])** *Let $\mathcal{H}_h$ be a $C^2$-hedgehog of $\mathbb{R}^3$. For every $\xi \in \mathbb{S}^2$, the projected algebraic area $v_2^\xi\,(h)$ is equal to the algebraic area of the plane hedgehog $\mathcal{H}_{h_\xi}$ of $P_\xi$, where $h_\xi$ is the restriction of $h$ to the great circle $\mathbb{S}_\xi^1 = \mathbb{S}^2 \cap P_\xi$, that is:*

$$v_2^\xi\,(h) = v_2\left(h_\xi\right).$$

This corollary could also have been established by reducing to the convex case or by using the Stokes formula. We will say that $\mathcal{H}_{h_\xi}$ is the ***projected hedgehog*** of $\mathcal{H}_h$ onto $P_\xi := \xi^\perp$.

### Consequences for Projective Hedgehogs

Since projective hedgehogs have an empty convex interior, another immediate consequence of Theorem 2.8.3 is the following.

**Corollary 2.8.6 ([93])** *Let $\mathcal{H}_h$ be a $C^2$-hedgehog of $\mathbb{R}^3$ that is projective (i.e., such that $h$ is odd). If a line cuts $\mathcal{H}_h$ transversely then it must encounter a hyperbolic region of $\mathcal{H}_h$.*

As we have $0 \leq -i_{h_n}(x) = n_h^{\xi}(x)^- - n_h^{\xi}(x)^+ \leq n_h^{\xi}(x)^-$ for all $x \in \Omega_\xi(h)$, we also deduce the following.

**Corollary 2.8.7 ([93])** *Let $\mathcal{H}_h$ be a $C^2$-projective hedgehog of $\mathbb{R}^3$. For every $\xi \in \mathbb{S}^2$,*

$$\left| v_2\left(h_\xi\right) \right| \leq \frac{1}{2} s_-(h),$$

*where $v_2\left(h_\xi\right)$ is the area of the plane hedgehog whose support function is the restriction of $h$ to $\mathbb{S}_\xi^1 = \mathbb{S}^2 \cap P_\xi$, and $s_-(h)$ the total area of hyperbolic regions of $\mathcal{H}_h$.*

## 2.9   More About Indexes and the Locus of Zeros

### 2.9.1   New Index and Transverse Orientation

In [102], the author introduced a new notion of index of a point $x \in \mathbb{R}^3 \backslash \mathcal{H}_h$ with respect to $\mathcal{H}_h$. This index, denoted by $j_h(x)$, is defined by:

$$j_h(x) := 1 - c_h(x),$$

where $c_h(x)$ denotes the number of connected components of $(h_x)^{-1}(\{0\})$ on $\mathbb{S}^2$, that is, the number of closed spherical curves formed by points $u \in \mathbb{S}^2$ such that $x$ belongs to the support hyperplane of $\mathcal{H}_h$ at $x_h(u)$. In certain respects, this $j_h$-index can play in $\mathbb{R}^3$ the same role as the Kronecker index does in $\mathbb{R}^2$ (compare the definition of $j_h(x)$ with the relationship between the Kronecker index of $x$ with respect to the plane hedgehog $\mathcal{H}_h$ and the number of zeros of the function $h_x(u) = h(u) - \langle x, u \rangle, \left(u \in \mathbb{S}^1\right)$). From the proof of Theorem 2.8.3, the index $j_h : x \mapsto j_h(x)$ remains constant on each connected component of $\mathbb{R}^3 \backslash \mathcal{H}_h$. In particular, $j_h$ is equal to 0 on the unbounded component of $\mathbb{R}^3 \backslash \mathcal{H}_h$. It is worth noting that the value of $j_h(x)$ must obviously decrease as $x$ transversally crosses $\mathcal{H}_h$ at a simple elliptic point from locally convex to locally concave side. Thus, if $\mathcal{H}_h$ is the boundary of a convex body $K$ of which $x$ is an interior point, we must have $j_h(x) = 1$, whereas $i_h(x) = -1$ or $i_h(x) = 1$ depending on whether $u$ points inward or outward from $K$ at $x_h(u) \in \mathcal{H}_h = \partial K$, $\left(u \in \mathbb{S}^2\right)$.

Now, the $j_h$-index corresponds to the transverse orientation of $\mathcal{H}_h$ that is such that whenever $x_h(u)$ is a simple regular point of $\mathcal{H}_h$, then the normal line to $\mathcal{H}_h$ at $x_h(u)$, is oriented in the direction that $j_h$ decreases by one unit. Contrary to the usual transverse orientation of $\mathcal{H}_h$, it is clear from its definition that this transverse orientation of $\mathcal{H}_h$ does not depend on the choice of the orientation of normal lines to $x_h\left(\mathbb{S}^2\right) = x_{\tilde{h}}\left(\mathbb{S}^2\right)$. We call it the ***absolute transverse orientation*** of $\mathcal{H}_h$. From the above, this absolute transverse orientation cannot change on an elliptic region (i.e., a region on which the Gauss curvature of $\mathcal{H}_h$ remains positive): indeed, the absolute transverse orientation is then simply given by the direction of convexity. On the other hand, it is worth noting that the absolute transverse orientation of a $C^2$–hedgehog $\mathcal{H}_h$ can be reversed along certain self-intersection curves of $\mathcal{H}_h$ (see paragraph below). From the above such a reversal of the absolute transverse orientation can only occur along self-intersection curves that are made of double hyperbolic points of $\mathcal{H}_h$. However, and this is a crucial point, such a reversal will not necessarily occur on any curve of hyperbolic double points of $\mathcal{H}_h$ but only on certain of them.

### The $\varepsilon_h$ Function on $\mathbb{S}^2$

The $\varepsilon_h$ function on $\mathbb{S}^2$ is a function with values in $\{-1, 0, 1\}$ whose sign at any regular point $u$ of $x_h : \mathbb{S}^2 \to \mathbb{R}^3$ indicates if the usual (i.e., relative) transverse orientation of $\mathcal{H}_h$ and its absolute one coincide or not at $x_h(u)$. At such a point $u \in \mathbb{S}^2$, we define $\varepsilon_h(u) \in \{-1, 1\}$ so that the unit vector

$$v_h(u) := \varepsilon_h(u)\, sgn\left[R_h(u)\right] u$$

direct the normal line of $\mathcal{H}_h$ at $x_h(u)$ when $\mathcal{H}_h$ is equipped with its absolute transverse orientation. Recall that $R_h$ denotes the curvature function of $\mathcal{H}_h$. If $x_h(u)$ is not a simple regular point of $\mathcal{H}_h$, then we put $\varepsilon_h(u) = 0$.

As noticed above, when $x_h(u)$ is an elliptic point of $\mathcal{H}_h$ (that is, when $R_h(u) > 0$), the normal vector $v_h(u)$ points at $x_h(u)$ to the side of $\mathcal{H}_h$ in which the tangent plane to $\mathcal{H}_h$ at $x_h(u)$ is located in the vicinity of $x_h(u)$. Therefore, the sign of $\varepsilon_h(u)$ does not change on the spherical image $\Omega$ of an elliptic region $x_h(\Omega)$ of $\mathcal{H}_h$. In the case of the spherical image $\Omega$ of a hyperbolic region $x_h(\Omega)$ of $\mathcal{H}_h$ the situation may be quite different. Indeed, in this second case, the sign of $\varepsilon_h(u)$ is likely to be reversed when we cross the spherical image of a curve of hyperbolic double points as proved by the example of a hedgehog version of the Steiner Roman surface with support function $h(x, y, z) := x\left(x^2 - 3y^2\right) + 2z^3$, $(x, y, z) \in \mathbb{S}^2 \subset \mathbb{R}^3$, which has been shown in Fig. 2.7. Recall that this hedgehog is projective. In this example, where the points of $x_h\left(\mathbb{S}^2\right)$ appearing as double are in fact quadruple, the index $j_h(x)$ is equal to $-2$ for any $x$ of a bounded component of $\mathbb{R}^3 \backslash \mathcal{H}_h$, and naturally equal to zero for any $x$ of the unbounded connected component of $\mathbb{R}^3 \backslash \mathcal{H}_h$. Therefore, in this example, the sign of $\varepsilon_h$ is reversed whenever we cross the spherical image of any of the hyperbolic double point curves.

Analogously to what we did in the plane, we can introduce a notion of convex interior for the hedgehogs of $\mathbb{R}^3$.

**Proposition 2.9.1** *For every $C^2$-hedgehog of $\mathbb{R}^3$,*

$$C_h = \left\{ x \in \mathbb{R}^3 \backslash \mathcal{H}_h \,|\, j_h(x) = 1 \right\}.$$

*is a convex subset of $\mathbb{R}^3$. We call $C_h$ the **convex interior** of $\mathcal{H}_h$.*

**Proof of Proposition 2.9.1** We can assume without loss of generality that the integral of $h$ over $\mathbb{S}^2$ is nonnegative since $C_h = C_{\widetilde{h}}$, where $\widetilde{h}(u) = -h(-u)$, $\left( u \in \mathbb{S}^2 \right)$. From the very definition of $j_h$, $C_h$ is the set of $x \in \mathbb{R}^3$ such that the function $u \mapsto h_x(u) - \langle x, u \rangle$ does not vanish on $\mathbb{S}^2$. It can of course be empty. If $x \in C_h$, then $h_x : \mathbb{S}^2 \to \mathbb{R}$ must remain positive on $\mathbb{S}^2$ by continuity of $h_x$ on $\mathbb{S}^2$ (the integral of $h_x$ over $\mathbb{S}^2$ is equal to the one of $h$, and thus nonnegative). $C_h$ can thus be written

$$C_h = \bigcap_{u \in \mathbb{S}^2} E_h^-(u),$$

where $E_h^-(u)$ is the open halfspace with inequation $\langle x, u \rangle < h(u)$. Therefore, $C_h$ is a convex of $\mathbb{R}^3$ as an intersection of convex subsets of $\mathbb{R}^3$.  ∎

**Definition 2.9.1** Let $\mathcal{H}_h$ be a $C^2$-hedgehog of $\mathbb{R}^2$. We define the $j_h$-interior (resp. $j_h$-exterior) of $\mathcal{H}_h$ in $\mathbb{R}^2$ to be the subset of $\mathbb{R}^2$ given by:

$$J_h := \left\{ x \in \mathbb{R}^2 \backslash \mathcal{H}_h \,|\, j_h(x) \neq 0 \right\}$$

$$\left( \text{resp. } E_j(\mathcal{H}_h) := \left\{ x \in \mathbb{R}^2 \backslash \mathcal{H}_h \,|\, j_h(x) = 0 \right\} \right).$$

For all $x \in \mathbb{R}^3 \backslash \mathcal{H}_h$, $j_h(x) = 1 - c_h(x) = 0$ implies $i_h(x) = r_h^+(x) - r_h^-(x) = 0$. Therefore $I_h \subset J_h$. This inclusion may be strict as shown by the example of non-trivial projective hedgehogs of $\mathbb{R}^3$: indeed, for such a hedgehog $\mathcal{H}_h^3$, we have $I_h = \emptyset$ and $J_h \neq \emptyset$.

**Remark** We have already seen in Sects. 2.3 and 2.4 that we can define *hedgehog polytopes*, also called *polyhedral hedgehogs* or virtual polytopes, which represent formal differences of polytopes in $\mathbb{R}^{n+1}$. We can naturally extend the two notions of indexes previously studied to certain classes of hedgehogs of $\mathbb{R}^3$ whose support function is not of class $C^2$ on $\mathbb{S}^2$, and in particular to hedgehog polytopes of $\mathbb{R}^3$ A large part of the notions and results related to these indexes can be extended to this polyhedral framework. In particular, the conclusion of Theorem 2.8.2 and the notions of interiors and exteriors corresponding to these indexes extend to hedgehog polytopes, and we still have $I_h \subset J_h$ for such a hedgehog $\mathcal{H}_h$ of $\mathbb{R}^3$.

### 2.9.2  New Notion of Volume and Geometrical Applications

Of course, this new notion of index also implies a new notion of (algebraic) volume in $\mathbb{R}^3$. The volume of $\mathcal{H}_h$ relative to our new index is defined by:

$$v_G\left(h\right) := \int_{\mathbb{R}^3 \setminus \mathcal{H}_h} j_h\left(x\right) d\lambda\left(x\right),$$

where $\lambda$ denotes the Lebesgue measure on $\mathbb{R}^3$. We call it the *geometric volume of* $\mathcal{H}_h$ *in* $\mathbb{R}^3$.

**Theorem 2.9.1 ([102])** *Let $\mathcal{H}_h$ be a $C^2$ projective hedgehog of $\mathbb{R}^3$. Then the following four properties hold:*

*(i)*  *For every $x \in \mathbb{R}^3 \setminus \mathcal{H}_h$, $j_h\left(x\right) = 1 - c_h\left(x\right) \leq 0$. Therefore, the geometric volume of $\mathcal{H}_h$ is non-positive: $v_G\left(h\right) \leq 0$;*

*(ii)*  *Let $x_h\left(u\right)$ be a simple elliptic point of $\mathcal{H}_h$ adherent to the $j_h$-exterior $E_j\left(\mathcal{H}_h\right)$. Then $\mathcal{H}_h$ turns its convexity towards $J_h$ at $x_h\left(u\right)$ (in other words, there exists a neighborhood of $x_h\left(u\right)$ in $\mathbb{R}^3$ in which the support plane with equation $\langle x, u \rangle = h\left(u\right)$ does not intersect $E_j\left(\mathcal{H}_h\right)$);*

*(iii)*  *The geometric volume of $\mathcal{H}_h$ is negative if $\mathcal{H}_h$ is not reduced to a single point.*

***Proof of Theorem 2.9.1***

*(i)*  Since $h_x$ is odd (and not identically equal to zero) on $\mathbb{S}^2$, it must change sign on $\mathbb{S}^2$, so that $c_h\left(x\right) \geq 1$.

*(ii)*  From *(i)*, as $x$ crosses $\mathcal{H}_h$ transversally at $x_h\left(u\right)$ in the direction of $E_j\left(\mathcal{H}_h\right)$, $j_h\left(x\right)$ must decrease from 0 to $-2$ (knowing that the $j_h$-index of a projective hedgehog $\mathcal{H}_h \subset \mathbb{R}^3$ takes its values in $2\mathbb{Z}$ since the parametrization $x_h$ describes the surface twice). In other words, $x$ is then crossing $\mathcal{H}_h$ transversally at $x_h\left(u\right)$ from locally convex to locally concave side.

*(iii)*  A nontrivial projective hedgehog $\mathcal{H}_h$ of $\mathbb{R}^3$ must have elliptic points (see Sect. 4.4) so that its $j_h$-index cannot be identically equal to 0 on $\mathbb{R}^3 \setminus \mathcal{H}_h$.

$\blacksquare$

**Remarks**

1.  It is not difficult to check that properties *(i)–(iii)* still hold for any hedgehog $\mathcal{H}_h$ of $\mathbb{R}^3$ whose support function $h$ satisfies

$$\int_{\mathbb{S}^2} h\left(u\right) d\sigma\left(u\right) = 0,$$

where $\sigma$ denotes the spherical Lebesgue measure on $\mathbb{S}^2$.

2.  Let $\mathcal{H}_h$ be such a hedgehog and assume that all its singularities are generic, $\left(h \in C^\infty\left(\mathbb{S}^2; \mathbb{R}\right)\right)$. Then no negative swallowtail of $\mathcal{H}_h$ can be seen from $E_j\left(\mathcal{H}_h\right)$. In other words, if a point $x_h\left(u\right)$ is a negative swallowtail of $\mathcal{H}_h$

belonging to the closure of $E_j\,(\mathcal{H}_h)$ then, near this point, the hyperbolic region to which it corresponds must lie in the complement of $E_j\,(\mathcal{H}_h)$.
3. Properties *(i)–(iii)* have of course to be compared with the corresponding properties of plane projective hedgehogs (for which, of course, the usual index $i_h$ replaces $j_h$).

## Application to the Minkowski Problem Extended to Hedgehogs

Let us see an immediate consequence that might be useful for studying the extension to $C^2$-hedgehogs of the classical Minkowski problem of prescribing the Gauss curvature of closed convex hypersurfaces. We will study this generalized Minkowski problem in the next sections, and in particular in Chap. 5 which will be entirely devoted to it. Minkowski's existence and uniqueness theorem is based on the following integral condition which is both necessary and sufficient for a positive continuous function $R\ :\ \mathbb{S}^n\ \to\ \mathbb{R}$ to be the curvature function of a convex hypersurface of class $C^2_+$ unique up to a translation:

$$\int_{\mathbb{S}^2} R\,(u)\,u\,d\sigma\,(u) = 0,$$

where $\sigma$ is the spherical Lebesgue measure on $\mathbb{S}^n$ Note that $R\ =\ 1/\kappa$, where $\kappa$ denotes the Gauss curvature of the convex hypersurface regarded as a function of the outer unit normal. As we will see later, the above integral condition, which can be seen as expressing the translation invariance of the volume, is still necessary for a continuous function $R$ to be the curvature function of some $C^2$-hedgehog, but it is no longer sufficient at all. Now, in the 3-dimensional case, we can consider the translation invariance of the geometric volume of $C^2$-hedgehogs. The geometric volume of a $C^2$-hedgehog $\mathcal{H}_h$ of $\mathbb{R}^3$ can be given by

$$v_G\,(h) := \int_{\mathbb{S}^2} \varepsilon_h\,(u)\,h\,(u)\,\frac{u}{\kappa_h(u)}\,d\sigma\,(u)$$

where $\sigma$ is the spherical Lebesgue measure on $\mathbb{S}^2$ and $\kappa_h$ the Gauss curvature of $\mathcal{H}_h$. By the translation invariance of the geometric volume, we obtain the following.

**Proposition 2.9.2** *Let $\mathcal{H}_h$ be a $C^2$-hedgehog of $\mathbb{R}^3$. Then we have*

$$\int_{\mathbb{S}^2} \varepsilon_h\,(u)\,\frac{u}{\kappa_h(u)}\,d\sigma\,(u) = 0,$$

*where $\sigma$ is the spherical Lebesgue measure on $\mathbb{S}^2$ and $\kappa_h$ the Gauss curvature.*

***Proof of Proposition 2.9.2*** For every $x \in \mathbb{R}^3$, consider the hedgehog with support function $h_x(u) := h(u) - \langle x, u \rangle$, $\left( u \in \mathbb{S}^2 \right)$. For all $x \in \mathbb{R}^3$, we have $x_{h_x}(u) = x_h(u) - x$ and in particular $\mathcal{H}_{h_x} = \mathcal{H}_h - \{x\}$. Therefore, we have:

$$\kappa_{h_x} = \kappa_h, \varepsilon_{h_x} = \varepsilon_h \quad \text{and} \quad v_G(h_x) = v_G(h).$$

Using these three equalities for every $x \in \mathbb{R}^3$, we obtain immediately:

$$\forall x \in \mathbb{R}^3, \int_{\mathbb{S}^2} \langle x, u \rangle \, \varepsilon_h(u) \, \frac{u}{\kappa_h(u)} \, d\sigma(u) = 0,$$

that is,

$$\left\langle x, \int_{\mathbb{S}^2} \varepsilon_h(u) \, \frac{u}{\kappa_h(u)} \, d\sigma(u) = 0 \right\rangle = 0,$$

which achieves the proof.                                                                ∎

We will see in Chap. 5 that a study of multiplicity of solutions in the Minkowski problem for hedgehogs should probably take into account these $\varepsilon_h$ functions.

# Chapter 3
# Volumes and Mixed Volumes

As already noticed in the introduction, the notion of mixed volumes, which forms the central part of the classical Brunn-Minkowski theory of convex bodies, arises naturally when one combines the two elementary notions of Minkowski addition and volume [167]. This notion is more precisely based on the following result in which $v_{n+1}$ stands for the $(n+1)$-dimensional volume and $s_n$ for the surface area measure (i.e., the surface area measure of order $n$):

**Theorem (See e.g., [167, Theorem 5.1.7])** *There is a nonnegative symmetric function $v : \left(\mathcal{K}^{n+1}\right)^{n+1} \to \mathbb{R}$ such that, for all $m \in \mathbb{N}$,*

$$v_{n+1}\left(\lambda_1 K_1 + \cdots + \lambda_m K_m\right) = \sum_{i_1,\dots,i_{n+1}=1}^{m} \lambda_{i_1} \cdots \lambda_{i_{n+1}} v\left(K_{i_1}, \dots, K_{i_{n+1}}\right)$$

*for arbitrary convex bodies $K_1, \dots, K_m \in \mathcal{K}^{n+1}$ and numbers $\lambda_1, \dots, \lambda_m \in \mathbb{R}_+$.*

*Further, there is a symmetric map $s$ from $\left(\mathcal{K}^{n+1}\right)^n$ into the space of finite Borel measures on $\mathbb{S}^n$, such that, for all $m \in \mathbb{N}$,*

$$s_n\left(\lambda_1 K_1 + \cdots + \lambda_m K_m, .\right) = \sum_{i_1,\dots,i_n=1}^{m} \lambda_{i_1} \cdots \lambda_{i_n} s\left(K_{i_1}, \dots, K_{i_n}\right)(.),$$

*where we write $s\left(K_{i_1}, \dots, K_{i_n}, .\right) := s\left(K_{i_1}, \dots, K_{i_n}\right)(.)$, for arbitrary convex bodies $K_1, \dots, K_m \in \mathcal{K}^{n+1}$ and numbers $\lambda_1, \dots, \lambda_m \in \mathbb{R}_+$.*

*The equality*

$$v\left(K_1, \dots, K_{n+1}\right) = \frac{1}{n+1} \int_{\mathbb{S}^n} h_{K_{n+1}}(u) \, s\left(K_1, \dots, K_n\right)(d\sigma(u)),$$

*where $h_{K_{n+1}}$ is the support function of $K_{n+1}$, holds for $K_1, \dots, K_{n+1} \in \mathcal{K}^{n+1}$.*

© The Author(s), under exclusive license to Springer Nature Switzerland AG 2026     67
Y. Martinez-Maure, *Hedgehog Theory*, Lecture Notes in Mathematics 2381,
https://doi.org/10.1007/978-3-032-04298-9_3

The symmetric maps $v : \left(\mathcal{K}^{n+1}\right)^{n+1} \to \mathbb{R}$ and $s$ are respectively called the ***the mixed volume*** and ***the mixed area measure***. The set $\mathcal{K}^{n+1}$ of convex bodies of $\mathbb{R}^{n+1}$, equipped with Minkowski addition and multiplication by nonnegative real numbers, forms a commutative semigroup, having the cancellation property, with scalar operator. But as we recalled earlier, it does not constitute a vector space since there is no subtraction in $\mathcal{K}^{n+1}$. Now hedgehogs (or equivalently formal differences of convex bodies) of $\mathbb{R}^{n+1}$ form a vector space $\mathcal{H}^{n+1}$ in which $\mathcal{K}^{n+1}$ is a cone that spans the entire space. It is thus natural to consider the multilinear extension of the mixed volume $v : \left(\mathcal{K}^{n+1}\right)^{n+1} \to \mathbb{R}$ to a symmetric $(n+1)$-linear form on $\mathcal{H}^{n+1}$. We still denote this extension by $v$. The notion of mixed volumes can thus be extended to hedgehogs with a few adaptations. In particular, areas and volumes have to be replaced by their algebraic versions, which can take negative values. Let us see how these notions can be introduced and interpreted for $C^2$-hedgehogs.

## 3.1  Volume and Surface Area

In this chapter, we are mainly interested in volumes and mixed volumes of hedgehogs in Euclidean vector space $\mathbb{R}^{n+1}$. In the following sections, we will be also interested in various volumetric questions related to different types of particular hedgehogs or convex bodies in $\mathbb{R}^{n+1}$, but also in questions about various extensions to geometric objects related or attached to hedgehogs (hedgehogs modelled only on a part of the unit sphere, multihedgehogs, focal hypersurfaces of hedgehogs, etc.). In the following chapters, we will also consider volumetric questions arising in other frameworks (complex hedgehogs, marginally trapped surfaces, hedgehogs of non-Euclidean spaces, etc.) as well as symplectic areas and volumes.

### *3.1.1  Surface Area Measure and Mixed Area*

**Surface Area Measure**

Let $\mathcal{H}_h$ be a $C^2$-hedgehog in $\mathbb{R}^{n+1}$. We have seen above that its so-called curvature function $R_h := 1/\kappa_h$ is given by $R_h(u) = \det\left[T_u x_h\right]$ for all $u \in \mathbb{S}^n$. Therefore, for all $u \in \mathbb{S}^n$, $|R_h(u)|\, d\sigma(u)$ is the area element around $x_h(u)$ on the hedgehog hypersurface $\mathcal{H}_h = x_h(\mathbb{S}^n)$ that corresponds to the area element $d\sigma(u)$ around $u$ on the sphere $\mathbb{S}^n$, where $\sigma$ is the spherical Lebesgue measure. For a $C^2$-hedgehog, it is more convenient to drop the absolute value and to consider instead $R_h(u)\, d\sigma(u)$, which is thus seen as the algebraic area element around $x_h(u) \in \mathcal{H}_h$ that corresponds to the area element $d\sigma(u)$ around $u \in \mathbb{S}^n$.

Naturally, we call the signed Borel measure defined by

$$s\,(h,\Omega) = \int_{\Omega} R_h\,(u)\,d\sigma\,(u)\,, \quad \text{for any Borel set } \Omega \subset \mathbb{S}^n$$

the (algebraic surface) area measure of $\mathcal{H}_h$, and hence the real number

$$s\,(h) = \int_{\mathbb{S}^n} R_h\,(u)\,d\sigma\,(u)$$

the (algebraic surface) area of $\mathcal{H}_h$. Note that $s\,(h)$ can be interpreted as the difference $s_+\,(h) - s_-\,(h)$, where $s_+\,(h)$ (resp. $s_-\,(h)$) denotes the total area of the smooth regions of $\mathcal{H}_h$ on which the Gauss curvature is positive (resp. negative).

**Mixed Curvature Function and Mixed Area**

**Proposition 3.1.1** *Let $H_{n+1}$ be the linear space of $C^2$-hedgehogs defined up to a translation in $\mathbb{R}^{n+1}$ and identified with their support function. Then the symmetric map*

$$R : H_{n+1}^n \longrightarrow C(\mathbb{S}^n; \mathbb{R})$$
$$(f_1, \ldots, f_n) \longmapsto R_{(f_1, \ldots, f_n)} = \frac{1}{n!} \sum_{k=1}^{n} (-1)^{n+k} \sum_{i_1 < \ldots < i_k} R_{(f_{i_1} + \ldots + f_{i_k})}$$

*is such that*

$$R_{(\lambda_1 h_1 + \ldots + \lambda_m h_m)} = \sum_{i_1, \ldots, i_n = 1}^{m} \lambda_{i_1} \ldots \lambda_{i_n} R_{(h_{i_1}, \ldots, h_{i_n})},$$

*for all $h_1, \ldots, h_m \in H_{n+1}$ and all $\lambda_1, \ldots, \lambda_m \in \mathbb{R}$.*

***Proof*** Following the definition of the mixed curvature function for convex bodies [167, p. 124], we obtain the existence of a symmetric map $R : H_{n+1}^n \to C(\mathbb{S}^n; \mathbb{R})$, $(f_1, \ldots, f_n) \longmapsto R_{(f_1, \ldots, f_n)}$ such that

$$R_{(\lambda_1 h_1 + \ldots + \lambda_m h_m)} = \sum_{i_1, \ldots, i_n = 1}^{m} \lambda_{i_1} \ldots \lambda_{i_n} R_{(h_{i_1}, \ldots, h_{i_n})},$$

for all $h_1, \ldots, h_m \in H_{n+1}$ and all $\lambda_1, \ldots, \lambda_m \in \mathbb{R}$. Next, we check that $R$ is given by

$$R_{(h_1, \ldots, h_n)} = \frac{1}{n!} \sum_{k=1}^{n} (-1)^{n+k} \sum_{i_1 < \ldots < i_k} R_{(h_{i_1} + \ldots + h_{i_k})},$$

following the reasoning used in [167] to express the mixed volume of strongly isomorphic polytopes in terms of Minkowski sums [167, Lemma 5.1.4, p. 277]. ∎

**Definition** This symmetric map $R : H^n_{n+1} \longrightarrow C(\mathbb{S}^n; \mathbb{R})$ is called the **mixed curvature function**, and the $n$-linear form $s : H^n_{n+1} \longrightarrow \mathbb{R}$ given by

$$s(h_1, \ldots, h_n) = \int_{\mathbb{S}^n} R_{(h_1, \ldots, h_n)}(p) d\sigma(p),$$

where $\sigma$ is the spherical Lebesgue measure, is called the **mixed (algebraic) area**.

Of course, if $h_1, \ldots, h_n$ are the respective support functions of $n$ convex bodies $K_1, \ldots, K_n$ of $\mathbb{R}^{n+1}$, then $s(h_1, \ldots, h_n)$ is nothing else but the mixed curvature function $s(K_1, \ldots, K_n)$ of $K_1, \ldots, K_n$.

### 3.1.2　Kronecker Index, Volume and Mixed Volume

Given a $C^2$-hedgehog $\mathcal{H}_h$ in $\mathbb{R}^{n+1}$, the Kronecker index of $x \in \mathbb{R}^{n+1} \setminus \mathcal{H}_h$ with respect to $\mathcal{H}_h$, say $i_h(x)$, is defined to be the degree of the map

$$\mathcal{U}_{(h,x)} : \mathbb{S}^n \to \mathbb{S}^n, \ u \longmapsto \frac{x_h(u) - x}{\|x_h(u) - x\|},$$

and interpreted as the algebraic intersection number of an oriented half-line with origin $x$ with the hypersurface $\mathcal{H}_h$ equipped with its transverse orientation (number independent of the oriented half-line for an open dense set of directions). We have already mentioned that, for all $x \in \mathbb{R}^{n+1} \setminus \mathcal{H}_h$, we have

$$i_h(x) = \frac{1}{\omega_n} \int_{\mathbb{S}^n} \frac{h(u) - \langle x, u \rangle}{\|x_h(u) - x\|^{n+1}} R_h(u) \, d\sigma(u),$$

where $R_h$ is the curvature function and $\sigma$ the spherical Lebesgue measure on $\mathbb{S}^n$.

For $n + 1 = 2$, the Kronecker index $i_h(x)$ is nothing but the winding number of $\mathcal{H}_h$ around $x$: it counts the total number of times that $\mathcal{H}_h$ winds around $x$. For instance, the index is equal to $-1$ at any interior point of the hedgehog represented on Fig. 2.4, since the curve winds once clockwise around the point. The (algebraic $(n + 1)$-dimensional) volume of a hedgehog $\mathcal{H}_h \subset \mathbb{R}^{n+1}$ can be defined by

$$v_{n+1}(h) := \int_{\mathbb{R}^{n+1} \setminus \mathcal{H}_h} i_h(x) \, d\lambda(x),$$

where $\lambda$ denotes the Lebesgue measure on $\mathbb{R}^{n+1}$, and it satisfies

$$v_{n+1}(h) = \frac{1}{n+1} \int_{\mathbb{S}^n} h(u) R_h(u) \, d\sigma(u),$$

just as in the particular case where $h$ is the support function of a convex body $K$ and $v_{n+1}(h)$ is the $n$-dimensional volume of $K$.

For instance, in the example of Fig. 2.4, the 2-dimensional volume (or algebraic area) $v_2(h)$ of $\mathcal{H}_h$ in $\mathbb{R}^2$ is equal to minus the area of the interior of the curve. Recall that this algebraic area $v_2(h)$ is also denoted by $a(h)$. As for convex bodies of class $C_+^2$, we introduce the mixed (algebraic $(n+1)$-dimensional) volume for $C^2$-hedgehogs to be the symmetric map $v : H_{n+1}^{n+1} \to \mathbb{R}$, $(h_1, \ldots, h_{n+1}) \mapsto v(h_1, \ldots, h_{n+1})$ given by

$$v(h_1, \ldots, h_{n+1}) = \frac{1}{n+1} \int_{\mathbb{S}^n} h_1(u) \, R_{(h_2, \ldots, h_{n+1})}(u) \, d\sigma(u),$$

where $R_{(h_2, \ldots, h_{n+1})}$ is the mixed curvature function of $\mathcal{H}_{h_2}, \ldots, \mathcal{H}_{h_{n+1}}$. Here again, if $h_1, \ldots, h_{n+1}$ are the respective support functions of $n+1$ convex bodies $K_1, \ldots, K_{n+1}$ of class $C_+^2$, then $v(h_1, \ldots, h_{n+1})$ is nothing else but the mixed volume $v(K_1, \ldots, K_{n+1})$ of $K_1, \ldots, K_{n+1}$.


## 3.2  Geometric Inequalities

### 3.2.1  *A Partial Extension of the Alexandrov-Fenchel Inequality to Hedgehogs and Some General Applications*

The classical Alexandrov-Fenchel inequality

$$v(H, K, L_3, \ldots, L_{n+1})^2 \geq v(H, H, L_3, \ldots, L_{n+1}) \, v(K, K, L_3, \ldots, L_{n+1}) \tag{3.2.1}$$

is a central result in the theory of mixed volumes. Here, $H, K, L_3, \ldots, L_{n+1}$ are convex bodies in $(n+1)$-dimensional real Euclidean vector space $\mathbb{R}^{n+1}$ and, $v$ denotes the mixed volume. Many geometric inequalities for convex bodies are consequences of (3.2.1) (see, e.g., [167, Chapter 7]). Connections with algebraic geometry have been discovered, which have led to new proofs of (3.2.1) via the Hodge index theorem [77, 181]. Equality holds in (3.2.1) if $K$ and $L$ are homothetic. But this is not the only equality case and until now the equality problem remains unsolved (see [167, Section 7.6] for a discussion). However, if $L_3, \ldots, L_{n+1}$ are convex bodies of class $C_+^2$, then this is the only equality case [167, Theorem 7.6.8].

In which follows, we will identify convex bodies and hedgehogs of $\mathbb{R}^{n+1}$ with their respective support functions regarded as functions on the unit sphere $\mathbb{S}^n$. Thus the classical Alexandrov-Fenchel inequality (3.2.1) will be rewritten

$$v(h, k; l)^2 \geq v(h, h; l) \, v(k, k; l),$$

where $h, k, l_3, \ldots, l_{n+1}$ denote the respective support functions of $H, K, L_3, \ldots,$ $L_{n+1}, l = (l_3, \ldots, l_{n+1})$ and $v(f, g; l) := v(f, g, l_3, \ldots, l_{n+1})$.

Recall that any real function $h$ of class $C^2$ on $\mathbb{S}^n$ is the support function of some hedgehog $\mathcal{H}_h$ in $\mathbb{R}^{n+1}$. Let $H_{n+1}$ denote the linear space of $C^2$-hedgehogs defined up to a translation in $\mathbb{R}^{n+1}$ and identified with their support function. Given $l_3, \ldots, l_{n+1}$ the support functions of $n-1$ convex bodies of class $C_+^2$, we consider the quadratic form

$$q : H_{n+1} \to \mathbb{R}, h \mapsto v(h, h; l) := v(h, h, l_3, \ldots, l_{n+1}),$$

and we denote by $b$ its associated bilinear form

$$b : H_{n+1}^2 \to \mathbb{R}, (f, g) \mapsto v(f, g; l) := v(f, g, l_3, \ldots, l_{n+1}).$$

In [92], the author gave the following partial extension of the Alexandrov-Fenchel inequality to hedgehogs under the assumption that $l_3, \ldots, l_{n+1}$ are the support functions of $n-1$ convex bodies of class $C_+^2$, $(n \geq 1)$.

**Theorem 3.2.1** *Let $\alpha \in H_{n+1}$ be such that $q(\alpha) := v(\alpha, \alpha; l) > 0$. Then*

$$b(\alpha, \beta)^2 \geq q(\alpha) q(\beta), \text{ that is } \quad v(\alpha, \beta; l)^2 \geq v(\alpha, \alpha; l) v(\beta, \beta; l),$$

*for any $\beta \in H_{n+1}$ and, the equality holds if and only if there exists $(\lambda, \mu) \in \mathbb{R}^2 \setminus \{(0, 0)\}$ such that $\lambda\alpha + \mu\beta = 0_{H_{n+1}}$.*

*By convenience, we will say that "$\mathcal{H}_\alpha$ and $\mathcal{H}_\beta$ are homothetic" if and only if "there exists $(\lambda, \mu) \in \mathbb{R}^2 \setminus \{(0, 0)\}$ such that $\lambda\alpha + \mu\beta = 0_{H_{n+1}}$". We start by establishing the following lemma.*

**Lemma 3.2.1** *Let $F$ be the linear subspace of $C^2(\mathbb{S}^n; \mathbb{R})$ given by*

$$F = \{f \in H_{n+1} \,|\, b(f, 1) = 0\}.$$

*If $h \in F \setminus \{0\}$, then $q(h) = v(h, h; l) < 0$.*

***Proof of Lemma 3.2.1*** Reasoning by absurd, we assume that $q(h) \geq 0$. Then the quadratic form $q$ is positive on the vector plane, say $V_h$, that is spanned by the functions 1 and $h$, so that

$$b(\alpha, \beta) \leq q(\alpha) q(\beta) \quad \text{for all } (\alpha, \beta) \in V_h^2$$

by the Cauchy-Schwarz inequality. Now, for all small enough $\varepsilon > 0$, $\alpha = 1$ and $\beta = 1 + \varepsilon h$ are the support functions of two non-homothetic convex bodies of class $C_+^2$, so that

$$b(\alpha, \beta) > q(\alpha) q(\beta)$$

by virtue of the Alexandrov-Fenchel inequality and its equality case if $l_3, \ldots, l_{n+1}$ are the support functions of $n - 1$ convex bodies of class $C_+^2$, which is clearly absurd.  ∎

***Proof of Theorem 3.2.1*** Let $A(t)$ be the second-degree polynomial given by:

$$A(t) := q(\beta + t\alpha) = q(\beta) + 2tb(\alpha, \beta) + t^2 q(\alpha).$$

From the assumption $q(\alpha) > 0$, we obviously deduce that $A(t) > 0$ for any large enough $t$. By this same assumption and Lemma 3.2.1, we also deduce that $b(\alpha, 1) \neq 0$. Now for $t = -b(\beta, 1)/b(\alpha, 1)$, we check by bilinearity of $b$ that $\beta + t\alpha \in F$ and hence $A(t) < 0$ unless $\beta + t\alpha = 0_{H_{n+1}}$. Therefore the reduced discriminant of $A$ (that is, $\Delta = b(\alpha, \beta)^2 - q(\alpha) q(\beta)$) is positive unless $\beta + t\alpha = 0_{H_{n+1}}$. To conclude, let us notice that if $\mathcal{H}_\alpha$ and $\mathcal{H}_\beta$ are homothetic, then we obviously have $b(\alpha, \beta)^2 = q(\alpha) q(\beta)$.  ∎

In this theorem, we have assumed that $l_3, \ldots, l_{n+1}$ are support functions of convex bodies of $\mathbb{R}^{n+1}$. We will see later that this condition is indeed necessary.

The following corollary is a straightforward consequence of Theorem 3.2.1.

**Corollary 3.2.1** *Let $k \in H_{n+1}$ be such that $q(k) > 0$, and let*

$$F_k = \{f \in H_{n+1} \mid b(f, k) = 0\}.$$

*The map $\sqrt{-q} : F_k \to \mathbb{R}_+, h \longmapsto \sqrt{-q(h)}$ is a norm associated with a scalar product on $F_k$. In particular, for all $(g, h) \in F_k^2$, we have the following inequalities:*

$$(i)\ \ \sqrt{-q(g+h)} \leq \sqrt{-q(g)} + \sqrt{-q(h)},$$
$$(ii)\ b(g, h)^2 \leq q(g) q(h)$$

*where equalities hold if and only if $\mathcal{H}_g$ and $\mathcal{H}_h$ are homothetic.*

Another immediate consequence of Theorem 3.2.1. is the following:

**Proposition 3.2.1** *Let $(f, g, h) \in H_{n+1}^3$ be such that $q(f) > 0$. Then*

$$2b(g, f)\, b(f, h)\, b(h, g) \geq q(g)\, b(h, f)^2 + q(h)\, b(g, f)^2,$$

*where equality holds if, and only if, $\mathcal{H}_g$ and $\mathcal{H}_h$ are homothetic.*

***Proof*** Indeed, the hedgehog $e = b(h, f)\, g - b(g, f)\, h$ is such that $b(e, f) = 0$ and the inequality stated above can be rewritten as $q(e) \leq 0$. Moreover, by Corollary 3.2.1, the equality $q(e) = 0$ holds if and only if $e = 0_{F_f}$, that is, if and only if $\mathcal{H}_g$ and $\mathcal{H}_h$ are homothetic.  ∎

Thanks to this proposition, we can prove a Brunn-Minkowski type inequality for two hedgehogs associated with $n - 1$ convex bodies in $\mathbb{R}^{n+1}$.

**Proposition 3.2.2** *Let $f \in H_{n+1}$ be such that $q(f) > 0$, and consider the following subset of $H_{n+1}$*

$$H_{n+1}^{+}(f) = \{k \in H_{n+1} | \, b(k, f) > 0 \text{ and } q(k) > 0\}.$$

*For all $(g, h) \in H_{n+1}^{+}(f)^2$, $g + h \in H_{n+1}^{+}(f)$ and*

$$\sqrt{q(g+h)} \geq \sqrt{q(g)} + \sqrt{q(h)},$$

*where equality holds if, and only if, $\mathcal{H}_g$ and $\mathcal{H}_h$ are homothetic.*

**Proof** Indeed, we have $b(g, h) > 0$ from Proposition 3.2.1, and it then follows from Theorem 3.2.1 that

$$q(g+h) = q(g) + q(h) + 2b(g, h) \geq \left(\sqrt{q(g)} + \sqrt{q(h)}\right)^2,$$

where equality holds if, and only if, $\mathcal{H}_g$ and $\mathcal{H}_h$ are homothetic.   ∎

This inequality admits a linear refinement under appropriate assumptions:

**Proposition 3.2.3** *Under assumptions of Proposition 3.2.2, if $\mathcal{H}_g$ and $\mathcal{H}_h$ are non-homothetic, then the inequality*

$$\sqrt{q(g+h)} \geq \sqrt{q(g)} + \sqrt{q(h)},$$

*admits the refinement*

$$2b(g, h) > q(g) + q(h),$$

*if, and only if, there exists $k \in H_{n+1}$ such that $q(k) > 0$ and $b(k, g) = b(k, h)$.*

**Proof** If the refinement holds (that is, if $\delta = g - h$ is such that $q(\delta) < 0$), then $k = b(\delta, g)\, h - b(\delta, h)\, g \in H_{n+1}$ is such that $b(k, g) = b(k, h)$, and $q(k) = (q(g)\,q(h) - b(g, h)^2)\, q(\delta) > 0$ by Theorem 3.2.1.

Conversely, if there exists $k \in H_{n+1}$ such that $q(k) > 0$ and $b(k, g) = b(k, h)$, then $\delta = g - h$ is such that $b(k, \delta) = 0$. By Proposition 3.2.1, we have then:

$$0 = 2b(\delta, 1)\, b(1, k)\, b(k, \delta) \geq q(\delta)\, b(k, 1)^2 + q(k)\, b(\delta, 1)^2.$$

Now since $q(k) > 0$ we have $b(k, 1)^2 \geq q(k)\,q(1) > 0$ by Theorem 3.2.1. Moreover, the case ($q(\delta) = 0$ and $b(\delta, 1) = 0$) is excluded by Lemma 3.2.1 since $\mathcal{H}_g$ and $\mathcal{H}_h$ are assumed to be non-homothetic. Consequently, $q(\delta) < 0$.   ∎

### 3.2.2 The 2 and 3-Dimensional Cases

**The 2-Dimensional Case**

For $n + 1 = 2$, the quadratic form $q : H_2 \to \mathbb{R}$, $h \mapsto v_2(h) := v(h, h)$ is given by

$$q(h) = \frac{1}{2} \int_0^{2\pi} h(\theta)(h + h'')(\theta)d\theta$$

$$= \frac{1}{2} \int_0^{2\pi} \left(h^2 - (h')^2\right)(\theta)d\theta, \quad \left(\theta \in \mathbb{S}^1 = \mathbb{R}/2\pi\mathbb{Z}\right),$$

and can be interpreted as the signed (or algebraic) area $a(h) := v_2(h)$ of $\mathcal{H}_h$. Its polar bilinear form $b : H_2^2 \to \mathbb{R}$, $(f, g) \mapsto v(f, g)$ is given by

$$b(g, h) = \frac{1}{2} \int_0^{2\pi} g(\theta)\left(h + h''\right)(\theta)\, d\theta = \frac{1}{2} \int_0^{2\pi} \left(gh - g'h'\right)(\theta)\, d\theta,$$

and can be thus be interpreted as the mixed signed (or algebraic) area of $\mathcal{H}_g$ and $\mathcal{H}_h$. By Theorem 3.2.1, we have:

**Corollary 3.2.2** *Let $(g, h) \in H_2^2$ be such that $a(g) > 0$ or $a(h) > 0$. Then, we have*

$$a(g, h)^2 \geq a(g)\, a(h),$$

*with equality if and only if $\mathcal{H}_g$ and $\mathcal{H}_h$ are homothetic.*

**Isoperimetric Inequality**

This corollary gives an extension to hedgehogs of the classical isoperimetric inequality $4\pi A \leq L^2$, where $A$ and $L$ are respectively the area and perimeter of a plane convex body $K$. Indeed, by setting $g = 1$, we immediately deduce the following result.

**Corollary 3.2.3** *For any hedgehog $h \in H_2$, we have*

$$a(h) \leq \frac{1}{4\pi} l(h)^2,$$

*where*

$$l(h) := \int_0^{2\pi} \langle x_h'(\theta), u'(\theta)\rangle(\theta)\, d\theta = \int_0^{2\pi} \left(h'' + h\right)(\theta)\, d\theta = \int_0^{2\pi} h(\theta)\, d\theta$$

*is interpreted as the **algebraic length** of $\mathcal{H}_h$. Equality holds if, and only if, $\mathcal{H}_h$ is a circle or a point.*

**A Refined Isoperimetric Inequality**

The following refinement gives an upper bound of the isoperimetric deficit of plane $C^3$-hedgehogs in terms of signed area of their evolute:

**Proposition 3.2.4** *For every $h \in C^3\left(\mathbb{S}^1; \mathbb{R}\right)$, we have*

$$0 \leq l\,(h)^2 - 4\pi a\,(h) \leq -4\pi a\,(\partial h),$$

*where $a\,(\partial h)$ is the algebraic area of the evolute of $\mathcal{H}_h$, which is the hedgehog with support function $\partial h \,:\, \mathbb{S}^1 = \mathbb{R}/2\pi\mathbb{Z} \to \mathbb{R}, \; \theta \mapsto h'\left(\theta - \frac{\pi}{2}\right)$. In each of above inequalities, equality holds if, and only if, $\mathcal{H}_h$ is a circle or a point.*

***Proof*** Recall that the ***evolute*** of a plane curve is the locus of all its centers of curvature or, equivalently, the envelope of its normal lines. In particular, the evolute of a plane hedgehog $\mathcal{H}_h \subset \mathbb{R}^2$ with support function $h \in C^3\left(\mathbb{S}^1; \mathbb{R}\right)$ is the locus of all its centers of curvature $c_h\,(\theta) := x_h\,(\theta) - R_h\,(\theta)\,u\,(\theta)$, where $R_h\,(\theta) := \det\left[T_{u(\theta)}x_h\right] = \left(h + h''\right)(\theta)$ is the curvature function of $\mathcal{H}_h$, and $u\,(\theta) := (\cos\theta, \sin\theta), \; \left(\theta \in \mathbb{S}^1 = \mathbb{R}/2\pi\mathbb{Z}\right)$. Equivalently, the evolute of $\mathcal{H}_h$ can be defined as the envelope of its 'normal lines' with equations $\langle x, u'\,(\theta)\rangle = h'\,(\theta)$, that is, as the hedgehog $\mathcal{H}_{\partial h}$ with support function $(\partial h)\,(\theta) := h'\left(\theta - \frac{\pi}{2}\right)$. Indeed, we have:

$$c_h\,(\theta) := h'\,(\theta)\,u'\,(\theta) + h''\,(\theta)\,u''\,(\theta) = x_{\partial h}\left(\theta + \frac{\pi}{2}\right).$$

Now, by virtue of the Cauchy-Schwarz inequality, we have

$$l\,(h)^2 = \left(\int_0^{2\pi} \left(h + h''\right)(\theta)\,d\theta\right)^2 \leq 2\pi \int_0^{2\pi} \left(h + h''\right)(\theta)^2\,d\theta$$

$$= 4\pi\left(a\,(h) - a\left(h'\right)\right),$$

which completes the proof since $a\left(h'\right) = a\,(\partial h)$.  ∎

**Remark** If we assume $h$ to be of class $C^4$, then we can also write that

$$0 \leq l\,(h)^2 - 4\pi a\,(h) \leq -4\pi a\,(\partial h) = -4\pi a\left(h, \partial^2 h\right),$$

where $\partial^2 h = -h''$ can be seen as the support function of the second evolute $\mathcal{H}_{\partial^2 h}$ of $\mathcal{H}_h$, and $a\left(h, \partial^2 h\right)$ as the mixed signed area of $\mathcal{H}_h$ and $\mathcal{H}_{\partial^2 h}$. Here, we have to recall that for all $h \in C^2\,(\mathbb{S}^n; \mathbb{R})$, the hedgehogs with support functions $h$ and $\widetilde{h}\,(u) = -h\,(-u), \; (u \in \mathbb{S}^n)$, have the same geometrical realizations since $x_{\widetilde{h}}\,(-u) = x_h\,(u)$ for all $u \in \mathbb{S}^n$. Thus here, $\theta \mapsto h''\,(\theta - \pi)$ and $\theta \mapsto -h''\,(\theta)$ can both be seen as support functions of the second evolute of $\mathcal{H}_h$ regarded itself as a hedgehog of $\mathbb{R}^2$.

## The 3-Dimensional Case

Let us start with an immediate consequence of Corollary 3.2.1. First note that for $n + 1 = 3$ and $l_3 = 1$, $q(h) = v(h, h, 1)$ is equal to one-third of

$$s(h) = \int_{\mathbb{S}^2} R_h d\sigma,$$

where $R_h(u) = \det[T_u x_h] = \left(R_h^1 R_h^2\right)(u)$ is the curvature function of $\mathcal{H}_h$, $R_h^1(u)$ and $R_h^2(u)$ denoting the principal radii of curvature of $\mathcal{H}_h$ at $x_h(u)$, $\left(u \in \mathbb{S}^2\right)$. As shown in Sect. 3.1, this integral $s(h)$ can be interpreted as the difference $s_+(h) - s_-(h)$, where $s_+(h)$ (resp. $s_-(h)$) denotes the total area of the smooth regions of $\mathcal{H}_h$ on which the Gauss curvature is positive (resp. negative). Now choosing $k = 1$, we can note that

$$F_1 := \left\{ h \in H_3 \;\middle|\; \int_{\mathbb{S}^2} R_{(1,h)} d\sigma = 0 \right\} = \left\{ h \in H_3 \;\middle|\; \int_{\mathbb{S}^2} h\, d\sigma = 0 \right\},$$

where $R_{(1,h)}(u) = \frac{1}{2} tr[T_u x_h] = \frac{1}{2}\left(R_h^1 + R_h^2\right)(u) = \left(h + \frac{\Delta h}{2}\right)(u)$ denotes the mean radius of curvature of $\mathcal{H}_h$, so that $F_1$ consists of all $h \in H_3$ whose '*integral mean curvature*', say $m(h)$, is equal to zero. Indeed, we have $R_{(1,h)} d\sigma = H dA$, where $H$ is the mean curvature of $\mathcal{H}_h$ and $dA = R_h d\sigma$ the (signed) area element around $x_h(u)$ on $\mathcal{H}_h$. Therefore:

*If the integral mean curvature of a $C^2$-hedgehog $\mathcal{H}_h$ of $\mathbb{R}^3$ is equal to zero, then its signed area $s(h) = s_+(h) - s_-(h)$ is negative unless $\mathcal{H}_h$ is reduced to a single point.*

## Quadratic Minkowskian Inequalities

Besides, note that Theorem 3.2.1 gives an extension to $C^2$-hedgehogs of the quadratic Minkowski inequality $4\pi S \le M^2$, where $S := v(k, k, 1)$ and $M := v(k, 1, 1)$ are respectively the surface area and integral mean curvature of a 3-dimensional convex body with support function $k$ in $\mathbb{R}^3$. Indeed, taking $\alpha = l_3 = 1$ and $\beta = h$ for $n + 1 = 2$ in Theorem 3.2.1 yields the following result.

**Proposition 3.2.5** *For every $h \in C^2\left(\mathbb{S}^2; \mathbb{R}\right)$, we have $4\pi s(h) \le m(h)^2$, where equality holds if, and only if, $\mathcal{H}_h$ is a sphere or a point.*

On the other hand, the quadratic Minkowski inequality $3MV \le S^2$, where $V$, $S$ and $M$ are the volume, area and integral of mean curvature of a 3-dimensional convex body $K$ in $\mathbb{R}^3$, does not extend to arbitrary $C^2$-hedgehogs of $\mathbb{R}^3$. For instance, the calculation shows that the hedgehog of $\mathbb{R}^3$ with support function given by

$$h(u) = 1 + \sqrt{\frac{5}{8}}\left(2x^2 - y^2 - z^2\right) \quad \text{for all } u = (x, y, z) \in \mathbb{S}^2 \subset \mathbb{R}^3$$

is such that $s\,(h) = 0$, $m\,(h) > 0$, $v\,(h) > 0$ and hence $3m\,(h)\,v\,(h) \geq s\,(h)^2$. By taking hedgehogs with support function of the form $f = h + r$, where $r \in \mathbb{R}_+^*$ is small enough, we can then deduce examples of hedgehogs such that

$$m\,(f) > 0,\, s\,(f) > 0,\, v\,(f) > 0 \text{ and } 3m\,(f)\,v\,(f) \geq s\,(f)^2.$$

Now the Minkowskian inequality $3MV \leq S^2$, which holds true for any 3-dimensional convex body with support function $k$ in $\mathbb{R}^3$, is a particular case of the Alexandrov-Fenchel inequality since it can be rewritten

$$v\,(1, 1, k)\,.v\,(k, k, k) \leq v\,(1, k, k)^2.$$

Therefore, in the statement of Theorem 3.2.1, it is necessary to assume that $l_3, \ldots, l_{n+1}$ are support functions of convex bodies of $\mathbb{R}^{n+1}$.

### 3.2.3   A Stability Estimate for the Alexandrov-Fenchel Inequality

For $k \in C^2\,(\mathbb{S}^n; \mathbb{R})$, we will write $k \in C_+^2\,(\mathbb{S}^n; \mathbb{R})$ to mean that $k : \mathbb{S}^n \to \mathbb{R}$ is the support function of a convex body the boundary of which is a hypersurface with positive Gauss curvature in $\mathbb{R}^{n+1}$. Theorem 3.2.1 gives a partial extension of the Alexandrov-Fenchel inequality to hedgehogs that can be reformulated as follows.

**Theorem 3.2.1** *Let $l = (l_3, \ldots, l_{n+1})$, where $l_3, \ldots, l_{n+1} \in C_+^2\,(\mathbb{S}^n; \mathbb{R})$, and let $f : \mathbb{S}^n \to \mathbb{R}$ be a $C^2$-function such that $v\,(f, f; l) > 0$. Then*

$$v\,(f, g; l)^2 \geq v\,(f, f; l)\,v\,(g, g; l)$$

*for any $g \in C^2\,(\mathbb{S}^n; \mathbb{R})$, and the equality holds if and only if there exists $(\lambda, \mu) \in \mathbb{R}^2 \setminus \{(0, 0)\}$ such that $\lambda f + \mu g$ be the support function of a point.*

For $v \in \mathbb{S}^n$, define $\sigma_v\,(u) := \frac{1}{2}\,|\langle u, v \rangle|$ for $u \in \mathbb{S}^n$: $\sigma_v$ is the support function of the unit segment $U\,(v)$ parallel to $v$ and centered at the origin. In [108], the author proved the following stability estimate for the Alexandrov-Fenchel inequality.

**Theorem 3.2.2** *For $h \in C^2\,(\mathbb{S}^n; \mathbb{R})$, $k \in C_+^2\,(\mathbb{S}^n; \mathbb{R})$ and $l = (l_3, \ldots, l_{n+1}) \in C_+^2\,(\mathbb{S}^n; \mathbb{R})^{n-1}$,*

$$v\,(h, k; l)^2 - v\,(h, h; l)\,v\,(k, k; l) \geq \frac{v\,(k, k; l)^2}{4}\,\left(M_{(h,k;l)} - m_{(h,k;l)}\right)^2,$$

*where $m_{(h,k;l)} := \min\limits_{v \in \mathbb{S}^n} \dfrac{v\,(h, \sigma_v; l)}{v\,(k, \sigma_v; l)}$ and $M_{(h,k;l)} := \max\limits_{v \in \mathbb{S}^n} \dfrac{v\,(h, \sigma_v; l)}{v\,(k, \sigma_v; l)}$.*

**Remark** Given $v \in \mathbb{S}^n$, denote by $v^\perp$ the vector subspace orthogonal to $v$. For any $f \in C^2(\mathbb{S}^n; \mathbb{R})$, we have

$$(n+1)\, v\left(f, \sigma_v; l_3, \ldots, l_{n+1}\right) = v_{v^\perp}\left(f^v; l_3^v, \ldots, l_{n+1}^v\right),$$

where $v_{v^\perp}$ is the $n$-dimensional mixed volume in $v^\perp$ and $f^v, l_3^v, \ldots, l_{n+1}^v$ the respective restrictions of $f, l_3, \ldots, l_{n+1}$ to $\mathbb{S}_v = \mathbb{S}^n \cap v^\perp$ (see later in Sect. 4.3).

Remind that if $f \in C^2(\mathbb{S}^n; \mathbb{R})$ is the support function of a convex body $K$, then $f^v \in C^2(\mathbb{S}_v; \mathbb{R})$ is the support function of the image of $K$ under orthogonal projection to $v^\perp$. The notion of mixed projection body extends to hedgehogs (see Sect. 4.3) and, if we denote by $\Pi_{(f;l)}$ the mixed projection hedgehog of the hedgehogs with support functions $f, l_3, \ldots, l_{n+1}$ and by $h_{\Pi_{(f;l)}}$ its support function, then the inequality of Theorem 3.2.2 can be rewritten in the form

$$v\left(h, k; l\right)^2 - v\left(h, h; l\right) v\left(k, k; l\right) \geq \frac{v\left(k, k; l\right)^2}{4} \mathcal{D}\left(\frac{h_{\Pi_{(h;l)}}}{h_{\Pi_{(k;l)}}}\right)^2,$$

where $\mathcal{D}\left(\dfrac{h_{\Pi_{(h;l)}}}{h_{\Pi_{(k;l)}}}\right)$ is the diameter of the image of $\mathbb{S}^n$ under $\dfrac{h_{\Pi_{(h;l)}}}{h_{\Pi_{(k;l)}}}$.

The proof is based on the study of equality cases in the extension of Theorem 3.2.1 to the case where $f \in C^2(\mathbb{S}^n; \mathbb{R}) \oplus \mathbb{R}\sigma_v$. It is inspired by the work of G. Bol [21] who proved the result for $n = 2$ and $l_3 = 1$. Unfortunately, Bol's work has apparently fallen into oblivion. This is perhaps due to the fact that Bol's proof contains a series of errors that make it difficult to understand. But fortunately it can be corrected and the approach can be adapted to this more general setting.

## Auxiliary Results

First, we fix $v \in \mathbb{S}^n$ and then we extend Theorem 3.2.1 by replacing $C^2(\mathbb{S}^n; \mathbb{R})$ by the real vector space, say $V(v)$, spanned by $C^2(\mathbb{S}^n; \mathbb{R})$ and $\sigma_v$ in $C(\mathbb{S}^n; \mathbb{R})$.

**Theorem 3.2.3** *Let $f$ be a function in $V(v)$ such that $v(f, f; l) > 0$. Then*

$$v(f, g; l)^2 \geq v(f, f; l) v(g, g; l) \tag{3.2.2}$$

*for all $g \in V(v)$ and, the equality holds if and only if there exists $(\lambda, \mu) \in \mathbb{R}^2 \setminus \{(0, 0)\}$ such that $\lambda f + \mu g$ is the support function of a point.*

***Proof of Theorem 3.2.3*** Let $q : V(v) \to \mathbb{R}$ be the quadratic form given by $q(h) := v(h, h; l)$. Denote by $b$ its polar form: $b(h, k) := v(h, k; l)$ for $(h, k) \in V(v)^2$. We start with an observation concerning the restriction of $q$ to the linear subspace $F(v)$ of $V(v)$ with equation $b(1, h) = 0$.

**Lemma 3.2.2** *If $h$ is in $F(v)$ and is not the support function of a point, then $q(h) := v(h, h; l) < 0$.*

***Proof of Lemma 3.2.2*** Such a function $h$ can be decomposed as $h = \gamma + \lambda \sigma_v$, where $\gamma \in C^2(\mathbb{S}^n; \mathbb{R})$ and $\lambda \in \mathbb{R}$. From Theorem 3.2.1, we may assume that $\lambda \neq 0$. Replacing $h$ by $-h$ if necessary, we may assume that $\lambda > 0$. Choose a number $\varepsilon > 0$ small enough so that $1 + \varepsilon h$ is the support function of a convex body. Such a number exists by Theorems 1.5.13 and 1.7.1 from [167]. Now, by Theorem 7.6.8 from [167], we know that equality holds in the classical Alexandrov-Fenchel inequality

$$v(H, K, L_3, \ldots, L_{n+1})^2 \geq v(H, H, L_3, \ldots, L_{n+1})\, v(K, K, L_3, \ldots, L_{n+1})$$

if and only if $H$ and $K$ are homothetic provided that $L_3, \ldots, L_{n+1}$ are smooth convex bodies. So, with our choice of $\varepsilon$, we must have $b(1, 1 + \varepsilon h)^2 > q(1)\, q(h)$.

If $q(h)$ was nonnegative, then the quadratic form $q$ would be positive semi-definite on the linear subspace $V_h$ of $V(v)$ spanned by 1 and $h$ so that we should have

$$b(\alpha, \beta)^2 \leq q(\alpha)\, q(\beta) \quad \text{for all } (\alpha, \beta) \in V_h^2$$

by the Cauchy-Schwarz inequality, which is contradictory.                                  ∎

***End of the Proof of Theorem 3.2.3*** Let $P$ be the degree 2 polynomial function given by

$$P(t) := q(g + tf) = q(g) + 2tb(f, g) + t^2 q(f) \text{ for } t \in \mathbb{R}.$$

Since $q(f) > 0$, $P(t) > 0$ for all large enough $t$. Furthermore, the lemma ensures that $b(1, f) \neq 0$ so that we may define

$$\tau := -\frac{b(1, g)}{b(1, f)}$$

and consider $g + \tau f$, which belongs to $F(v)$. Thus, by the lemma, $P(\tau) < 0$ unless $g + \tau f$ is the support function of a point. By considering the discriminant of $P$, we deduce that $b(f, g)^2 > q(f)\, q(g)$ unless $g + \tau f$ is the support function of a point. Finally, note that if there exists $(\lambda, \mu) \in \mathbb{R}^2 \setminus \{(0, 0)\}$ such that $\lambda f + \mu g$ is the support function of a point, then $b(f, g)^2 = q(f)\, q(g)$.          ∎

Next, we deduce the following.

**Corollary 3.2.4** *Let $f$ be a function in $V(v)$ such that $v(f, f; l) > 0$. If $g$ is any function in $V(v)$ such that $v(f, g; l) = v(g, g; l) = 0$, then the hedgehog $\mathcal{H}_g$ is reduced to a point.*

*In other words:*

**Corollary 3.2.5** *Let $f \in V(v)$. If there exists a hedgehog not reduced to a point with support function $g \in V(v)$ such that $v(f, g; l) = v(g, g; l) = 0$, then $v(f, f; l) \le 0$.*

***Proof of Corollary 3.2.4*** It follows from assumptions that we are in an equality case of (3.2.1). So, by Theorem 3.2.3, there exists $(\lambda, \mu) \in \mathbb{R}^2 \setminus \{(0, 0)\}$ such that $\lambda f + \mu g$ is the support function of a point. Since $\mathcal{H}_{\lambda f + \mu g}$ is a point, $v(\lambda f + \mu g, \lambda f + \mu g; l) = 0$. Developing by multilinearity and using assumptions, we deduce that $\lambda^2 v(f, f; l) = 0$. Since $v(f, f; l) > 0$, $\lambda = 0$ and hence $\mathcal{H}_{\mu g}$ is reduced to a point. Now $\mu \neq 0$ since $(\lambda, \mu) \neq (0, 0)$. Therefore, $\mathcal{H}_g$ is reduced to a point. ∎

***Proof of Theorem 3.2.2*** Finally, we apply Corollary 3.2.5 to

$$g := \sigma_v \quad \text{and} \quad f := h - \lambda k, \quad \text{where} \quad \lambda := \frac{v(h, \sigma_v; l)}{v(k, \sigma_v; l)}.$$

Let us check that all the assumptions of Corollary 3.2.5 are then satisfied. Of course, $\mathcal{H}_g$ is not reduced to a point since it is a unit segment $U(v)$. Since the mixed volume $v : V(v)^{n+1} \to \mathbb{R}$ is linear in each of its arguments, we have $v(f, g; l) = v(h, \sigma_v; l) - \lambda v(k, \sigma_v; l) = 0$. Applying formula (5.77) from [167, p. 302], we obtain

$$(n + 1) v(\sigma_v, \sigma_v; l) = v_{v\perp}\left(U(v)^v, L_3^v, \ldots, L_{n+1}^v\right),$$

where $v_{v\perp}$ denotes the $n$-dimensional mixed volume in the linear subspace orthogonal to $v$ and, $U(v)^v, L_3^v, \ldots, L_{n+1}^v$ the respective images of $U(v)$, $L_3$, …, $L_{n+1}$ under orthogonal projection to this subspace, and thus $v(g, g; l) = 0$ since $U(v)^v = \{0\}$.

Hence by Corollary 3.2.5, we have

$$v(h - \lambda k, h - \lambda k; l) \le 0.$$

After replacing $\lambda$ by its value and rearranging, we obtain

$$v(h, k; l)^2 - v(h, h; l) v(k, k; l) \ge \left(v(h, k; l) - \frac{v(h, \sigma_v; l)}{v(k, \sigma_v; l)} v(k, k; l)\right)^2.$$

Using the inequality $a^2 + b^2 \ge \frac{1}{2}(a - b)^2$ with $a := v(h, k; l) - m_{(h,k;l)} v(k, k; l)$ and $b := v(h, k; l) - M_{(h,k;l)} v(k, k; l)$, we deduce that

$$v(h, k; l)^2 - v(h, h; l) v(k, k; l) \ge \frac{v(k, k; l)^2}{4} \left(M_{(h,k;l)} - m_{(h,k,l)}\right)^2.$$

∎

For the study of several particular cases and a comparison with a stability result by R. Schneider, and independently by P.R. Goodey and H. Groemer, we refer the reader to [108].

### 3.2.4   A Remark on the Missing Boundary of the Blaschke Diagram

In this subsubsection, we briefly mention without proof an application of our results to the missing boundary of the Blaschke diagram. The interested reader will find details and proofs in [92]. The **Blaschke diagram** is the image of the set of convex bodies of $\mathbb{R}^3$ that are not reduced to a single point under the map associating to each such convex body $K$ the point $(x, y) \in \mathbb{R}^2_+$ given by

$$x = \frac{4\pi S}{M^2} \quad \text{and} \quad y = \frac{48\pi^2 V}{M^3},$$

where $V$, $S$ and $M$ respectively denote the 3-dimensional volume, the surface area, and the integral of the mean curvature of $K$. We know from the following Minkowskian inequalities

$$4\pi S \leq M^2, \quad 3MV \leq S^2 \quad \text{and} \quad 48\pi^2 V \leq M^3,$$

that $x \leq 1$, $y \leq 1$ and $y \leq x^2$, but a part of the boundary of the Blaschke diagram must correspond to an unknown inequality of the form $V \geq f(S, M)$. This problem raised by Blaschke [18] still today remains unsolved. The reader may refer to the studies of this problem presented by Hadwiger [62], Bieri [17] and Sangwine-Yager [162]. Using the solution of the Christoffel problem provided by Firey [46], we prove the following proposition.

**Proposition 3.2.6 ([92])**   *Let $g \in C^3\left(\mathbb{S}^2; \mathbb{R}\right)$ be the support function of a convex body of $\mathbb{R}^3$. There exists a $C^3$-hedgehog $\mathcal{H}_f$, which is unique up to translation, such that $R_{(1,f)} = R_g$.*

We will say that this unique $C^3$-hedgehog $\mathcal{H}_f$ is associated to the convex hedgehog $\mathcal{H}_g$.

**Theorem 3.2.4 ([92])**   *Let $g \in C^3\left(\mathbb{S}^2; \mathbb{R}\right)$ be the support function of a convex body of $\mathbb{R}^3$ such that the associated hedgehog of $\mathcal{H}_g$ satisfies the condition $s(f) > 0$. Then we have*

$$v(g) \geq \frac{s(g)^2}{3(m(g) + \sqrt{m(g)^2 - 4\pi s(g)})},$$

*and the equality holds if, and only if, $\mathcal{H}_g$ is a sphere.*

*In other words, the convex body with support function g is then such that*

$$V \geq \frac{S^2}{3(M + \sqrt{M^2 - 4\pi S})} = \frac{SM}{12\pi}\left(1 - \sqrt{1 - \frac{4\pi S}{M^2}}\right),$$

*that is,*

$$y \geq \frac{x^2}{1 + \sqrt{1 - x}} = x(1 - \sqrt{1 - x}).$$

# Chapter 4
# Special Convex Bodies, Hedgehogs and Multihedgehogs

This section is devoted to a series of special convex bodies, hedgehogs, and multihedgehogs, which are also called $N$-hedgehogs, $(N \in \mathbb{N}^*)$: an envelope of a family of cooriented planes of $\mathbb{R}^{n+1}$ will be called an $N$-***hedgehog*** if, for an open dense set of $u \in \mathbb{S}^n$, it has exactly $N$ cooriented support planes with normal vector $u$. Thus, ordinary $C^2$-hedgehogs are merely 1-hedgehogs. Figure 4.1 shows a planar 2-hedgehog. We will extend to hedgehogs a series of classical notions for convex bodies. For instance, we will start by extending the notion of width to hedgehogs. As an application of our study, we will give an example of a noncircular algebraic curve of constant width whose equation is relatively simple, which answers a problem raised by S. Rabinowitz (Sect. 4.1.2). In passing, we will study various concepts related to convex bodies. In particular, we will study the relationship between planar projective hedgehogs (which are the planar hedgehogs of constant width 0) and Zindler curves (which are those planar closed curves such that all chords that divide the curve perimeter—or area—in a half, have the same length) in Sect. 4.2.2. We will then rely on a notion of symplectic area to introduce and study Zindler-type surfaces in $\mathbb{R}^4$. Section 4.5 will aim to motivate the development of a Brunn-Minkowski theory for minimal hedgehogs or multihedgehogs by continuing the pioneering works by R. Langevin, G. Levitt, H. Rosenberg and E. Toubiana [81, 82, 159].

This will also be an opportunity to discover a first series of applications of hedgehog theory to analysis. In Sect. 4.3, we will consider the cosine transform, which associates to any continuous function $f : \mathbb{S}^n \to \mathbb{R}$ the map $T_f : \mathbb{R}^{n+1} \to \mathbb{R}$ defined by

$$T_f(x) := \int_{\mathbb{S}^n} |\langle x, v \rangle| \, f(v) \, d\sigma(v),$$

where $\langle ., . \rangle$ is the standard inner product and $\sigma$ the spherical Lebesgue measure. We will prove that the cosine transform, which often appears in convex geometry, is a bounded linear operator from $C(\mathbb{S}^n; \mathbb{R})$ to $C^2(\mathbb{S}^n; \mathbb{R})$. It follows that the boundaries of zonoids (resp. generalized zonoids) whose generating measures



Y. Martinez-Maure, *Hedgehog Theory*, Lecture Notes in Mathematics 2381,
https://doi.org/10.1007/978-3-032-04298-9_4

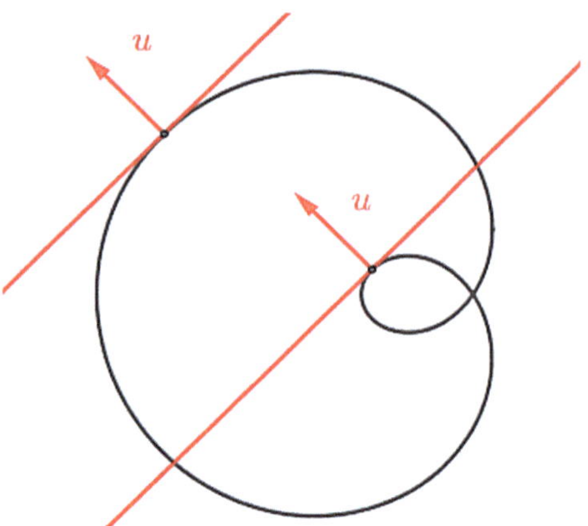


**Fig. 4.1**  A plane 2-hedgehog

have a continuous density with respect to $\sigma$ can be considered as $C^2$-hedgehogs. We will study such hedgehogs. Recall that zonotopes are the Minkowski sums of line segments, and that zonoids are (necessarily centrally symmetric) convex bodies that are the limit, in the sense of the Hausdorff metric, of a sequence of zonotopes. Zonoids play an important role in various areas such as the theory of vector measures, Banach space theory or stochastic geometry. We will obtain a local property of zonoids whose generating measures have a continuous density with respect to $\sigma$. We then define projection hedgehogs (resp. mixed projection hedgehogs) and interpret their support functions in terms of $n$-dimensional volume (resp. mixed volume). Finally, this study will lead us to consider the extension of the Minkowski problem (in differential geometry, the one of the existence, uniqueness and regularity of a closed convex hypersurface with preassigned curvature function) to hedgehogs. The classical Minkowski problem played an important role in the development of the theory of elliptic Monge-Ampère equations. The study of its extension to hedgehogs will be the subject of Chap. 5. In Sect. 4.4, we will study the existence of a nontrivial $C^2$-hedgehog in $\mathbb{R}^3$ that is hyperbolic (i.e., with an everywhere nonpositive curvature function), in order to determine the validity of the characterization of the 2-sphere conjectured by A.D. Alexandrov. This question amounts to studying a partial differential inequation. We will prove this Alexandrov conjecture in some particular cases, such as the case when the surface is assumed to be of constant width, and give a counterexample in the general case. In passing, we will consider the discrete version of hyperbolic hedgehogs. After a brief presentation of hedgehog polytopes (also called polyhedral hedgehogs) in $\mathbb{R}^3$, we will introduce two notions of hyperbolicity (weak and strong hyperbolicity) for hedgehog polytopes of $\mathbb{R}^3$ and give examples. Our example of a strongly hyperbolic polytope is obtained by a discretization of our counterexample to Alexandrov's conjecture. In Sect. 4.6, we will give a geometric proof of the Sturm-Hurwitz theorem in the framework of planar multihedgehogs. We will take this opportunity to present a series of geometric consequences and inequalities. We will end Chap. 4 by a detailed study of planar general hedgehogs (i.e., Minkowski differences of arbitrary convex bodies of $\mathbb{R}^2$). Our way of introducing general hedgehogs (proceeding by induction on $n$ and replacing support sets by 'support hedgehogs') makes clear that a perfect understanding of planar hedgehogs is a prerequisite to a study of general hedgehogs of $\mathbb{R}^{n+1}$. In particular, we will: ($i$) study their length measures and solve the extension of the Christoffel-Minkowski Problem to plane hedgehogs; ($ii$) characterize support functions of plane convex bodies among support functions of plane hedgehogs and support functions of plane hedgehogs among continuous functions; ($iii$) study the mixed area of hedgehogs in $\mathbb{R}^2$ and give an extension of the classical Minkowski inequality (and thus of the isoperimetric inequality) to hedgehogs.

## 4.1  Convex Bodies and Hedgehogs of Constant Width

In this subsection, we introduce and study the notion of a hedgehog of constant width in $\mathbb{R}^{n+1}$. In Chap. 7, we will consider this notion for non-Euclidean hedgehogs, and in particular for hedgehogs of the hyperbolic space $\mathbb{H}^{n+1}$.

### 4.1.1  Introduction

Let $K$ be a convex body of $(n+1)$-Euclidean vector space $\mathbb{R}^{n+1}$, that is, let $K \in \mathcal{K}^{n+1}$. The **width function** $w_K : \mathbb{S}^n \to \mathbb{R}$ of $K$ is defined by

$$w_K(u) = h_K(u) + h_K(-u) \quad \text{for all } u \in \mathbb{S}^n,$$

where $h_K : \mathbb{S}^n \longrightarrow \mathbb{R}$ is the support function of $K$. For all $\in \mathbb{S}^n$, the number $w_K(u)$ is called the width of $K$ in the direction $u$: $w_K(u)$ is simply the distance between the two support hyperplanes of $K$ orthogonal to $u$. The width of a closed convex hypersurface of $\mathbb{R}^{n+1}$ is defined to be the width of the body that it bounds. Of course, when $w_K$ is constant the convex body $K$ and its boundary $\partial K$ are said to be of constant width. Obvious bodies of constant width are Euclidean balls, but there are many others. In fact, every convex body $K$ of $\mathbb{R}^{n+1}$ whose support function $h_K$ is of the form $f + r$, where $f$ is an odd function on $\mathbb{S}^n$ and $r$ a constant, is a convex body of constant width. Besides, focusing on boundaries of convex bodies, we can state that any $C_+^2$-closed convex hypersurface of constant width $2r$, say $\mathcal{H}_{f+r}$, in $\mathbb{R}^{n+1}$ can be regarded as the parallel hypersurface at distance $r$ to its 'projective part' $\mathcal{H}_f$, which is a projective hedgehog. Indeed, we have;

$$\forall u \in \mathbb{S}^n, x_{f+r}(u) = x_f(u) + ru.$$

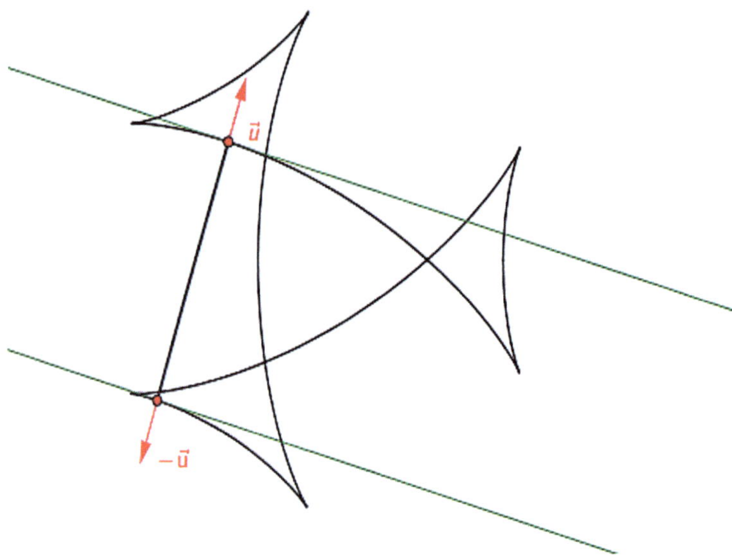


**Fig. 4.2**  A plane hedgehog of constant width: $\mathcal{H}_h$, where $h(\theta) = 2 + \cos(3\theta)$

The theory of bodies of constant width is a popular and fascinating topic, which has benefited from the works of a large number of scientists and mathematicians. It is now a well-established field of classical convex geometry. Various other names can be found in the literature to refer to convex bodies of constant width. For instance, these bodies are also called orbiforms in dimension 2, or spheroforms in dimension 3. For a recent book on bodies of constant width and their relations with various parts of mathematics, we refer the reader to [120].

The notion of being of constant width extends naturally to hedgehogs. A hedgehog $\mathcal{H}_h$ of $\mathbb{R}^{n+1}$ is said to be *of constant width* if the signed distance $w_h(u) :=$ $h(u) + h(-u)$ between the two cooriented support hyperplanes orthogonal to $u \in \mathbb{S}^n$ does not depend on the direction $u$ (Fig. 4.2).

In particular, projective hedgehogs of $\mathbb{R}^{n+1}$ (that is, hedgehogs $\mathcal{H}_h \subset \mathbb{R}^{n+1}$ with an odd support function $h : \mathbb{S}^n \to \mathbb{R}$) are exactly the hedgehogs of $\mathbb{R}^{n+1}$ that are of zero constant width. Of course, projective hedgehogs owe their name to the fact that the natural parametrization $x_h : \mathbb{S}^n \to \mathbb{R}$ of a $C^2$-hedgehog can be defined on the projective space $\mathbb{R}P^n = \mathbb{S}^n/(\text{antipodal map})$. Indeed, for all $f \in C^2(\mathbb{S}^n; \mathbb{R})$, we have:

$$(\mathcal{H}_h \text{ is projective}) \Longleftrightarrow (\forall u \in \mathbb{S}^n, h(-u) = -h(u))$$
$$\Longleftrightarrow (\forall u \in \mathbb{S}^n, x_h(-u) = x_h(u)).$$

Recall that we have already briefly touched on the concept of a projective hedgehog, first in Sect. 2.2.2 when we were interested in hedgehog versions of the classical models of the projective plane in $\mathbb{R}^3$ in response to a question raised by

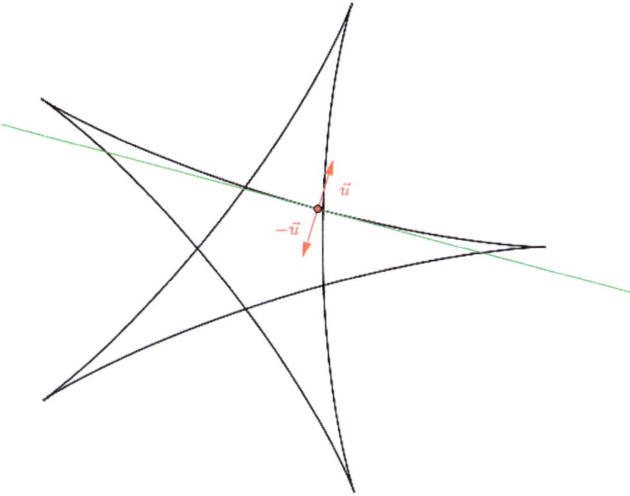

**Fig. 4.3**  A hedgehog is projective if, and only if, it has 0 constant width

Hilbert and Cohn-Vossen, then in Sect. 2.5 when we gave an example of a fractal projective hedgehog with a $C^1$-support function, and finally in Sects. 2.8 and 2.9 by presenting some of their geometric properties.

In this subsection, we will not review the classical results of the general theory of bodies of constant width, for which we refer the reader to [120]. We will restrict ourselves to parts of the theory that extend to hedgehogs as well as to the new results that hedgehogs have made it possible to discover (Fig. 4.3).

### 4.1.2   Noncircular Algebraic Curves of Constant Width: An Answer to Rabinowitz

A disc has the property that it can be rotated between two fixed parallel lines without losing contact with either line. It has been known for a long time that there are many other plane convex bodies with the same property. Such plane convex bodies are called plane convex bodies 'of constant width' or 'orbiforms'. Their boundaries are of course called 'plane convex curves of constant width'. A classical noncircular example is the famous Reuleaux triangle [192] which is made of three circular arcs. But a noncircular plane convex curve of constant width can be smooth, and not having any circular arc in its boundary. As we just saw above, the notion of a convex body of constant width extends to higher dimension.

In this subsubsection, we are essentially interested in noncircular algebraic curves of constant width. Rabinowitz [155] found that the zero set of the following polynomial $P \in \mathbb{R}[X, Y]$ forms a noncircular algebraic curve of constant width in $\mathbb{R}^2$:

$$
P(x, y) := \left(x^2 + y^2\right)^4 - 45 \left(x^2 + y^2\right)^3 - 41283 \left(x^2 + y^2\right)^2
$$
$$
+ 7950960 \left(x^2 + y^2\right) + 16 \left(x^2 - 3y^2\right)^3 + 48 \left(x^2 + y^2\right) \left(x^2 - 3y^2\right)^2
$$
$$
+ x \left(x^2 - 3y^2\right) \left(16 \left(x^2 + y^2\right)^2 - 5544 \left(x^2 + y^2\right) + 266382\right) - 720^3.
$$

Then, he raised the following open questions: "*The polynomial curve found is pretty complicated. Can it be put in simpler form? Our polynomial is of degree 8. Is there one with lower degree? What is the lowest degree polynomial whose graph is a noncircular curve of constant width?*". A couple of years ago, Bardet and Bayen [14, Cor. 2.1] proved that the degree of $P$, that is 8, is the minimum possible degree for a noncircular plane **convex** curve of constant width. Here, we emphasize the convexity assumption because it is implicit in the statement of Corollary 2.1 in [14]. In this subsubsection taken from [113], we provide additional answers to **Rabinowitz's open questions**. First, we recall the notion of a plane hedgehog curve of constant width, and we notice that in this setting, we can find algebraic curves of constant width much simpler. Second, we give an example of a noncircular algebraic curve of constant width whose equation is simpler than the one of Rabinowitz. Finally, we notice that we can deduce from it (relatively) simple examples in higher dimensions.

For a presentation of the main results intended for the general public with additional illustrations and animations, we refer the reader to [112].

**Plane Algebraic Hedgehogs of Constant Width**

Here, we will follow more or less [91].

**Definition 4.1.1** For any smooth function $h : \mathbb{S}^1 = \mathbb{R} \backslash 2\pi \mathbb{Z} \to \mathbb{R}, \theta \mapsto h(\theta)$, we let $\mathcal{H}_h$ denote the envelope of the family of lines given by

$$
x \cos \theta + y \sin \theta = h(\theta), \tag{4.1.1}
$$

where $(x, y)$ are the coordinates in the canonical basis of the Euclidean vector space $\mathbb{R}^2$. We say that $\mathcal{H}_h$ is the plane hedgehog with support function $h$, and that $\mathcal{H}_h$ is projective if $h(\theta + \pi) = -h(\theta)$ for all $\theta \in \mathbb{S}^1$.

Partial differentiation of (4.1.1) yields

$$
- x \sin \theta + y \cos \theta = h'(\theta). \tag{4.1.2}
$$

From (4.1.1) and (4.1.2), the parametric equations for $\mathcal{H}_h$ are

$$\begin{cases} x = h(\theta)\cos\theta - h'(\theta)\sin\theta \\ y = h(\theta)\sin\theta + h'(\theta)\cos\theta. \end{cases}$$

The family of lines $(D(\theta))_{\theta\in\mathbb{S}^1}$, of which $\mathcal{H}_h$ is the envelope, is the family of 'support lines' of $\mathcal{H}_h$. Suppose that $\mathcal{H}_h$ has a well-defined tangent line at the point $(x, y)$, say $T$. Then $T$ is the support line with Eq. (4.1.1): the unit vector $u(\theta) = (\cos\theta, \sin\theta)$ is normal to $T$ and $h(\theta)$ may be interpreted as the signed distance from the origin to $T$.

A plane hedgehog is thus simply a plane envelope that has exactly one oriented support line in each direction. A singularity-free plane hedgehog is simply a convex curve. A plane hedgehog is projective if it has exactly one nonoriented support line in each direction

Now, we can define the width, say $w_h(\theta)$, of such a plane hedgehog $\mathcal{H}_h$ in the direction $u(\theta)$ to be the signed distance between the two support lines of $\mathcal{H}_h$ that are orthogonal to $u(\theta)$, that is,

$$w_h(\theta) = h(\theta) + h(\theta + \pi).$$

Thus plane projective hedgehogs are hedgehogs of constant width 0, and the condition that a plane hedgehog $\mathcal{H}_h$ is of constant width $2r$ is simply that its support function $h$ has the form $f + r$, where $f$ is the support function of a projective hedgehog. Here are three examples of plane hedgehogs: $(a)$ a convex hedgehog of constant width; $(b)$ a hedgehog with four cusps; $(c)$ a plane projective hedgehog which is a hypocycloid with three cusps (Fig. 4.4).

**Proposition 4.1.1** *The plane projective hedgehog $\mathcal{H}_h$ with support function $h :$ $\mathbb{S}^1 \to \mathbb{R}$, $\theta \mapsto \sin(3\theta)$ is a noncircular algebraic curve of constant width 0 with equation*

$$\left(x^2 + y^2\right)^2 + 18\left(x^2 + y^2\right) - 8y\left(y^2 - 3x^2\right) = 27.$$

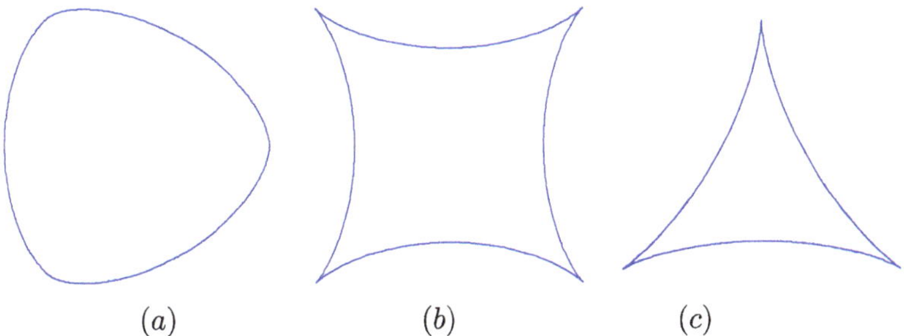

(a)      (b)      (c)

**Fig. 4.4** (**a**) $h(\theta) = 10 + \cos(3\theta)$. (**b**) $h(\theta) = \cos(2\theta)$. (**c**) $h(\theta) = \sin 3\theta$

***Proof*** We already know that $\mathcal{H}_h$ is a noncircular curve of constant width 0. From the parametric equations for $\mathcal{H}_h$, we deduce that

$$x^2 + y^2 = h(\theta)^2 + h'(\theta)^2 = \sin^2(3\theta) + 9\cos^2(3\theta) = 5 + 4\cos(6\theta).$$

Now, $h : \mathbb{S}^1 \to \mathbb{R}, \theta \mapsto \sin(3\theta)$ is the restriction of the polynomial $-y\left(y^2 - 3x^2\right)$ to the unit circle $\mathbb{S}^1$, and the linearization of $-y\left(y^2 - 3x^2\right)$ as a trigonometric function of $\theta$ gives

$$-y\left(y^2 - 3x^2\right) = -12 - 14\cos(6\theta) - \cos(12\theta) = -11 - 14\cos(6\theta) - 2\cos^2(6\theta).$$

From the above two equations, we deduce easily that

$$\left(x^2 + y^2\right)^2 + 18\left(x^2 + y^2\right) - 8y\left(y^2 - 3x^2\right) = 27,$$

∎

## A Noncircular Algebraic Curve of Constant Width Whose Equation Is Relatively Simple

Any hedgehog whose support function $h : \mathbb{S}^1 \to \mathbb{R}$ is of the form $h(\theta) = r - \sin(3\theta)$, for some constant $r$, is a hedgehog of constant width $2r$. Such a function $h : \mathbb{S}^1 \to \mathbb{R}$ is the support function of a convex body if and only if $\left(h + h''\right)(\theta) = r - 8\sin(3\theta) \geq 0$ for all $\theta \in \mathbb{S}^1$, that is if and only if $r \geq 8$. We choose $r = 8$ in order to be 'as closed as possible' to the previous example.

**Theorem 4.1.1** *The plane hedgehog $\mathcal{H}_h$ with support function $h : \mathbb{S}^1 \to \mathbb{R}, \theta \mapsto 8 - \sin(3\theta) = 4\sin^3\theta - 3\sin\theta + 8$ is a noncircular convex algebraic curve of constant width* 16 *with equation*

$$\left(\left(x^2 + y^2\right)^2 + 8y\left(y^2 - 3x^2\right)\right)^2 + 432y\left(y^2 - 3x^2\right)\left(351 - 10\left(x^2 + y^2\right)\right)$$

$$= 567^3 + 28\left(x^2 + y^2\right)^3 + 486\left(x^2 + y^2\right)\left(67\left(x^2 + y^2\right) - 567 \times 18\right).$$

$$(4.1.3)$$

***Proof*** The parametric equations for $\mathcal{H}_h$ are equivalent to:

$$\begin{cases} x = -8\left(\sin^3(\theta) - 1\right)\cos(\theta) \\ y = -2\cos(2\theta) - \cos(4\theta) + 8\sin(\theta). \end{cases}$$

Expanding $x$ and $y$ in terms of $c = \cos\theta$ and $s = \sin\theta$, we obtain after simplification:

$$\begin{cases} x = -8\left(s^3 - 1\right)c \\ y = -3 + 4s\left(2 + 3s - 2s^3\right). \end{cases}$$

Squaring the first equation and substituting in $c^2 = 1 - s^2$ gives us the following system of equations in the three unknowns $x$, $y$, and $s$:

$$\begin{cases} 64\left(1 - s^2\right)\left(s^3 - 1\right)^2 - x^2 = 0 \\ -3 + 4s\left(2 + 3s - 2s^3\right) - y = 0. \end{cases}$$

We then eliminate $s$ by computing the resultant of the polynomials

$$A\left(s\right) = 64\left(1 - s^2\right)\left(s^3 - 1\right)^2 - x^2 \quad \text{and} \quad B\left(s\right) = -3 + 4s\left(2 + 3s - 2s^3\right) - y$$

with Mathematica, and find after simplification that:

$$\left(\left(x^2 + y^2\right)^2 + 8y\left(y^2 - 3x^2\right)\right)^2 + 432y\left(y^2 - 3x^2\right)\left(351 - 10\left(x^2 + y^2\right)\right)$$

$$= 567^3 + 28\left(x^2 + y^2\right)^3 + 486\left(x^2 + y^2\right)\left(67\left(x^2 + y^2\right) - 567 \times 18\right).$$

Here, it is important to note that our example has another advantage over Stanley Rabinowitz's. The coordinates of the points of the Rabinowitz curve are indeed such that $P\left(x, y\right) = 0$. But they are not the only ones! The set of points of the plane whose coordinates are such that $P\left(x, y\right) = 0$ also contains isolated points that are not on the curve with equation $P\left(x, y\right) = 0$ [143]. Now, in our example, we are lucky that the points that correspond to these isolated points are also located on our curve [143] (Fig. 4.5). ∎

## Higher Dimension

As we already mentioned, the notion of a hedgehog of constant width can of course be extended to higher dimension. Each of the above two examples of algebraic curves of constant width admits an axis of symmetry in $\mathbb{R}^2$. By rotating it around such an axis, we deduce immediately an example of algebraic surface of revolution that is of constant width in $\mathbb{R}^3$. More precisely, the algebraic surface with equation.

$$\left(x^2 + y^2 + z^2\right)^2 + 18\left(x^2 + y^2 + z^2\right) \quad 8z\left(z^2 - 3\left(x^2 + y^2\right)\right) = 27$$

**Fig. 4.5** The noncircular
convex curve of constant
width 16 with Eq. (4.1.3)

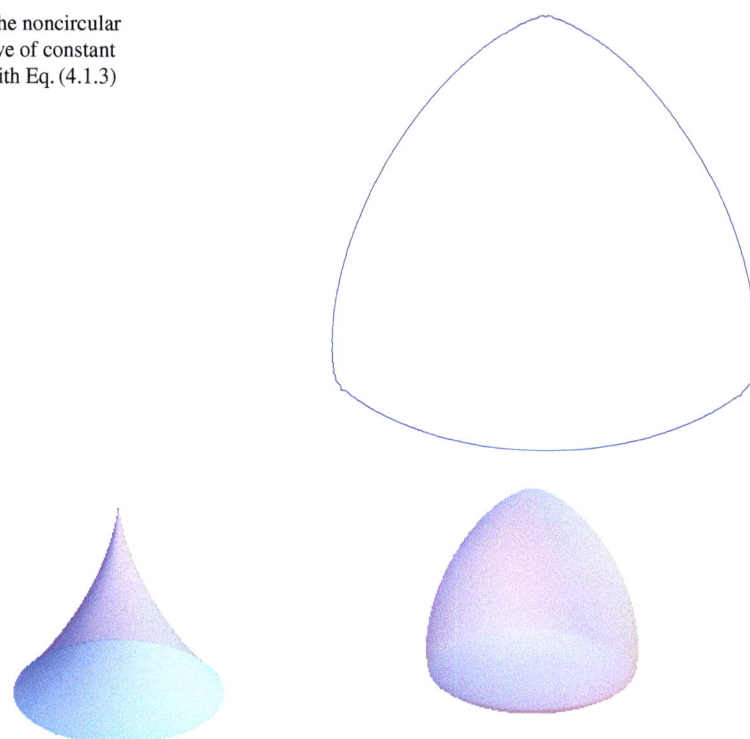

**Fig. 4.6** Our two algebraic surfaces of constant width

is a 'projective hedgehog' of revolution and a surface of constant width 0 in $\mathbb{R}^3$ (see
Fig. 4.6, left), and the algebraic surface with equation

$$\left(\left(x^2 + y^2 + z^2\right)^2 + 8z\left(z^2 - 3\left(x^2 + y^2\right)\right)\right)^2$$
$$+ 432z\left(z^2 - 3\left(x^2 + y^2\right)\right)\left(351 - 10\left(x^2 + y^2 + z^2\right)\right)$$
$$= 567^3 + 28\left(x^2 + y^2 + z^2\right)^3$$
$$+ 486\left(x^2 + y^2 + z^2\right)\left(67\left(x^2 + y^2 + z^2\right) - 567 \times 18\right)$$

is a convex surface of constant width 16 in $\mathbb{R}^3$ (see Fig. 4.6, right).

There are several methods to explicitly find algebraic constant width bodies, even
without being of revolution (see, e.g., [120, Section 8.5]).

### 4.1.3   *Hedgehogs of Constant Relative Width*

More generally, given an arbitrary norm $\|.\|$ on $\mathbb{R}^{n+1}$, we may define the (signed) width of a $C^2$ hedgehog $\mathcal{H}_h$ of $\mathbb{R}^{n+1}$ relative to the centered (i.e., 0-symmetric) convex body $K = \{x \in \mathbb{R}^{n+1} \,|\, \|x\| \leq 1\}$, also called the $K$-**width** of $\mathcal{H}_h$, in the direction $u$ to be the (signed) $K$-distance (i.e., the distance relative to $\|.\|$) between the two cooriented support hyperplanes orthogonal to $u$, that is, by

$$w_K(h, u) := 2\frac{w_h(u)}{w_{h_K}(u)},$$

where $h_K : \mathbb{S}^n \to \mathbb{R}$ is the support function of $K$.

The following proposition relates plane $C^2$ hedgehogs of zero relative mean width to the subspaces $F_k$ considered in Corollary 3.2.1 in the case $n + 1 = 2$, that is, to the subspaces $F_k := \{h \in H_2 \,|\, a\,(h, k) = 0\}$, where $a$ denotes the mixed area, and $k \in H_2$ has positive area.

**Proposition 4.1.2** *Let $\|.\|$ be an arbitrary norm on $\mathbb{R}^2$, and let $K := \{x \in \mathbb{R}^2 \,|\, \|x\| \leq 1\}$. There exists a plane convex $C^2$ hedgehog $\mathcal{H}_k$ such that the subspace $F_k$ that is a-orthogonal to $k$ is the subspace of $H_2$ constituted by all plane $C^2$ hedgehogs $\mathcal{H}_h$ of zero mean $K$-width, that is, by all $C^2$ hedgehogs satisfying the condition*

$$\int_0^{2\pi} w_K(h, \theta)d\theta = 0 \quad \text{or, equivalently,} \quad \int_0^{2\pi} \frac{h\,(\theta)}{h_K\,(\theta)}d\theta = 0.$$

***Proof of Proposition 4.1.2*** Since $K$ is a plane centered convex body of $\mathbb{R}^2$, its support function $h_K : \mathbb{S}^1 = \mathbb{R}\backslash 2\pi\mathbb{Z} \to \mathbb{R}$ is $\pi$-periodic, so that the general solution $k$ of the differential equation $y + y'' = 1/h_K$, namely

$$k(\theta) = \left(\int \frac{\cos\theta}{h_K(\theta)}\,d\theta\right)\sin\theta - \left(\int \frac{\sin\theta}{h_K(\theta)}\,d\theta\right)\cos\theta,$$

is a $2\pi$-periodic $C^2$-function that defines (up to a translation) a hedgehog $\mathcal{H}_k$ that is convex because $R_k := k + k'' = 1/h_K > 0$. Moreover, for any $h \in H_2$, we have

$$a\,(h, k) = \frac{1}{2}\int_0^{2\pi} h\,(\theta)\,(k + k'')\,(\theta)\,d\theta = \frac{1}{2}\int_0^{2\pi} \frac{h\,(\theta)}{h_K\,(\theta)}d\theta,$$

so that

$$a\,(h, k) = 0 \quad \text{if, and only if,} \quad \int_0^{2\pi} w_K(h, \theta)d\theta = 0.$$

## 4.2   Concepts Related to Constant Width

### 4.2.1   Equichordal Points

Let $K$ be a convex body in $(n + 1)$-dimensional Euclidean space $\mathbb{R}^{n+1}$ and let $S$ be its boundary $\partial K$. An interior point $o$ of $K$ is called an **equichordal point** of $S$ if all chords of $S$ passing through $o$ have the same length. In 1916, Fujiwara proved that a plane convex curve cannot have more than two equichordal points, and raised the problem of whether there exists a plane convex curve with exactly two equichordal points [51]. Independently, Blaschke, Rothe and Weitzenböck posed the same equichordal point problem a year later [20]. Wirsing proved (assuming its existence) that such a curve must be analytic [194]. Petty and Crotty proved the existence of Minkowski spaces of arbitrary dimension in which there are convex hypersurfaces with exactly two equichordal points [145]. But, despite its elementary formulation, the equichordal point problem remained unsolved for a long time. Indeed, it was only in 1996 that Rychlik managed to solve this difficult problem in the negative in a long article using methods of advanced complex analysis and algebraic geometry [160].

In Sect. 2.7, we recalled the notion of pedal hypersurface (with respect to the origin $o$) of any $C^2$-hedgehog of $\mathbb{R}^{n+1}$ with non-vanishing support function. If the $C^2$-hedgehog is of constant width, then we see immediately that its pedal hypersurface $\mathcal{P}\left(\mathcal{H}_h\right)$ has $o$ as an equichordal point, but $\mathcal{P}\left(\mathcal{H}_h\right)$ is not necessarily convex. Conversely, we have the following:

**Proposition 4.2.1** *If $S$ is a closed convex hypersurface (that is, the boundary of a convex body $K$) of class $C_+^2$ in $\mathbb{R}^{n+1}$ with an equichordal point $o$, then $S$ is the pedal hypersurface (with respect to $o$) of a hedgehog of constant width $\mathcal{P}^{-1}\left(S\right)$. We will then say that this hedgehog $\mathcal{P}^{-1}\left(S\right)$ is the negative pedal of $S$ (with respect to $o$).*

***Proof*** Let us assume without loss of generality that $o$ is the origin. From the assumptions, the hypersurface $S$ is starlike and admits a parametrization of the form

$$X : \mathbb{S}^n \to S, u \longmapsto \rho\left(u\right) u,$$

where $\rho > 0$ denotes its radial function with respect to the origin. Since $\rho$ is equal to $1/g$ where $g$ is the support function of the polar body $K^o$, which is also of class $C_+^2$ (see Sect. 2.5 Higher regularity and curvature in [167]), this radial function $\rho$ is of class $C^2$ on $\mathbb{S}^n$. Now the condition that $S$ have the origin as an equichordal point is simply that $\rho$ be of the form $h + r$, where $r$ is a constant and $h$ is an odd $C^2$-function on $\mathbb{S}^n$ (that is, the support function of a projective hedgehog). Since we have $x_{h+r}\left(u\right) = \rho\left(u\right) u + \left(\nabla h\right)\left(u\right)$ for all $u \in \mathbb{S}^n$, we see that $S$ is the pedal hypersurface of $\mathcal{H}_{h+r}$ with respect to $o$. Furthermore, this hedgehog $\mathcal{H}_{h+r}$ is of constant width $2r$.    ∎

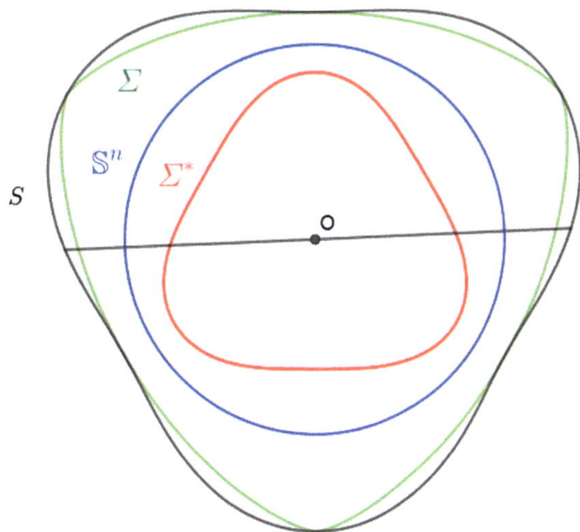

**Fig. 4.7** Illustration for the proof of Proposition 4.2.2



**Proposition 4.2.2** *If S is a closed convex hypersurface (that is the boundary of a convex body) of class $C_+^2$ in $\mathbb{R}^{n+1}$ with an equichordal point o, then $\mathcal{P}^{-1}(S)$, the negative pedal of S (with respect to o), is convex if and only if the hypersurface obtained from S by inversion with respect to $\mathbb{S}^n$ is convex.*

***Proof*** We can of course assume without loss of generality that $o$ is the origin. Let $\Sigma$ and $\Sigma^*$ denote respectively the negative pedal hypersurface $\mathcal{P}^{-1}(S)$ and the hypersurface obtained from $S$ by inversion with respect to $\mathbb{S}^n$. By polarity, we know that if $\Sigma$ or $\Sigma^*$ is convex, then $\Sigma$ and $\Sigma^*$ are the boundaries of two respective polar bodies $K$ and $K^o$ (see Sect. 2.7). This achieves the proof.    ∎

See Fig. 4.7 for an illustration of the proof.

**Proposition 4.2.3** *Let $\mathcal{H}_h$ in $\mathbb{R}^n$ be a $C^2$-hedgehog of constant width whose support function $h$ does not vanish on $\mathbb{S}^n$. Then the pedal hedgehog of $\mathcal{H}_h$ with respect to the origin o is a smooth hypersurface with the origin as an equichordal point. Furthermore, $\mathcal{P}(\mathcal{H}_h)$ is convex if and only if $1/h$ is the support function of a convex body.*

***Proof*** The pedal hedgehog $\mathcal{P}(\mathcal{H}_h)$ can be parametrized by

$$X : \mathbb{S}^n \to S, u \longmapsto h(u)\, u,$$

and has the origin as an equichordal point since $u \mapsto h(u) + h(-u)$ is constant. Furthermore, we have $\mathcal{P}(\mathcal{H}_h) = (\mathcal{I} \circ \mathcal{P})(\mathcal{H}_{1/h})$, where $\mathcal{I}$ denotes the inversion with respect to $\mathbb{S}^n$ and $\mathcal{P}$ the map that assigns to each hedgehog (with non-vanishing support function), its pedal surface with respect to $o$. Therefore, if $\mathcal{P}(\mathcal{H}_h)$ or $\mathcal{H}_{1/h}$ is convex, then $\mathcal{P}(\mathcal{H}_h)$ and $\mathcal{H}_{1/h}$ are the boundaries of two polar bodies. This achieves the proof.    ∎

Note that these questions are also related to equireciprocal points of convex bodies (e.g., see [79]).

### 4.2.2  Zindler Curves in $\mathbb{R}^2$ and Zindler-Type Hypersurfaces in $\mathbb{R}^4$

**Introduction**  The geometric concept of a Zindler curve was discovered a century ago by K. Zindler [198] and later named by H. Auerbach [10]. *Zindler curves* are those plane closed curves such that all chords that divide the curve length (or area) into halves, have the same length. The most simple examples of Zindler curves are circles but K. Zindler discovered other examples and gave a method to construct such curves. Since these curves are also the boundaries of figures with half the density of water that float in water in equilibrium in any position [24], they are used to study the floating body problem in two dimensions. They also serve as solutions to other well-known problems, such as the tire-track problem or the problem of determining the trajectory of an electron in a perpendicular parabolic magnetic field (see, e.g., [180] and [188]).

There are many known generalizations of Zindler curves, such as in normed planes [118, 119] or in non-Euclidean geometries [157, 178].

There exists a geometric duality relationship between (a huge class of) Zindler curves and convex curves of constant width through their association with their middle hedgehog. Let us recall that the *middle hedgehog* of a convex curve of constant width is the hedgehog curve that is the locus of the midpoints of all its diameters. This duality appears when a right-angle rotation is applied to the constant-length chords defining either curve type. For convex curves of constant width, the constant-length chords are the diameters, which are double normals. For Zindler curves the constant length chords divide the perimeter into equal halves. For the construction of Zindler curves by a right-angle rotation of the diameters of a convex curve of constant width around their midpoint, see [60, 120] or [156] (Fig. 4.8).

The same idea led to generalize the concept of a Zindler curve to curves in higher dimensions. On the one hand, Hoschek extended the concept of a Zindler curve to curves of $\mathbb{R}^3$ by turning double-normals of a closed transnormal space curve of constant width [68, 69]. The resulting spatial curve has properties analogous to those of its planar counterparts. Also using transnormal curves as starting point, Wegner generalized results of Hoschek to higher dimensions [187]. On the other hand, Wunderlich developed a kinematic principle based on the family of tangent lines to the midpoint curve that allows the construction of spatial Zindler curves without using spatial curves of constant width [195]. Pottmann generalized these results on Zindler curves in $\mathbb{R}^n$ [152]. In this generalization, the Zindler curves can simply be generated since the midpoints of the constant length chords are situated on the striction curve of the main chord-surface.

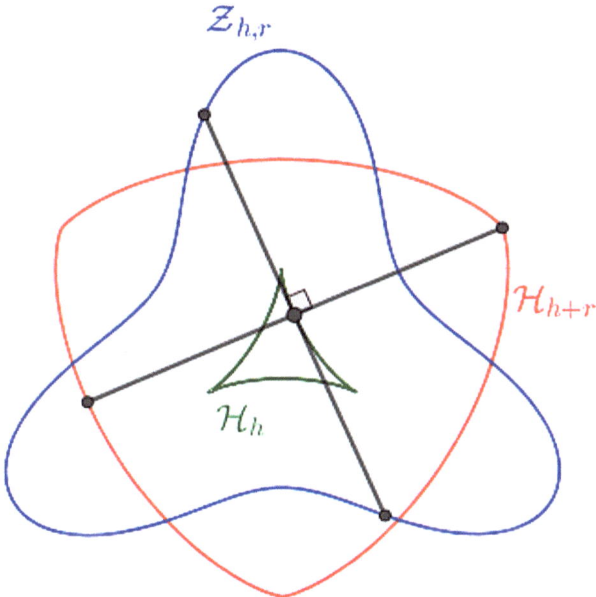


**Fig. 4.8** A Zindler curve $\mathcal{Z}_{h,r}$ constructed by rotating double-normals of a constant width curve $\mathcal{H}_{h+r}$; the midpoint travels along the hedgehog $\mathcal{H}_h$. Here, $h(\theta) = \sin 3\theta$ and $r = 8$

The notion of a hypersurface of constant width has been widely studied: see [120] and references therein. The concept of a hedgehog hypersurface of constant width in $\mathbb{R}^{n+1}$ is also well-known, see Sect. 4.1, including the notion of a projective hedgehog as those hypersurfaces of constant width 0. However, to the author's knowledge, no suitable notion of a Zindler surface of $\mathbb{R}^3$ or, more generally of a Zindler hypersurface of $\mathbb{R}^{n+1}$ with $n \geq 2$, has been proposed so far.

The present subsubsection is based on a joint work of the author with D. Rochera [116]. It aims to introduce a definition for Zindler-type hypersurfaces in $\mathbb{R}^4$, generalizing planar Zindler curves while preserving analogous properties. Since our method relies on a symplectic structure, it requires an even dimension. Consequently, this technique cannot be applied to find a suitable notion of a Zindler surface in $\mathbb{R}^3$, leaving this as an open problem.

**Properties of Planar Zindler Curves**  We begin by summarizing the main properties of planar Zindler curves. Later, we will show how some of them can be generalized to the Zindler-type hypersurfaces that we will introduce in $\mathbb{R}^4$.

Middle hedgehogs of convex curves of constant width are projective hedgehogs. As already noticed, we can construct Zindler curves from projective hedgehog curves regarded as their middle hedgehogs. However, not all Zindler curves arise from a middle hedgehog: see e.g., Mampel's example in [89]. In this subsubsection, we are only interested in Zindler curves associated with a curve of constant width and, thus, that can be generated from a projective hedgehog. The midpoint curve is the envelope of the halving chords [37]:

**Proposition 4.2.4** *Let $z : t \longmapsto z(t)$ be a $C^1$-regular parametric curve, and for all $t$, let $c(t)$ be the vector defining its halving chord at $z(t)$. Denote by $m$ the curve generated by the midpoints of the halving chords. Then $z$ is a Zindler curve if and only if $c'(t)$ is orthogonal to $m'(t)$ for every $t$.*

Note that in Pottmann's generalization of Zindler curves in $\mathbb{R}^n$ [152], this condition is required in the definition.

Recall that any convex curve with a $C^2$ support function is a hedgehog. $C^2$-hedgehog curves of constant width $2r$ are those $C^2$-hedgehogs $\mathcal{H}_h$ whose support functions $h$ are such that $h(\theta) + h(\theta + \pi) = 2r$ for all $\theta \in \mathbb{S}^1 = \mathbb{R}/2\pi\mathbb{Z}$. $C^2$-hedgehogs of constant width $0$ are exactly $C^2$ projective hedgehogs. Here, we will only consider $C^2$-hedgehogs. Auerbach was the first who identified the relationship between Zindler curves and curves of constant width [10]. He proved that Zindler curves have associated curves of constant width of equal area. Conversely, curves of constant width have associated Zindler curves of equal area [89]. This can be elegantly proved using Holditch's theorem. For an introduction to Holditch's theorem, we refer the reader to [153] or [128].

**Proposition 4.2.5** *If a pair of curves, consisting of a convex curve of constant width and a Zindler curve, are associated through a $C^2$ projective hedgehog, then they have the same area.*

**Proof** Let $\mathcal{H}_h$ be a projective $C^2$-hedgehog of $\mathbb{R}^2$. Now, let $r \in \mathbb{R}_+^*$ be such that $x_{h+r} : \mathbb{S}^1 = \mathbb{R}/2\pi\mathbb{Z} \to \mathbb{R}^2$ and $z_{h,r} : \mathbb{S}^1 = \mathbb{R}/2\pi\mathbb{Z} \to \mathbb{R}^2, \theta \mapsto x_h(\theta) + ru'(\theta) = h(\theta)u(\theta) + \left(h'(\theta) + r\right)u'(\theta)$ are a convex curve of constant width $2r$ and its associated Zindler curve, respectively, for chords of length $2r$. Recall that for $f \in C^2\left(\mathbb{S}^1; \mathbb{R}\right)$, $a(f)$ denotes the area of $\mathcal{H}_f$. By Holditch's theorem, we have that $a(h+r) - a(h) = \pi r^2$ and $a\left(\mathcal{Z}_{h,r}\right) - a(h) = \pi r^2$, where $a\left(\mathcal{Z}_{h,r}\right)$ denotes the area of the Zindler curve $\mathcal{Z}_{h,r} := z_{h,r}\left(\mathbb{S}^1\right)$. Therefore $a\left(\mathcal{Z}_{h,r}\right) = a(h+r)$.    ∎

Note that this can also be seen as a particular case of swept-out areas by bicycle tire-track curves (see e.g., [47]).

**Proposition 4.2.6** *All the Zindler curves generated from a $C^2$-projective hedgehog are regular.*

**Proof** Let $\mathcal{H}_h$ be a projective $C^2$-hedgehog of $\mathbb{R}^2$. Given $r \in \mathbb{R}_+^*$, the corresponding Zindler curve $\mathcal{Z}_{h,r}$ can be parametrized by:

$$z_{h,r} : \mathbb{S}^1 = \mathbb{R}/2\pi\mathbb{Z} \to \mathbb{R}^2, \theta \mapsto x_h(\theta) + ru'(\theta) = h(\theta)u(\theta) + \left(h'(\theta) + r\right)u'(\theta),$$

where $u(\theta) := (\cos\theta, \sin\theta)$. Since

$$z'_{h,r}(\theta) = \left(h + h''\right)(\theta)u'(\theta) - ru(\theta) \quad \text{for all } \theta \in \mathbb{S}^1,$$

we have that

$$\left\| z'_{h,r}\left(\theta\right) \right\| = \sqrt{\left(h+h''\right)\left(\theta\right)^2 + r^2} > 0 \quad \text{for all } \theta \in \mathbb{S}^1,$$

so that $z_{h,r}$ is regular. $\blacksquare$

A well-known result concerns the angles that the halving chords of a Zindler curve form with the tangent vectors at their endpoints. It can be stated as follows (see e.g., [195]).

**Proposition 4.2.7** *The halving chords of a Zindler curve make equal angles with tangent vectors to the curve at both corresponding endpoints.*

**Proof** For all $\theta \in \mathbb{S}^1$, the endpoints $z_1\left(\theta\right)$ and $z_2\left(\theta\right)$ of the halving chord $[z_1\left(\theta\right), z_2\left(\theta\right)]$ of a Zindler curve $\mathcal{Z}_{h,r} = z_{h,r}\left(\mathbb{S}^1\right)$ can be described from the middle hedgehog $\mathcal{H}_h := x_h\left(\mathbb{S}^1\right)$ by

$$\begin{cases} z_1\left(\theta\right) = x_h\left(\theta\right) - ru'\left(\theta\right) \\[2mm] z_2\left(\theta\right) = x_h\left(\theta\right) + ru'\left(\theta\right) \end{cases},$$

where $u\left(\theta\right) := \left(\cos\theta, \sin\theta\right)$. Now, since we have

$$\begin{cases} z'_1\left(\theta\right) = \left(h+h''\right)\left(\theta\right)u'\left(\theta\right) + ru\left(\theta\right) \\[2mm] z'_2\left(\theta\right) = \left(h+h''\right)\left(\theta\right)u'\left(\theta\right) - ru\left(\theta\right) \end{cases},$$

for all $\theta \in \mathbb{S}^1$, we deduce that $\left\langle z_2\left(\theta\right) - z_1\left(\theta\right), z'_1\left(\theta\right)\right\rangle = \left\langle z_2\left(\theta\right) - z_1\left(\theta\right), z'_2\left(\theta\right)\right\rangle$, and the result follows. $\blacksquare$

The following result gives a nice geometric relationship between the evolute of the middle hedgehog $\mathcal{H}_h$ and the Zindler curve $\mathcal{Z}_{h,r}$.

**Proposition 4.2.8** *Let $\mathcal{H}_h$ be a projective $C^2$-hedgehog of $\mathbb{R}^2$. Recall that its evolute is the projective hedgehog with support function $\partial h : \mathbb{S}^1 = \mathbb{R}/2\pi\mathbb{Z} \to \mathbb{R}$, $\theta \mapsto h'\left(\theta - \frac{\pi}{2}\right)$. Let $r \in \mathbb{R}^*_+$, and let $\mathcal{Z}_{h,r}$ be the Zindler curve parametrized by $z_{h,r} : \mathbb{S}^1 = \mathbb{R}/2\pi\mathbb{Z} \to \mathbb{R}^2$, $\theta \mapsto x_h\left(\theta\right) + ru'\left(\theta\right)$, where $u\left(\theta\right) := \left(\cos\theta, \sin\theta\right)$: $\mathcal{Z}_{h,r}$ corresponds to the hedgehog curve of constant width $\mathcal{H}_{h+r}$. Then, for all $\theta \in \mathbb{S}^1$, the vector $z_{h,r}\left(\theta\right) - x_{\partial h}\left(\theta\right)$ has the same length as $z'_{h,r}\left(\theta\right)$, and it is orthogonal to $\mathcal{Z}_{h,r}$ at $z_{h,r}\left(\theta\right)$.*

**Proof** The evolute of $\mathcal{H}_h$, which we defined as the locus of the centers of curvature of $\mathcal{H}_h$, can be parametrized by

$$c_h : \mathbb{S}^1 \to \mathbb{R}^2, \theta \mapsto c_h\left(\theta\right) := x_h\left(\theta\right) - R_h\left(\theta\right)u\left(\theta\right) = h'\left(\theta\right)u'\left(\theta\right) - h''\left(\theta\right)u\left(\theta\right),$$

where $R_h = h + h''$ is the curvature function of $\mathcal{H}_h$. On another side, the Zindler curve $\mathscr{Z}_{h,r}$ can be parametrized by each of the two parametrizations $z_1$ and $z_2$ described in the proof of Proposition 4.2.7. From this, we deduce immediately that

$$
\begin{cases}
z_1\left(\theta\right) - c_h\left(\theta\right) = \left(h + h''\right)\left(\theta\right)u\left(\theta\right) - ru'\left(\theta\right) = -Jz_1'\left(\theta\right) \\[2ex]
z_2\left(\theta\right) - c_h\left(\theta\right) = \left(h + h''\right)\left(\theta\right)u\left(\theta\right) + ru'\left(\theta\right) = -Jz_2'\left(\theta\right)
\end{cases},
$$

where $J : \mathbb{R}^2 \to \mathbb{R}^2$ is the linear complex structure given by $J\left(a, b\right) = \left(-b, a\right)$ for all $\left(a, b\right) \in \mathbb{R}^2$, so that $Ju\left(\theta\right) = u'\left(\theta\right)$ and $Ju'\left(\theta\right) = -u\left(\theta\right)$ for all $u \in \mathbb{S}^1$. This completes the proof since the vectors $-Jz_1'\left(\theta\right)$ and $-Jz_2'\left(\theta\right)$ are orthogonal to $\mathscr{Z}_{h,r}$ at $z_1\left(\theta\right)$ and $z_2\left(\theta\right)$, respectively, and have the same length as $z_1'\left(\theta\right)$ and $z_2'\left(\theta\right)$, respectively. ∎

Let us observe that Proposition 4.2.8 provides a geometric method to construct Zindler curves directly from the evolute of any projective hedgehog: see Fig. 4.9. Let $\mathcal{H}_h$ be a projective $C^2$-hedgehog of $\mathbb{R}^2$ oriented by its canonical basis. Consider its evolute, which is parametrized by $c_h : \mathbb{S}^1 \to \mathbb{R}^2, \theta \mapsto c_h\left(\theta\right) := x_h\left(\theta\right) - R_h\left(\theta\right)u\left(\theta\right) = x_{\partial h}\left(\theta + \frac{\pi}{2}\right)$. For all $\theta \in \mathbb{S}^1 = \mathbb{R}/2\pi\mathbb{Z}$, there exists a circle centered

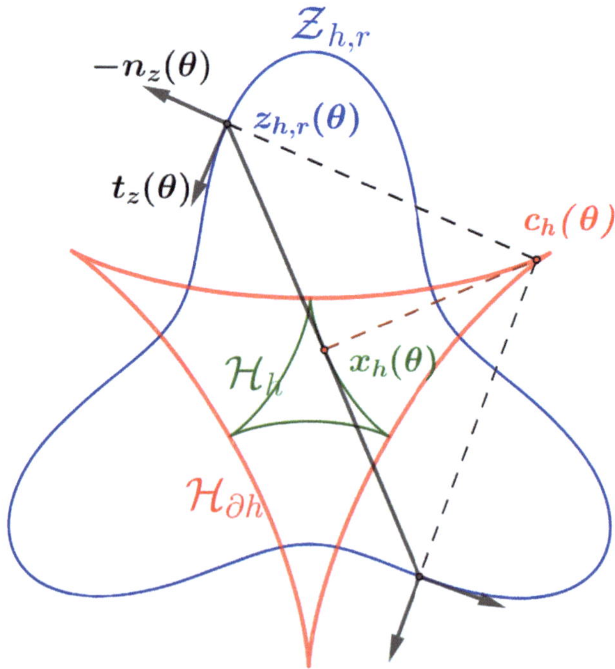


Fig. 4.9 A plane projective hedgehog $\mathcal{H}_h$, its evolute $\mathcal{H}_{\partial h}$, and the associated Zindler curve $\mathscr{Z}_{h,r}$ with tangent and normal unit vectors $\mathbf{t}_z$ and $\mathbf{n}_z$, respectively

at $c_h(\theta)$ that cuts the support line of $\mathcal{H}_h$ at $x_h(\theta)$ in points $z_1(\theta)$ and $z_2(\theta)$ such that $[z_1(\theta), z_2(\theta)]$ has length $2r$. Then $z_1 : \mathbb{S}^1 \to \mathbb{R}^2$ and $z_2 : \mathbb{S}^1 \to \mathbb{R}^2$ are parametrizations of a Zindler curve. Furthermore, the Frenet frame of $\mathcal{Z}_i := z_i(\mathbb{S}^1)$ at $z_i(\theta)$ is given by

$$\left( \frac{J\big(z_i(\theta) - c(\theta)\big)}{\big\| z_i(\theta) - c(\theta) \big\|}, \frac{c(\theta) - z_i(\theta)}{\big\| z_i(\theta) - \varepsilon(\theta) \big\|} \right), (i \in \{1, 2\}).$$

**Geometric Preliminaries in $\mathbb{R}^4$**   Here, we aim to present the linear space in which we will work, along with essential concepts from hedgehog theory and symplectic geometry that are crucial to our objectives. We will come back to this in much more detail in Chap. 6 when we will study real hedgehogs in $\mathbb{R}^{2n}$ endowed with a linear complex structure.

**Symplectic Area of a Closed Curve**

Let us equip $V := \mathbb{R}^4$ with a linear complex structure $J$ that is compatible with the standard Euclidean metric $\langle ., . \rangle$ on $\mathbb{R}^4$, that is, a linear isometry $J : V \to V$ that is such that $J^2 = -Id_V$. Then we denote by $\omega$ the associated Kähler form, that is, the alternating 2-form given by $\omega(X, Y) := \langle JX, Y \rangle$. Note that $\langle X, Y \rangle = \langle JX, JY \rangle = \omega(X, JY)$ so that we retrieve $\langle ., . \rangle$ from $\omega$. The *symplectic area of a closed curve* $\gamma \cong \mathbb{R}/2\pi\mathbb{Z} : \mathbb{S}^1 \to V$ is defined by

$$\mathcal{A}(\gamma) := \int_\gamma \alpha, \tag{4.2.1}$$

where $\alpha$ is the 1-form given by $(\alpha)_x(dx) = \frac{1}{2}\omega(x, dx)$, which is such that $d\alpha = \omega$. Notice that the integral (4.2.1) does not depend on the orientation of the curve $\gamma$ (as if we change the orientation of $\gamma$, the 1-form $\alpha$ is changed into its opposite). In other words, the *symplectic area of* $\gamma$, which is sometimes called the *action of* $\gamma$, can be written in the form

$$\mathcal{A}(\gamma) = \frac{1}{2} \int_0^{2\pi} \omega\big(\gamma(\theta), \gamma'(\theta)\big) d\theta.$$

Geometrically, the symplectic area of $\gamma$ can be interpreted as the sum of the algebraic areas of its projections onto a pair of orthogonal vector planes. The interested reader can find more details in [121–123] or [27].

**Geometry of $\mathbb{R}^4$ as a Kähler Vector Space**

The reader can refer to [111, 182] and [123] as primary sources for more background information on what follows.

Identify $\mathbb{R}^4$ with the quaternion algebra $\mathbb{H}$ and thus the unit sphere $\mathbb{S}^3$ with the set $\mathbb{S}_{\mathbb{H}}^1$ of unit quaternions. Denote by $\mathbb{S}^2$ the set $\mathbb{S}_{\mathbb{H}}^1 \cap \mathrm{Im}\,(\mathbb{H})$ of pure unit quaternions. To any pure unit quaternion $v$, associate the linear complex structure $J_v : \mathbb{R}^4 \to \mathbb{R}^4$, $x \longmapsto vx$. In other words, for any $v \in \mathbb{S}^2$, consider the Kähler vector space $\left(\mathbb{R}^4, J_v, \omega_v\right)$, where $\omega_v$ denotes the associated Kähler form (i.e. the alternating 2-form $\omega_v\,(X, Y) = \langle J_v X, Y\rangle$, where $\langle \cdot, \cdot \rangle$ is the standard Euclidean metric on $\mathbb{R}^4$). To any $v \in \mathbb{S}^2$, there corresponds a **Hopf fibration** and a **Hopf flow** on $\mathbb{S}^3 \cong \mathbb{S}_{\mathbb{H}}^1$ leaving the Hopf fibration invariant, namely the Hopf flow induced on $\mathbb{S}^3$ by the vector field $X_v\,(u) := J_v\,(u)$, that is the Hopf flow $\left\{(\phi_v)_\theta\right\}_{\theta \in \mathbb{S}^1}$ given by $(\phi_v)_\theta\,(u) := (\cos\theta)\,u + (\sin\theta)\,vu$, for any $u \in \mathbb{S}^3$.

For every $(u, v) \in \mathbb{S}^3 \times \mathbb{S}^2$, let $\mathbb{S}_{u,v}^1$ be the oriented geodesic of $\mathbb{S}^3$ through $u$ in the direction of $J_v(u)$. This oriented **Hopf circle** of $\mathbb{S}^3$ in $\left(\mathbb{R}^4, J_v\right)$ can be regarded as the unit circle of the vector plane $\mathbb{C}(u, v) := \mathbb{R}u + \mathbb{R}J_v\,(u)$ oriented by $(u, J_v(u))$. Conversely, any oriented vector plane $\xi$ in $\mathbb{R}^4$ determines an oriented unit circle $\mathbb{S}_\xi^1 = \mathbb{S}^3 \cap \xi$ and a pure unit quaternion $v_\xi$ that is such that: for all $u \in \mathbb{S}_\xi^1$, $T_u\mathbb{S}_\xi^1$ is oriented by the unit vector $J_{v_\xi}\,(u)$. Given a $C^2$-hedgehog $\mathcal{H}_h$ in $\mathbb{R}^4$, let us consider the symplectic area of $x_h : \mathbb{S}_{u,v}^1 \to \mathbb{R}^4$ in the Kähler vector space $\left(\mathbb{R}^4, J_v, \omega_v\right)$.

**Symplectic Area of the Image of the Hopf Circle $\mathbb{S}_{u,v}^1$ Under $x_h$**

Let $\mathcal{H}_h$ be any $C^2$-hedgehog in $\mathbb{R}^4$, and let $s_{u,v}(h)$ denote the symplectic area of the curve $x_h : \mathbb{S}_{u,v}^1 \to \mathbb{R}^4$ in $\left(\mathbb{R}^4, J_v, \omega_v\right)$:

$$s_{u,v}\,(h) = \int_{x_h\left(\mathbb{S}_{u,v}^1\right)} \alpha_v,$$

where $\alpha_v$ is the 1-form given by $(\alpha_v)_x\,(dx) = \tfrac{1}{2}\omega_v(x, dx) = \tfrac{1}{2}\langle x, (-J_v)\,(dx)\rangle$.

The symplectic area of $x_h : \mathbb{S}_{u,v}^1 \to \mathbb{R}^4$ is the sum of the algebraic areas of its projections onto the planes $\mathbb{C}(u, v)$ and $\mathbb{C}(u, v)^\perp$.

Note that the orientation of $\mathbb{S}_{u,v}^1$ (or the one of the plane that contains this Hopf circle) is not relevant for the value of the symplectic area. The author proved the following proposition (see Chap. 6):

**Proposition 6.6.5** *For all $h \in C^\infty\left(\mathbb{S}^3; \mathbb{R}\right)$ and $(u, v) \in \mathbb{S}^3 \times \mathbb{S}^2$,*

$$s_{u,v}\,(h) = \frac{1}{2}\int_0^{2\pi} \langle x_h\,(u_\theta)\,, R_h\,(u_\theta, v)\,u_\theta\rangle\,d\theta,$$

*where $u_\theta = (\cos\theta)\, u + (\sin\theta)\, J_v(u)$ and $R_h\,(u_\theta, v) = -v\left(T_{u_\theta} x_h\right)(J_v\,(u_\theta))\,\overline{u_\theta}$, with $\overline{u_\theta}$ being the quaternion conjugate of $u_\theta$.*

## Decomposition of Hedgehogs (See Chap. 6)

Let $(v, w)$ be any couple of pure unit quaternions that are orthogonal when they are regarded as vectors of $\mathbb{R}^4$. The quadruple $(1, v, w, vw)$ is then a direct orthonormal basis of $\mathbb{H} \cong \mathbb{R}^4$. For any $C^2$-hedgehog $\mathcal{H}_h$ in $\mathbb{R}^4$ and, for any $u \in \mathbb{S}^3$, we have the following decompositions:

$$
\begin{aligned}
x_h(u) &= h(u)\, u + \nabla h(u)\\
&= h(u)\, u + \langle \nabla h(u), vu\rangle\, vu + \langle \nabla h(u), wu\rangle\, wu + \langle \nabla h(u), vwu\rangle\, vwu\\
&= h(u)\, u + \partial_v h(vu)\, vu + \partial_w h(wu)\, wu + \partial_{vw} h(vwu)\, vwu\\
&= (h(u) + \partial_v h(vu)\, v + \partial_w h(wu)\, w + \partial_{vw} h(vwu)\, vw)\, u,
\end{aligned}
$$

where, for any pure unit quaternion $q$,

$$
\partial_q h : \mathbb{S}^3 \to \mathbb{R}, u \longmapsto \left\langle \nabla h\left(-J_q\,(u)\right), u\right\rangle
$$

is the support function of the so-called evolute of $\mathcal{H}_h$ in the Kähler vector space $\left(\mathbb{R}^4, J_q, \omega_q\right)$, see Chap. 6).

**Zindler-Type Hypersurfaces in $\mathbb{R}^4$**   Let $\mathcal{H}_h$ be a $C^2$ projective hedgehog in $\mathbb{R}^4$. Recall that a $C^2$-hedgehog $\mathcal{H}_h$ of $\mathbb{R}^4$ is said to be projective if, and only if, $h(u) + h(-u) = 0$ for all $u \in \mathbb{S}^3$, and thus $x_h(-u) = x_h(u)$ for all $u \in \mathbb{S}^3$. For all $r \in \mathbb{R}$, the $C^2$-hedgehog of $\mathbb{R}^4$ with support function $h + r$ is then of constant width $2r$ (this means that, for all $u \in \mathbb{S}^3$, the signed distance between the two cooriented support hyperplanes that are orthogonal to the line $\mathbb{R}u$ is equal to $2r$). Furthermore, if $r \in \mathbb{R}_+^*$ is large enough then $\mathcal{H}_{h+r}$ is necessarily of class $C_+^2$.

**Definition 4.2.1**   Let $\mathcal{H}_h$ be a $C^2$ projective hedgehog in $\mathbb{R}^4$ and let $r \in \mathbb{R}_+^*$ be such that $\mathcal{H}_{h+r}$ is a convex hypersurface of constant width. Given any pure unit quaternion $v \in \mathbb{S}^2$, the hypersurface $\mathcal{Z}_{h,r}^v$ of $\mathbb{R}^4$ that is parametrized by

$$
\begin{aligned}
z_{h,r}^v : \mathbb{S}^3 &\to \mathbb{R}^4\\
u &\mapsto x_h\,(u) + rvu.
\end{aligned}
$$

will be called the *v-Zindler hypersurface associated with $\mathcal{H}_{h+r}$*.

Note that the hypersurface $\mathcal{Z}_{h,r}^v = z_{h,r}^v\left(\mathbb{S}^3\right)$ is fibrated by the smooth curves that are the image of the Hopf circles $\mathbb{S}_{u,v}^1$ under $z_{h,r}^v : \mathbb{S}^3 \to \mathbb{R}^4$. The following theorem outlines key properties that make $\mathcal{Z}_{h,r}^v$ a Zindler-type hypersurface.

**Theorem 4.2.1** *Let $\mathcal{H}_h$ be a $C^2$ projective hedgehog in $\mathbb{R}^4$ and let $r \in \mathbb{R}_+^*$ be such that $\mathcal{H}_{h+r}$ is a convex hypersurface of constant width $2\,r$. For any $v \in \mathbb{S}^2$, let $\mathcal{Z}_{h,r}^v$ be the associated $v$-Zindler hypersurface. The following properties hold.*

1. *For all $u \in \mathbb{S}^3$, the chord $\left[ z_{h,r}^v(-u), z_{h,r}^v(u) \right]$ has length $2\,r$, and $x_h(u)$ is its midpoint.*
2. *$\mathcal{Z}_{h,r}^{u,v} := z_{h,r}^v\left(\mathbb{S}_{u,v}^1\right)$ is regular and, for all $\theta \in \mathbb{R}/2\pi\mathbb{Z}$, $\left[z_{h,r}^v(-u_\theta), z_{h,r}^v(u_\theta)\right]$ is a perimeter halving chord of $\mathcal{Z}_{h,r}^{u,v}$, where $u_\theta := (\cos\theta)\,u + (\sin\theta)\,vu \in \mathbb{S}_{u,v}^1$.*
3. *The halving chords $\left[z_{h,r}^v(-u_\theta), z_{h,r}^v(u_\theta)\right]$ make equal angles with tangent vectors to $\mathcal{Z}_{h,r}^{u,v}$ at both corresponding endpoints.*
4. *The orthogonal projection of the curve $\mathcal{Z}_{h,r}^{u,v}$ onto the plane $\mathbb{C}(u,v) := \mathbb{R}u + \mathbb{R}J_v(u)$ oriented by $(u, J_v(u))$ is the Zindler curve of $\mathbb{C}(u,v)$ parametrized by*

$$z : \mathbb{R}/2\pi\mathbb{Z} \to \mathbb{C}(u,v)$$
$$\theta \mapsto x_{h|_{\mathbb{S}_{u,v}^1}}(\theta) + rvu_\theta.$$

*where $\mathcal{H}_{h_{|\mathbb{S}_{u,v}^1}}$ is the projective hedgehog of $\mathbb{C}(u,v)$ whose support function is the restriction of $h$ to $\mathbb{S}_{u,v}^1$.*

***Proof***

1. Let $u \in \mathbb{S}^3$. It follows from the definition of $z_{h,r}^v$ that

$$\frac{z_{h,r}^v(-u) + z_{h,r}^v(u)}{2} = x_h(u).$$

Moreover, since $x_h(-u) = x_h(u)$, we have:

$$\left\| z_{h,r}^v(u) - z_{h,r}^v(-u) \right\| = \| 2rvu \| = 2r.$$

2. First, note that for all $\theta \in \mathbb{R}/2\pi\mathbb{Z}$, $u_{\theta+\pi} = -u_\theta \in \mathbb{S}_{u,v}^1$. We need to demonstrate that the length of each part of $\mathcal{Z}_{h,r}^{u,v}$ connecting $z_{h,r}^v(-u_\theta)$ and $z_{h,r}^v(u_\theta)$ is equal to half the total length of $\mathcal{Z}_{h,r}^{u,v}$. The map $\theta \mapsto z_{h,r}^v(\theta) := z_{h,r}^v(u_\theta)$ is such that for all $\theta$,

$$\left(z_{h,r}^v\right)'(\theta) = \left(T_{u_\theta}x_h\right)(vu_\theta) + rv(vu_\theta) = \left(T_{u_\theta}x_h\right)(vu_\theta) - ru_\theta,$$

where $T_{u_\theta}x_h$ denotes the tangent map of $x_h$ at $u_\theta$. Since $\mathcal{H}_h$ is projective,

$$\left(z_{h,r}^v\right)'(\theta + \pi) = \left(T_{u_{\theta+\pi}}x_h\right)(vu_{\theta+\pi}) - ru_{\theta+\pi}$$
$$= \left(T_{-u_\theta}x_h\right)(-vu_\theta) + ru_\theta = \left(T_{u_\theta}x_h\right)(vu_\theta) + ru_\theta.$$

Therefore,

$$\left\| \left(z_{h,r}^{v}\right)'(\theta + \pi) \right\| = \sqrt{\left\| \left(T_{u_\theta} x_h\right)(vu_\theta)\right\|^2 + r^2} = \left\| \left(z_{h,r}^{v}\right)'(\theta) \right\|.$$

In other words, the map $\theta \mapsto \left\| \left(z_{h,r}^{v}\right)'(\theta) \right\|$ is $\pi$-periodic, and hence

$$\int_{\theta}^{\theta+\pi} \left\| \left(z_{h,r}^{v}\right)'(t) \right\| dt = \int_{0}^{\pi} \left\| \left(z_{h,r}^{v}\right)'(t) \right\| dt = \frac{L}{2},$$

where $L$ is the length of $\mathcal{Z}_{h,r}^{u,v}$. Moreover, the curve $\mathcal{Z}_{h,r}^{u,v}$ is indeed regular since we have $\left\| \left(z_{h,r}^{v}\right)'(\theta) \right\| \geq r > 0$ for all $\theta$.

3. For all $\theta$, $z_{h,r}^{v}(u_\theta) - z_{h,r}^{v}(-u_\theta) = 2rvu_\theta$ is orthogonal to $u_\theta$. So

$$\left\langle z_{h,r}^{v}(u_\theta) - z_{h,r}^{v}(-u_\theta), \left(z_{h,r}^{v}\right)'(\theta) \right\rangle = \left\langle z_{h,r}^{v}(u_\theta) - z_{h,r}^{v}(-u_\theta), \left(z_{h,r}^{v}\right)'(\theta + \pi) \right\rangle.$$

4. By the decomposition of a hedgehog given above, we have

$$x_f(u) = f(u)u + \partial_v f(vu) vu \in \mathbb{C}(u, v)$$

where $f = h_{|\mathbb{S}_{u,v}^1}$. Therefore, the orthogonal projection of $\mathcal{Z}_{h,r}^{u,v}$ onto $\mathbb{C}(u, v)$ can be parametrized by

$$z : \mathbb{S}^1 = \mathbb{R}/2\pi\mathbb{Z} \to \mathbb{C}(u, v), \theta \mapsto x_{h_{|\mathbb{S}_{u,v}^1}}(\theta) + rvu_\theta.$$

Note that $z\left(\mathbb{S}^1\right)$ is a planar Zindler curve since the projective hedgehog $\mathcal{H}_{h_{|\mathbb{S}_{u,v}^1}}$ of $\mathbb{C}(u, v)$ is such that the support line at $x_{h_{|\mathbb{S}_{u,v}^1}}(\theta)$ is directed by $J_v(u_\theta) = vu_\theta$ for all $\theta$. Observe that the length $r$ is large enough to have a Zindler curve since $\mathcal{H}_{h_{|\mathbb{S}_{u,v}^1} + r}$, which is the orthogonal projection of $\mathcal{H}_{h+r}$ onto $\mathbb{C}(u, v)$, is a planar convex curve of constant width $2r$.

■

**Remark** The curve $\mathcal{Z}_{h,r}^{u,v} = z_{h,r}^{v}\left(\mathbb{S}_{u,v}^1\right)$ from Theorem 4.2.1 is a spatial Zindler curve in the sense of Pottmann [152]. Indeed, $z_{h,r}^{v}$ can be written in the form

$$\theta \mapsto z_{h,r}^{v}(\theta) = x_h(u_\theta) + rvu_\theta$$

where $\theta \mapsto m(\theta) = x_h(u_\theta)$ is the striction line of the ruled surface generated by the halving chords with directions $e(\theta) = vu_\theta$, $(\theta \in \mathbb{R}/2\pi\mathbb{Z})$. This is due to the fact that $m'(\theta)$ is orthogonal to $e'(\theta)$ for all $\theta$. In addition, we have $m(\theta + \pi) = m(\theta)$ and $e(\theta + \pi) = -e(\theta)$ for all $\theta$. Therefore, we can say that the Zindler hypersurface $\mathcal{Z}_{h,r}^{v}$ is fibrated by space Zindler curves in the sense of Pottmann.

The following theorem demonstrates that the space Zindler curve $\mathcal{Z}_{h,r}^{u,v}$ on the Zindler hypersurface $\mathcal{Z}_{h,r}^{v}$ has an associated space curve of constant width that lies on the corresponding hypersurface of constant width $\mathcal{H}_{h+r}$.

Several generalizations of constant width curves to $\mathbb{R}^3$ exist in the literature, though they are not necessarily equivalent. Fujiwara was the first to introduce such curves [49]. He also introduced a more restrictive class of curves in [49] and later in [50]. Subsequently, some other authors proposed similar definitions based on his work (see, e.g., [186], [23, p. 147] or [172]).

Fujiwara's original definition of a spatial curve of constant width can be extended to $\mathbb{R}^n$ as follows.

**Definition 4.2.2 (Fujiwara)** Let $C$ be a closed and regular curve in $\mathbb{R}^n$. For a given point $A \in C$, let $d(A, P)$ denote the shortest distance between the tangent line at $A$ and the tangent line at another point $P \in C$. The *width of $C$ with respect to $A$* is defined as

$$M_A = \max_{P \in C} d\,(A, P)\,.$$

The curve $C$ is said to be of constant width if $M_A$ does not depend on $A \in C$.

It can be proved that a sufficient condition for $C$ to be of constant width is the following:

**Proposition 4.2.9** *If there exists a diffeomorphism $P \rightleftarrows P'$ between points of $C$ such that $\left[PP'\right]$ are maximal chords which are double-normal (orthogonal to the tangents to $C$ at the endpoints $P$ and $P'$) and such that as a point $M$ on $C$ moves from $P$ to $P'$ according to an orientation, the point $M'$ moves from $P'$ to $P$ with the same orientation, then the curve $C$ is of constant width.*

We will use this proposition to prove the following theorem.

**Theorem 4.2.2** *Let $\mathcal{H}_h$ be a $C^2$ projective hedgehog in $\mathbb{R}^4$ and let $r > 0$ be such that $\mathcal{H}_{h+r}$ is a convex and regular hypersurface of constant width $2r$. For all $v \in \mathbb{S}^2$, let $\mathcal{Z}_{h,r}^{v} = z_{h,r}^{v}\left(\mathbb{S}^3\right)$ be its associated $v$-Zindler hypersurface. Then, for all $(u, v) \in \mathbb{S}^3 \times \mathbb{S}^2$, $x_{h+r}(\mathbb{S}_{u,v}^1)$ is a space curve of constant width that has the space Zindler curve $z_{h,r}^{v}(\mathbb{S}_{u,v}^1)$ as its associated: the vectors determined by the endpoints of the corresponding constant length chords are related through the linear complex structure $J_v : \mathbb{R}^4 \to \mathbb{R}^4$, $x \longmapsto vx$.*

***Proof*** Let $\alpha : \mathbb{S}_{u,v}^1 \to \mathbb{R}^4, u_\theta \mapsto \alpha(\theta) := \alpha(u_\theta)$ be defined by $\alpha(\theta) = x_{h+r}(u_\theta) = x_h(u_\theta) + ru_\theta$, where $u_\theta := (\cos\theta)u + (\sin\theta)vu \in \mathbb{S}_{u,v}^1$, for all $\theta \in \mathbb{R}/2\pi\mathbb{Z}$. First, note that $\alpha$ is regular since $x_{h+r}$ is regular.

For $\theta \in \mathbb{R}/2\pi\mathbb{Z}$, $\alpha(\theta)$ and $\alpha(\theta+\pi)$ are unequivocally associated through a diffeomorphism. In addition, for all $\theta \in \mathbb{R}/2\pi\mathbb{Z}$, we have

$$\alpha'(\theta) = \left(T_{u_\theta}x_h\right)(vu_\theta) + rvu_\theta,$$

$$\alpha'(\theta+\pi) = \left(T_{u_\theta}x_h\right)(vu_\theta) - rvu_\theta,$$

and

$$\alpha(\theta) - \alpha(\theta+\pi) = 2ru_\theta.$$

since $\mathcal{H}_h$ is projective. Thus, the chords $[\alpha(\theta), \alpha(\theta+\pi)]$ are all double-normal and have constant length:

$$\|\alpha(\theta) - \alpha(\theta+\pi)\| = 2r \quad \text{for all } \theta \in \mathbb{R}/2\pi\mathbb{Z}.$$

Since $\alpha\left(\mathbb{S}^1_{u,v}\right)$ lies on a hypersurface of constant width $2r$, these chords are also maximal. Therefore, we can conclude by Proposition 4.2.2 that the curve $\alpha\left(\mathbb{S}^1_{u,v}\right)$ is of constant width.

Finally, observe that if the chords $[\alpha(\theta), \alpha(\theta+\pi)]$ of constant length $2r$ are rotated through the pure unit quaternion $v$, we obtain the halving chords $\left[z^v_{h,r}(u_\theta), z^v_{h,r}(u_{\theta+\pi})\right]$ of constant length $2r$ of $\mathcal{Z}^{u,v}_{h,r} = z^v_{h,r}\left(\mathbb{S}^1_{u,v}\right)$:

$$z^v_{h,r}(u_\theta) - z^v_{h,r}(u_{\theta+\pi}) = 2rvu_\theta = v\left(\alpha(\theta) - \alpha(\theta+\pi)\right).$$

Thus, $\alpha\left(\mathbb{S}^1_{u,v}\right)$ is associated with $z^v_{h,r}\left(\mathbb{S}^1_{u,v}\right)$ through the pure unit quaternion $v$.  ∎

Note that the space curve of constant width considered in Theorem 4.2.2 is not transnormal. Indeed, the vectors $\alpha'(\theta)$ and $\alpha'(\theta+\pi)$ are not collinear and, therefore, they do not share the same normal hyperplane at the corresponding points $\alpha(\theta)$ and $\alpha(\theta+\pi)$.

The previous results can be complemented with the following theorem, which establishes that the symplectic areas of these pairs of associated curves (constituted by a space curve of constant width $x_{h+r}(\mathbb{S}^1_{u,v})$ and a space Zindler curve $z^v_{h,r}(\mathbb{S}^1_{u,v})$) are equal. This result serves as a generalization of Proposition 4.2.5.

**Theorem 4.2.3** *Let $\mathcal{H}_h$ be a $C^2$ projective hedgehog in $\mathbb{R}^4$ and let $r > 0$ be such that $\mathcal{H}_{h+r}$ is a convex hypersurface of constant width $2r$. For all $v \in \mathbb{S}^2$, let $\mathcal{Z}^v_{h,r} = z^v_{h,r}\left(\mathbb{S}^3\right)$ be its associated $v$-Zindler hypersurface. Then, for all $(u, v) \in \mathbb{S}^3 \times \mathbb{S}^2$, the symplectic area of $z^v_{h,r}(\mathbb{S}^1_{u,v})$ and the symplectic area $s_{u,v}(h+r)$ of $x_{h+r}(\mathbb{S}^1_{u,v})$ are equal. More precisely, both symplectic areas are equal to $s_{u,v}(h) + \pi r^2$.*

***Proof*** By definition, the symplectic area of $z_{h,r}^v(\mathbb{S}_{u,v}^1)$ in the Kähler vector space $(\mathbb{R}^4, J_v, \omega_v)$ is

$$\int_{z_{h,r}^v(\mathbb{S}_{u,v}^1)} \alpha_v = \frac{1}{2}\int_0^{2\pi} \left\langle z_{h,r}^v(\theta), (-J_v)\left(\left(z_{h,r}^v\right)'(\theta)\right)\right\rangle d\theta$$

$$= \frac{1}{2}\int_0^{2\pi} \left\langle x_h(u_\theta) + rvu_\theta, (-J_v)\left((T_{u_\theta}x_h)(vu_\theta) - ru_\theta\right)\right\rangle d\theta$$

$$= \frac{1}{2}\int_0^{2\pi} \left\langle x_h(u_\theta) + rvu_\theta, (-J_v)\left((T_{u_\theta}x_h)(vu_\theta)\right) + rvu_\theta\right\rangle d\theta.$$

By bilinearity of $\langle., .\rangle$ we then deduce that

$$\int_{z_{h,r}^v(\mathbb{S}_{u,v}^1)} \alpha_v = s_{u,v}(h) + \pi r^2 + \frac{r}{2}\int_0^{2\pi} \langle x_h(u_\theta), vu_\theta\rangle\, d\theta$$

$$- \frac{r}{2}\int_0^{2\pi} \left\langle vu_\theta, v(T_{u_\theta}x_h)(vu_\theta)\right\rangle d\theta.$$

Now,

$$\int_0^{2\pi} \langle x_h(u_\theta), vu_\theta\rangle\, d\theta = \int_0^{2\pi} \langle h(u_\theta)u_\theta + \nabla h(u_\theta), vu_\theta\rangle\, d\theta$$

$$= \int_0^{2\pi} \langle \nabla h(u_\theta), vu_\theta\rangle\, d\theta = 0,$$

since

$$\langle \nabla h(u_{\theta+\pi}), vu_{\theta+\pi}\rangle = -\langle \nabla h(u_\theta), vu_\theta\rangle \quad \text{for all } \theta \in \mathbb{R}/2\pi\mathbb{Z},$$

by the fact that $\mathcal{H}_h$ is projective. Since $J_v : \mathbb{R}^4 \to \mathbb{R}^4$, $x \mapsto vx$ is an isometry, we also have

$$\int_0^{2\pi} \left\langle vu_\theta, v(T_{u_\theta}x_h)(vu_\theta)\right\rangle d\theta = \int_0^{2\pi} \left\langle u_\theta, (T_{u_\theta}x_h)(vu_\theta)\right\rangle d\theta = 0.$$

Therefore

$$\int_{z_{h,r}^v(\mathbb{S}_{u,v}^1)} \alpha_v = s_{u,v}(h) + \pi r^2.$$

Now, by Proposition 6.6.5 above, we have

$$s_{u,v}\,(h+r) = \frac{1}{2}\int_0^{2\pi} \langle x_{h+r}\,(u_\theta)\,,\,R_{h+r}\,(u_\theta,v)\,u_\theta\rangle\,d\theta,$$

$$= \frac{1}{2}\int_0^{2\pi} \langle x_h\,(u_\theta) + r u_\theta\,,\,R_h\,(u_\theta,v)\,u_\theta + r\,u_\theta\rangle\,d\theta,$$

and by bilinearity of $\langle .,.\rangle$, we then deduce that

$$s_{u,v}\,(h+r) = s_{u,v}\,(h) + \pi r^2 + \frac{r}{2}\int_0^{2\pi} \langle x_h\,(u_\theta)\,,\,u_\theta\rangle\,d\theta$$

$$+ \frac{r}{2}\int_0^{2\pi} \langle u_\theta\,,\,R_h\,(u_\theta,v)\,u_\theta\rangle\,d\theta$$

The first integral is zero by the fact that $\mathcal{H}_h$ is projective:

$$\int_0^{2\pi} \langle x_h\,(u_\theta)\,,\,u_\theta\rangle\,d\theta = \int_0^{2\pi} \langle h\,(u_\theta)\,u_\theta + \nabla h\,(u_\theta)\,,\,u_\theta\rangle\,d\theta$$

$$= \int_0^{2\pi} h\,(u_\theta)\,d\theta = 0,$$

Since $J_v : \mathbb{R}^4 \to \mathbb{R}^4$, $x \mapsto vx$ is an isometry, we also have

$$\int_0^{2\pi} \langle u_\theta\,,\,R_h\,(u_\theta,v)\,u_\theta\rangle\,d\theta = \int_0^{2\pi} \langle u_\theta\,,\,-v\,(T_{u_\theta}x_h)\,(vu_\theta)\rangle\,d\theta$$

$$= \int_0^{2\pi} \langle vu_\theta\,,\,(T_{u_\theta}x_h)\,(vu_\theta)\rangle\,d\theta = 0,$$

by the fact that

$$\langle vu_{\theta+\pi}\,,\,(T_{u_{\theta+\pi}}x_h)\,(vu_{\theta+\pi})\rangle = \langle -vu_\theta\,,\,-(T_{u_\theta}x_h)\,(-vu_\theta)\rangle,$$

for all $\theta \in \mathbb{R}/2\pi\mathbb{Z}$. Therefore,

$$s_{u,v}\,(h+r) = s_{u,v}\,(h) + \pi r^2.$$

## 4.3  Projection Bodies, Zonoids, Generalized Zonoids and Hedgehogs

Unless explicitly stated otherwise, the results of this subsection are essentially taken from [94].[1] Recall that the so-called ***cosine transform***, which associates to any continuous function $f \in C\left(\mathbb{S}^n; \mathbb{R}\right)$ the map $T_f : \mathbb{R}^{n+1} \to \mathbb{R}$ defined by

$$T_f\left(x\right) := \int_{\mathbb{S}^n} |\langle x, v \rangle|\, f\left(v\right) d\sigma\left(v\right),$$

where $\langle ., . \rangle$ is the standard inner product and $\sigma$ the spherical Lebesgue measure, often appears in convex geometry. For instance, every sufficiently smooth support function of a convex body arises as a cosine transform of a continuous function. In this section, we prove that for any $R \in C\left(\mathbb{S}^n; \mathbb{R}\right)$, the map

$$h : \mathbb{S}^n \to \mathbb{R}, u \mapsto \int_{\mathbb{S}^n} |\langle u, v \rangle|\, R(v) d\sigma(v),$$

is of class $C^2$. It follows that the boundaries of zonoids (resp. generalized zonoids) whose generating measure have a continuous density with respect to $\sigma$ can be considered as $C^2$-hedgehogs. We deduce the following local property for such zonoids: Given a zonoid $K$ whose generating measure has a continuous density with respect to $\sigma$, if one of the principal radii of curvature of its boundary is zero at $u \in \mathbb{S}^n$, then all the other ones are also zero at $u$. We give a formula for the curvature function of the hedgehog defined by $h$ and we deduce a necessary and sufficient condition for $h$ being the support function of a convex body of class $C^2_+$. We define projection hedgehogs (resp. mixed projection hedgehogs) and interpret their support functions in terms of $n$-dimensional volume (resp. mixed volume). Finally, this study leads us to consider the extension of the classical Minkowski problem to hedgehogs.

### 4.3.1  Introduction and Statement of Results

Recall that Minkowski sums of line segments form  a type of polytopes called ***zonotopes*** (see Fig. 4.10). A ***zonoid*** is a (necessarily centrally symmetric) convex body that is the limit, in the sense of the Hausdorff metric, of a sequence of zonotopes. Zonoids play an important role in various areas such as the theory of vector measures, Banach space theory or stochastic geometry. After recalling basic facts on zonoids, we present a study of zonoids and their generalizations based on

---

[1] This article was published in Advances in Math. 158, Y. Martinez-Maure, *Hedgehogs and zonoids*, pp. 1–17. Copyright Elsevier 2001.

**Fig. 4.10** Zonotope of $\mathbb{R}^3$ that is the sum of the line segments connecting $(0, 0, 0)$ to $(1, 0, 0)$, $(0, 1, 0)$, $(0, 1, 1)/\sqrt{2}$, $(0, -1, 1)/\sqrt{2}$, $(1, 0, 1)/\sqrt{2}$, $(-1, 0, 1)/\sqrt{2}$

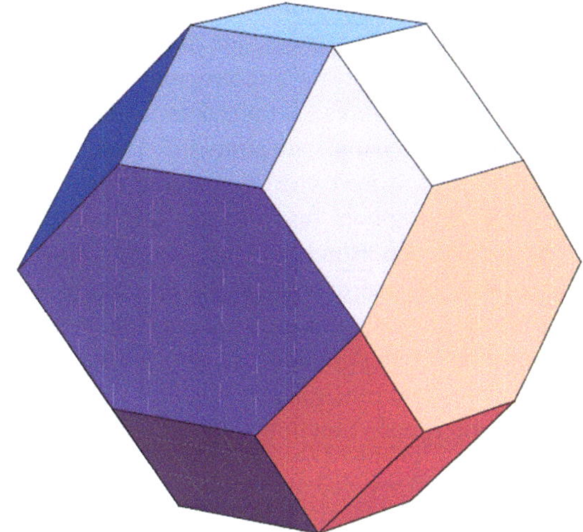

hedgehogs. For further results on zonoids and related topics, we refer the reader to the survey of Schneider and Weil [170].

Remind that the support function of a convex body $K$ of $\mathbb{R}^{n+1}$ is defined by

$$h_K(u) = \max_{v \in K} \langle u, v \rangle \quad \text{for } u \in \mathbb{R}^{n+1},$$

where $\langle ., . \rangle$ denotes the standard inner product. It is well-known [167, Theorem 3.5.3] that a centered (i.e., symmetric with respect to the origin) convex body $K \subset \mathbb{R}^{n+1}$ is a zonoid if and only if its support function $h_K$ can be represented in the form

$$h_K(u) = \int_{\mathbb{S}^n} |\langle u, v \rangle| \, d\mu(v), \tag{4.3.1}$$

where $\mu$ is a nonnegative even Borel measure on the unit sphere $\mathbb{S}^n$ of $\mathbb{R}^{n+1}$. Here $\mu$ is called even if it is invariant under reflection in the origin.

Projection bodies are an important class of zonoids. Given a convex body $K$ in $\mathbb{R}^{n+1}$, the **projection body** of $K$ is the centered zonoid $\Pi K$ with support function

$$h_{\Pi K}(u) = \frac{1}{2} \int_{\mathbb{S}^n} |\langle u, v \rangle| \, dS(K, v) = \int_{\mathbb{S}^n} |\langle u, v \rangle| \, d\rho(v),$$

where $S(K, .)$ is the surface area measure of $K$ and $\rho$ is the Borel measure defined by $\rho(\Omega) = \frac{1}{4}(S(K, \Omega) + S(K, -\Omega))$ for any Borel set $\Omega \subset \mathbb{S}^n$. For any $u \in \mathbb{S}^n$, $h_{\Pi K}(u)$ is the **brightness** of $K$ in the direction of $u$, that is the $n$-dimensional volume of the orthogonal projection of $K$ onto the linear subspace orthogonal to $u$ [167, p. 302]; this explains the expression 'projection body'. From the general case of

Minkowski's existence theorem [167, Theorem 8.2.2], any $(n + 1)$-dimensional centered zonoid is a projection body. In this section, we extend the notion of projection body (and its interpretation in terms of brightness) to hedgehogs and we consider the Minkowski problem for hedgehogs.

Generalized zonoids are simply defined by extending the integral representation (4.3.1) to signed measures: a centered convex body $K$ of $\mathbb{R}^{n+1}$ is a **generalized zonoid** if its support function $h_K$ can be represented in the form (4.3.1) with $\mu$ a signed even Borel measure on $\mathbb{S}^n$. Such a representation is unique [167, Theorem 3.5.4] and $\mu$ is called the generating measure of $K$.

The present section is mostly interested in generalized zonoids whose generating measures have a continuous density with respect to spherical Lebesgue measure. For any even real function $h$ of class $C^k$ on $\mathbb{S}^n$, where $k \geq n+3$ is even, there exists an even continuous function $R$ on $\mathbb{S}^n$ such that

$$h(u) = \int_{\mathbb{S}^n} |\langle u, v \rangle| \, R(v) d\sigma(v), \tag{4.3.2}$$

where $\sigma$ is the spherical Lebesgue measure [167, Theorem 3.5.4]. Consequently, any centered convex body with sufficiently smooth support function is a generalized zonoid of this type. Note that Schneider [165] has obtained explicit examples of smooth centered convex bodies that are not zonoids. Given any even continuous real function $R$ on $\mathbb{S}^n$, Lindquist [85, Theorem 1] has proved that (4.3.2) defines a support function of a convex body if and only if

$$\int_S \langle v, x \rangle^2 \, R(v) d\sigma_S(v) \geq 0, \tag{4.3.3}$$

for all $(n-1)$-dimensional great sphere $S \subset \mathbb{S}^n$ with spherical Lebesgue measure $\sigma_S$ and for all $x \in S$. Our first theorem states that (4.3.2) defines a support function of a $C^2$-hedgehog for any continuous real function $R$ on $\mathbb{S}^n$. As a corollary, we deduce a criterion for $h$ being the support function of a convex body of class $C^2_+$ (i.e., whose boundary is a convex hypersurface of class $C^2$ with positive Gauss curvature). We also deduce a new proof and the geometrical significance of **Lindquist's criterion**. Our second result gives a local property of zonoids whose generating measures have a continuous density with respect to spherical Lebesgue measure. Recall that Weil [189] has shown that a strictly local characterization of zonoids cannot exist.

Remind that if a hedgehog $\mathcal{H}_h$ of $\mathbb{R}^{n+1}$ is not necessarily a convex hypersurface, its natural parametrization $x_h : \mathbb{S}^n \to \mathcal{H}_h, u \mapsto h(u)u + (\nabla h)(u)$ can be interpreted as the inverse of its Gauss map, in the sense that: at each regular point $x_h(u)$ of $\mathcal{H}_h$, $u$ is a normal vector to $\mathcal{H}_h$. As we already said, for computations it is often convenient to extend $h$ to $\mathbb{R}^{n+1} \setminus \{0\}$ as a positively 1-homogeneous function (i.e., such that $h(tu) = th(u)$ for all $t > 0$). Then the second differential of $h$ at $u \in \mathbb{S}^n$, considered as a bilinear form on $\mathbb{R}^{n+1}$, satisfies

$$d^2 h_u(u, u) = 0 \text{ and } d^2 h_u(x, x) = \langle x, T_u x_h(x) \rangle \text{ for all } x \in T_u \mathbb{S}^n,$$

so that its eigenvalues are 0 and the principal radii of curvature of $\mathcal{H}_h$ at $u$ [167, Corollary 2.5.2].

The following observation relates zonoids to hedgehogs:

*For any continuous real function $R$ on $\mathbb{S}^n$, Relation  (4.3.2) defines the support function of a hedgehog $\mathcal{H}_h$.*

This assertion is an immediate consequence of our first theorem:

**Theorem 4.3.1** *Let $R$ be a continuous real function on $\mathbb{S}^n$. Then, the positively 1-homogeneous function*

$$h : \mathbb{R}^{n+1} \to \mathbb{R}, \ u \longmapsto \int_{\mathbb{S}^n} |\langle u, v \rangle| \, R(v) d\sigma(v),$$

*is of class $C^2$ on $\mathbb{R}^{n+1} \setminus \{0\}$.*

*Its second differential at $u$, considered as a bilinear form on $\mathbb{R}^{n+1}$, is given by*

$$d^2 h_u(x, y) = \frac{2}{\|u\|} \int_{S_u} \langle v, x \rangle \, \langle v, y \rangle \, R(v) d\sigma_u(v),$$

*where $S_u$ is the unit sphere of the n-dimensional subspace of $\mathbb{R}^{n+1}$ that is orthogonal to $u$ and $\sigma_u$ is the spherical Lebesgue measure on $S_u$.*

A direct consequence of Theorem 4.3.1 is that the cosine transform $T$ : $C(\mathbb{S}^n; \mathbb{R}) \to C^2(\mathbb{S}^n; \mathbb{R})$, $f \mapsto T_f$ is a bounded linear operator from $C(\mathbb{S}^n; \mathbb{R})$ to $C^2(\mathbb{S}^n; \mathbb{R})$, which is an essential tool in the proof of Ivaki's fixed points results [70, 71].

Theorem 4.3.1 has various geometric applications related to zonoids. We begin by the above-mentioned criterion of Lindquist. As is well-known, a necessary and sufficient condition that a positively 1-homogeneous function $h$ : $\mathbb{R}^{n+1} \to \mathbb{R}$ of class $C^2$ on $\mathbb{R}^{n+1} \setminus \{0\}$ should be the support function of a convex body is that

$$d^2 h_u(x, x) \geq 0 \text{ for all } u \in \mathbb{S}^n \text{ and } x \in S_u$$

Therefore, *Theorem 4.3.1 provides a new proof of Lindquist's criterion.* Moreover, for such a function $h$, $d^2 h_u(x, x)$ has a geometric significance in the linear plane $\Pi$ spanned by $u \in \mathbb{S}^n$ and $x \in S_u$. Indeed, a computation shows that $d^2 h_u(x, x)$ is the (principal) radius of curvature at $u$ of the hedgehog of $\Pi$ whose support function is the restriction of $h$ to $\mathbb{S}^n \cap \Pi$. For $h$ convex, this radius of curvature $R_h(u, x)$ is called the ***tangential radius of curvature*** of $\mathcal{H}_h$ at $x_h(u)$ in the direction $x$ [167, p. 126]. Extending this definition to hedgehogs, we can state *the geometric significance of Lindquist's criterion:*

*Lindquist's condition asserts that all tangential radii of curvature are nonnegative.*

Theorem 4.3.1 also provides information on the boundary structure of zonoids (resp. generalized zonoids). By Theorem 4.3.1, *the boundary of a generalized zonoid whose generating measure has a continuous density with respect to spherical*

*Lebesgue measure is a hedgehog.* In particular, its principal radii of curvature are everywhere defined as functions of the outer unit normal vector.

We know that a characterization of zonoids by a strictly local criterion is not possible [189]. However, using Theorem 4.3.1 we prove the following local property:

**Theorem 4.3.2** *Let $K$ be a zonoid whose generating measure has a continuous density with respect to spherical Lebesgue measure. If one of its principal radii of curvature is zero at $u$, then all its principal radii of curvature are zero at $u$.*

Thus, *Theorem 4.3.2 yields a class of centered convex bodies with smooth support function that are not zonoids.* For example, the centered convex body $K$ of $\mathbb{R}^3$ with support function given by

$$h_K(v) = 2 - \langle u, v \rangle^2 \text{ for } v \in \mathbb{S}^2,$$

where $u \in \mathbb{S}^2$ is fixed, is not a zonoid. Indeed, on the great circle $S_u$, one of its principal radii of curvature is zero whereas the other one is nonzero.

*Theorem 4.3.1 also allows us to extend the notion of projection body to hedgehogs.* First consider a convex body $K$ whose boundary is a hedgehog $\mathcal{H}_f$ of $\mathbb{R}^{n+1}$ with everywhere positive radii of curvature. Its surface area measure $S(K, .)$ has the product $R_f$ of the principal radii of curvature (that is, the reciprocal Gauss curvature) as a density with respect to spherical Lebesgue measure $\sigma$ (see [167, p. 545]), so that the support function of the projection body $\Pi K$ is given by

$$h_{\Pi K}(u) = \frac{1}{2} \int_{\mathbb{S}^n} |\langle u, v \rangle| \, R_f(v) \, d\sigma(v).$$

As the reciprocal Gauss curvature $R_f$ is a continuous function, it follows from Theorem 4.3.1 that the boundary of $\Pi K$ is a hedgehog. Now, for any hedgehog, the product of the principal radii of curvature is a continuous function. Consequently, Theorem 4.3.1 permits us to state:

**Proposition 4.3.1 (and Definition)** *Let $\mathcal{H}_f$ be a hedgehog of $\mathbb{R}^{n+1}$ and let $R_f$ be the product of its principal radii of curvature. Then, the function*

$$h_f : \mathbb{S}^n \to \mathbb{R}, \ u \longmapsto \frac{1}{2} \int_{\mathbb{S}^n} |\langle u, v \rangle| \, R_f(v) d\sigma(v),$$

*is the support function of a hedgehog $\Pi_f$.*
*We call $\Pi_f$ the **projection hedgehog** of $\mathcal{H}_f$.*

Our following proposition asserts that the signed distance from the origin to the oriented support hyperplane of $\Pi_f$ with unit normal vector $u$ is equal to the $n$-dimensional volume of the hedgehog whose support function is the restriction of $f$ to $S_u = \mathbb{S}^n \cap u^\perp$, where $u^\perp$ is the $n$-dimensional subspace of $\mathbb{R}^{n+1}$ that is orthogonal to $u$:

**Proposition 4.3.2** *Let $\mathcal{H}_f$ be a hedgehog of $\mathbb{R}^{n+1}$. Then, the support function $h_f$ of its projection hedgehog $\Pi_f$ is such that:*

$$h_f(u) = v_n(f_u) \text{ for all } u \in \mathbb{S}^n,$$

*where $\mathcal{H}_{f_u}$ is the hedgehog of $u^\perp$ whose support function $f_u$ is the restriction of $f$ to $S_u = \mathbb{S}^n \cap u^\perp$.*

Using Theorem 4.3.1, we can also extend the notion of mixed projection body introduced by E. Lutwak [88] to hedgehogs. Indeed, as the mixed curvature function of $n$ hedgehogs is a continuous function on $\mathbb{S}^n$, we can state the following generalization of Proposition 4.3.1:

**Proposition 4.3.3 (and Definition)** *Let $\mathcal{H}_{f_1}, \ldots, \mathcal{H}_{f_n}$ be $n$ hedgehogs of $\mathbb{R}^{n+1}$. Then, the function*

$$h_{(f_1,\ldots,f_n)} : \mathbb{S}^n \to \mathbb{R}, \ u \longmapsto \frac{1}{2}\int_{\mathbb{S}^n} |\langle u, v\rangle| \ R_{(f_1,\ldots,f_n)}(v)d\sigma(v),$$

*is the support function of a hedgehog $\Pi_{(f_1,\ldots,f_n)}$.*

*We call $\Pi_{(f_1,\ldots,f_n)}$ the **mixed projection hedgehog** of $\mathcal{H}_{f_1}, \ldots, \mathcal{H}_{f_n}$.*

Naturally, we have the following generalization of Proposition 4.3.2:

**Proposition 4.3.4** *Let $\mathcal{H}_{f_1}, \ldots, \mathcal{H}_{f_n}$ be $n$ hedgehogs of $\mathbb{R}^{n+1}$. Then, the support function $h_{(f_1,\ldots,f_n)}$ of their mixed projection hedgehog is such that*

$$h_{(f_1,\ldots,f_n)}(u) = v_n(f_1^u, \ldots, f_n^u) \text{ for all } u \in \mathbb{S}^n,$$

*where $f_i^u$ is the restriction of $f_i$ to $S_u = \mathbb{S}^n \cap u^\perp$ for $i \in \{1, \ldots, n\}$.*

By integration, it follows that

$$s(f_1, \ldots, f_n) = \frac{1}{\kappa_n}\int_{\mathbb{S}^n} v_n(f_1^u, \ldots, f_n^u)d\sigma(u),$$

where $\kappa_n$ is the Lebesgue measure of the unit ball in $\mathbb{R}^n$. As a particular case, we have the extension of ***Kubota's formula*** to hedgehogs:

$$v_{n+1}^m(f) = \frac{1}{(n+1)\,\kappa_n}\int_{\mathbb{S}^n} v_n^m(f_u)d\sigma(u),$$

where $v_{n+1}^m(f) = v_{n+1}(\underbrace{f, \ldots, f}_{m}, \underbrace{1, \ldots, 1}_{n-m+1})$ and $v_n^m(f_u) = v_n(\underbrace{f_u, \ldots, f_u}_{m}, \underbrace{1, \ldots, 1}_{n-m})$.

The following proposition ensures that every centered (i.e., symmetric with respect to the origin) hedgehog with sufficiently smooth support function is a mixed projection hedgehog.

**Proposition 4.3.5** *Let $\mathcal{H}_h$ be a hedgehog whose support function can be represented in the form*

$$h(u) = \int_{\mathbb{S}^n} |\langle u, v \rangle|\, R(v) d\sigma(v), \tag{4.3.2}$$

*where $R$ is an even continuous function on $\mathbb{S}^n$ (that is the case if $h$ is an even function of class $C^k$ on $\mathbb{S}^n$, where $k \geq n + 3$ is even). Then, there exists a hedgehog $\mathcal{H}_f$ for which $\mathcal{H}_h$ is the mixed projection hedgehog $\Pi_{(1,\dots,1,f)}$.*

By Proposition 4.3.5, *any generalized zonoid whose generating measure has a continuous density with respect to spherical Lebesgue measure* (and thus, any centered convex body with sufficiently smooth support function) *can be viewed as a mixed projection hedgehog.*

As representation (4.3.2) is unique [167, Theorem 3.5.4], the question whether $\mathcal{H}_h$ is the projection hedgehog of some hedgehog whose curvature function is even boils down to **the Minkowski Problem for hedgehogs**, that is:

**Extended Minkowski Problem** *What are necessary and sufficient conditions for a real continuous function $R$ on the sphere $\mathbb{S}^n$ to be the curvature function (that is, the inverse $\frac{1}{\kappa}$ of the Gauss curvature $\kappa$) of some $C^2$-hedgehog?*

This extension of Minkowski's problem to hedgehogs, very difficult and still largely open, will be the subject of our next section. But it would be a shame not to mention it briefly here to answer the question whether $\mathcal{H}_h$ is the projection hedgehog of some hedgehog whose curvature function is even.

In differential geometry, the classical Minkowski problem is that of the existence (and uniqueness up to translations) of a closed convex hypersurface with Gauss curvature prescribed as a function of the outer unit normal vector. It is well-known (see the survey by Gluck [54]]) that a positive continuous function $K$ on $\mathbb{S}^n$ which satisfies the condition

$$(c) \quad \int_{\mathbb{S}^n} \frac{u}{K(u)}\, d\sigma(u) = 0$$

is the Gauss curvature (in the sense of Gauss' definition) of a unique (not necessarily $C^2$-smooth) closed convex hypersurface $H$. Condition $(c)$ is necessary by the fact that the vector area of a closed hypersurface $H$ is equal to zero:

$$\int_H v(x) d S(x) = 0,$$

where $v$ is the Gauss map and $S$ the surface area measure on $H$.

In the present case, the condition

$$\int_{\mathbb{S}^n} u R(u) d\sigma(u) = 0$$

is still necessary (for instance from the translation invariance of the volume) but not sufficient to ensure that a hedgehog with curvature function $R$ exists. For instance, a negative continuous function on $\mathbb{S}^2$ that satisfies this integral condition cannot be the curvature function of a hedgehog (since there is no compact surface with negative Gauss curvature in $\mathbb{R}^3$). In fact, this integral condition is not sufficient even if $R$ is a smooth function which changes sign cleanly on $\mathbb{S}^2$:

**Proposition 4.3.6**   *There exists a smooth real function $R$ on $\mathbb{S}^2$ that is not the curvature function of a hedgehog although it satisfies the following two conditions:*

*(i)   $R$ is even and thus such that*

$$\int_{\mathbb{S}^2} u R(u) d\sigma(u) = 0;$$

*(ii)   $R$ changes sign cleanly on $\mathbb{S}^2$.*
*Here 'cleanly' means that $dR(u) \neq 0$ if $R(u) = 0$.*

It follows that *a hedgehog $\mathcal{H}_h$ whose support function can be represented in the form (4.3.2) is not necessarily the projection hedgehog of a centered (i.e., symmetric with respect to the origin) hedgehog.*

Our Theorem 4.3.1 shows that the cosine transform of a continuous function $f : \mathbb{S}^n \to \mathbb{R}$ is a $C^2$ function on $\mathbb{R}^{n+1} \setminus \{0\}$. Let us notice to conclude that Y. Lonke obtained the following generalization [86]: The *$L^p$-cosine transform of an even continuous function $f : \mathbb{S}^n \to \mathbb{R}$ is defined to be the map $T_f^p : \mathbb{R}^{n+1} \to \mathbb{R}$ given by*

$$T_f^p(x) := \int_{\mathbb{S}^n} |\langle x, v \rangle|^p f(v) \, d\sigma(v).$$

*If $p$ is not an even integer then all the partial derivatives of even order of $T_f^p$ up to order $p + 1$ (including $p + 1$ is $p$ is an odd integer) exist and are continuous on $\mathbb{R}^{n+1} \setminus \{0\}$.*

### 4.3.2   *Further Remarks and Proof of Results*

We begin by the following lemma:

**Lemma 4.3.1**   *Let $h$ be as in Theorem 4.3.1. Then, $h$ is a function of class $C^1$ on $\mathbb{R}^{n+1} \setminus \{0\}$ and we have*

$$(grad \, h)(x) = \int_{\mathbb{S}^n} sgn(\langle u, x \rangle) \, u R(u) d\sigma(u),$$

*for all $x \in \mathbb{R}^{n+1} \setminus \{0\}$, where $sgn$ is the signum function.*

***Proof*** It suffices to prove that the directional derivatives

$$h'(x; y) = \lim_{t \downarrow 0} \frac{h(x + ty) - h(x)}{t}$$

exist and are given by

$$h'(x; y) = \left\langle \int_{\mathbb{S}^n} sgn\left(\langle u, x \rangle\right) u R(u) d\sigma(u) , \ y \right\rangle,$$

for $x \in \mathbb{R}^{n+1} \setminus \{0\}$ and $y \in \mathbb{R}^{n+1}$.

Using positive 1-homogeneity of $h$, the proof boils down to the case where $x$ and $y$ are linearly independent unit vectors. In this case, we have

$$h'(x; y) = \lim_{t \downarrow 0} \frac{1}{t} \int_{\mathbb{S}^n} \left(|\langle u, x + ty \rangle| - |\langle u, x \rangle|\right) R(u) d\sigma(u)$$

$$= \lim_{t \downarrow 0} \left[ \int_{P_t} sgn\left(\langle u, x \rangle\right) \langle u, y \rangle R(u) d\sigma(u) \right.$$

$$\left. - \frac{2}{t} \int_{N_t} |\langle u, x \rangle| R(u) d\sigma(u) + \int_{N_t} |\langle u, y \rangle| R(u) d\sigma(u) \right]$$

where $P_t = \{u \in \mathbb{S}^n \,|\, \langle u, x + ty \rangle \langle u, x \rangle > 0\}$ and $N_t = \mathbb{S}^n \setminus P_t$. Now, we see easily that $|\langle u, x \rangle| \leq t$ for all $u \in N_t$, so that

$$\left| \frac{1}{t} \int_{N_t} |\langle u, x \rangle| R(u) d\sigma(u) \right| \leq \max_{u \in \mathbb{S}^n} |R(u)| \sigma(N_t).$$

We deduce that

$$\lim_{t \downarrow 0} \frac{1}{t} \int_{N_t} |\langle u, x \rangle| R(u) d\sigma(u) = 0$$

since $\lim_{t \downarrow 0} \sigma(N_t) = 0$. Of course, we similarly obtain

$$\lim_{t \downarrow 0} \int_{N_t} |\langle u, y \rangle| R(u) d\sigma(u) = 0.$$

Thus, as $f(u) = sgn\left(\langle u, x \rangle\right) \langle u, y \rangle R(u)$ satisfies

$$\left| \int_{\mathbb{S}^n} f(u) d\sigma(u) - \int_{P_t} f(u) d\sigma(u) \right| = \left| \int_{N_t} f(u) d\sigma(u) \right|$$

$$\leq \max_{u \in \mathbb{S}^n} |R(u)| \sigma(N_t),$$

we finally obtain

$$h'(x; y) = \lim_{t \downarrow 0} \int_{P_t} f(u)d\sigma(u) = \int_{\mathbb{S}^n} f(u)d\sigma(u),$$

which completes the proof.  ∎

***Proof of Theorem 4.3.1***  First prove that the directional derivatives

$$h''(u; x, y) = \lim_{t \downarrow 0} \frac{h'(u + ty; x) - h'(u; x)}{t}$$

exist and are given by

$$h''(u; x, y) = \frac{2}{\|u\|} \int_{S_u} \langle v, x \rangle \langle v, y \rangle R(v)d\sigma_u(v),$$

for $u \in \mathbb{R}^{n+1} \setminus \{0\}$ and $x, y \in \mathbb{R}^{n+1}$.

Using positive 1-homogeneity of $h$, we see easily that the proof boils down to the case where $u \in \mathbb{S}^n$ and where $x, y$ belong to an orthonormal basis of $T_u\mathbb{S}^n$. Let us choose an orthonormal basis $(e_1, \ldots, e_{n+1})$ of $\mathbb{R}^{n+1}$ such that $e_{n+1} = u$ and $e_i = x$, $e_j = y$ for some $i, j \in \{1, \ldots, n\}$. We write $v_k$ for the $k$th coordinate of $v \in \mathbb{R}^{n+1}$: $v_k = \langle v, e_k \rangle$. We have

$$h''(u; e_i, e_j) = \lim_{t \downarrow 0} \frac{1}{t} \int_{\mathbb{S}^n} \left( sgn \left( \langle u, v \rangle + tv_j \right) - sgn \left( \langle u, v \rangle \right) \right) v_i R(v)d\sigma(v)$$

$$= \lim_{t \downarrow 0} \frac{2}{t} \int_{S(t)} F(v)d\sigma(v),$$

where $F(v) = sgn(v_j)v_i R(v)$ and $S(t) = \left\{ v \in \mathbb{S}^n \,\middle|\, \langle u, v \rangle \left( \langle u, v \rangle + tv_j \right) < 0 \right\}$.

Denote by $v'$ the vector in $S_u$ with the same direction as the orthogonal projection of $v$ into $u^\perp$. Fix $\varepsilon > 0$. By uniform continuity of $v \longmapsto v_i R(v)$ on $\mathbb{S}^n$, there exists $\delta > 0$ such that $0 < t \leq \delta$ implies

$$\left| F(v') - F(v) \right| = \left| v'_i R(v') - v_i R(v) \right| \leq \varepsilon \text{ for all } v \in S(t),$$

and hence

$$\left| \int_{S(t)} F(v')d\sigma(v) - \int_{S(t)} F(v)d\sigma(v) \right| \leq \varepsilon \sigma(S(t)).$$

As $\sigma(S(t)) = \frac{\arctan(t)}{\pi} \sigma(\mathbb{S}^n)$, it follows that

$$h''(u; e_i, e_j) = \lim_{t \downarrow 0} \frac{2}{t} \int_{S(t)} F(v')d\sigma(v). \qquad (4.3.4)$$

Now, we express the coordinates of $v$ in hyperspherical coordinates:

$$
\begin{cases}
v_{n+1} = \cos\theta_1 \\
v_n = \sin\theta_1 \cos\theta_2 \\
\vdots \\
v_2 = \sin\theta_1 \ldots \sin\theta_{n-1}\cos\theta_n \\
v_1 = \sin\theta_1 \ldots \sin\theta_{n-1}\sin\theta_n
\end{cases},
$$

where $0 \leq \theta_i \leq \pi$ for $1 \leq i \leq n-1$ and $0 \leq \theta_n \leq 2\pi$. The surface area element $d\sigma(v)$ on $\mathbb{S}^n$ becomes

$$
\begin{aligned}
d\sigma(v) &= (\sin\theta_1)^{n-1}(\sin\theta_2)^{n-2}\ldots(\sin\theta_{n-1})\,d\theta_1\ldots d\theta_n \\
&= \left((\sin\theta_1)^{n-1}\,d\theta_1\right)d\sigma_p(v')
\end{aligned}
$$

where $d\sigma_u(v') = (\sin\theta_2)^{n-2}\ldots(\sin\theta_{n-1})\,d\theta_2\ldots d\theta_n$ is the element of surface area at $v'$ on $S_u$. As for $\langle u, v\rangle + tv_j = 0$, we have $v = (\cos\alpha)v' + (\sin\alpha)u$ with $\alpha = \arctan(-tv_j')$, we obtain from (4.3.4):

$$
h''(u; e_i, e_j) = \lim_{t\downarrow 0}\frac{2}{t}\int_{S_u} F(v)\phi_t(v)\,d\sigma_u(v),
$$

where $\phi_t(v) = \displaystyle\int_0^{\arctan(t|v_j|)}(\cos\alpha)^{n-1}\,d\alpha$.

Now, we can see easily that

$$
0 \leq t\,|v_j| - \phi_t(u) \leq \frac{n+1}{2}t^3,
$$

so that

$$
\begin{aligned}
h''(u; e_i, e_j) &= 2\int_{S_u} F(v)\,|v_j|\,d\sigma_u(v) \\
&= 2\int_{S_u} v_i v_j R(v)\,d\sigma_u(v),
\end{aligned}
$$

that is,

$$
h''(u; x, y) = 2\int_{S_u}\langle v, x\rangle\,\langle v, y\rangle\,R(v)\,d\sigma_u(v).
$$

Thus, for all $u \neq 0$, the continuous linear operator

$$
h''(u) : \mathbb{R}^{n+1} \longrightarrow \mathbb{R}^{n+1},\ x \longmapsto h''(u)x = \frac{2}{\|u\|}\int_{S_u}\langle v, x\rangle\,vR(v)\,d\sigma_u(v),
$$

is such that $h''(u; x, y) = \langle h''(u)x, y \rangle$. As $u \longmapsto h''(u)$ is a continuous map from $\mathbb{R}^{n+1} \setminus \{0\}$ to the space $L(\mathbb{R}^{n+1})$ of continuous linear operators from $\mathbb{R}^{n+1}$ to itself, it follows that the function $h$ is of class $C^2$ on $\mathbb{R}^{n+1} \setminus \{0\}$.  ∎

We have the following corollary of Theorem 4.3.1:

**Corollary 4.3.1** *Let $h$ be as in Theorem 4.3.1. Then, the curvature function $R_h$ of $\mathcal{H}_h$ is given by*

$$R_h(u) = \frac{2^n}{n!} \int_{S_u} \cdots \int_{S_u} D(v_1, \ldots, v_n)^2 R(v_1) \ldots R(v_n) d\sigma_u(v_1) \ldots d\sigma_u(v_n),$$

$$(4.3.5)$$

*where $D(v_1, \ldots, v_n)$ denotes the n-dimensional volume of the parallelepiped spanned by $v_1, \ldots, v_n$. Thus, the function $h$ is the support function of a convex body of class $C^2_+$ if and only if the right-hand side of (4.3.5) is positive for all $u \in \mathbb{S}^n$.*

***Proof*** Let $(e_1, \ldots, e_{n+1})$ be an orthonormal basis of $\mathbb{R}^{n+1}$ such that $e_{n+1} = u$. We know that the curvature function $R_h$ is given by

$$\begin{aligned} R_h(u) &= \det \left( h''(u; e_i, e_j) \right)_{i,j=1}^n \\ &= \sum_{\sigma \in \mathfrak{S}_n} \varepsilon(\sigma) h''(u; e_{\sigma(1)}, e_1) \ldots h''(u; e_{\sigma(n)}, e_n), \end{aligned}$$

where $\mathfrak{S}_n$ is the symmetric group and $\varepsilon(\sigma)$ the sign of $\sigma \in \mathfrak{S}_n$. Hence, for all $\tau \in \mathfrak{S}_n$, we have

$$(\tau) \qquad \begin{aligned} R_h(u) &= \varepsilon(\tau) \det \left( h''(u; e_i, e_{\tau(j)}) \right)_{i,j=1}^n \\ &= \varepsilon(\tau) \sum_{\sigma \in \mathfrak{S}_n} \varepsilon(\sigma) h''(u; e_{\sigma(1)}, e_{\tau(1)}) \ldots h''(u; e_{\sigma(n)}, e_{\tau(n)}). \end{aligned}$$

Now, Theorem 4.3.1 yields

$$h''(u; e_{\sigma(i)}, e_{\tau(i)}) = 2 \int_{S_u} v_{i,\sigma(i)} v_{i,\tau(i)} d\sigma_u(v_i),$$

for $i = 1, \ldots, n$, where $v_{i,j} = \langle v_i, e_j \rangle$ $(1 \leq j \leq n)$. Replacing, we can rewrite $(\tau)$ as

$$R_h(u)$$

$$= 2^n \int_{S_u} \cdots \int_{S_u} \varepsilon(\sigma) v_{1,\tau(1)} \ldots v_{n,\tau(n)} \Delta_v R(v_1) \ldots R(v_n) d\sigma_u(v_1) \ldots d\sigma_u(v_n),$$

where $\Delta_v = \det \left( v_{j,i} \right)_{i,j=1}^n = \sum_{\sigma \in \mathfrak{S}_n} \varepsilon(\sigma) v_{1,\sigma(1)} \ldots v_{n,\sigma(n)}.$

Adding equalities $(\tau)$, we obtain

$$R_h(u) = \frac{2^n}{n!} \int_{S_u} \dots \int_{S_u} D(v_1, \dots, v_n)^2 R(v_1) \dots R(v_n) d\sigma_u(u_1) \dots d\sigma_u(v_n),$$

since $D(v_1, \dots, v_n) = |\Delta_v|$. $\blacksquare$

***Proof of Theorem 4.3.2*** Let $h$ be the support function of $K$. By hypothesis, $h$ can be represented in the form

$$h(u) = \int_{\mathbb{S}^n} |\langle u, v \rangle| R(v) d\sigma(v),$$

where $R$ is a nonnegative even continuous function on $\mathbb{S}^n$. Therefore, by Theorem 4.3.1 $h$ is of class $C^2$ on $\mathbb{R}^{n+1} \setminus \{0\}$ and the tangential radii of curvature of $\mathcal{H}_h$ are given by

$$R_h(u, x) = 2 \int_{S_u} \langle v, x \rangle^2 R(v) d\sigma_u(v).$$

By continuity and nonnegativity of $R$ on $\mathbb{S}^n$, it follows that if one of the tangential radii of curvature of $\mathcal{H}_h$ at $x_h(u)$ is zero, then the restriction of $R$ to $S_u$ is the null function and hence all tangential radii of curvature of $\mathcal{H}_h$ at $x_h(u)$ are zero.

As the principal radii of curvature of $K$ at $u$ are tangential radii of curvature of $\mathcal{H}_h$ at $x_h(u)$, Theorem 4.3.2 follows. $\blacksquare$

***Proof of Proposition 4.3.4 (and Thus of Proposition 4.3.2)*** The result is well-known when $f_1, \dots, f_n$ are support functions of convex bodies [88]. Using the fact that the Minkowski sum of any hedgehog with a large enough sphere is the boundary of a convex body, Proposition 4.3.4 follows immediately by $n$-linearity. $\blacksquare$

***Proof of Proposition 4.3.5*** First note that the mixed curvature function $R_{(1,\dots,1,f)}$ is the mean radius of curvature of $\mathcal{H}_f$:

$$R_{(1,\dots,1,f)} = \frac{1}{n} \sum_{k=1}^{n} R_f^k,$$

where $R_f^1, \dots, R_f^n$ are the principal radii of curvature of $\mathcal{H}_f$. To see this, it suffices to compare the coefficients of $\lambda^{n-1}$ in the following expressions of the curvature function of $H_{f+\lambda}$, where $\lambda$ is a constant:

$$R_{f+\lambda} = \sum_{k=0}^{n} C_n^k \lambda^k R_{(\underbrace{1,\dots,1}_{k},\underbrace{f,\dots,f}_{n-k})},$$

from Proposition 3.1.1, and

$$R_{f+\lambda} = \prod_{k=1}^{n} \left( R_f^k + \lambda \right),$$

from the fact that the principal radii of curvature of $\mathcal{H}_{f+\lambda}$ are $R_f^1 + \lambda, \ldots, R_f^n + \lambda$.

Now, the study of Christoffel's problem given by Firey [46] shows the existence of a hedgehog $\mathcal{H}_f$ whose mean radius of curvature $R_{(1,\ldots,1,f)}$ is equal to $R$, which completes the proof. ∎

Our proof of Proposition 4.3.6 requires the following lemma:

**Lemma 4.3.2** *Let $\mathcal{H}_f$ be a hedgehog of $\mathbb{R}^3$ and let $u \in \mathbb{S}^2$. If the curvature function $R_f$ of $\mathcal{H}_f$ is positive on the great circle $S_u = \mathbb{S}^2 \cap u^{\perp}$, then*

$$\int_{\mathbb{S}^2} |\langle u, v \rangle|\, R_f(v)\, d\sigma(v) > 0.$$

***Proof*** By positivity of $R_f$ on $S_u$, at any point of $x_f(S_u)$ the tangential radii of curvature of $\mathcal{H}_f$ are all nonzero. In particular, the hedgehog of $u^{\perp}$ whose support function $f_u$ is the restriction of $f$ to $S_u$ is nonsingular (in other words, $\mathcal{H}_{f_u}$ is a nonsingular convex curve of $u^{\perp}$). Consequently, its 2-dimensional volume $v_2(f_u)$ is positive. Now, Proposition 4.3.2 yields

$$v_2(f_u) = \frac{1}{2} \int_{\mathbb{S}^2} |\langle u, v \rangle|\, R_f(v)\, d\sigma(v),$$

which completes the proof. ∎

***Proof of Proposition 4.3.6*** Consider the real smooth function $R$ defined by

$$R(v) = 1 - 2 \langle u, v \rangle^2 \ \text{ for } v \in \mathbb{S}^2,$$

where $u \in \mathbb{S}^2$ is fixed. This function satisfies obviously conditions $(i)$ and $(ii)$, but Lemma 4.3.2 ensures that it cannot be the curvature function of some hedgehog. Indeed, $R$ is positive on $S_u$ whereas an elementary computation shows that

$$\int_{\mathbb{S}^2} |\langle u, v \rangle|\, R(v)\, d\sigma(v) = 0.$$

∎

## 4.4 Hyperbolic Hedgehogs in $\mathbb{R}^3$

Let $\mathcal{H}_h$ be a $C^k$-hedgehog in $\mathbb{R}^3$, with $k \in \mathbb{N} \cup \{\infty\}$ and $k \geq 2$. If $\mathcal{H}_h$ does not have any singularity, then $\mathcal{H}_h$ has an everywhere positive Gaussian curvature (in other words, all its points are elliptic). Indeed, it is of course impossible for $\mathcal{H}_h$ to have an everywhere negative curvature function (that is, to have only hyperbolic points) since there is no compact surface in $\mathbb{R}^3$ with negative Gaussian curvature. But if we allow $\mathcal{H}_h$ to have singularities, is it possible for $\mathcal{H}_h$ to have no elliptic point at all without being trivial (that is, reduced to a single point)? In other words:

**Problem 4.4.1** Does there exist a nontrivial $C^k$-hedgehog $\mathcal{H}_h$ of $\mathbb{R}^3$ with an everywhere nonpositive curvature function $R_h : \mathbb{S}^2 \to \mathbb{R}$, $u \mapsto R_h(u) := 1/\kappa_h(u) = \det[T_u x_h]$, $(k \in \mathbb{N} \cup \{\infty\}$ and $k \geq 2)$?

If such a $C^k$-hedgehog $\mathcal{H}_h$ exists in $\mathbb{R}^3$, we will say that $\mathcal{H}_h$ is a **hyperbolic $C^k$-hedgehog** *of* $\mathbb{R}^3$. As we will see later, this existence problem is of great importance in the theory of PDE's in $\mathbb{R}^3$. But this problem first appeared in connection with a famous characterization of the 2-sphere conjectured by A. D. Alexandrov in the mid 1930s, and disproved by the author by resorting to the geometric study of hedgehogs of $\mathbb{R}^3$ [95]. Let us start by seeing how the conjecture could be reduced to the above problem of existence of hyperbolic hedgehogs and how the construction of such a hedgehog makes it possible to build a counterexample to A. D. Alexandrov's conjecture.

### 4.4.1 $C^2$-Counter-Example to a Conjectured Characterization of the 2-Sphere

**Introduction** The following characterization of the 2-sphere had long been conjectured.

*Conjecture* $(C)$ *([1] and [2, p. 352])* If $S$ is a closed convex surface of class $C_+^2$ of $\mathbb{R}^3$ (i.e., a $C^2$-surface of $\mathbb{R}^3$ with an everywhere positive Gaussian curvature) whose principal curvature $k_1$ and $k_2$ satisfy the following inequality

$$(k_1 - c)(k_2 - c) \leq 0,$$

with some constant $c > 0$, *then* $S$ must be a sphere of radius $1/c$.

For analytical surfaces, this had been established by A.D. Alexandrov himself [1, 2] and H. Münzner [131]. Conjecture $(C)$ had also been verified for surfaces of revolution [130], and more generally for all those that admit a circular orthogonal projection [78]. But after these works, no progress was made in the direction of the general case for almost thirty years. Since the study of Conjecture $(C)$ amounts to comparing the surface $S$ to a sphere of radius $r := 1/c$, such as the sphere $S\left(0_{\mathbb{R}^3}; r\right)$

with center at the origin and radius $r$, and since these surfaces can be regarded as hedgehogs, it is natural from the hedgehog point of view to decompose the surface $S$ into the sum

$$S = S\left(0_{\mathbb{R}^3}; r\right) + \left(S - S\left(0_{\mathbb{R}^3}; r\right)\right)$$

in order to reduce the comparison between these hedgehogs $S$ and $S\left(0_{\mathbb{R}^3}; r\right)$ to the study of the hedgehog $\mathcal{H} = S - S\left(0_{\mathbb{R}^3}; r\right)$. Notice that $(k_1 - c)(k_2 - c) \leq 0$ is equivalent to $(R_1 - r)(R_2 - r) \leq 0$, where $R_1$, $R_2$ are the principal radii of curvature of $S$ at a point (regarded as functions of the unit normal). Now, we know that for all $u \in \mathbb{S}^2$, $R_1(u)$, $R_2(u)$ are the eigenvalues of the tangent map $T_u x_f :$ $T_u \mathbb{S}^2 \to T_u \mathbb{S}^2$, where $f$ is the support function of $S = \mathcal{H}_f$, and thus $R_1(u) - r$, $R_2(u) - r$ are the principal radii of curvature of the hedgehog $\mathcal{H} = S - S\left(0_{\mathbb{R}^3}; r\right)$ (i.e., the eigenvalues of the tangent map $T_u x_h : T_u \mathbb{S}^2 \to T_u \mathbb{S}^2$, where $h$ is the support function of $\mathcal{H} = S - S\left(0_{\mathbb{R}^3}; r\right)$). Solving Conjecture $(C)$ then amounts to knowing whether this hedgehog $\mathcal{H}_h = \mathcal{H}$ must necessarily be reduced to a single point under the assumption $R_h(u) = \det[T_u x_h] \leq 0$ for all $u \in \mathbb{S}^2$. In other words, we are directly led to reformulate Conjecture $(C)$ in the following form.

*Conjecture* $(H)$  If $\mathcal{H}_h$ is a $C^2$-hedgehog of $\mathbb{R}^3$ whose curvature function is nonpositive all over the unit sphere $\mathbb{S}^2$, then $\mathcal{H}_h$ is reduced to a single point.

Note that conjectures $(C)$ and $(H)$ are in fact equivalent. In particular, if $\mathcal{H}$ is any counterexample to Conjecture $(H)$ (that is, if $\mathcal{H}$ is a hyperbolic $C^2$-hedgehog of $\mathbb{R}^3$), then for any sphere $\Sigma$ with a large enough radius, $S = \Sigma + \mathcal{H}$ is a counterexample to Conjecture $(C)$. In [95], the author constructed a counterexample to Conjecture $(H)$ given by an explicit formula, and thus disproved Conjecture $(C)$ by proving the following.

**Theorem 4.4.1 ([95])** *There exists a nontrivial $C^2$-hedgehog $\mathcal{H}_h$ of $\mathbb{R}^3$ whose curvature function $R_h := \det[T_u x_h] = h^2 + h\Delta_2 h + \Delta_{22} h \leq 0$ is nonpositive all over the unit sphere $\mathbb{S}^2$.*

As we will see below, the crucial fact was the existence of a (non compact) cross-cap whose Gauss map is one-to-one and whose curvature function is defined and nonpositive on $\mathbb{S}^2$ minus a semi-great circle. By gluing four cross-caps together with a central part, we constructed a closed surface to which we were able to give an appropriate saddle shape to obtain a nontrivial hedgehog whose curvature function is nonpositive all over the unit sphere $\mathbb{S}^2$ (see Fig. 4.11).

We will recall at the very end of this subsubsection the detailed construction method of this hyperbolic $C^2$-hedgehog $\mathcal{H}_h$, as well as its analytical expression. Such a $C^2$-hedgehog is a counterexample to Conjecture $(H)$, since it is not reduced to a single point. Furthermore by adding a large enough sphere to it, we obtain a counterexample to Conjecture $(C)$.

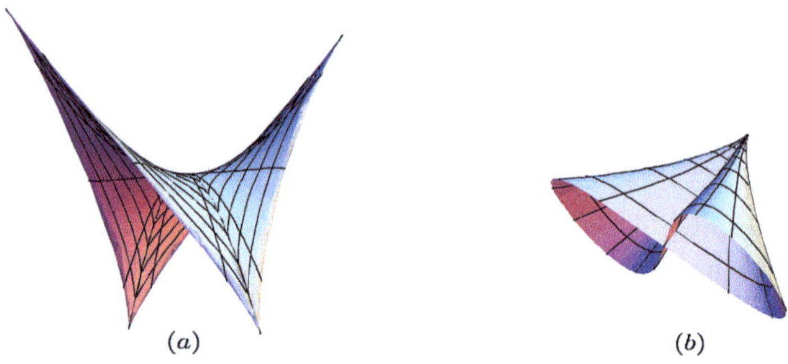

Fig. 4.11  (a) A counterexample to $(H)$; (b) a hedgehog cross cap

Note that two years later, the author presented a polytopal version of this counterexample to Conjecture $(H)$, namely an example of a 'strongly hyperbolic hedgehog polytope' in $\mathbb{R}^3$ [97]: see the next subsubsection.

From an analytical point of view, it is worth mentioning that disproving conjectures $(C)$ and $(H)$ was also equivalent, by considering $C^2$-functions on $\mathbb{S}^2$ as support functions of $C^2$-hedgehogs of $\mathbb{R}^3$, to proving the following statement which is the analytical formulation of Theorem 4.4.1.

**Analytical Reformulation of Theorem 4.4.1 [95]** *There exists a nonlinear function $\varphi : \mathbb{R}^3 \to \mathbb{R}$ whose restriction to $\mathbb{S}^2$, say $h$, is $C^2$ and satisfies the partial differential inequation*

$$h^2 + h\Delta_2 h + \Delta_{22} h \leq 0,$$

*where $\Delta_2$ and $\Delta_{22}$ are respectively the spherical Laplacian and the **Monge-Ampère operator**, that is, the sum and the product of the eigenvalues of the Hessian of $h = \varphi_{|\mathbb{S}^2}$.*

Indeed, this last statement also invalidates a conjecture formulated by D. Koutroufiotis and L. Nirenberg in 1973 [78]. Note in passing that in his paper [78], D. Koutroufiotis gave a proof of Alexandrov's conjecture in the case that the surface $S$ is assumed to be an ovaloid of class $C^3$ with a circular enveloping cylinder in some direction.

Incidentally, we also note that our counterexample of conjecture $(H)$ has made it possible to highlight a subtle error in a 'proof' of Conjecture $(C)$ that had been proposed by A.V. Pogorelov in 1998–1999 [149, 150]: see for instance [141]. In this paper, G. Panina gave new and $C^\infty$-smooth counterexamples to A. D. Alexandrov's conjecture. However, she moved in the opposite direction: she started by constructing hyperbolic hedgehog polytopes, and then she made use of smoothening techniques to construct $C^\infty$-smooth counterexamples. A direct consequence of her method is that the resulting new counterexamples to Conjecture $(H)$ admit many other singularities than the endpoints of their cross-caps: indeed, contrary to our counterexample, for each of these new counterexamples there are necessarily open regions of $\mathbb{S}^2$ on which the principal radii of curvature $R_1$, $R_2$ satisfy $R_1 R_2 = 0$ but not $R_1 = R_2 = 0$ (contrary to what was initially required by A.D. Alexandrov). In our counterexample to Conjecture $(H)$, there are exactly four disjoint geodesic semicircles of $\mathbb{S}^2$ (corresponding to the endpoints of the 4 cross-caps) on which $R_1 R_2 = 0$, and along them we have $R_1 = R_2 = 0$. At all the other points of $\mathbb{S}^2$, the curvature function is negative; in other words, *apart from the 4 endpoints of the 4 cross-caps, the hedgehog surface is regular and has negative curvature*: see below. The '4 semicircles theorem' by J. A. Gálvez and P. Mira gives a converse of our result: if $\mathcal{H}_h$ is any hyperbolic $C^2$-hedgehog of $\mathbb{R}^3$, then its principal radii of curvature $R_1$, $R_2$ both vanishes along four disjoint geodesic semicircles of $\mathbb{S}^2$ [52]. If, in addition to the condition $R_1 = R_2 = 0$ whenever $R_1 R_2 = 0$, the support function of the hedgehog surface is required to be $C^\infty$, or even only $C^3$, then the Alexandrov's conjecture remains open.

We refer the reader to [52], and to the book [133] by Nadirashvili, Tkachev and Vladut, for a reformulation of Alexandrov's conjecture in terms of positively 1-homogeneous solutions of a linear elliptic equation in $\mathbb{R}^3$, and for proofs of special cases. In particular, Han, Nadirashvili and Yuan proved in [63] that the Alexandrov conjecture holds in the uniformly elliptic case.

**Study of Two Particular Cases of Alexandrov's Conjecture**   We are going to give a new proof of Conjecture $(C)$ for analytical surfaces (as recalled above, in this particular case, Conjecture $(C)$ had been established by A.D. Alexandrov himself [1, 4] and H. Münzner [131]), and a proof of Conjecture $(C)$ in the case that the convex surface $S$ is assumed to be of constant width. Of course, in both cases the 'hedgehog approach' will consist in considering the corresponding particular cases of Conjecture $(H)$. Let $\mathcal{H}_h$ be a $C^2$-hedgehog of $\mathbb{R}^3$ whose curvature function $R_h$ is nonpositive all over the unit sphere $\mathbb{S}^2$. We will consider its images under orthogonal projections onto linear planes to obtain informations on the geometry of $\mathcal{H}_h$.

For any $v \in \mathbb{S}^2$, let $h_v$ be the restriction of $h$ to the great circle $\mathbb{S}_v^1 = \mathbb{S}^2 \cap v^\perp$, where $v^\perp$ is the linear plane orthogonal to $v$. We know that $h_v$ is the support function of the hedgehog $\mathcal{H}_{h_v}$ that is the image of $x_h\left(\mathbb{S}_v^1\right)$ under the orthogonal projection onto $v^\perp$:

$$\mathcal{H}_{h_v} = \pi_v \left[ x_h \left( \mathbb{S}_v^1 \right) \right],$$

where $\pi_v$ is the orthogonal projection onto the plane $v^\perp$. Then, the index of a point $x \in v^\perp - \mathcal{H}_{h_v}$ with respect to $\mathcal{H}_{h_v}$ (i.e., the winding number of $\mathcal{H}_{h_v}$ around $x$) gives us information on the curvature of $\mathcal{H}_h$ on the line $\{x\} + \mathbb{R}v$. For every $v \in \mathbb{S}^2$, $P_v$ will denote the oriented plane vector of $\mathbb{R}^3$ with unit normal vector $v$, and $S_v^+$ the half unit sphere given by $\langle u, v \rangle \geq 0$, $\left( u \in \mathbb{S}^2 \right)$. In Sect. 2.8, we saw the following theorem.

**Theorem 2.8.3**   *Let $\mathcal{H}_h$ be a $C^2$-hedgehog of $\mathbb{R}^3$, and let $x$ be a regular value of the map $x_h^v = \pi_v \circ x_h : \mathbb{S}^2 \to P_v$. The index of $x \in P_v \backslash \mathcal{H}_{h_v}$ with respect to $\mathcal{H}_{h_v}$ is* *given by*

$$i_{h_v}(x) = n_h^v(x)^+ - n_h^v(x)^-,$$

*where $n_h^v(x)^+$ (resp. $n_h^v(x)^-$) is the number of $u \in \mathbb{S}_v^+$ such that $x_h(u)$ is an elliptic (resp. a hyperbolic) point of $\mathcal{H}_h$ lying on the line $\{x\} + \mathbb{R}v$.*

In the present case, where the curvature function $R_h$ is assumed to be nonpositive all over the unit sphere $\mathbb{S}^2$, this result implies that:

- For all $v \in \mathbb{S}^2$, we have $i_{h_v} \leq 0$ on $P_v \backslash \mathcal{H}_{h_v}$;
- If $\mathcal{H}_h$ is not reduced to a single point then, for all $v \in \mathbb{S}^2$, there exists some $x$ in $P_v \backslash \mathcal{H}_{h_v}$ such that $i_{h_v}(x) < 0$

Now, the index $i_{h_v}$ is necessarily nonnegative if $\mathcal{H}_{h_v}$ is a circle or a point. This proves immediately that *Conjecture (C) is true if we assume that the surface S admits a circular enveloping cylinder in some direction $v \in \mathbb{S}^2$.*

In order to state the following corollary, we have to introduce the notion of a point of a $C^2$-hedgehog of $\mathbb{R}^3$. To every $C^2$-hedgehog $\mathcal{H}_h$ of $\mathbb{R}^3$ is associated a pseudometric defined on $\mathbb{S}^2$ by:

$$\forall (p, q) \in \left(\mathbb{S}^2\right)^2, \quad \rho_h(p, q) = \inf\left(\left\{L(x_h \circ \gamma) \,\middle|\, \gamma \in C_{pq}\right\}\right),$$

where $C_{pq}$ is the set of all piecewise $C^1$ paths joining $p$ to $q$ on $\mathbb{S}^2$, and $L(x_h \circ \gamma)$ is the length of $x_h \circ \gamma$. Let $\mathcal{R}_h$ be the equivalent relation defined on $\mathbb{S}^2$ by

$$\forall (p, q) \in \left(\mathbb{S}^2\right)^2, \quad p \, \mathcal{R}_h \, q \iff (\rho_h(p, q) = 0),$$

and let $\mathbb{S}^2 / \mathcal{R}_h$ be the corresponding quotient space:

$$\mathbb{S}^2 / \mathcal{R}_h = \left\{[p] \,\middle|\, p \in \mathbb{S}^2\right\},$$

where $[p]$ denotes the equivalent class of $p$. We will define a point of $\mathcal{H}_h$ as a point of this quotient space $\mathbb{S}^2 / \mathcal{R}_h$ endowed with the metric given by:

$$\forall (p, q) \in \left(\mathbb{S}^2\right)^2, \quad d_h([p], [q]) = \rho_h(p, q).$$

But in practice, for the sake of simplicity, we will generally not distinguish $[p]$ from its geometric realization $x_h(p)$ in $\mathbb{R}^3$.

**Corollary 4.4.1** *Let $v \in \mathbb{S}^2$. If $\mathcal{H}_h$ is a $C^2$-hedgehog whose curvature function $R_h$ is nonpositive all over $\mathbb{S}^2$, then:*

(i) *For all $x \in v^{\perp} \backslash \mathcal{H}_{h_v}$, there exist exactly $-2i_{h_v}(x)$ points of $\mathcal{H}_h$ whose geometric realizations lie on the line $\{x\} + \mathbb{R}v$;*

*(ii) For all $x \in \mathcal{H}_{h_v}$, every point of $\mathcal{H}_h$ whose geometric realization lies on the line $\{x\} + \mathbb{R}v$ is the equivalent class of a point of the circle $\mathbb{S}_v^1 = \mathbb{S}^2 \cap v^{\perp}$.*

*Given a $C^2$-hedgehog $\mathcal{H}_h$ of $\mathbb{R}^{n+1}$, we will say that $s \in x_h\left(\mathbb{S}^n\right)$ is an extremal point of $\mathcal{H}_h$ in direction $v \in \mathbb{S}^n$ if $\langle x_h\left(u\right), v \rangle \leq \langle s, v \rangle$ for all $u \in \mathbb{S}^n$.*

**Lemma 4.4.1** *Let $\mathcal{H}_h$ be a $C^2$-hedgehog of $\mathbb{R}^2$, and let $s$ be an extremal point of $\mathcal{H}_h$ in direction $v \in \mathbb{S}^1$ such that $s \notin \{x_h\left(-v\right), x_h\left(v\right)\}$. On $\mathbb{S}^1$, the connected components of $x_h^{-1}\left(s\right)$ are separated from each other by $\{-v, v\}$.*

***Proof of Lemma 4.4.1*** It suffices to prove that if $\Gamma$ is a subarc of $\mathbb{S}^1$ with distinct endpoints $\alpha, \beta \in x_h^{-1}\left(s\right)$ such that $\Gamma \cap \{-v, v\} = \varnothing$, then $\Gamma \subset x_h^{-1}\left(s\right)$. Let $\Gamma$ be such a subarc of $\mathbb{S}^1$.

We may assume without loss of generality that $s = (0, 0)$ and $v = (0, 1)$. For $u\left(\theta\right) = \left(\cos\theta, \sin\theta\right) \in \mathbb{S}^1 \backslash \{-v, v\}$, the support line of $\mathcal{H}_h$ with unit normal vector $u\left(\theta\right)$ then cuts the linear line $v^{\perp}$ at the point of coordinates $\left(x\left(\theta\right), 0\right)$, where $x\left(\theta\right) = h\left(\theta\right) / \cos\theta$. If $u\left(\theta\right) \in \Gamma \backslash \{\alpha, \beta\}$ tends to $\alpha$ or $\beta$, then $x\left(\theta\right)$ tends to $0$, and on the other hand

$$x'\left(\theta\right) = \frac{\langle x_h\left(\theta\right), v \rangle}{\cos^2\theta} \leq \frac{\langle s, v \rangle}{\cos^2\theta} = 0 \quad \text{for all } u\left(\theta\right) \in \Gamma \backslash \{\alpha, \beta\},$$

since $s$ is an extremal point of $\mathcal{H}_h$ in direction $v \in \mathbb{S}^1$. Therefore, $x\left(\theta\right) = 0$ for all $u\left(\theta\right) \in \Gamma \backslash \{\alpha, \beta\}$, so that $h = 0$ on $\Gamma$, which implies $\Gamma \subset x_h^{-1}\left(s\right)$. $\blacksquare$

**Theorem 4.4.2** *Let $\mathcal{H}_h$ be a hyperbolic $C^2$-hedgehog of $\mathbb{R}^3$ (i.e. a nontrivial $C^2$-hedgehog of $\mathbb{R}^3$ whose curvature function $R_h$ is nonpositive all over $\mathbb{S}^2$). Then, there exists $s \in x_h\left(\mathbb{S}^2\right)$ that is an extremal point of $\mathcal{H}_h$ in some direction $v \in \mathbb{S}^2$, and such that $x_h^{-1}\left(s\right)$ avoids some half great circle joining $-v$ to $v$ on $\mathbb{S}^2$.*

***Proof of Lemma 4.4.2*** Since $\mathcal{H}_h$ is a hyperbolic $C^2$-hedgehog, it is easy to prove by considering its convex hull that there exists a direction $v$ in which $\mathcal{H}_h$ admits two distinct extremal points $s_1$ and $s_2$ such that $x_h\left(-v\right)$ and $x_h\left(v\right)$ do not belong to the plane with equation $\langle x, v \rangle = \langle s_i, v \rangle$, $(i \in \{1, 2\})$. Let $u = (s_2 - s_1) / \|s_2 - s_1\|$, and $\mathbb{S}_u^1 = \mathbb{S}^2 \cap u^{\perp}$, where $u^{\perp}$ denotes the linear plane orthogonal to $u$. Consider the image of $x_h\left(\mathbb{S}_u^1\right)$ under the orthogonal projection onto $u^{\perp}$, that is, the plane hedgehog of $u^{\perp}$ whose support function $h_u$ is the restriction of $h$ to $\mathbb{S}_u^1$. The common projection of $s_1$ and $s_2$ on $u^{\perp}$, say $s$, is an extremal point of $\mathcal{H}_{h_u}$ in direction $v$, and $s \notin \left\{x_{h_u}\left(-v\right), x_{h_u}\left(v\right)\right\}$. No connected component of $x_{h_u}^{-1}\left(s\right)$ can meet both $x_h^{-1}\left(s_1\right)$ and $x_h^{-1}\left(s_2\right)$ because $\mathcal{H}_h$ cannot contain the line segment $[s_1, s_2]$ (if $[s_1, s_2]$ was included in $\mathcal{H}_h$, then the orthogonal projection onto the plane $w^{\perp}$, where $w \in \mathbb{S}_u^1 \cap \mathbb{S}_v^1$, would send $x_h\left(\mathbb{S}_w^1\right)$ to a plane hedgehog containing some non trivial line segment, which is impossible). Now, $x_h^{-1}\left(s_1\right)$ and $x_h^{-1}\left(s_2\right)$ must meet $x_{h_u}^{-1}\left(s\right)$ by Corollary 4.4.1, and the connected component of $x_{h_u}^{-1}\left(s\right)$ are separated from each other by $\{-v, v\}$ on $\mathbb{S}_u^1$ by Lemma 4.4.1. Therefore, $x_h^{-1}(s_1)$ (resp. $x_h^{-1}(s_2)$) does not meet one of the two half great circles joining $-v$ to $v$.  ∎

**Theorem 4.4.3** *Let $\mathcal{H}_h$ be a $C^2$-hedgehog of $\mathbb{R}^3$ whose curvature function $R_h$ is nonpositive all over $\mathbb{S}^2$. Then, $\mathcal{H}_h$ is reduced to a single point if one of the following two conditions is satisfied:*

*(i)  $\mathcal{H}_h$ is analytic $\left(\text{that is, the support function } h : \mathbb{S}^2 \to \mathbb{R} \text{ is analytic}\right)$;*
*(ii)  $\mathcal{H}_h$ is projective $\left(\text{that is, } h(-u) = -h(u) \text{ for all } u \in \mathbb{S}^2\right)$.*

**Proof of Theorem 4.4.3** Let $s$ be an extremal point of $\mathcal{H}_h$ in some direction $v \in \mathbb{S}^2$, and let $[m]$ be a point of $\mathcal{H}_h$ of which $s$ is the geometric realization $x_h(m)$ in $\mathbb{R}^3$. From Theorem 4.4.2, it suffices to prove that $x_h^{-1}(s)$ meets each half great circle joining $-v$ to $v$ on $\mathbb{S}^2$. Now, from Corollary 4.4.1, $[m]$ meets each great circle passing through $v$ on $\mathbb{S}^2$. If $\mathcal{H}_h$ is projective, then $x_h^{-1}(s)$ is invariant under the antipodal map and the result follows.

Now, assume that $\mathcal{H}_h$ is analytic and not reduced to a single point. Since the function $\phi : \mathbb{S}^2 \to \mathbb{R}$, $u \mapsto \langle x_h(u), u \rangle$ is then analytic, each connected component of $\phi^{-1}(\langle s, v \rangle)$ is either a point, or a simple closed arc, or an embedded connected graph all of whose vertices have an even degree [179]. Let $C$ denote the connected component of $\phi^{-1}(\langle s, v \rangle)$ that contains $[m]$. As $[m]$ meets any great circle passing through $v$ and meets at most once any half great circle joining $-v$ to $v$ on $\mathbb{S}^2$, by virtue of Lemma 4.4.1, either $C$ meets $\{-v, v\}$, or $C$ is a simple closed arc separating $-v$ and $v$ on $\mathbb{S}^2$. Now, $C \subset x_h^{-1}(s)$, since otherwise a projected hedgehog of $\mathcal{H}_h$ would contain a non-trivial segment, which is impossible. Therefore, $x_h^{-1}(s)$ meets each half great circle joining $-v$ to $v$ on $\mathbb{S}^2$, which is in contradiction with Theorem 4.4.2.  ∎

**Corollary 4.4.2** *If $S$ is analytic, or if $S$ is of constant width, then Conjecture $(C)$ is true.*

**Proof of Corollary 4.4.2** This corollary is an immediate consequence of Theorem 4.4.3 in the case when $S$ is analytic. If $S$ is of constant width, then we know that $h = f - r$ has the form $g + k$, where $g$ is an odd function on $\mathbb{S}^2$ and $k$ is a real constant. We then have

$$R_h = R_g + 2k R_{(1,g)} + k^2,$$

where $R_{(1,g)}$ is the mean radius of curvature of $\mathcal{H}_g$. Now, if we take the inequality $R_h \leq 0$ at two antipodal singular points of $x_g$ (such a pair $\{-u, u\}$ of antipodal points, for which $R_h(-u) = R_h(u) = 0$, exists since $\mathcal{H}_g$ is projective), it appears that $k^2 \leq 0$, so that $\mathcal{H}_h$ is projective since $h = g$.  ∎

**The Explicit Construction of Our Counterexample to Conjecture** $(H)$  Neither the methods that we have just seen above, nor any of the works prior to our paper [95] can make it possible to take into account the possible existence of singularities

**Fig. 4.12** A cross-cap hyperbolic hedgehog

of cross-cap type for hedgehogs. As we can easily check, the map

$$X : \mathbb{R}^2 \to \mathbb{R}^3, \ (u, v) \longmapsto (x, y, z) = r^4 \, (u, 1, uv) \,,$$

where $r = \sqrt{u^2 + v^2}$, parametrizes a cross-cap of $\mathbb{R}^3$ which can be regarded as a hyperbolic hedgehog $\mathcal{H}_h$ (modelled on the sphere $\mathbb{S}^2$ minus the semicircle given by $z = 0$ and $x \leq 0$): see Fig. 4.12. At the source $\mathbb{S}^2$, the singular locus of this hedgehog $\mathcal{H}_h$ is the semicircle given by $y = 0$ and $x \geq 0$. This semicircle is the spherical representation of the extreme point of $\mathcal{H}_h$ in the direction of $n = (0, -1, 0)$. It does not meet the geodesic semicircles of $\mathbb{S}^2$ joining $-n$ to $n$ through the hemisphere of $\mathbb{S}^2$ given by $x < 0$. Therefore, the method that we have applied above to treat the special cases of Conjecture $(H)$ cannot take into account such singularities. The regular part of the cross-cap hyperbolic hedgehog $\mathcal{H}_h$ in $\mathbb{R}^3$ is the portion $y > 0$ of the algebraic surface with equation

$$x^4 y^5 - \left( x^4 + y^2 z^2 \right)^2 = 0.$$

This surface can be obtained by welding (gluing) the graph of the function of two variables $f(x, y) = (x/y) \sqrt{y^{5/2} - x^2}$ with its symmetric with respect to the plane $z = 0$. This example shows, in passing, that a non-analytical hedgehog can admit an analytic parametrization.

The existence and the construction of this example of a cross-cap hyperbolic hedgehog $\mathcal{H}_h$ in $\mathbb{R}^3$ suggest us that we try to assemble conveniently 4 cross-cap hyperbolic hedgehogs in order to construct a nontrivial hyperbolic $C^2$-hedgehog modelled on the entire unit sphere $\mathbb{S}^2$. Let us begin by welding (gluing) the graph of the function of two variables

$$f(x, y) = \frac{xy}{1 - x^4 - y^4}$$

$$\times \sqrt{\left(1 - x^4 - y^4\right)^{\frac{5}{2}} - 25x^2 y^2 \left(x^8 + y^8 + 3\left(x^4 + y^4 - x^4 y^4\right) + 1\right)^{\frac{1}{2}}},$$

where $(x, y) \in D = \left\{ (u, v) \in \mathbb{R}^2 \,\middle|\, |u|^{\frac{4}{3}} + |v|^{\frac{4}{3}} \leq 1 \right\}$, with its symmetric with respect to the plane $z = 0$. Note that the expression under the radical vanishes precisely on the boundary of $D$. The resulting surface is formed by 4 cross-caps, but it still admits pieces with positive curvature. In order to eliminate these elliptic regions, let us add to $f$ a function $g : D \to \mathbb{R}$ of the form $g(x, y) = a\left(x^2 - y^2\right) + b\left(x^4 - y^4\right)$, with $a \neq 0$ and $a + 6b = 0$, so that the graph of $g$ has negative curvature except at the singular points of its boundary. More precisely, let us consider the one parameter family of surfaces $(\mathcal{S}_t)_{t \in \mathbb{R}_+^*}$ given by:

$$X_t : D \times \{-1, 1\} \longrightarrow \mathbb{R}^3$$
$$(x, y, \varepsilon) \longmapsto \left(x, y, \left(x^2 - y^2\right) - \tfrac{1}{6}\left(x^4 - y^4\right) + t\varepsilon f(x, y)\right).$$

Computer calculations prove the existence of an interval of $t$-values for which the surface $\mathcal{S}_t$ has negative Gauss curvature. In this interval, we take $t = 1/12$. The surface $\mathcal{S}_{1/12}$ is a portion of an algebraic surface. It is symmetric with respect to the planes $x = 0$ and $y = 0$, and also with respect to the lines $x = y = 0$, and, $z = 0$ and $x - y = 0$ (resp. $x + y = 0$). Since $\mathcal{S}_{1/12}$ is the welding (gluing) of two graphs above $D$, seen in the $(x, y)$-plane identified with $\mathbb{R}^2$, and since the boundary of $D$ is the plane hedgehog with support function $(u, v) \longmapsto uv\left(u^4 + v^4\right)^{-\frac{1}{4}}$, $(u, v) \in \mathbb{S}^1 \subset \mathbb{R}^2$, the negativity of the Gauss curvature (and thus the local injectivity of the Gauss map) implies that $\mathcal{S}_{1/12}$ can be regarded as a hedgehog $\mathcal{H}_h$. Let us see what we can say about the differentiability class of $\mathcal{H}_h$. The computer calculations show that the singular locus of $\mathcal{H}_h$ is formed on $\mathbb{S}^2$ by the four following disjoint geodesic semicircles: $(i)$ $3x + 4z = 0$, $y \geq 0$ ; $(ii)$ $3y - 4z = 0$, $x \geq 0$ ; $(iii)$ $3x - 4z = 0$, $y \leq 0$ ; $(iv)$ $3y + 4z = 0$, $x \leq 0$. We can say right away that $h$ is $C^\infty$ outside this singular locus, and $C^1$ on $\mathbb{S}^2$ (recall that $x_h : \mathbb{S}^2 \to \mathbb{R}^3$ is the restriction to $\mathbb{S}^2$ of the gradient of the positively 1-homogenous extension $\varphi(v) = \|v\| h(v/\|v\|)$ of $h$ to $\mathbb{R}^3 \setminus \{0\}$). We then prove that $h$ is $C^2$ on $\mathbb{S}^2$ by noting that, for all $u \in \mathbb{S}^2$, the eigenvalues of the Hessian of $\varphi$ at $u$, which are 0 and the principal radii of curvature $R_h^1(u)$, $R_h^2(u)$ of $\mathcal{H}_h$ at $x_h(u)$, tend to 0 when $u$ tends to the singular locus. Indeed, the calculations prove that $R_h(u) = R_h^1(u) R_h^2(u)$ and $R_{(1,h)}(u) = \left(R_h^1(u) + R_h^2(u)\right)/2$ tend to 0 when $u$ tends to the singular locus. Fig. 4.13 shows the hyperbolic hedgehog given by our construction.

**Remark 4.4.1** The support function $h$ of our counterexample to Conjecture $(H)$ is the subject of a conjecture by C. Mooney in his study of the question whether Lipschitz minimizers of $\int F(\nabla u) \, dx$ in $\mathbb{R}^3$ are $C^1$ when $F$ is strictly convex [129, Conjecture 2.4].

**Fig. 4.13** A hyperbolic $C^2$-hedgehog

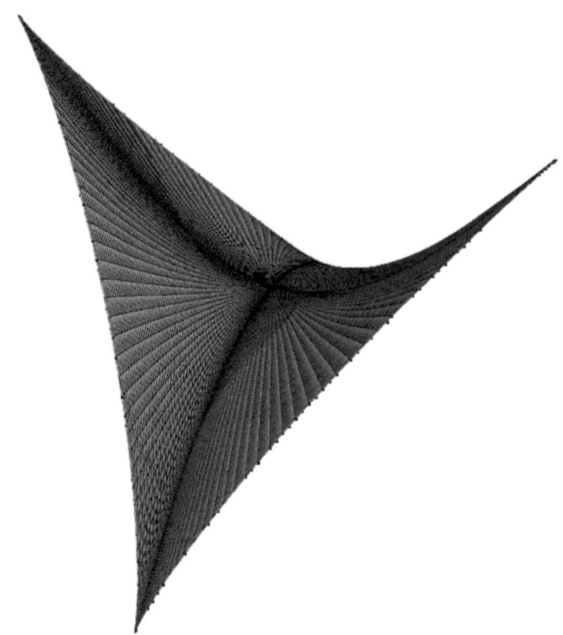

### 4.4.2   *Weak and Strong Hyperbolicity for Hedgehog Polytopes of $\mathbb{R}^3$*

We have already seen in Sects. 2.3 and 2.4 that we can define *hedgehog polytopes*, also called *polyhedral hedgehogs* or virtual polytopes, which represent formal differences of polytopes in $\mathbb{R}^{n+1}$. In this subsubsection, we are interested in the discrete version of hyperbolic hedgehogs in $\mathbb{R}^3$, that is, in *weak and strong hedgehog polytopes*. The results of this subsubsection are those of [97].

**Hedgehog Polytopes in $\mathbb{R}^3$**   We define a **convex polyhedron** of $\mathbb{R}^{n+1}$ to be the intersection of finitely many closed halfspaces in $\mathbb{R}^{n+1}$, and a **polytope** of $\mathbb{R}^{n+1}$ to be a bounded convex polyhedron of $\mathbb{R}^{n+1}$. Alternatively, a polytope of $\mathbb{R}^{n+1}$ may be defined to be the convex hull of finitely many points in $\mathbb{R}^{n+1}$. We now define a **hedgehog polytope** of $\mathbb{R}^{n+1}$ to be any hedgehog of $\mathbb{R}^{n+1}$ that can be represented as the difference $K - L$ of two nonempty polytopes $K$ and $L$ of $\mathbb{R}^{n+1}$. In other words, a polytope hedgehog of $\mathbb{R}^{n+1}$ is any hedgehog of $\mathbb{R}^{n+1}$ whose support function $h : \mathbb{S}^n \to \mathbb{R}$ belongs to the linear space generated by support functions of nonempty polytopes in $C\left(\mathbb{S}^n; \mathbb{R}\right)$, that is, whose (extended) support function $h : \mathbb{R}^{n+1} \setminus \{0\} \to \mathbb{R}$ is continuous, and piecewise linear (with respect to some division of $\mathbb{R}^{n+1}$ into polyhedral cones having the origin as their common vertex). As we said, we are now interested in hedgehog polytopes of $\mathbb{R}^3$.

Let $\mathcal{H}_h$ be a hedgehog polytope of $\mathbb{R}^3$. For all $u \in \mathbb{S}^2$, we denote by $\mathcal{F}_h^u$ the union of the support hedgehog $\mathcal{H}_h^u = \{h(u)u\} + H_h^u$ and its interior $\mathcal{I}_h^u := \{h(u)u\} + \left\{ x \in u^{\perp} \smallsetminus H_h^u \middle| i_{h'(u;.)}(x) \neq 0) \right\}$. Now, for any $u \in \mathbb{S}^2$ such that the interior $\mathcal{I}_h^u$ of the support hedgehog $\mathcal{H}_h^u$ is nonempty, we say that $\mathcal{F}_h^u$ is the face of $\mathcal{H}_h$ with unit normal vector $u$. The algebraic area of $\mathcal{H}_h$ is then defined to be the sum of the algebraic areas of all the faces of $\mathcal{H}_h$, that is, by

$$s(h) = \sum_{u \in \mathcal{N}(h)} a_u(h),$$

where $a_u(h) = a\left(h'(u;.)\right)$ is the algebraic area of the face $\mathcal{F}_h^u$, and $\mathcal{N}(h) := \left\{ u \in \mathbb{S}^2 \middle| \mathcal{I}_h^u \neq \varnothing \right\}$. Given any $u \in \mathcal{N}(h)$, every $h(u)u + x \in \mathcal{I}_h^u$ will be regarded as a point with multiplicity $\left| i_{h'(u;.)}(x) \right|$ of $\mathcal{F}_h^u$, at which the normal line to $\mathcal{H}_h$ is directed by $sgn\left( i_{h'(u;.)}(x) \right) u$, $sgn(.)$ being the signum function. This allows us to consider $\mathcal{H}_h$ as a transversely oriented surface of $\mathbb{R}^3$ the algebraic volume of which is defined by

$$v(h) := \int_{\mathbb{R}^3 \smallsetminus \mathcal{H}_h} i_h(x)\, d\lambda(x),$$

where $i_h(x)$ is the algebraic intersection number of an oriented half-line with origin $x$ with the surface $\mathcal{H}_h$ equipped with its transverse orientation (number independent of the oriented half-line for an open dense set of directions), and $\lambda$ is the Lebesgue measure on $\mathbb{R}^3$. This algebraic volume is then given by

$$v(h) = \frac{1}{3} \sum_{u \in \mathcal{N}(h)} h(u)\, a_u(h)$$

for every hedgehog polytope $\mathcal{H}_h$ of $\mathbb{R}^3$.

**Weak and Strong Hyperbolic Polytopes** Let $\mathcal{H}_h$ be a hedgehog polytope of $\mathbb{R}^3$, and let $\mathcal{S}_h$ be its spherical representation (i.e. the graph constituted by all the points $u \in \mathbb{S}^2$ such that $h : \mathbb{R}^3 \smallsetminus \{0\} \to \mathbb{R}$ is not linear in the vicinity of $u$ in $\mathbb{R}^3$). Assume that for any couple $(u, v) \in \mathcal{N}(h)^2$ of vertices of $\mathcal{S}_h$, we have $u + v \neq 0$. Then we will say that $\mathcal{H}_h$ is **weakly hyperbolic** if all its support hedgehogs $\mathcal{H}_h^u$ are of negative index, that is, if:

$$\forall u \in \mathcal{N}(h),\, \forall y = h(u)u + x \in \mathcal{I}_h^u,\, i_{h'(u;.)}(x) < 0.$$

A way of conceiving examples of weakly hyperbolic polytopes is to discretize example of $C^2$ hedgehogs, the elliptic regions of which are bounded by a single cuspidal edge, by reducing each of these elliptic regions into a single edge. Edges coming from an elliptic region then have the particularity of being represented by a great arc (an arc of great circle) of length greater than $\pi$ on the sphere $\mathbb{S}^2$. We will say of such an edge that it is *elliptical*. Figure 4.14a represents the

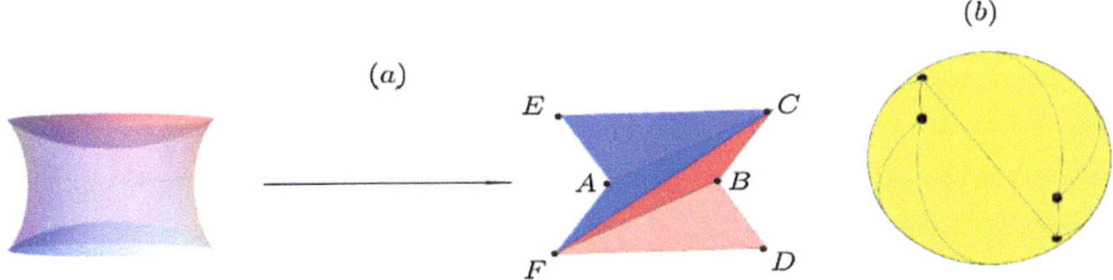

**Fig. 4.14** A weakly hyperbolic polytope obtained by discretization (**a**), and its spherical representation (**b**)

discretization of a hedgehog of revolution (with one hyperbolic region and two elliptic ones) into a weakly hyperbolic polytope (with 6 vertices, 8 faces and 12 edges of which two are elliptic). Let $(x, y, z)$ be the standard coordinates in $\mathbb{R}^3$. On Fig. 4.14a, this weakly hyperbolic polytope is represented from a point of the line with equations $x = y = 0$. Its 6 vertices are the points $A$, $B$, $C$, $D$, $E$, $F$ of $\mathbb{R}^3$ with respective coordinates $(1, 0, 0)$, $(-1, 0, 0)$, $(2, 3/2, 3/2)$, $(2, -3/2, -3/2), (-2, 3/2, -3/2), (-2, -3/2, 3/2)$, and its 8 faces are triangles $ACE$, $AED$, $ADF$, $AFC$, $BCE$, $BED$, $BDF$, $BFC$. This weakly hyperbolic polytope is symmetric with respect to three orthogonal lines including the axis of revolution. Figure 4.14b gives its spherical representation. A very simple example of a weakly hedgehog polytope is given by the hyperbolic tetrahedron (see, e.g., [142]).

**Definition 4.4.1** A weakly hyperbolic polytope $\mathcal{H}_h$ of $\mathbb{R}^3$ is said to be **strongly hyperbolic** if it does not have any elliptic edge, that is, if all its edges are represented on $\mathcal{S}_h$ by a great arc with length less than $\pi$.

**Theorem 4.4.4** *There exists a strongly hyperbolic hedgehog polytope in* $\mathbb{R}^3$.

An example of a strongly hyperbolic polytope (with 14 vertices, 24 faces, and 36 edges) can indeed be obtained by a discretization of the $C^2$ hyperbolic hedgehog constructed above in order to give a counterexample to Alexandrov's conjecture. This example is composed of a central part (Fig. 4.15a) and 4 discrete cross-caps (Fig. 4.15b). Our example of a strongly hyperbolic polytope of $\mathbb{R}^3$ is symmetric with respect to two orthogonal planes (with respective Cartesian equations $x = 0$ and $y = 0$) and two orthogonal lines (with respective equations $x - y = z = 0$ and $x + y = z = 0$). Figure 4.15c gives its spherical representation on $\mathbb{S}^2$ seen from a point of the line with equations $x = y = 0$.

## 4.5  Minimal $N$-hedgehogs and Brunn-Minkowski Theory

The aim of this subsection is to motivate the development of a Brunn-Minkowski theory for minimal surfaces. In 1985, R. Langevin, G. Levitt and H. Rosenberg introduced a sum operation in the set $W$ of complete minimal surfaces in $\mathbb{R}^3$ of finite

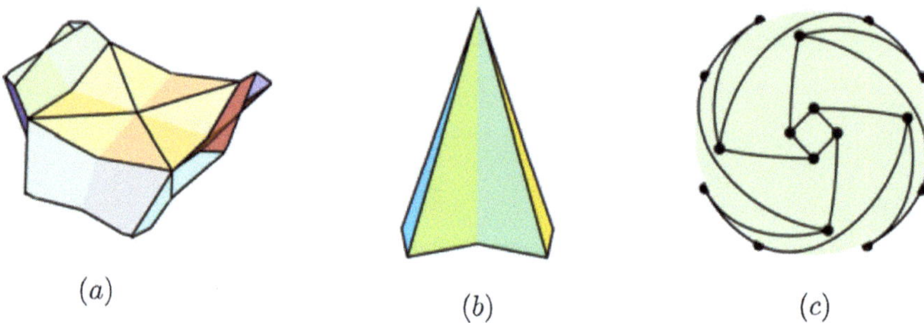



**Fig. 4.15** Discretization of our counterexample to Alexandrov's conjecture

total curvature and with finitely many branch points [82]. In 1988, H. Rosenberg and
E. Toubiana studied this sum operation and noticed that points of $\mathbb{R}^3$ and surfaces
of $W$ of total (absolute) curvature $4\pi$ (the so-called ***minimal hedgehogs***) constitute
a real vector space [159]. In this subsection, we: $(i)$ give geometric inequalities
for minimal surfaces of $\mathbb{R}^3$; $(ii)$ study the relation between support functions
and Enneper-Weierstrass representation; $(iii)$ introduce and studies a new type of
addition for minimal surfaces; $(iv)$ extend notions and techniques from the classical
Brunn-Minkowski theory to minimal surfaces. Two characterizations of the catenoid
among minimal hedgehogs are given.

The notion of $C^2$-hedgehog of $\mathbb{R}^3$ can be extended by considering hedgehogs
whose support functions are defined (and $C^2$) only on some spherical domain $\Omega \subset
\mathbb{S}^2$. Among hedgehogs defined on the unit sphere $\mathbb{S}^2$ punctured at a finite number of
points, we can consider those that are minimal, that is, those whose mean curvature
$H$ is zero at all the smooth points. ***The condition for a hedgehog $\mathcal{H}_h$ of $\mathbb{R}^3$ to be
minimal*** is simply that its principal radii of curvature $R_1 := \lambda_1 + h$ and $R_2 := \lambda_2 + h$,
where $\lambda_1 \leq \lambda_2$ denote the eigenvalues of the Hessian $\nabla^2 h$, satisfy $R_1 + R_2 = 0$,
that is, that its support function $h$ satisfies the equation

$$\triangle_S h + 2h = 0,$$

where $\triangle_S$ is the spherical Laplace operator on $\mathbb{S}^2$. In other words, a minimal
hedgehog $\mathcal{H}_h$ (modelled on $\mathbb{S}^2$ punctured at a finite number of points) is a trivial
hedgehog (i.e., a point) or a (possibly branched) minimal surface with total curvature
$-4\pi$ that is parametrized by the inverse of its Gauss map.

A first study of minimal hedgehogs had been given by H. Rosenberg and E.
Toubiana [159]. Concerning linear structures on the collections of minimal surfaces
in $\mathbb{R}^3$ and $\mathbb{R}^4$, the reader is also referred to the paper by A. Small [174]. Unless
explicitly states otherwise, the results of this section are essentially taken from [99].

### 4.5.1 *Geometric Inequalities for Minimal N-hedgehogs in $\mathbb{R}^3$*

In this subsection, we are mainly interested in the extension to minimal surfaces of notions and techniques from the Brunn-Minkowski theory. The idea of developing a Brunn-Minkowski theory for minimal surfaces of $\mathbb{R}^3$ arises naturally from the fact that a (reversed) Brunn-Minkowski type inequality holds for minimal hedgehogs.

Let $K$ be the closure of a (nonempty) connected open subset of $\mathbb{S}^2$ and let $\mathcal{H}_k$ be a minimal hedgehog modelled on $K$. Then, the area of $x_k(K)$ is finite and given by

$$Area\,[x_k\,(K)] = -\int_K R_k\,d\sigma,$$

where $\sigma$ is the spherical Lebesgue measure on $\mathbb{S}^2$ and $R_k$ the curvature function of $\mathcal{H}_k$, that is, $1/\kappa_k$, where $\kappa_k$ is the Gauss curvature of $\mathcal{H}_k$ (regarded as a function of the unit normal). Now, if $\mathcal{H}_l$ is another minimal hedgehog modelled on $K$, then

$$\sqrt{A(k+l)} \leq \sqrt{A(k)} + \sqrt{A(l)}, \tag{4.5.1}$$

where $A(h) = Area\,[x_h\,(K)]$. In fact, we can regard the set of hedgehogs modelled (up to a translation) on $K$ as a real vector space endowed with a prehilbertian structure for which the norm is given by the square root of the area. Consider the set of support functions (of a minimal hedgehog) modelled on $K$ and identify two such functions $k$ and $l$ when $x_k(K)$ and $x_l(K)$ are translates of each other. Then, the quotient set $\mathcal{H}(K)$ inherits a real vector space structure and we obtain the following result.

**Theorem 4.5.1** *The map $\sqrt{A} : \mathcal{H}(K) \to \mathbb{R}_+, h \longmapsto \sqrt{Area\,[x_h\,(K)]}$, is a norm associated with a scalar product $A : \mathcal{H}(K)^2 \to \mathbb{R}$, which may be interpreted as an algebraic mixed area:*

$$\forall\,(k,l) \in \mathcal{H}(K)^2,\ A\,(k,l) := (Mixed\,Area)\,[x_k\,(K)\,, x_l\,(K)]$$

*By the Cauchy-Schwarz inequality we have:*

$$A(k,l)^2 \leq A(k).A(l). \tag{4.5.2}$$

**Corollary 4.5.1** *Consequently, the area $A : \mathcal{H}(K) \to \mathbb{R}_+, h \longmapsto Area\,[x_h(K)]$, is a strictly convex map, and thus for any nonempty convex subset $\mathcal{K}$ of $\mathcal{H}(K)$, the problem of minimizing $A$ over $\mathcal{K}$ has at most one optimal solution.*

**Remark 4.5.1** Inequality (4.5.1) (resp. (4.5.2)) has to be compared with the following Brunn-Minkowski inequality (resp. Minkowski inequality). For any pair $(K, L)$ of convex bodies of $\mathbb{R}^3$, we have (see e.g., [167])

$$\sqrt{A\,(K+L)} \geq \sqrt{A\,(K)} + \sqrt{A\,(L)}$$

and

$$A (K, L)^2 \geq A (K) \cdot A (L),$$

where $A (H)$ (resp. $A (K, L)$) is the surface area (resp. the mixed surface area) of the convex body $H \subset \mathbb{R}^3$ (resp. of the pair $(K, L)$).

**Remark 4.5.2** Let $\mathcal{H} (\mathbb{S}^2)$ be the real vector space of support functions of minimal hedgehogs defined (up to a translation) on the unit sphere punctured at a finite number of points. To each $h \in \mathcal{H} (\mathbb{S}^2)$ let us assign the positive Borel measure $\mu_h$ defined on $\mathbb{S}^2$ by

$$\forall \Omega \in \mathcal{B} \left(\mathbb{S}^2\right), \mu_h (\Omega) = - \int_\Omega R_h \, d\sigma,$$

where $\mathcal{B} (\mathbb{S}^2)$ denotes the $\sigma$-algebra of Borel subsets of $\mathbb{S}^2$. Then, we notice that the map

$$m : \mathcal{H} (\mathbb{S}^2) \to \left\{ \sqrt{\mu} \, | \mu \text{ is a positive Borel measure on } \mathbb{S}^2 \right\}$$
$$h \longmapsto \sqrt{\mu_h},$$

satisfies the following properties:

(i)  $\forall h \in \mathcal{H} (\mathbb{S}^2), m (h) = 0 \iff h = 0_{\mathcal{H}(\mathbb{S}^2)};$

(ii)  $\forall \lambda \in \mathbb{R}, \forall h \in \mathcal{H} (\mathbb{S}^2), m (\lambda h) = |\lambda| m (h);$

(iii)  $\forall (k, l) \in \mathcal{H} (\mathbb{S}^2)^2, m (k + l) \leq m (k) + m (l).$

**Remark 4.5.3** Let $\mathcal{H}_k$ and $\mathcal{H}_l$ be two hedgehogs whose support function is defined (and $C^2$) on some spherical domain $\Omega \subset \mathbb{S}^2$. On this domain, we can define their mixed curvature function by

$$R_{(k,l)} := \frac{1}{2} (R_{k+l} - R_k - R_l).$$

The symmetric map $(\alpha, \beta) \mapsto R_{(\alpha,\beta)}$ is bilinear on the vector space of hedgehogs modelled on $\Omega$ (see Sect. 3.1). Given any $u \in \Omega$, the polynomial function $P_u (t) = R_{k+tl} (u)$ thus satisfies $P_u (t) = R_k (u) + 2t R_{(k,l)} (u) + t^2 R_l (u)$ for all $t \in \mathbb{R}$.

When $k$ and $l$ are the support functions of two convex bodies of class $C_+^2$, $P_u (t)$ must have a zero, so that

$$R_{(k,l)} (u)^2 \geq R_k (u) . R_l (u)$$

and hence

$$\sqrt{R_{k+l} (u)} \geq \sqrt{R_k (u)} + \sqrt{R_l (u)},$$

by noticing that $R_{(k,l)} > 0$.

When $\mathcal{H}_k$ and $\mathcal{H}_l$ are minimal hedgehogs, $P_u\,(t)$ is nonpositive on $\mathbb{R}$, so that

$$R_{(k,l)}\,(u)^2 \leq R_k\,(u)\,.R_l\,(u)$$

and hence

$$\sqrt{-R_{k+l}\,(u)} \leq \sqrt{-R_k\,(u)} + \sqrt{-R_l\,(u)}.$$

Note that $A\,(k,l) = \int_K R_{(k,l)}d\sigma$ for all $(k,l) \in \mathcal{H}\,(K)^2$ and that inequality (4.5.2) can be deduced from the inequality $R_{(k,l)}{}^2 \leq R_k.R_l$.

Inequality (4.5.1) can be extended to some **asymptotic areas of embedded ends** in $\mathbb{R}^3$. The (possibly branched) complete minimal surfaces of finite nonzero total curvature in $\mathbb{R}^3$ can be regarded as '**multihedgehogs**' provided they have only a finite number of branch points [159]: the (possibly singular) envelope of a family of cooriented planes of $\mathbb{R}^3$ is called an $N$-**hedgehog** if, for an open dense set of $u \in \mathbb{S}^2$, it has exactly $N$ cooriented support planes with normal vector $u$. Hedgehogs with a $C^2$ support function are merely 1-hedgehogs.

It is known that embedded ends of a minimal surface of $\mathbb{R}^3$ are planar or of catenoid type (i.e., asymptotic to a planar or catenoid end). More precisely, each embedded end is the graph (over the exterior of a bounded region in an $(x_1, x_2)$-plane orthogonal to the limiting normal at the end) of a function of the form

$$u\,(x_1, x_2) = a \ln\,(r) + b + \frac{cx_1 + dx_2}{r^2} + O\left(\frac{1}{r^2}\right), \text{ where } r = \sqrt{x_1^2 + x_2^2},$$

with $a = 0$ when the end is planar [171].

Let $E$ be an embedded planar end of a minimal surface of $\mathbb{R}^3$ and let $P$ be its asymptotic plane. Define the asymptotic area of $E$ by

$$A_s\,[E] = \iint_\Delta \left(\sqrt{1 + u_{x_1}\,(x_1, x_2)^2 + u_{x_2}\,(x_1, x_2)^2} - 1\right) dx_1 dx_2 \in [0, +\infty],$$

where $u : \Delta \rightarrow \mathbb{R}$, $(x_1, x_2) \mapsto u\,(x_1, x_2)$ is the function whose graph is equal to $E$. Given any increasing sequence $(K_n)$ of compact subsets of $P$ such that $K_n \rightarrow \Delta$, $A_s\,[E]$ may be interpreted as the limit of

$$Area\left[\pi^{-1}\,(K_n) \cap E\right] - Area\,[K_n],$$

where $\pi$ denotes the orthogonal projection onto the asymptotic plane.

**Theorem 4.5.2**  *The asymptotic area of every embedded planar end of a minimal surface $S$ of $\mathbb{R}^3$ is finite.*

Note that hedgehogs never have planar ends: if an end is planar, then the limiting normal at the end is a branch point of the Gauss map so that the surface cannot be a hedgehog (see e.g., [81]). Let $E$ be an embedded planar end of a minimal $N$-hedgehog, where $N \geq 2$. After a rotation, we may assume the limiting normal at the end is $n = (0, 0, -1)$. Then $E$ admits a Weierstrass representation $(g\,(z)\,,\,f\,(z)dz)$ of the form

$$g(z) = z^N \text{ and } f(z) = \frac{\alpha}{z^2} + \sum_{k=0}^{+\infty} c_k z^k,$$

where $\alpha$ is nonzero [81]. The reader who is not familiar with the Weierstrass representation of minimal surfaces in $\mathbb{R}^3$ will find an introduction to it in the next subsubsection. Given $r \in \,]0, 1[$, the pieces of minimal $N$-hedgehogs defined (up to a translation) by a parametrization of the form

$$X_f : D = \{z \in \mathbb{C} \,|\, 0 < |z| \leq r\} \rightarrow \mathbb{R}^3$$

$$z = x + iy \mapsto \mathrm{Re}\left( \int \frac{1}{2} f(z) \left(1 - z^{2N}\right) dz,\right.$$

$$\left. \int \frac{i}{2} f(z) \left(1 + z^{2N}\right) dz, \int f(z) z^N dz \right),$$

where $f\,(z) = \frac{\alpha}{z^2} + \sum_{k=0}^{+\infty} c_k z^k$ ($\alpha$ may be 0), constitute a real vector space $(E_N, +, .)$, where addition is defined by $X_{f_1} + X_{f_2} = X_{f_1 + f_2}$ and scalar multiplication by $\lambda.X_f = X_{\lambda f}$. Let us denote by $S_f$ the surface parametrized by $X_f : D \rightarrow \mathbb{R}^3$.

**Theorem 4.5.3** *For every $S_f \in E_N$, define $A_s\,(f)$ by*

$$A_s\,(f) := \iint_D (1 - \langle N(z), n\rangle) \left\| \left(\frac{\partial X_f}{\partial x} \times \frac{\partial X_f}{\partial y}\right)(z) \right\| dxdy,$$

*where* $N(z) = \frac{2}{|z|^{2N}+1}\left(\mathrm{Re}\left(z^N\right), \mathrm{Im}\left(z^N\right), \frac{|z|^{2N}-1}{2}\right)$ *is the unit normal at* $X_f\,(z)$ *if* $\left(\frac{\partial X_f}{\partial x} \times \frac{\partial X_f}{\partial y}\right)(z) \neq 0$ *and where* $D$ *is identified with* $\left\{(x, y) \in \mathbb{R}^2 \,\middle|\, 0 < \sqrt{x^2 + y^2} \leq r\right\}$.

(i) *If $S_f$ is an embedded planar end, then $A_s\,(f)$ is its asymptotic area $A_s\left[S_f\right]$.*
(ii) *The map $\sqrt{A_s} : E_N \rightarrow R_+$, $S_f \longmapsto \sqrt{A_s\,(f)}$ is a norm associated with a scalar product (which may be interpreted as a **mixed algebraic asymptotic area**).*

***Proof of Theorem 4.5.3***

(i) If $S_f$ is an embedded planar end, then $A_s\,(f)$ is its asymptotic area $A_s\left[S_f\right]$ for $\langle N(z), n\rangle \left\| \left(\frac{\partial X_f}{\partial x} \times \frac{\partial X_f}{\partial y}\right)(z) \right\| dxdy$ is the area of the orthogonal projection, onto the asymptotic plane, of the element of area $\left\| \left(\frac{\partial X_f}{\partial x} \times \frac{\partial X_f}{\partial y}\right)(z) \right\| dxdy$ on the end.

(ii) We know that (see, e.g., [137]):

$$\forall z = x + iy \in D, \ \left\|\left(\frac{\partial X_f}{\partial x} \times \frac{\partial X_f}{\partial y}\right)(z)\right\| = \left(|f(z)| \frac{\left(1 + |z|^{2N}\right)}{2}\right)^2,$$

so that

$$A_s(f) = \iint_D (1 - \langle N(z), n\rangle) \left\|\left(\frac{\partial X_f}{\partial x} \times \frac{\partial X_f}{\partial y}\right)(z)\right\| dx dy$$

$$= \iint_D |f(z)|^2 |z|^{2N} \frac{1 + |z|^{2N}}{2} dx dy.$$

Consequently, $\sqrt{A_s} : E_N \to \mathbb{R}_+$ is a norm associated with the scalar product given by

$$A_s(f_1, f_2) = \iint_D \mathrm{Re}\left[f_1(z) \overline{f_2(z)}\right] |z|^{2N} \frac{1 + |z|^{2N}}{2} dx dy.$$

$\blacksquare$

### 4.5.2   Addition of Minimal Surfaces and Enneper-Weierstrass Representation

It is well known that any minimal surface $S$ of $\mathbb{R}^3$ (possibly with isolated branch points) can be locally represented in the form

$$\begin{cases} X_1(x, y) = \frac{1}{2} \mathrm{Re}\left[\int_{z_0}^{z} \left(1 - g(\zeta)^2\right) f(\zeta) d\zeta\right] + c_1 \\ X_2(x, y) = \frac{1}{2} \mathrm{Re}\left[\int_{z_0}^{z} i\left(1 + g(\zeta)^2\right) f(\zeta) d\zeta\right] + c_2 \\ X_3(x, y) = \mathrm{Re}\left[\int_{z_0}^{z} g(\zeta) f(\zeta) d\zeta\right] + c_3, \end{cases} \tag{4.5.3}$$

where $f(z)$ is an arbitrary holomorphic function on an open simply connected neighborhood $\mathcal{U}$ of $z_0 \in \mathbb{C}$ and $g(z)$ an arbitrary meromorphic function on $\mathcal{U}$ such that, at each pole of order $n$ of $g(z)$, $f(z)$ has a zero of order at least $2n$, the integral being taken along any path connecting $z_0$ to $z = x + iy \in \mathbb{C}$ in $\mathcal{U}$, and naturally, $c_1, c_2$ and $c_3$ denote real constants. Recall that (see e.g., [137])

$$N(z) := \frac{\left(\frac{\partial X}{\partial x} \times \frac{\partial X}{\partial y}\right)(x, y)}{\left\|\left(\frac{\partial X}{\partial x} \times \frac{\partial X}{\partial y}\right)(x, y)\right\|}$$

$$= \frac{2}{|g(z)|^2 + 1} \left( \mathrm{Re}\left[g(z)\right], \mathrm{Im}\left[g(z)\right], \frac{|g(z)|^2 - 1}{2} \right),$$

is the (unit) normal to the surface at $X(x, y) = (X_1(x, y), X_2(x, y), X_3(x, y))$ and $g(z)$ its image under the stereographic projection $\sigma : \mathbb{S}^2 - \{(0, 0, 1)\} \to \mathbb{C}, (x, y, t) \mapsto \frac{x+iy}{1-t}$. Thus, $X : \mathcal{U} \to \mathbb{R}^3, z = x+iy \mapsto (X_1(x, y), X_2(x, y), X_3(x, y))$ is a hedgehog (that is, $X$ can be interpreted as the inverse of the stereographic projection of its Gauss map) if and only if $g(z) = z$. The simplest choice of 'Weierstrass data' $(g(z), f(z)\,dz) = (z, dz)$ gives **Enneper's surface**. Recall that this surface and the **catenoid**, which is given by $(g(z), f(z)\,dz) = \left(z, dz/z^2\right)$, are the only two complete regular minimal surfaces that are hedgehogs (see e.g., [137]).

Representation (4.5.3) can be generalized to generate all minimal surfaces of $\mathbb{R}^3$: if $S \subset \mathbb{R}^3$ is a minimal surface (possibly with isolated branch points), $M$ its Riemann surface and $g = \sigma \circ N : M \to \mathbb{C} \cup \{\infty\}$ the stereographic projection (from the north pole) of its Gauss map, then $S$ can be represented in the form (4.5.3) for some holomorphic function $f$ on $M$ and some fixed $z_0 \in M$.

Given any two (possibly branched) minimal surfaces $S_1$ and $S_2$ modelled (up to a translation) by Weierstrass data $(g(z), f_1(z)\,dz)$ and $(g(z), f_2(z)\,dz)$ on a Riemann surface $M$ (and thus sharing the same 'Gauss map' $g(z)$), we can define their sum $S_1 + S_2$ as the (possibly branched) minimal surface given (up to a translation) by $(g(z), (f_1(z) + f_2(z))\,dz)$. For any minimal surface $S$ modelled (up to a translation) by Weierstrass data $(g(z), f(z)\,dz)$ on $M$ and for any complex number $\lambda$, we can define the minimal surface $\lambda S$ as the minimal surface given (up to a translation) by $(g(z), \lambda f(z)\,dz)$. Of course, in order for $z \mapsto \mathrm{Re}\left[\int \phi_\lambda(z)\,dz\right]$ to be well defined on $M$, where

$$\phi_\lambda(z) := \lambda f(z) \left( \frac{1}{2}\left(1 - g(z)^2\right), \frac{i}{2}\left(1 + g(z)^2\right), g(z) \right),$$

we need that no component of $\phi_\lambda$ has a real period on $M$, that is

$$Period_\gamma\,[\phi_\lambda] := \mathrm{Re} \oint_\gamma \phi_\lambda(z)\,dz = 0_{\mathbb{R}^3},$$

for all closed curves $\gamma$ on $M$, but in the case when this period condition is not satisfied, we may consider the minimal surface $\lambda S$ modelled on the universal covering space of $M$ (i.e., $\mathbb{C}$ or the open unit disc). By hypothesis, $\phi_1$ has no real period on $M$ since $S$ is modelled on $M$. It follows that for any $\lambda \in \mathbb{R}$ the surface $\lambda S$ is also modelled on $M$ (since $\phi_\lambda$ clearly has no real period on $M$ if $\lambda \in \mathbb{R}$). Thus, minimal surfaces modelled (up to a translation) by Weierstrass data $(g(z), f(z)\,dz)$ on a common Riemann surface $M$ and sharing the same 'Gauss map' $g(z)$ constitute a real vector space $E_M$ (which can be identified with the space of all holomorphic functions $f(z)$ having a zero of order at least $2n$ at each pole of order $n$ of $g(z)$ and satisfying

$$Period_\gamma\left[f\left(\frac{1}{2}\left(1-g^2\right),\frac{i}{2}\left(1+g^2\right),g\right)\right]=0_{\mathbb{R}^3}$$

for all closed curves $\gamma$ on $M$).

Recall that: $(i)$ the associate surfaces to a minimal surface $S$ modelled (up to a translation) by Weierstrass data $(g(z),f(z)dz)$ on a Riemann surface $M$ are the surfaces $S_\theta=e^{i\theta}S$ given (up to a translation) by $\left(g(z),e^{i\theta}f(z)dz\right)$, where $\theta\in\left[0,\frac{\pi}{2}\right]$; and $(ii)$ the conjugate surface $S^*$ to $S$ is the associate surface $S_{\frac{\pi}{2}}$. Clearly, $S^*$ and $S_\theta$ are (locally) parametrized by $X^*(z)=-\operatorname{Im}\left[\int\phi(z)dz\right]$ and $X_\theta=(\cos\theta)X-(\sin\theta)X^*$, where $\phi:=f\left(\frac{1}{2}\left(1-g^2\right),\frac{i}{2}\left(1+g^2\right),g\right)$ and $X(z):=\operatorname{Re}\left[\int\phi(z)dz\right]$. In other words, we have $S_\theta=(\cos\theta)S-(\sin\theta)S^*$, where the surfaces are modelled on the universal covering space of $M$ in the case when $\phi$ has a real period on $M$.

**Remark 4.5.4** Every hedgehog $\mathcal{H}_h\subset\mathbb{R}^{n+1}$ has a unique representation in the form

$$\mathcal{H}_h=\mathcal{H}_c+\mathcal{H}_p, \tag{4.5.4}$$

where $\mathcal{H}_c$ is centered (that is, centrally symmetric with center at the origin) and $\mathcal{H}_p$ projective (that is, modelled on $\mathbb{P}^n(\mathbb{R})=\mathbb{S}^n/\{Id,-Id\}$). This representation is given by

$$h=c+p,$$

where

$$c(u)=\frac{1}{2}(h(u)+h(-u))\quad\text{and}\quad p(u)=\frac{1}{2}(h(u)-h(-u)).$$

In the same way, every minimal hedgehog $\mathcal{H}_h$ of $\mathbb{R}^3$ has a unique representation in the form (4.5.4). If $\mathcal{H}_h$ is given by Weierstrass data $(z,f(z)dz)$, then $\mathcal{H}_c$ and $\mathcal{H}_p$ are given (up to a translation) by the following decomposition of $f(z)$:

$$f(z)=f_c(z)+f_p(z),$$

where

$$f_c(z)=\frac{1}{2}\left(f(z)+\overline{\frac{1}{z^4}f\left(\frac{-1}{\overline{z}}\right)}\right)\quad\text{and}\quad f_p(z)=\frac{1}{2}\left(f(z)-\overline{\frac{1}{z^4}f\left(\frac{-1}{\overline{z}}\right)}\right)$$

**Fig. 4.16** Central symmetrization of Enneper's surface

(see [175] for the determination of $f_p(z)$). Let us consider the case of Enneper's surface, whose support function is given by

$$h(u) = \frac{\left(x^2 - y^2\right)(2r - t)}{2(r - t)^2},$$

where $r = \sqrt{x^2 + y^2 + t^2}$ and $u = (x, y, t) \in \mathbb{S}^2 \subset \mathbb{R}^3$. In this case, we obtain

$$c(u) = \frac{x^2 - y^2}{\left(x^2 + y^2\right)^2} \quad \text{and} \quad p(u) = \frac{t\left(x^2 - y^2\right)\left(2r^2 + x^2 + y^2\right)}{2\left(x^2 + y^2\right)^2}$$

(resp. $f_c(z) = \frac{1}{2}\left(1 + \frac{1}{z^4}\right)$ and $f_p(z) = \frac{1}{2}\left(1 - \frac{1}{z^4}\right)$) and we notice that: $(i)$ $\mathcal{H}_c$ has 5 planes of symmetry (with equations $x = 0$, $y = 0$, $z = 0$ $x + y = 0$ and $x - y = 0$), 4 curves of double points lying on the plane $z = 0$, and 4 branch points (namely $\left(1/\sqrt{2}, 1/\sqrt{2}, 0\right)$ and the points deduced from it by symmetry); $(ii)$ $\mathcal{H}_p$ is **Henneberg's surface** (which is thus the 'projective part' of Enneper's surface). Figure 4.16 shows the **central symmetrization of Enneper's surface.**

### 4.5.3 Relation Between Enneper-Weierstrass Representation and Support Function

**Theorem 4.5.4** *Let* $X : \mathcal{U} \ni z_0 \to \mathbb{R}^3, z \mapsto \mathrm{Re}\left[\int_{z_0}^{z} \phi(\zeta)\, d\zeta\right]$, *where* $\phi(z) :=$ $f(z)\left(\frac{1}{2}\left(1 - g(z)^2\right), \frac{i}{2}\left(1 + g(z)^2\right), g(z)\right)$, *be the Weierstrass representation of* *a piece of minimal surface (possibly with isolated branch points) such that*

$$N : \mathcal{U} \to N\,(\mathcal{U}) \subset \mathbb{S}^2, z \mapsto N\,(z)$$

$$= \frac{2}{|g\,(z)|^2 + 1} \left( \mathrm{Re}\,[g\,(z)]\,,\, \mathrm{Im}\,[g\,(z)]\,,\, \frac{|g\,(z)|^2 - 1}{2} \right)$$

*is a diffeomorphism of $\mathcal{U}$ onto $N\,(\mathcal{U})$. Then $X\,(\mathcal{U})$ can be regarded as a hedgehog $\mathcal{H}_h$ whose parametrization $x_h : N\,(\mathcal{U}) \to \mathcal{H}_h \subset \mathbb{R}^3$ is given by $x_h = \nabla\varphi$, where $\varphi : v \mapsto \|v\|\,h\,(v/\|v\|)$ is the positively 1-homogeneous extension of $h$ to $\{ tu \,|\, u \in N\,(\mathcal{U})$ and $t \in \mathbb{R}_+^* \}$ .Given $g(z)$, the support function $h$ and the holomorphic function $f$ are related by*

$$\phi\,(z) = \frac{2g'\,(z)}{1 + |g\,(z)|^2}\,(L_\varphi)_{N(z)}\,\left(\overline{v_g\,(z)}\right),  \tag{4.5.5}$$

*where $(L_\varphi)_{N(z)}$ is the endomorphism of $\mathbb{C}^3$ that is represented in the standard basis by the Hessian matrix $(Hess\,\varphi)_{N(z)}$ of $\varphi$ at $N\,(z)$ and $v_g\,(z) = (1, i, g\,(z))$, so that*

$$f\,(z) = \frac{2g'\,(z)}{\left(1 + |g\,(z)|^2\right)^2}\,\left[{}^t\overline{V_g\,(z)}.\,(Hess\,\varphi)_{N(z)}.\overline{V_g\,(z)}\right],$$

*where $V_g\,(z)$ is the column matrix ${}^t v_g\,(z)$.*

**Proof of Theorem 4.5.4** For all $z = x + iy \in \mathcal{U}$ we have

$$X\,(z) = x_h\,[N\,(z)] = (\nabla\varphi)\,[N\,(z)]\,,$$

and thus

$$X_\xi\,(z) = (L_\varphi)_{N(z)}\,\left(N_\xi\,(z)\right),$$

where $N_\xi\,(z) = \frac{\partial}{\partial\xi}\,[N\,(x + iy)]$, $X_\xi\,(z) = \frac{\partial}{\partial\xi}\,[X\,(x + iy)]$, and $\xi = x$ or $y$. Note that

$$N_\xi\,(z) = \frac{2}{1 + |g\,(z)|^2}\,\left[(P_\xi, Q_\xi, P P_\xi + Q Q_\xi)\,(z) - \left(P P_\xi + Q Q_\xi\right)\,(z)\,N\,(z)\right],$$

where $g\,(z) = P\,(x, y) + i Q\,(x, y)$, $P_\xi = \frac{\partial P}{\partial\xi}$ and $Q_\xi = \frac{\partial Q}{\partial\xi}$. As $\varphi$ is positively 1-homogeneous, we have

$$\left(L_\varphi\right)_{N(z)}\,(N\,(z)) = 0,$$

and we thus get

$$X_\xi(z) = \frac{2}{1 + |g(z)|^2} \left(L_\varphi\right)_{N(z)} \left[\left(P_\xi, Q_\xi, PP_\xi + QQ_\xi\right)(z)\right].$$

Now, direct calculation gives

$$\begin{cases} \mathrm{Re}\left[\frac{2g'(z)}{1+|g(z)|^2}\left(L_\varphi\right)_{N(z)}\overline{\left(v_g(z)\right)}\right] \\ \quad = \frac{2}{1+|g(z)|^2}\left(L_\varphi\right)_{N(z)}\left[\left(P_x, Q_x, PP_x + QQ_x\right)(z)\right] \\[2mm] \mathrm{Im}\left[\frac{2g'(z)}{1+|g(z)|^2}\left(L_\varphi\right)_{N(z)}\overline{\left(v_g(z)\right)}\right] \\ \quad = -\frac{2}{1+|g(z)|^2}\left(L_\varphi\right)_{N(z)}\left[\left(P_y, Q_y, PP_y + QQ_y\right)(z)\right], \end{cases}$$

so that

$$\phi(z) = X_x(z) - iX_y(z) = \frac{2g'(z)}{1 + |g(z)|^2}\left(L_\varphi\right)_{N(z)}\overline{\left(v_g(z)\right)}.$$

∎

Let $\mathcal{H}_h$ be a minimal hedgehog of $\mathbb{R}^3$ defined by Weierstrass data $(z, f(z)\,dz)$ on the sphere $\mathbb{S}^2$ punctured at a finite number of points. From (4.5.5), it follows that

$$f(z) = \frac{2}{z\left(1 + |z|^2\right)}\left[\left(\nabla\varphi_t\right)(N(z)).\overline{V_g(z)}\right],$$

where $\varphi_t$ is the partial derivative of $\varphi$ with respect to the third coordinate in the standard basis of $\mathbb{R}^3$ and $\nabla\varphi_t = \left(\varphi_{xt}, \varphi_{yt}, \varphi_{t^2}\right)$ is its gradient. Changing the orientation of the normal, this gives $\tilde{f}(z) = \frac{2}{z(1+|z|^2)}\left[\left(\nabla\tilde{\varphi}_t\right)(N(z)).\overline{V_g(z)}\right]$, where $\tilde{f}(z) = -\frac{1}{z^4}\overline{f\left(\frac{-1}{\bar{z}}\right)}$ and $\tilde{\varphi}(u) = -\varphi(-u)$. Noting that $N\left(\frac{-1}{\bar{z}}\right) = -N(z)$ and comparing $f\left(\frac{-1}{\bar{z}}\right)$ with $\tilde{f}(z)$, we get easily

$$\varphi_{t^2}(N(z)) = \mathrm{Re}\left[z^2 f(z)\right].$$

Now, inflection points of level curves of a hedgehog $\mathcal{H}_h \subset \mathbb{R}^3$ (with a support function of class $C^\infty$) are given by

$$\varphi_{t^2}(u) = 0, \ \nabla\varphi_t(u) \neq 0 \text{ and } R_h(u) \neq 0,$$

where $\varphi(u) = \|u\| h(u/\|u\|)$. By 'inflection point' of a level curve $\mathcal{C} \subset \mathcal{H}_h$ we mean a point where $\mathcal{C}$ has a contact of order $\geq 2$ with its tangent line. Therefore we have:

**Corollary 4.5.2** *Let $\mathcal{H}_h$ be a nontrivial minimal hedgehog of $\mathbb{R}^3$ defined by Weierstrass data $(z, f(z)\,dz)$ on the unit sphere $\mathbb{S}^2$ punctured at a finite number of points. The inflection points of level curves of $\mathcal{H}_h$ are given by*

$$\mathrm{Re}\left[z^2 f(z)\right] = 0, \ z \neq 0 \text{ and } f(z) \neq 0.$$

*It follows easily that the hedgehog $\mathcal{H}_h$ is necessarily a catenoid if it is complete and if no level curve of $\mathcal{H}_h$ has an inflection point.*

### 4.5.4  Orthogonal-Projection Techniques

Theorem 2.8.3 enables us to obtain information on an ordinary $C^2$-hedgehog of $\mathbb{R}^3$ by considering its image under orthogonal projections onto planes. We have the following analogue for minimal hedgehogs:

**Theorem 4.5.5** *Let $\mathcal{H}_h \subset \mathbb{R}^3$ be a complete minimal hedgehog modelled on $\mathbb{S}^2$ punctured at a finite number of points $e_1, \ldots, e_n$ (corresponding to its ends) and let $u \in \mathbb{S}^2$ be such that $\mathbb{S}_u^1 \subset \mathbb{S}^2 - \{e_1, \ldots, e_n\}$. Then, for any regular value $x \in u^\perp - \mathcal{H}_{h_u}$ of the map $x_h^u = \pi_u \circ x_h : \mathbb{S}^2 - \{e_1, \ldots, e_n\} \to u^\perp$, we have*

$$i_{h_u}(x) + N_h^u(x)^+ = \sum_{e_k \in \mathbb{S}_u^+} d(e_k),$$

*where $\mathbb{S}_u^+ \subset \mathbb{S}^2$ is the halfsphere defined by $\langle u, v \rangle > 0$, $N_h^u(x)^+$ the number of $v \in \mathbb{S}_u^+ - \{e_j \,|\, \langle e_j, u \rangle > 0\}$ such that $x_h(v) \in \{x\} + \mathbb{R}u$ and $d(e_k)$ the winding number of the end with limiting normal $e_k$. Replacing $u$ by $-u$, it follows that*

$$i_{h_u}(x) + N_h^u(x)^- = \sum_{e_k \in \mathbb{S}_u^-} d(e_k),$$

*where $\mathbb{S}_u^- \subset \mathbb{S}^2$ is the halfsphere defined by $\langle u, v \rangle < 0$ and $N_h^u(x)^-$ the number of $v \in \mathbb{S}_u^- - \{e_j \,|\, \langle e_j, u \rangle < 0\}$ such that $x_h(v) \in \{x\} + \mathbb{R}u$. Consequently,*

$$i_{h_u}(x) = \frac{1}{2}\left(N(h) - N_h^u(x)\right),$$

*where $N_h^u(x) = N_h^u(x)^- + N_h^u(x)^+$ is the number of $v \in \mathbb{S}^2 - \{e_1, \ldots, e_n\}$ such that $x_h(v) \in \{x\} + \mathbb{R}u$ and $N(h)$ the total spinning of $\mathcal{H}_h$, that is, $N(h) = \sum_{k=1}^n d(e_k)$.*

***Proof of Theorem 4.5.5*** It suffices to prove the relation

$$i_{h_u}(x) + N_h^u(x)^+ = \sum_{e_k \in \mathbb{S}_u^+} d(e_k),$$

for any regular value $x \in u^\perp - \mathcal{H}_{h_u}$ of $x_h^u = \pi_u \circ x_h : \mathbb{S}^2 - \{e_1, \ldots, e_n\} \to u^\perp$.

Let $(x_1, x_2, x_3)$ be the standard coordinates in $\mathbb{R}^3$. Without loss of generality, we can identify $u^\perp$ to the plane with equation $x_3 = 0$ (and thus with the Euclidean vector plane $\mathbb{R}^2$) and assume that $x$ is its origin $0_{\mathbb{R}^2}$. The index $i_{h_u}(x)$ is the winding number of $\mathcal{H}_{h_u}$ around $x \in u^\perp - \mathcal{H}_{h_u}$. It is given by

$$i_{h_u}(x) = \frac{1}{2\pi} \int_{\mathcal{H}_{h_u}} \omega,$$

where $\omega$ is the closed 1-form defined by $\omega_{(x_1,x_2)} = \frac{x_1 dx_2 - x_2 dx_1}{x_1^2 + x_2^2}$ on $\mathbb{R}^2 - \{0_{\mathbb{R}^2}\}$. This index $i_{h_u}(x)$ can also be regarded as the winding number of $x_h(\mathbb{S}_u^1)$ around the oriented line, say $D_x(u)$, passing through $x$ and directed by $u$. In other words, $i_{h_u}(x)$ is given by

$$i_{h_u}(x) = \frac{1}{2\pi} \int_{x_h(\mathbb{S}_u^1)} \omega,$$

which can be checked by an easy calculation. Writing $\Sigma_u^+ = \mathbb{S}_u^+ - \{e_j \,|\, e_j \in \mathbb{S}_u^+\}$, we then have

$$i_{h_u}(x) = \frac{1}{2\pi} \int_{\partial S} \omega,$$

where $S$ denotes the surface $x_h\left[\Sigma_u^+\right]$ equipped with its transverse orientation. Let $\{f_1, \ldots, f_L\}$ be the set consisting of all $e_j$ such that $\langle e_j, u \rangle > 0$, i.e., $e_j \in \mathbb{S}_u^+$. Since $x$ is a regular value of the map $x_h^u = \pi_u \circ x_h : \mathbb{S}^2 - \{e_1, \ldots, e_n\} \to u^\perp$, there exists a small closed disc, say $D$, centered at $x$ whose inverse image under $\left(x_h^u\right)^+ : \mathbb{S}_u^+ - \{f_1, \ldots, f_L\} \to u^\perp, v \mapsto x_h^u(v)$ is empty or admits a partition of the form

$$\left[\left(x_h^u\right)^+\right]^{-1}(D) = \bigcup_{k=1}^{K} D_k,$$

where $K = N_h^u(x)^+$ and $D_k$ is such that the map $\pi_u \circ x_h$ defines a diffeomorphism from $D_k$ onto $D$ for all $k \in \{1, \ldots, K\}$. As $f_1, \ldots, f_L$ are limiting normals at ends of the complete minimal hedgehog $\mathcal{H}_h$, there exist small disjoint spherical discs

$\Delta_1, \ldots, \Delta_L$ punctured at $f_1, \ldots, f_L$ that are disjoint from $\mathbb{S}_u^1$ and from each $D_k$ $(1 \leq k \leq K)$. Now, Stokes's formula gives

$$\int_{\partial S} \omega = \sum_{k=1}^{K} \int_{\partial S_k} \omega + \sum_{l=1}^{L} \int_{\partial \Sigma_l} \omega,$$

where $S_k$ (resp. $\Sigma_l$) denotes the surface $x_h \, (D_k)$ (resp. $x_h \, (\Delta_l)$) equipped with its transverse orientation. As $\mathcal{H}_h$ is a (possibly branched) minimal surface, the maps $x_h : D_k \rightarrow S_k$ are orientation reversing and thus the orthogonal projections of the oriented curves $\partial S_k$ into the $(x_1, x_2)$-plane have winding number $-1$ around $x$. Consequently,

$$\sum_{k=1}^{K} \int_{\partial S_k} \omega = -N_h^u \, (x)^+ .$$

To complete the proof, it suffices to notice that we have also

$$\sum_{l=1}^{L} \int_{\partial \Sigma_l} \omega = \sum_{l=1}^{L} d \, (f_l) = \sum_{e_k \in \mathbb{S}_u^+} d \, (e_k) ,$$

from the definition of the winding number of an end.                                   ∎


## 4.6  Plane $N$-hedgehogs and the Sturm-Hurwitz Theorem

Sturm theory is closely related to the geometry of curves. In particular, Sturm-type oscillation theorems enable us to minorate the number of certain special points or concurrent lines for different types of closed curves. The central result of Sturm oscillation theory is the famous Sturm-Hurwitz theorem.

**The Sturm-Hurwitz Theorem**  *Every continuous real function of the form* $h \, (\theta) = \sum_{n \geq N} (a_n \cos n\theta + b_n \sin n\theta)$ *has at least as many zeros on the circle* $\mathbb{S}^1 = \mathbb{R}/2\pi\mathbb{Z}$ *as its first nonvanishing harmonics, that is*

$$\# \left( \left\{ \theta \, (\mathrm{mod} \, 2\pi) \in \mathbb{S}^1 \, | h \, (\theta) = 0 \right\} \right) \geq 2N.$$

We know many proofs of this theorem. The interested reader can for instance find five proofs of the Sturm-Hurwitz theorem in [139, Appendice 8.1]. For the history and importance of the Sturm-Hurwitz theorem, we refer the reader to the papers by V.I. Arnold and in particular to [9]. In that paper, V.I. Arnold explained why we can consider Sturm theory as a Morse theory extended to higher order derivative and laments that: "There are many proofs of this Sturm theorem but all of them are incomprehensible. Of course, I can reproduce them but you get no intuition from those proofs". In the present subsection, we will give a geometrical interpretation and a new proof of the Sturm-Hurwitz theorem for $C^2$ functions by the way of Rolle's theorem, and by considering the $2N\pi$-periodic function $h(\theta/N)$ as the support function of an '$N$-hedgehog' $\mathcal{H}_h$ of $\mathbb{R}^2$. $N$-hedgehogs are defined in the same way as hedgehogs, as envelopes of families of cooriented lines having exactly $N$ cooriented support lines with a given unit normal $u \in \mathbb{S}^1$. Plane 1-hedgehogs are just plane hedgehogs.

The geometrical interpretation of the Sturm-Hurwitz theorem for $C^2$-functions will be the following.

**Multihedgehog Version of the Sturm-Hurwitz Theorem**   *If $\mathcal{H}_h$ is an $N$-hedgehog of $\mathbb{R}^2$ such that*

$$h(N\theta) = \sum_{n=N}^{+\infty} (a_n \cos n\theta + b_n \sin n\theta),$$

*for some sequences of real numbers $(a_n)$ and $(b_n)$, then $\mathcal{H}_h$ has no 'positive area' (that is, for all $x \in \mathbb{R}^2 \backslash \mathcal{H}_h$, the winding number $i_h(x)$ of $\mathcal{H}_h$ around $x$ is nonpositive).*

Of course, we hope that hedgehogs and multihedgehogs can constitute a useful geometric tool to obtain other results related to Sturm theory, in particular in higher dimensions. In Chap. 9, we will show through a study of the concurrent normals conjecture that this hope is justified by first new results making use of hedgehogs. The results of the present subsection are mainly extracted from [98].

We mention in passing that the author also proved the following "*Sturm-type comparison theorem by a geometric study of multihedgehogs*" in the Illinois J. Math. 52 (2008), 981–993:

**Theorem**   *Let $h$ be a real $2N\pi$-periodic function of class $C^2$ on $\mathbb{R}$, ($N \in \mathbb{N}^*$). The number $S \in \mathbb{N} \cup \{\infty\}$ of zeros of $h + h''$ in $[0, 2N\pi[$ satisfies*

$$n_h \leq S + 4N - 2,$$

*where $n_h \in \mathbb{N} \cup \{\infty\}$ is the number of zeros of $h$ in $[0, 2N\pi[$.*

### 4.6.1   Plane N-hedgehogs

In the Euclidean vector plane $\mathbb{R}^2$, $N$-hedgehogs are defined in the same way as hedgehogs, except that their support functions are $2N\pi$-periodic instead of being $2\pi$-periodic, ($N \in \mathbb{N}^*$). The integer $N$ is just the number of full rotations of the coorienting normal vector $u(\theta) = (\cos\theta, \sin\theta)$ when $\theta$ varies from 0 to $2N\pi$. Therefore, an $N$-hedgehog $\mathcal{H}_h$ of $\mathbb{R}^2$ has exactly $N$ cooriented support lines with a given normal vector $u \in \mathbb{S}^1$ (counted with their multiplicity). So plane 1-hedgehogs are simply hedgehogs. Let us recall basic definitions.

Let $N \in \mathbb{N}^*$. For every $2N\pi$-periodic $C^2$-function $h : \mathbb{R} \to \mathbb{R}$, let us consider the envelope of the family of lines with equation

$$\langle x, u(\theta) \rangle = h(\theta), \tag{4.6.1}$$

where $\langle ., . \rangle$ denotes the standard scalar product on $\mathbb{R}^2$, and $u(\theta) = (\cos\theta, \sin\theta)$, $(\theta \in [0, 2N\pi[)$. This envelope is called the $N$-hedgehog with support function $h$ and denoted by $\mathcal{H}_h$. Partial differentiation of ($4.6.1$) with respect to $\theta$ gives

$$\langle x, u'(\theta) \rangle = h'(\theta), \tag{4.6.2}$$

From ($4.6.1$) and ($4.6.2$), the natural parametrization of $\mathcal{H}_h$ is:

$$x_h : [0, 2N\pi] \longrightarrow \mathbb{R}^2, \ \theta \mapsto x_h(\theta) = h(\theta)u(\theta) + h'(\theta)u'(\theta).$$

Differentiation of $x_h$ gives

$$x'_h(\theta) = R_h(\theta)(\theta)u'(\theta) \quad \text{for all } \theta \in \mathbb{R}$$

where $R_h := h + h''$ is the so-called *curvature function* of $\mathcal{H}_h$. Note that $R_h(\theta)$ is well defined for all $\theta \in \mathbb{R}$ and that $R_h(\theta) = 0$ if and only if $x_h(\theta)$ is a singular point of $\mathcal{H}_h$.

Here are some examples of '*multihedgehogs*' of $\mathbb{R}^2$: for every $n \geq 2$, the **hypocycloid** (resp. the **epicycloid**) with support function $h_n(\theta) = \sin(n\theta)$ (resp. $e_n(\theta) = \sin((n-1)\theta/n))$ is a 1-hedgehog (resp. an $n$-hedgehog) with $2n$ (resp. $2(n-1)$) cusps (cf. Fig. 4.17); when $n$ is odd, the cusps of the hypocycloid are counted twice because $x_{h_n}(\theta)$ travels twice $\mathcal{H}_{h_n}$ when $\theta$ varies from 0 to $2\pi$.

The (algebraic) area of an $N$-hedgehog $\mathcal{H}_h$ of $\mathbb{R}^2$ can be defined by

$$a(h) = \frac{1}{2} \int_0^{2N\pi} h(\theta)(h + h'')(\theta)\, d\theta = \frac{1}{2} \int_0^{2N\pi} \left(h^2 - (h')^2\right)(\theta)\, d\theta$$

and interpreted as the integral

$$a(h) = \int_{\mathbb{R}^2 \setminus \mathcal{H}_h} i_h(x)\, d\lambda(x),$$

where $i_h(x)$ denotes the winding number of $\mathcal{H}_h$ around $x$ and $\lambda$ the Lebesgue measure on $\mathbb{R}^2$. Note that the winding number $i_h(x)$ can also be seen as the algebraic intersection number of almost every oriented half-line with origin $x$ with $\mathcal{H}_h$ equipped with its transverse orientation (this number is independent of the oriented half-line for an open dense set of directions). For the hypocycloid with three cusps with support function $h_3$, the algebraic area is equal to $-4\pi$, that is

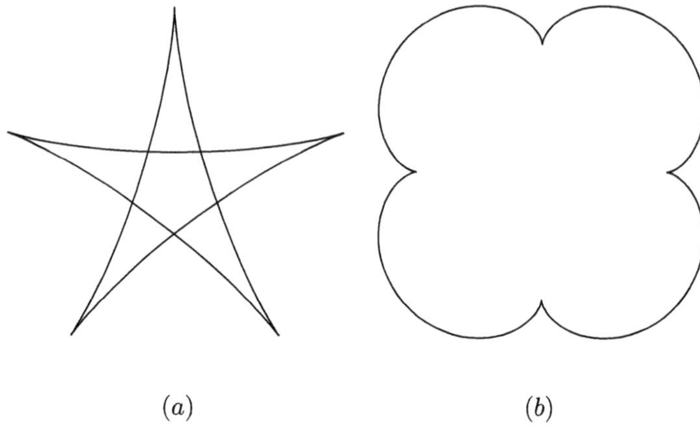

**Fig. 4.17** Examples of multihedgehogs. (**a**) The hypocycloid $\mathcal{H}_{h_5}$. (**b**) The epicycloid $\mathcal{H}_{e_3}$

to $-\,2\,area(D)$, where $D$ is the domain delimited by $\mathcal{H}_{h_3}$; the minus sign comes from the fact that $D$ is concave at the regular points of its boundary and the factor 2 from the fact that $x_{h_3}(\theta)$ travels the curve twice when $\theta$ varies from 0 to $2\pi$. For the epicycloid with four cusps with support function $e_3$ (see Fig. 4.17b), the algebraic area is equal to $(5/6)\pi$.

Note that if the expansion in a Fourier series of $h(N\theta)$ is given by $h(N\theta) = \sum_{n\geq 0}(a_n\cos n\theta + b_n\sin n\theta)$, then

$$a(h) = Na_0^2\pi + \frac{\pi}{2N}\sum_{n=1}^{+\infty}\left(N^2 - n^2\right)\left(a_n^2 + b_n^2\right). \tag{4.6.3}$$

The (algebraic) area defines a quadratic form on the linear space, say $L_N$, of $N$-hedgehogs defined up to a translation in $\mathbb{R}^2$ and identified with their support functions. Its polar form

$$a : L_N^2 \to \mathbb{R}$$
$$(h, k) \mapsto a(h, k)$$

can be interpreted as a mixed algebraic area; for all $(h, k) \in L_N^2$, the mixed algebraic area of $\mathcal{H}_h$ and $\mathcal{H}_k$ is thus given by:

$$a(h, k) = \frac{1}{2}\int_0^{2N\pi} h(\theta)\left(k + k''\right)(\theta)\,d\theta = \frac{1}{2}\int_0^{2N\pi}\left(hk - h'k'\right)(\theta)\,d\theta.$$

The (algebraic) length of an $N$-hedgehog $\mathcal{H}_h$ of $\mathbb{R}^2$ is defined by

$$l(h) := 2a(1, h) = \int_0^{2N\pi}\left(h + h''\right)(\theta)\,d\theta = \int_0^{2N\pi} h(\theta)\,d\theta,$$

whereas its absolute length is given by

$$L(h) = \int_0^{2N\pi} \left|\left(h + h''\right)(\theta)\right| \, d\theta,$$

since $x'_h(\theta) = R_h(\theta)(\theta)u'(\theta)$   for all $\theta \in \mathbb{R}$.

### 4.6.2   Multihedgehog Version of the Sturm-Hurwitz Theorem

The above multihedgehog version of the Sturm-Hurwitz theorem is based on the following relationship between the index $i_h(x)$ of a point $x \in \mathbb{R}^2 \setminus \mathcal{H}_h$ with respect to $\mathcal{H}_h$ and the number of zeros of $h_x(\theta) = h(\theta) - \langle x, u(\theta) \rangle$, where $u(\theta) = (\cos\theta, \sin\theta)$. Here, the index $i_h(x)$ is of course the winding number of $\mathcal{H}_h$ around $x$ in $\mathbb{R}^2$.

**Theorem 4.6.1 ([98])** *For every $N$-hedgehog $\mathcal{H}_h$ of $\mathbb{R}^2$ with a $C^2$ support function, we have:*

$$\forall x \in \mathbb{R}^2 \setminus \mathcal{H}_h, \; i_h(x) = N - \frac{1}{2}n_h(x), \qquad (4.6.4)$$

*where $n_h(x)$ is the number of cooriented support lines through $x$ (i.e., the number of zeros of $h_x : [0, 2N\pi[ \to \mathbb{R}, \theta \longmapsto h(\theta) - \langle x, u(\theta) \rangle$, where $u(\theta) = (\cos\theta, \sin\theta))$. Note that relationship (4.6.4) allows us to define $i_h(x) \in \mathbb{Z} \cup \{-\infty\}$ for any $x \in \mathbb{R}^2$.*

***Proof of Theorem 4.6.1*** We will slightly abbreviate the proof for it is very similar to the one of Theorem 2.8.1, which corresponds to the case $N = 1$. The index $i_h(x)$ can be defined as the degree of the map

$$\mathcal{U}_{(h,x)} : S_N \to \mathbb{S}^n, \; \theta \longmapsto \frac{x_h(\theta) - x}{\|x_h(\theta) - x\|},$$

where $S_N := \mathbb{R} \setminus 2N\pi\mathbb{Z}$. Since $x_{h_x}(\theta) = x_h(\theta) - x$ for all $\theta \in S_N$, we may assume without loss of generality that $x$ is the origin $0_{\mathbb{R}^2} = (0, 0)$. A straightforward computation then gives

$$i_h\left(0_{\mathbb{R}^2}\right) = \frac{1}{2\pi}\int_0^{2N\pi} \frac{h(\theta)\left(h+h''\right)(\theta)}{h(\theta)^2 + h'(\theta)^2} \, d\theta$$

$$= N + \frac{1}{2\pi}\int_{\Gamma_h} \omega,$$

where $\Gamma_h$ is the oriented curve of $\mathbb{R}^2$ that is parametrized by

$$\gamma_h : [0, 2N\pi] \to \mathbb{R}^2, \; \theta \longmapsto \left(h\left(\theta\right), h'\left(\theta\right)\right),$$

and $\omega$ the 1-differential form given by

$$\omega(x_1, x_2) = \frac{x_1 dx_2 - x_2 dx_1}{x_1^2 + x_2^2} \quad \text{for all} \quad (x_1, x_2) \neq (0, 0) \text{ in } \mathbb{R}^2.$$

We then deduce that

$$\frac{1}{2\pi} \int_{\Gamma_h} \omega = -\frac{1}{2} n_h \left(0_{\mathbb{R}^2}\right)$$

in the same way we did it in the proof of Theorem 2.8.1.                    ∎

**Corollary 4.6.1** *Let $\mathcal{H}_h$ be an $N$-hedgehog of $\mathbb{R}^2$ for which $h\left(N\theta\right)$ is a trigonometric polynomial of degree $\leq N$. Then the map $i_h : \mathbb{R}^2 \backslash \mathcal{H}_h \to \mathbb{Z}$ is everywhere nonnegative on $\mathbb{R}^2 \backslash \mathcal{H}_h$, and it is identically equal to 0 on $\mathbb{R}^2 \backslash \mathcal{H}_h$ if and only if $\mathcal{H}_h$ is reduced to a single point.*

*So the mixed area $a$ is a scalar product on the linear space $P_N$ of all $N$-hedgehogs defined up to a translation in $\mathbb{R}^2$ (this point only means that the harmonic of order $n$ remains unspecified) the support functions of which satisfy the same condition as $h$. Thus for all $(h_1, h_2) \in P_N^2$, we have*

$$\sqrt{a(h_1 + h_2)} \leq \sqrt{a(h_1)} + \sqrt{a(h_2)}, \tag{4.6.5}$$

*and*

$$a(h_1, h_2)^2 \leq a(h_1)\, a(h_2). \tag{4.6.6}$$

*Equality holds in (4.6.5) (resp. (4.6.6)) if and only if $\mathcal{H}_{h1}$ and $\mathcal{H}_{h2}$ are linearly dependent in $P_N$.*

***Proof of Corollary 4.6.1*** The second part of the statement follows simply from (4.6.3). But the whole statement is a corollary of Theorem 4.6.1 since any trigonometric polynomial of degree $\leq N$ has at most $2N$ zeros in $[0, 2\pi[$.                    ∎

Note that inequalities (4.6.5) and (4.6.6) are **analogues of the Brunn-Minkowski and Minkowski ones (with reversed inequality signs)**. In particular, (4.6.6) yields a **reverse isoperimetric inequality** for $k = 1$:

$$a(h) \geq \frac{1}{4N\pi} l\left(h\right)^2, \tag{4.6.7}$$

where equality holds if and only if $\mathcal{H}_h$ is a circle travelled $N$ times.

Our proof of the multihedgehog version of the Sturm-Hurwitz theorem will rely on the following proposition. Recall that the evolute of a plane curve is the locus of all its centers of curvature or, equivalently, the envelope of its normal lines.

**Proposition 4.6.1** *Let $\mathcal{H}_h \subset \mathbb{R}^2$ be an $N$-hedgehog the support function of which is of class $C^3$. The evolute of $\mathcal{H}_h$ is the $N$-hedgehog with support function $(\partial h)(\theta) = h'\left(\theta - \frac{\pi}{2}\right)$. Moreover for every $x \in \mathbb{R}^2 \setminus \mathcal{H}_h$ we have $n_h(x) \le n_{\partial h}(x)$ and hence $i_{\partial h}(x) \le i_h(x)$.*

***Proof of Proposition 4.6.1*** For all $\theta \in \mathbb{R}$, the center of curvature of $\mathcal{H}_h$ at $x_h(\theta)$ is given by $c_h(\theta) = x_h(\theta) - R_h(\theta)u(\theta)$, so that $c_h\left(\theta - \frac{\pi}{2}\right) = x_{\partial h}(\theta)$. Moreover, since from Rolle's theorem any two distinct zeros of $h_x(\theta) := h(\theta) - \langle x, u(\theta)\rangle$ must be separated by a zero of $h_x'(\theta) := h'(\theta) - \langle x, u'(\theta)\rangle = (\partial h)\left(\theta + \frac{\pi}{2}\right) - \langle x, u\left(\theta + \frac{\pi}{2}\right)\rangle = (\partial h)_x\left(\theta + \frac{\pi}{2}\right)$, we have $n_h(x) \le n_{\partial h}(x)$.   ∎

For every $N$-hedgehog $\mathcal{H}_h$ of $\mathbb{R}^2$, we define the positive (resp. negative) area to be the integral

$$a_+(h) = \int_{\mathcal{P}_h} i_h(x)\,d\lambda(x) \quad \left(\text{resp. } a_-(h) = \int_{\mathcal{N}_h} |i_h(x)|\,d\lambda(x)\right),$$

where $\mathcal{P}_h$ (resp. $\mathcal{N}_h$) is the set of all $x \in \mathbb{R}^2$ such that $i_h(x) > 0$ (resp. $i_h(x) < 0$). Under the assumptions of Proposition 4.6.1, we immediately have

$$\mathcal{P}_{\partial h} \subset \mathcal{P}_h \quad \text{and} \quad \mathcal{N}_{\partial h} \supset \mathcal{N}_h,$$

and hence

$$a_+(\partial h) \le a_+(h) \quad \text{and} \quad a_-(\partial h) \ge a_-(h).$$

**Proof of the Multihedgehog Version of the Sturm-Hurwitz Theorem** Let $V_N$ be the linear space of $N$-hedgehogs $\mathcal{H}_h$ (defined up to a translation in $\mathbb{R}^2$ and identified with their support functions) for which the expansion in a Fourier series of $h(N\theta)$ has the form $h(N\theta) = \sum_{n \ge N}(a_n \cos n\theta + b_n \sin n\theta)$. Let us denote by $\partial^{-1}$ the endomorphism of $V_N$ that assigns to each $N$-hedgehog with support function

$$h(\theta) = \sum_{n \ge N}\left(a_n \cos\frac{n\theta}{N} + b_n \sin\frac{n\theta}{N}\right),$$

the involute (or evolvent) of $\mathcal{H}_h$ that has algebraic length 0:

$$\partial^{-1} : V_N \longrightarrow V_N$$

$$h(\theta) - \sum_{n \ge N}\left(a_n \cos\tfrac{n\theta}{N} + b_n \sin\tfrac{n\theta}{N}\right) \longmapsto \left(\partial^{-1}h\right)(0),$$

where

$$\left(\partial^{-1}h\right)(\theta) = \sum_{n \geq N} \frac{N}{n} \left(a_n \sin \frac{n}{N}\left(\theta + \frac{\pi}{2}\right) - b_n \cos \frac{n}{N}\left(\theta + \frac{\pi}{2}\right)\right).$$

Naturally, we have $\partial\left(\partial^{-1}h\right) = h$ for all $h \in V_N$. Given any $h \in V_\theta$, we define the sequence $(h_n)_{n\in\mathbb{N}}$ of $V_N$ by

$$h_0 = h \quad \text{and} \quad h_{n+1} = \partial^{-1}h_n \text{ for all } n \in \mathbb{N}.$$

Since $\left(\mathcal{P}_{h_n}\right)$ is increasing from Proposition 4.6.1, in order to ensure that $\mathcal{P}_h = \varnothing$ (i.e., that $\mathcal{H}_h$ has no positive area) it suffices to prove that the sequence of hedgehogs $\left(\mathcal{H}_{h_n}\right)_{n\in\mathbb{N}}$ converges to $\{(a_N, b_N)\}$ (i.e., to $\mathcal{H}_{h_\infty}$ where $h_\infty(\theta) = a_N \cos\theta + b_N \sin\theta$) in the sense of the Hausdorff metric. In other words, it suffices to prove that $\|x_h(\theta) - (a_N, b_N)\| = \left\|x_h(\theta) - x_{h_\infty}(\theta)\right\| = \left\|x_{h-h_\infty}(\theta)\right\|$ converges uniformly on $\mathbb{R}$ to 0. To this aim, we first note that

$$(h_n - h_\infty)(\theta) = \sum_{k \geq N+1} \left(\frac{N}{k}\right)^n \left(a_k \cos\theta_{n,k} + b_k \sin\theta_{n,k}\right),$$

where $\theta_{n,k} = \frac{k}{N}\left(\theta + n\frac{\pi}{2}\right) - n\frac{\pi}{2}$. Since $h$ is of class $C^2$, we have $a_k, b_k = O\left(\frac{1}{k^2}\right)$, and we can then deduce that

$$\sup_{\theta\in\mathbb{R}} \left\|x_{(h_n-h_\infty)}(\theta)\right\| \leq 2M \left(\frac{N}{N+1}\right)^{n-1},$$

where $M = \sum_{k \geq N+1} \left(|a_k| + |b_k|\right)$.

This completes the proof.

See below Fig. 4.18 for an illustration of this proof.

**Corollary 4.6.2** *Let $\mathcal{H}_h$ be an $N$-hedgehog of $\mathbb{R}^2$ for which $h(N\theta)$ has the form $h(N\theta) = \sum_{n \geq N} (a_n \cos n\theta + b_n \sin n\theta)$. Then the map $i_h : \mathbb{R}^2\backslash\mathcal{H}_h \to \mathbb{Z}$ is everywhere nonpositive, and it is identically equal to 0 on $\mathbb{R}^2\backslash\mathcal{H}_h$ if and only if $\mathcal{H}_h$ is reduced to a single point. So the opposite $- a$ of the mixed area $a$ is a scalar product on the linear space $V_N$ of all $N$-hedgehogs defined up to a translation in $\mathbb{R}^2$ (this point only means that the harmonic of order $n$ remains unspecified). the support functions of which satisfy the same condition as $h$. Thus for all $(h, k) \in V_N^2$, we have*

$$\sqrt{-a(h+k)} \leq \sqrt{-a(h)} + \sqrt{-a(k)}, \tag{4.6.8}$$

*and*

$$a(h, k)^2 \leq a(h)\,a(k). \tag{4.6.9}$$

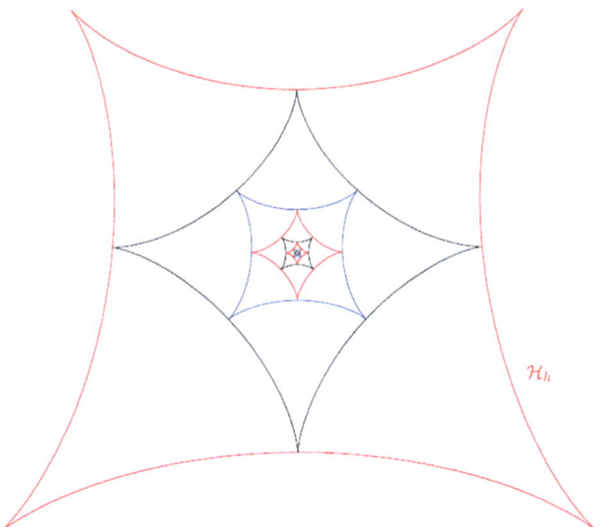

**Fig. 4.18** $\mathcal{H}_h$ and its iterates by $\partial^{-1} : V_N \to V_N$

*Equality holds in (4.6.8) (resp. (4.6.9)) if and only if $\mathcal{H}_h$ and $\mathcal{H}_k$ are linearly dependent in $V_N$.*

***Proof of Corollary 4.6.2*** Taking into account Theorem 4.6.1, Corollary 4.6.2 is an immediate consequence of the multihedgehog version of the Sturm-Hurwitz theorem. ∎

### 4.6.3   Some Immediate Consequences for Hedgehogs

Let us see some immediate consequences of the Sturm-Hurwitz theorem that are of interest for the study of hedgehogs. Let $h : \mathbb{S}^1 = \mathbb{R}/2\pi\mathbb{Z} \to \mathbb{R}$ be the support function of a $C^2$ projective hedgehog of $\mathbb{R}^2$ (i.e., $h$ is $C^2$, and such that $h(\theta + \pi) = -h(\theta)$ for all $\theta \in \mathbb{R}$). Since the curvature function $R_h := h + h''$ of $\mathcal{H}_h$ is then also an odd function on $\mathbb{S}^1$ (i.e., $R_h(\theta + \pi) = -R_h(\theta)$ for all $\theta \in \mathbb{R}$), the Fourier expansion of $R_h$ contains only harmonics of odd order. Since the differential operator $d^2/d\theta^2 + 1$ kills the first harmonics, we then see that the Fourier expansion of $R_h$ starts with the third harmonics. Thus, we have proved the following statement since the singularities of $\mathcal{H}_h$ correspond to the values of $\theta \pmod{2\pi} \in \mathbb{S}^1 = \mathbb{R}/2\pi\mathbb{Z}$ for which $R_h(\theta) = 0$.

**Corollary 4.6.3**  *A $C^2$ projective hedgehog of $\mathbb{R}^2$ has at least six singularities at the source $\mathbb{S}^1$, that is, at least three singularities on the projective line $\mathbb{R}P$ if we identify antipodal points of $\mathbb{S}^1$.*

From Corollary 4.6.3, we can now deduce the following result.

**Corollary 4.6.4** *A plane $C^3$ hedgehog of constant width $2r$, $(r \in \mathbb{R})$, has at least six vertices.*

**Proof of Corollary 4.6.4** The condition that a plane $C^2$ hedgehog $\mathcal{H}_h$ is of constant width $2r$ (i.e., such that the signed distance $h(\theta) + h(\theta + \pi)$ between two parallel support lines is constant, equal to $2r$) is simply that its support function has the form $h = f + r$, where $f$ is the support function of a $C^2$ projective hedgehog. Here we assume that $h$ (and hence $f$) is of class $C^3$.

A *vertex* of $\mathcal{H}_h$ is a point of at least third order tangency of $\mathcal{H}_h$ with a circle, that is, a point where the radius of curvature of $\mathcal{H}_h$ has a stationary value. In other words, the vertices of $\mathcal{H}_h$ correspond to the values of $\theta$ (mod $2\pi$) $\in \mathbb{S}^1 = \mathbb{R}/2\pi\mathbb{Z}$ for which $\left( f' + f''' \right)(\theta) = 0$. Since these values correspond to the singularities of the $C^2$ projective hedgehog with support function $f'$, the theorem follows from the previous corollary. ∎

From Corollary 4.6.3, we can deduce by duality the following particular case of the **classical Möbius theorem**, which states that *any simple noncontractible smooth curve in the projective plane has at least three inflection points.*

**Corollary 4.6.5** *Let $C$ be a closed simple smooth curve of $\mathbb{S}^2$ that is everywhere transverse to the meridians. If $C$ is invariant under the antipodal map then $C$ has at least six inflection points.*

Here, an inflection point is simply a zero of the geodesic curvature, that is, a point of at least second order tangency of the curve with a great circle of $\mathbb{S}^2$.

**Proof of Corollary 4.6.5** By virtue of assumptions, $\mathcal{C}$ admits a parametrization of the form

$$\gamma_h : \mathbb{S}^1 = \mathbb{R}/2\pi\mathbb{Z} \longrightarrow \mathbb{S}^2 \subset \mathbb{R}^2 \times \mathbb{R}$$
$$\theta \,(\mathrm{mod}\, 2\pi) \longmapsto \frac{1}{\sqrt{1+h(\theta)^2}}(\cos\theta, \sin\theta, h(\theta)),$$

where $h : \mathbb{S}^1 \to \mathbb{R}$ is the support function of a $C^2$ projective hedgehog $\mathcal{H}_h$ of $\mathbb{R}^2$, so that $h(\theta + \pi) = -h(\theta)$ for all $\theta \in \mathbb{R}$. If $\theta_0$ is a zero of $h + h''$, we observe that the spherical curve parametrized by $\gamma_g$, where $g(\theta) = h(\theta_0)\cos(\theta - \theta_0) + h'(\theta_0)\sin(\theta - \theta_0)$, is a great circle whose order of tangency with $C$ at $\gamma_h(\theta_0)$ is at least equal to two, since

$$g(\theta_0) = h(\theta_0), \; g'(\theta_0) = h'(\theta_0), \; g''(\theta_0) = -g(\theta_0) = -h(\theta_0) = h''(\theta_0.)$$

Hence, the zeros of $h + h''$ correspond to inflection points of $C$, and the theorem follows from Corollary 4.6.3. ∎

## 4.7  Plane General Hedgehogs and Their Support Functions

The results of the present subsection are essentially extracted from [100]. Note that plane general hedgehogs (i.e., Minkowski differences of arbitrary convex bodies of $\mathbb{R}^2$) do not only constitute a real linear space $(\mathcal{H}^2, +, .)$ but also a commutative and associative $\mathbb{R}$-algebra: see Sect. 10.1 for their convolution. We will start by recalling below why a detailed study of planar hedgehogs is a prerequisite of the theory. We have already seen the usefulness of $C^2$-hedgehogs. Note that planar general hedgehogs have also proven useful for the study of convex bodies. For instance, R. Schneider proved that a typical convex body (in the sense of Baire category) has infinitely many convexity points by proving that the **middle hedgehog** (i.e., the projective part) of a typical convex body has infinitely many exposed points (recall that a point of a convex body $K$ is said to be a convexity point of $K$ if the union of $K$ and its reflection in the point is convex) [168].

### 4.7.1  Construction of Plane General Hedgehogs From Their Support Functions

First, recall that any $C^2$-hedgehog of $\mathbb{R}^{n+1}$ can be regarded as the Minkowski difference $K - L$ of two convex bodies (or of two hypersurfaces by considering their boundaries) $K, L$ of class $C^2_+$ in $\mathbb{R}^{n+1}$ (see Sect. 2.2). Now, let us briefly recall how we proceeded to introduce general hedgehogs (Minkowski differences $K - L$ of two arbitrary convex bodies $K, L \in \mathcal{K}^{n+1}$) in $\mathbb{R}^{n+1}$. Let $\mathcal{K}_{n+1}$ denote the set of support functions of convex bodies of $\mathbb{R}^{n+1}$ and $\mathcal{H}_{n+1}$ be the subspace spanned by $\mathcal{K}_{n+1}$ in the linear space of continuous real functions on $\mathbb{R}^{n+1}$. Note that $\mathcal{H}_{n+1}$ can be identified with the real vector space of formal differences of convex bodies of $\mathbb{R}^{n+1}$. To any $h \in \mathcal{H}_{n+1}$, we associated as follows a geometric realization by induction on $n + 1 \in \mathbb{N}$:

For any $h \in \mathcal{H}_1$, we define the hedgehog with support function $h$ as the oriented segment $\mathcal{H}_h = [-h(-1), h(1)]$. Given $n \geq 1$, we assume that the notion of hedgehog with support function $h \in \mathcal{H}_k$ has already been defined for $k = n$. Then for any $h \in \mathcal{H}_{n+1}$, we associate to each $u \in \mathbb{S}^n$, the hedgehog $H_h^u$ of $u^\perp$ with support function $h'(u; v) = \lim_{t \downarrow 0} [h(u + tv) - h(u)]/t$. We denote by $\mathcal{H}_h$ and call (general) **hedgehog with support function $h$**, the datum of all the support hedgehogs $\mathcal{H}_h^u = \{h(u)u\} + H_h^u$, where $u \in \mathbb{S}^n$.

The above definition makes clear that a good understanding of plane hedgehogs is a prerequisite to a study of general hedgehogs of $\mathbb{R}^{n+1}$. That is why we devote this subsection to case $n + 1 = 2$. The set $\mathcal{K}_2$ of support functions of convex bodies of $\mathbb{R}^2$ spans a subspace $\mathcal{H}_2$ in the linear space of continuous real functions on $\mathbb{R}^2$:

$$\mathcal{H}_2 = \{h_K - h_L \,|(h_K, h_L) \in \mathcal{K}_2 \times \mathcal{K}_2\}.$$

Each $h = h_{\mathcal{K}} - h_{\mathcal{L}} \in \mathcal{H}_2$ is the *support function* of a hedgehog $\mathcal{H}_h$, which can be regarded as the geometrical realization of the formal difference $\mathcal{K} - \mathcal{L}$. This hedgehog $\mathcal{H}_h$ is obtained by associating to each $u \in \mathbb{S}^1$, the oriented segment $\sigma_h(u) = \{h(u)u\} + \left[-h'\left(u; -u^{\top}\right), h'\left(u; u^{\top}\right)\right]$, where $u^{\top} \in \mathbb{S}^1$ is the unit vector such that $\left(u, u^{\top}\right)$ is a direct orthonormal frame of $\mathbb{R}^2$, which we assume equipped with the canonical orientation. See Figs. 2.2 and 2.8 for illustrations in the polygonal case.

The main result of this subsection will be that plane general hedgehogs can always be regarded as rectifiable closed curves having exactly one cooriented support line in each direction. We know that every $h \in C^2\left(\mathbb{S}^1; \mathbb{R}\right)$ defines a hedgehog $\mathcal{H}_h$ of $\mathbb{R}^2$ that can be seen as the envelope of the family of cooriented lines $(L_u)_{u \in \mathbb{S}^1}$ with equation $\langle x, u \rangle = h(u)$. We already noticed that it is still possible to associate such an envelope $\mathcal{H}_h$ to every $h \in C^1\left(\mathbb{S}^1; \mathbb{R}\right)$, but it no longer true that such a hedgehog can always be interpreted as a difference of two plane convex bodies when $h$ is only $C^1$. As we saw in Sect. 2.5, such "hedgehogs" can even be nowhere differentiable fractal curves of infinite length.

**Reminders on Plane Convex Bodies**  Let us recall how the boundary of a convex body $\mathcal{K}$ of $\mathbb{R}^2$ is determined by its support function $h = h_{\mathcal{K}} : \mathbb{R}^2 \to \mathbb{R}$. Its boundary, say $\partial \mathcal{K}$, is constituted of the union of all its support sets $\mathcal{K}_u = \{x \in \mathcal{K} \mid \langle x, u \rangle = h_{\mathcal{K}}(u)\}$, where $u \in \mathbb{S}^1$:

$$\partial \mathcal{K} = \bigcup_{u \in \mathbb{S}^1} \mathcal{K}_u.$$

For every $u \in \mathbb{S}^1$, the convexity of $h : \mathbb{R}^2 \to \mathbb{R}$ implies the existence of

$$h'(u; v) = \lim_{t \downarrow 0} \frac{h(u + tv) - h(u)}{t},$$

for all $v \in \mathbb{R}^2$, and $h'(u; .)$ is the support function of the support set with normal vector $u$, so that

$$\mathcal{K}_u = \left\{x \in \mathbb{R}^2 \,\middle|\, \forall v \in \mathbb{R}^2, \langle x, v \rangle \leq h'(u; v)\right\}.$$

Now, let us see the support function $h$ as a $2\pi$-periodic function $p$ on $\mathbb{R}$: $p(\theta) = h(u(\theta))$, where $u(\theta) = (\cos\theta, \sin\theta)$. The left derivative $p'_l$ and the right derivative $p'_r$ of $p$ are everywhere defined on $\mathbb{R}$, and they are respectively given by $p'_l(\theta) = -h'\left(u(\theta); -u'(\theta)\right)$ and $p'_r(\theta) = h'\left(u(\theta); u'(\theta)\right)$. The calculation gives $h'\left(u\left(\theta\right); \alpha\, u\left(\theta\right) + \beta\, u'\left(\theta\right)\right) = \alpha\, p\left(\theta\right) + \varepsilon\, \beta\, h'\left(u\left(\theta\right); \varepsilon\, u'\left(\theta\right)\right)$, where $\varepsilon = sgn(\beta)$. Thus: $\forall \theta \in J = [0, 2\pi[$,

$$\mathcal{K}_{u(\theta)} = \left\{ x \in \mathbb{R}^2 \,\middle|\, \forall v \in \mathbb{R}^2,\ \langle x, v \rangle \le h'\left(u\left(\theta\right); v\right) \right\}$$

$$= \left\{ x \in \mathbb{R}^2 \,\middle|\, \exists t \in \left[ p'_l\left(\theta\right), p'_r\left(\theta\right) \right],\ x = p\left(\theta\right) u\left(\theta\right) + t\, u'\left(\theta\right) \right\}.$$

Therefore, the boundary of $\mathcal{K}$ is constituted of the union of all the segments $\sigma_h(\theta) = [x_h^-(\theta), x_h^+(\theta)]$, $(\theta \in J = [0, 2\pi[)$, where $x_h^-(\theta) = p(\theta)\, u(\theta) + p'_l(\theta)\, u'(\theta)$ and $x_h^+(\theta) = p(\theta)\, u(\theta) + p'_r(\theta)\, u'(\theta)$:

$$\partial \mathcal{K} = \bigcup_{\theta \in J} \sigma_h\left(\theta\right).$$

Recall that this boundary $\partial \mathcal{K}$ is a rectifiable simple closed curve. For proofs and more details, we refer the reader to [167].

**Plane General Hedgehogs Are Rectifiable Curves**  Let us begin by proving that for any $h \in \mathcal{H}_2$, the segments $\sigma_h(\theta) = [x_h^-(\theta), x_h^+(\theta)]$, $(\theta \in J)$, are well defined and still constitute a rectifiable (but not necessarily simple) closed curve. The proof is based on the following proposition.

**Proposition 4.7.1**  *For all $h \in \mathcal{H}_2$, the following four properties are satisfied:*

$(i)$  *the function $p = h \circ u$, where $u\left(\theta\right) = (\cos\theta, \sin\theta)$, is Lipschitzian on $\mathbb{R}$;*

$(ii)$  *the function $p$ admits a left derivative $p'_l$ (resp., a right derivative $p'_r$) that is continuous from the left (resp., from the right) on $\mathbb{R}$, and we have*

$$\forall \theta \in \mathbb{R},\ p'_l\left(\theta\right) = \lim_{\alpha \to \theta^-} p'_r\left(\alpha\right) \quad and \quad p'_r\left(\theta\right) = \lim_{\alpha \to \theta^+} p'_l\left(\alpha\right);$$

$(iii)$  *the family $\left(\left(p'_r - p'_l\right)\left(\theta\right)\right)_{\theta \in J = [0, 2\pi[}$ is absolutely summable;*

$(iv)$  *the functions $p'_l$ and $p'_r$ are of bounded variation on $I = [0, 2\pi]$.*

***Sketch of the Proof***  Note that it is sufficient to check the result for $h \in \mathcal{K}_2$. Let us continue with the function $h = h_{\mathcal{K}}$ and with the notations we introduced above.

$(i)$  Property $(i)$ follows from the convexity of $h$, cf. [167, Theorem 1.5.3].

$(ii)$  As $h_\theta(t) = h(u(\theta) + t\, u'(\theta))$ is convex on $\mathbb{R}$, $h_\theta$ admits a left derivative $(h_\theta)'_l$ that is continuous from the left and a right derivative $(h_\theta)'_r$ that is continuous from the right and such that $(h_\theta)'_l \le (h_\theta)'_r$, cf. [167, Theorem 1.5.4]. Property $(ii)$ follows by expressing $(h_\theta)'_l$ and $(h_\theta)'_r$ in terms of the functions $p$, $p'_l$ and $p'_r$.

$(iii)$  Property $(iii)$ results from the fact that $\partial K$ is a rectifiable simple closed curve that is constituted of the union of all the segments $\sigma_h(\theta) = [x_h^-(\theta), x_h^+(\theta)]$, $(\theta \in J = [0, 2\pi[)$, the relative interiors of which are pairwise disjoint.

($iv$)  As for every increasing sequence $(\theta_i)_{i\geq 0}$ of $I$, the segments $\sigma_h(\theta_i)$ are placed in the increasing order of subscripts on the anticlockwise oriented curve $\partial K$, it also follows that $x_h^-$ and $x_h^+$ are of bounded variation on $I$. Using ($i$), we then deduce that the maps $p_l'\,u'$ and $p_r'\,u'$ are also of bounded variation on $I$. Noting that the functions $p_l'$ and $p_r'$ are necessarily bounded, we at last deduce Property ($iv$).

∎

Given any $h \in \mathcal{H}_2$, let us consider the union of all the segments $\sigma_h(\theta) = [x_h^-(\theta), x_h^+(\theta)]$, $(\theta \in J = [0, 2\pi[)$. These segments, which are well defined from Property ($ii$), make up the image of the map

$$x_h : D_{x_h} \to \mathbb{R}^2, \ (\theta, t) \longmapsto p(\theta)\,u(\theta) + tu'(\theta),$$

where $u(\theta) = (\cos\theta, \sin\theta)$ and $p(\theta) = h(\cos\theta, \sin\theta)$ for all $\theta \in \mathbb{R}$ and where $D_{x_h} = \left\{(\theta, t) \in J \times \mathbb{R} \,\middle|\, (t - p_l'(\theta))(t - p_r'(\theta)) \leq 0\right\} \cup \left\{(2\pi, p_l'(2\pi))\right\}$. Note that the points $A = (0, p_l'(0))$ and $B = (2\pi, p_l'(2\pi))$ satisfy $x_h(A) = x_h(B)$.

Let us show how $x_h(D_{x_h})$ can be seen as a closed curve of $\mathbb{R}^2$. To this aim, let us equip the set $D_{x_h}$ with the metric defined by

$$d((\theta_1, t_1), (\theta_2, t_2))$$

$$= \begin{cases} |t_1 - t_2| \ \text{if } \theta_1 = \theta_2 \\[2mm] \theta_j - \theta_i + \left|p_r'(\theta_i) - t_i\right| + s(\theta_i, \theta_j) + \left|t_j - p_l'(\theta_j)\right| \ \text{if } \theta_i < \theta_j \end{cases},$$

where $s(\theta_i, \theta_j) = \sum\limits_{\theta_i < \alpha < \theta_j} \left|(p_r' - p_l')(\alpha)\right|$, (see property ($iii$)).

Let us observe that this metric is such that

($i$)  The map $d_h : D_{x_h} \to \mathbb{R}, M \longmapsto d(A, M)$ is an isometry from $D_{x_h}$ onto $I_h = [0, F_h]$, where $F_h = d_h(B) = 2\pi + \sum_{\theta \in J}\left|(p_r' - p_l')(\theta)\right|$ (there is no particular difficulty in proving this point);

($ii$)  The map $x_h$ is continuous on $D_{x_h}$ (reduce this point to the continuity of $D_{x_h} \to \mathbb{R}, (\theta, t) \longmapsto t$ and make use of Property ($ii$)).

This allows us to define the map

$$\gamma_h = x_h \circ d_h^{-1} : I_h \to \mathbb{R}^2, \lambda \longmapsto x_h(\theta(\lambda), t(\lambda)), \ where \ (\theta(\lambda), t(\lambda)) = d_h^{-1}(\lambda),$$

and to assert that it is continuous and such that $\gamma_h(I_h) = x_h(D_{x_h})$. The curve of $\mathbb{R}^2$ that it defines is closed since $x_h(A) = x_h(B)$ and we can state the following.

**Theorem 4.7.1**  *For every $h \in \mathcal{H}_2$, the map $\gamma_h : I_h \to \mathbb{R}^2$ defines a closed curve of $\mathbb{R}^2$ whose geometric realization $\gamma_h(I_h)$ is the union of the segments $\sigma_h(\theta) = \left[x_h^-(\theta), x_h^+(\theta)\right]$, $(\theta \in J = [0, 2\pi[)$.*

**Definition 4.7.1**   For every $h \in \mathcal{H}_2$, the closed curve of $\mathbb{R}^2$ that is defined by $\gamma_h : I_h \to \mathbb{R}^2$ is denoted by $\mathcal{H}_h$ and called hedgehog with support function $h$. Regular parts of $\mathcal{H}_h$ are assumed to be equipped with the transverse orientation for which the unit normal at $\gamma_h(\lambda) = x_h(\theta(\lambda), t(\lambda))$ is $u(\theta(\lambda))$. For $\theta \in J$, the oriented segment $\sigma_h(\theta) = \left[x_h^-(\theta), x_h^+(\theta)\right]$ is called support hedgehog of $\mathcal{H}_h$ in direction $u(\theta)$.

**Theorem 4.7.2**   *For every $h \in \mathcal{H}_2$, the hedgehog $\mathcal{H}_h$ is a rectifiable curve of $\mathbb{R}^2$. Let us denote its length by $L(h)$. This length satisfies the inequality*

$$L(h) \geq \int_0^{F_h} \left\| \gamma_h'(\lambda) \right\| d\lambda,$$

*where $\|.\|$ is the Euclidean norm on $\mathbb{R}^2$.*

**Proof**   Given any partition $\sigma = (\lambda_0, \ldots, \lambda_n)$ of $I_h = [0, F_h]$, let $L(\sigma, h)$ denote the sum

$$\sum_{i=0}^{n-1} \left\| \gamma_h(\lambda_{i+1}) - \gamma_h(\lambda_i) \right\| = \sum_{i=0}^{n-1} \left\| x_h(\theta(\lambda_{i+1}), t(\lambda_{i+1})) - x_h(\theta(\lambda_i), t(\lambda_i)) \right\|.$$

Recall that the length of $\mathcal{H}_h$ can be defined by

$$L(h) = \sup_{\sigma} L(\sigma, h),$$

where the supremum is taken over all partitions of $I_h$. By definition, the curve $\mathcal{H}_h$ is a rectifiable if and only if it has a finite length (which means analytically that the components of $\gamma_h : I_h \to \mathbb{R}^2$ are functions of bounded variation on $I_h$). In this case, the derivative $\gamma_h'$ exists almost everywhere on $I_h$ and the inequality

$$L(h) \geq \int_0^{F_h} \left\| \gamma_h'(\lambda) \right\| d\lambda$$

holds. Consequently, it suffices to prove the existence of some real constant $C$ such that $L(\sigma, h) \leq C$ for every partition $\sigma$ of $I_h$. Put $u(\theta) = (\cos\theta, \sin\theta)$ and $p(\theta) = h(u(\theta))$ for all $\theta \in \mathbb{R}$. Writing $p(\alpha)u(\alpha) - p(\beta)u(\beta)$ under the form $p(\alpha)(u(\alpha) - u(\beta)) + (p(\alpha) - p(\beta))u(\beta)$ and noting that $(\theta(\lambda_i))_{i=0}^n$ is an increasing sequence of $I$, we get at once

$$(A) \qquad \sum_{i=0}^{n-1} \left\| p(\theta(\lambda_{i+1}))u(\theta(\lambda_{i+1})) - p(\theta(\lambda_i))u(\theta(\lambda_i)) \right\|$$

$$\leq 2\pi(p_h + q_h),$$

where $q_h$ is the best Lipschitz constant of $p : \mathbb{R} \to \mathbb{R}$ and $p_h = \sup_{\theta \in I} |p(\theta)|$. Similarly, we get

$$(B) \qquad \sum_{i=0}^{n-1} \left\| t(\lambda_{i+1}) \, u'(\theta(\lambda_{i+1})) - t(\lambda_i) \, u'(\theta(\lambda_i)) \right\|$$

$$\leq 2\pi \, t_h + \sum_{i=0}^{n-1} |t(\lambda_{i+1}) - t(\lambda_i)|,$$

where $t_h = \sup\left(\left\{ |t| \, \big| \, \exists \theta \in I, (\theta, t) \in D_{x_h} \right\}\right)$. Besides, using properties $(iii)$ and $(iv)$ we get

$$s_h := \sup_{\sigma} \sum_{i=0}^{n-1} |t(\lambda_{i+1}) - t(\lambda_i)| < +\infty,$$

where the supremum is taken over all partitions of $I_h$. As $L(\sigma, h)$ is less or equal to the sum of left-hand sides of inequalities $(A)$ and $(B)$, we now conclude that we can take $C = 2\pi \, (p_h + q_h + t_h) + s_h$. $\blacksquare$

**Remarks**

1. For every $h \in \mathcal{H}_2$, let $X_h : [0, L(h)] \to \mathcal{H}_h \subset \mathbb{R}^2$, $s \mapsto X_h(s) = \left( X_h^1(s), X_h^2(s) \right)$ be the parametrization by arclength of $\mathcal{H}_h$ (for which the length of every subarc $X_h : [0, L] \to \mathcal{H}_h$ $(0 \leq L \leq L(h))$ is equal to $L$). This parametrization $X_h = \left( X_h^1, X_h^2 \right)$ is such that $X_h^1$ and $X_h^2$ are absolutely continuous (and thus almost everywhere differentiable) on $[0, L(h)]$ and there exists an increasing map $s \mapsto \theta_s$ of $[0, L(h)]$ into $J$ such that, for almost every $s \in [0, L(h)]$, $X_h(s) \in \sigma_h(\theta_s)$ and $X_h'(s) = \varepsilon_h(s) \, u'(\theta_s)$, where $\varepsilon_h(s) \in \{-1, 1\}$.
2. Note that the length of $\mathcal{H}_h$ can be interpreted in terms of the 1-dimensional (outer) Hausdorff measure $\Lambda_1$ in $\mathbb{R}^2$:

$$L(h) = \int_{\mathbb{R}^2} n_h(x) \, d\Lambda_1(x)$$

where $n_h(x)$ is the number of $\lambda \in [0, F_h]$ such that $\gamma_h(\lambda) = x$; see [135, pp. 125–126].
3. In definitions and results following Proposition 4.7.1, we can replace $\mathcal{H}_2$ by the linear subspace consisting of all functions of $C\left(\mathbb{S}^1; \mathbb{R}\right)$ that satisfy properties $(i)$–$(iv)$. We will see later that this subspace is nothing but $\mathcal{H}_2$.

**Proposition 4.7.2** *For every $h \in \mathcal{H}_2$, Proposition 4.7.1 ensures that $p = h \circ u$ admits a left derivative $p_l'$ and a right derivative $p_r'$ on $\mathbb{R}$. These left and right derivatives of $p$ admit a common derivative at almost every $\theta \in \mathbb{R}$. We will simply denote it by $p''(\theta)$.*

***Proof*** From property $(iv)$, $p'_l$ and $p'_r$ are of bounded variation on $I$, and thus almost everywhere differentiable on $\mathbb{R}$. Now, property $(i)$ ensures that $p$ is Lipschitzian and thus almost everywhere differentiable on $\mathbb{R}$. Therefore, $p'_l$ and $p'_r$ coincide almost everywhere on $\mathbb{R}$, so that their derivatives must also coincide almost everywhere on $\mathbb{R}$. ∎

**Proposition 4.7.3** *For every $h \in \mathcal{K}_2$, the function $p(\theta) = h(u(\theta))$ satisfies the following two properties:*

*(v)* $p'_l(\theta) \leq p'_r(\theta)$ *for all $\theta \in \mathbb{R}$;*
*(vi)* $(p + p'')(\theta) \geq 0$ *for almost every $\theta \in \mathbb{R}$.*

***Proof*** This proposition is an immediate consequence of the following characterization, which is due to M. Kallay [76]: $\forall h \in C(\mathbb{S}^1; \mathbb{R})$,

$$(h \in \mathcal{K}_2) \Longleftrightarrow \left( \forall \theta \in I, \forall \alpha \in \left[0, \frac{\pi}{2}\right], \ p(\theta + \alpha) + p(\theta - \alpha) \geq 2\,p(\theta) \cos \alpha \right),$$

where $p(\theta) = h(u(\theta))$. We just have to observe that:

$$p''(\theta) = \lim_{\alpha \to 0} \frac{p(\theta + \alpha) + p(\theta - \alpha) - 2p(\theta)}{\alpha^2}.$$

∎

As we will see in the next subsubsection (cf. Theorem 4.7.6), properties $(v)$ and $(vi)$ do not characterize support functions of convex bodies in $\mathcal{H}_2$.

### 4.7.2 Length and Area Measures of Plane General Hedgehogs

In the previous subsubsection, we saw that every formal difference of two convex bodies of $\mathbb{R}^2$ can be seen as a (transversely oriented) rectifiable curve, which we called a (general) hedgehog. In the present subsection, we will introduce and study the notions of length measure and mixed area for these hedgehogs of $\mathbb{R}^2$.

Whereas the length measure $L(\mathcal{C}, .)$ of a convex curve $\mathcal{C}$ of $\mathbb{R}^2$ is defined as a (positive) Borel measure on $\mathbb{S}^1$, the length measure of a hedgehog $\mathcal{H}_h$ of $\mathbb{R}^2$ will be defined as a (possibly signed) Borel measure $l_h$ on $\mathbb{S}^1$ in order that the map $\mathcal{H}_h \mapsto l_h$ be linear. This algebraic length measure will of course be interpreted and studied from a geometrical point of view. In the same way, the area of a hedgehog $\mathcal{H}_h$ of $\mathbb{R}^2$ will be defined as an algebraic area in order to extend the mixed area $A : \mathcal{K}^2 \times \mathcal{K}^2 \to \mathbb{R}$, $(\mathcal{K}, \mathcal{L}) \longmapsto A(\mathcal{K}, \mathcal{L})$ to a symmetric bilinear form $a : \mathcal{H}^2 \times \mathcal{H}^2 \to \mathbb{R}$. The area of a general hedgehog $\mathcal{H}_h$ of $\mathbb{R}^2$ will be interpreted as the integral over $\mathbb{R}^2 \setminus \mathcal{H}_h$ of the winding number $i_h(x)$ of $\mathcal{H}_h$ with respect to $x \in \mathbb{R}^2 \setminus \mathcal{H}_h$. We will see in the next subsubsection that the extended bilinear form $a : \mathcal{H}^2 \times \mathcal{H}^2 \to \mathbb{R}$ satisfies a partial extension of the Minkowski inequality $A(\mathcal{K}, \mathcal{L}) \geq$

$A(\mathcal{K}) \cdot A(\mathcal{L})$ which leads to a natural extension of the isoperimetric inequality to plane general hedgehogs. On the way, we will solve the Christoffel-Minkowski problem for plane general hedgehogs by giving a necessary and sufficient condition for a (possibly signed) Borel measure on $\mathbb{S}^1$ to be the length measure of a general hedgehog. Moreover, we will characterize support functions of plane convex bodies among support functions of plane general hedgehogs, and support functions of plane general hedgehogs among continuous functions.

Let us begin by recalling some basic facts concerning the area measure of order 1 of plane convex bodies, that is, the length measure of plane convex curves. We will use notations of the previous subsubsection, and $\mathcal{B}\left(\mathbb{S}^1\right)$ will denote the $\sigma$-algebra of Borel subsets of $\mathbb{S}^1$. The area measure of order 1 of a convex body $\mathcal{K}$ of $\mathbb{R}^2$ (that is, the length measure of its boundary $\partial\mathcal{K}$) is the (positive) Borel measure $S_1(\mathcal{K}, .)$ defined as follows: $(i)$ if $\mathcal{K}$ is contained in a line, then

$$\forall \Omega \in \mathcal{B}\left(\mathbb{S}^1\right), \; S_1(\mathcal{K}, \Omega) := \sum_{u \in \Omega} Length\left[\sigma_{h_K}(u)\right];$$

$(ii)$ if $\mathcal{K}$ is not contained in a line, then

$$\forall \Omega \in \mathcal{B}\left(\mathbb{S}^1\right), \; S_1(\mathcal{K}, \Omega) := \Lambda_1\left[\bigcup_{u \in \Omega} \sigma_{h_K}(u)\right],$$

where $\Lambda_1$ denotes the 1-dimensional (outer) Hausdorff measure in $\mathbb{R}^2$. This area measure of order 1 determines $\mathcal{K}$ up to a translation. More precisely, we have the following existence and uniqueness result for plane convex bodies with prescribed area measure of order 1.

**Theorem 4.7.3 (See e.g. [167, Theorem 8.3.1])** *Let $m$ be a finite positive Borel measure on $\mathbb{S}^1$. If $m$ satisfies*

$$(C) \qquad \int_{\mathbb{S}^1} u\, dm(u) = 0_{\mathbb{R}^2},$$

*then $m$ is the length measure of a unique (up to translations) convex body of $\mathbb{R}^2$.*

Recall that integral condition $(C)$ is necessary from the translation invariance of the area of $\mathcal{K}$. Let us recall the following formula for the perimeter of a plane convex body.

**Theorem 4.7.4 (Barbier 1860 [13])** *Let $\mathcal{K}$ be a convex body of $\mathbb{R}^2$. The perimeter of $\mathcal{K}$, that is, the length $L(\partial\mathcal{K}) := S_1\left(\mathcal{K}, \mathbb{S}^1\right)$ of its boundary $\partial\mathcal{K}$, is given by*

$$L(\partial\mathcal{K}) = \int_0^{2\pi} p(\theta)\, d\theta,$$

*where $p(\theta) = h_{\mathcal{K}}(\cos\theta, \sin\theta)$.*

*Remind that if the restriction of $h = h_{\mathcal{K}}$ to $\mathbb{S}^1$ is of class $C^2$, then :*

*(i) $\partial \mathcal{K}$ can be parametrized by*

$$x_h : \mathbb{S}^1 \to \partial \mathcal{K} \subset \mathbb{R}^2, \ u(\theta) = (\cos\theta, \sin\theta) \longmapsto x_h(\theta)$$
$$= p(\theta) \, u(\theta) + p'(\theta) \, u'(\theta),$$

*where $p(\theta) = h(u(\theta))$;*

*(ii) $x_h$ is of class $C^1$ on $\mathbb{S}^1$ and we have: $\forall \theta \in J$, $x_h'(\theta) = (p + p'')(\theta) \, u'(\theta)$;*

*(iii) $R_h : \mathbb{S}^1 \to \mathbb{R}$, $u(\theta) \longmapsto R_h(\theta) := (p + p'')(\theta)$ is nonnegative, and $R_h(\theta)$ can be interpreted as the (principal) radius of curvature of $\mathcal{H}_h$ at $x_h(\theta)$. Therefore, in this case the length measure of $\partial \mathcal{K}$ is given by*

$$\forall \Omega \in \mathcal{B}\left(\mathbb{S}^1\right), \ S_1(\mathcal{K}, \Omega) = \int_\Omega R_h \, d\sigma,$$

*where $\sigma$ is the circular Lebesgue measure on $\mathbb{S}^1$.*

**Algebraic Length Measure**

**Definition 4.7.2** Let $\mathcal{H}_h$ be a hedgehog of $\mathbb{R}^2$. For every $\Omega \in \mathcal{B}\left(\mathbb{S}^1\right)$, we put

$$l_h(\Omega) := S_1(\mathcal{K}, \Omega) - S_1(\mathcal{L}, \Omega),$$

where $\mathcal{K}$ and $\mathcal{L}$ are two convex bodies of $\mathbb{R}^2$ such that $h = h_{\mathcal{K}} - h_{\mathcal{L}}$, that is, two convex bodies of which $\mathcal{H}_h$ is the difference. As $h_{\mathcal{K}}$ and $S_1(\mathcal{K}, .)$ depend linearly on $\mathcal{K} \in \mathcal{K}^2 \subset \mathcal{H}^2$, this definition does not depend on the choice of $(\mathcal{K}, \mathcal{L}) \in \mathcal{K}^2 \times \mathcal{K}^2$. This signed measure $l_h$ is called the algebraic length measure of $\mathcal{H}_h$. Naturally, if $h$ is the support function of a convex body $\mathcal{K} \in \mathcal{K}^2$, then $l_h(.) = S_1(\mathcal{K}, .)$.

**Definition 4.7.3** Let $\mathcal{H}_h$ be a hedgehog of $\mathbb{R}^2$. The algebraic length of $\mathcal{H}_h$ is defined by $l(h) := l_h(\mathbb{S}^1)$. By the definition of $l_h$, it follows from the above Barbier's theorem that $l(h)$ is given by

$$l(h) = \int_0^{2\pi} p(\theta) \, d\theta,$$

where $p(\theta) = h(\cos\theta, \sin\theta)$.

We will see later in this subsubsection how $l_h$ and $l(h)$ can be interpreted from a geometrical point of view. The following generalization of Theorem 4.7.3 solves **the Christoffel-Minkowski problem** for general hedgehogs of the Euclidean plane.

**Theorem 4.7.5**  *Let $m$ be a (possibly signed) Borel measure on $\mathbb{S}^1$. If $m$ satisfies*

$$\int_{\mathbb{S}^1} u \, dm \, (u) = 0_{\mathbb{R}^2},$$

*then $m$ is the (algebraic) length measure of a unique (up to translations) hedgehog $\mathcal{H}_h$ of $\mathbb{R}^2$.*

***Proof* Existence**  Let $m = m^+ - m^-$ be the Jordan decomposition of $m$:

$$m^+ = \frac{1}{2} \left( |m| + m \right) \text{ and } m^- = \frac{1}{2} \left( |m| - m \right),$$

where $|m|$ is the total variation measure of $m$. From the assumption, there exists some $(a, b) \in \mathbb{R}^2$, such that:

$$\int_{\mathbb{S}^1} u \, dm^+ \, (u) = \int_{\mathbb{S}^1} u \, dm^- \, (u) = (a, b) \,.$$

Let $m_f$ be the Borel measure given by: $\forall \Omega \in \mathcal{B} \left( \mathbb{S}^1 \right)$, $m_f \left( \Omega \right) = \int_\Omega f \, (u) \, d\sigma \, (u)$, where $\sigma$ is the circular Lebesgue measure and $f \, (u \, (\theta)) := c - \frac{1}{\pi} \left( a \cos \theta + b \sin \theta \right)$, $c$ being some constant larger than $\frac{1}{\pi} \sqrt{a^2 + b^2}$. The Borel measures defined by $\mu = m^+ + m_f$ and $\nu = m^- + m_f$ are positive and such that

$$m = \mu - \nu \quad \text{and} \quad \int_{\mathbb{S}^1} u \, d\mu \, (u) = \int_{\mathbb{S}^1} u \, d\nu \, (u) = 0_{\mathbb{R}^2}.$$

Thus, from Theorem 4.7.3, there exists $(\mathcal{K}, \mathcal{L}) \in \left( \mathcal{K}^2 \right)^2$, such that $\mu = S_1 \, (\mathcal{K}, .)$ and $\nu = S_1 \, (\mathcal{L}, .)$, so that

$$m = \mu - \nu = S_1 \, (\mathcal{K}, .) - S_1 \, (\mathcal{L}, .) = l_h,$$

where $h = h_\mathcal{K} - h_\mathcal{L}$.

**Uniqueness up to Translations**  Assume that $\tilde{h} = h_{\widetilde{\mathcal{K}}} - h_{\widetilde{\mathcal{L}}} \in \mathcal{H}_2$ is such that $l_h = l_{\tilde{h}}$, where $\left( \widetilde{\mathcal{K}}, \widetilde{\mathcal{L}} \right) \in \left( \mathcal{K}^2 \right)^2$. We then have

$$S_1 \, (\mathcal{K}, .) - S_1 \, (\mathcal{L}, .) = S_1 \left( \widetilde{\mathcal{K}}, . \right) - S_1 \left( \widetilde{\mathcal{L}}, . \right) \,.$$

As $S_1$ is Minkowski linear, we thus have

$$S_1 \left( \widetilde{\mathcal{K}} + \mathcal{L}, . \right) = S_1 \left( \widetilde{\mathcal{K}}, . \right) + S_1 \, (\mathcal{L}, .) = S_1 \, (\mathcal{K}, .) + S_1 \left( \widetilde{\mathcal{L}}, . \right) = S_1 \left( \mathcal{K} + \widetilde{\mathcal{L}}, . \right).$$

From Theorem 4.7.3, it follows that $\widetilde{\mathcal{K}} + \mathcal{L}$ and $\mathcal{K} + \widetilde{\mathcal{L}}$ are translates of each other, so that $h_{\widetilde{\mathcal{K}}+\mathcal{L}} - h_{\mathcal{K}+\widetilde{\mathcal{L}}}$ is a linear form $\phi$ on $\mathbb{R}^2$. As $h_{\mathcal{K}+\widetilde{\mathcal{L}}} = h_{\mathcal{K}} + h_{\widetilde{\mathcal{L}}}$ and $h_{\widetilde{\mathcal{K}}+\mathcal{L}} = h_{\widetilde{\mathcal{K}}} + h_{\mathcal{L}}$, it follows that $\widetilde{h} = h + \phi$, which completes the proof.

$\blacksquare$

**Curvature Function**

**Definition 4.7.4** Let $\mathcal{H}_h$ be a hedgehog of $\mathbb{R}^2$, and let $\sigma$ denote the circular Lebesgue measure on $\mathbb{S}^1$. From the Lebesgue decomposition theorem, there is a unique pair of mutually singular measures $l_h^a$ and $l_h^s$ such that

$$l_h = l_h^a + l_h^s,$$

where $l_h^a$ is absolutely continuous with respect to $\sigma$ and $l_h^s$ mutually singular with $\sigma$. Furthermore, from the Radon-Nykodym theorem, there is a unique $R_h \in L^1(\sigma)$ such that

$$\forall \Omega \in \mathcal{B}\left(\mathbb{S}^1\right), \; l_h^a(\Omega) = \int_\Omega R_h \, d\sigma.$$

$R_h \in L^1(\sigma)$ is called the **curvature function** of $\mathcal{H}_h$. For the sake of simplicity, we will often consider $R_h$ as a function of $\theta \in \mathbb{R}$ by putting $R_h(\theta) = R_h(u(\theta))$.

Note that this curvature function of $\mathcal{H}_h$ is such that

$$R_h(\theta) = \lim_{\alpha \downarrow 0} \frac{l_h^a([\theta, \theta + \alpha])}{\alpha},$$

for almost every $\theta \in \mathbb{R}$. Of course, if the restriction of $h$ to $\mathbb{S}^1$ is a function of class $C^2$ then $R_h(\theta) = (p + p'')(\theta)$, where $p(\theta) = h(u(\theta))$, for every $\theta \in \mathbb{R}$, and $R_h(\theta)$ can be interpreted as the (principal) radius of curvature of $\mathcal{H}_h$ at $x_h(\theta)$.

**Absolute Length Measure**

**Definition 4.7.5** For every hedgehog $\mathcal{H}_h$ of $\mathbb{R}^2$, let $L_h$ denote the total variation of $l_h$, that is the (positive) Borel measure $|l_h|$ defined by:

$$\forall \Omega \in \mathcal{B}\left(\mathbb{S}^1\right), \; L_h(\Omega) = \sup_{(\Omega_i)_{i \in \mathfrak{I}} \in P(\Omega)} \sum_{i \in \mathfrak{I}} |l_h(\Omega_i)|,$$

where $P(\Omega)$ denotes the set of all partitions $(\Omega_i)_{i \in \mathfrak{I}}$ of $\Omega$. This Borel measure $L_h$ is called the **absolute length measure** of $\mathcal{H}_h$. Naturally, if $h \in \mathcal{K}_2$, then $L_h$ is the length measure $S_1(\mathcal{K}, .)$, where $\mathcal{K}$ is the convex body with support function $h$.

**Remarks**

1. As it will be checked at the end of the subsubsection, for every $h \in \mathcal{H}_2$, $L_h\left(\mathbb{S}^1\right)$ is the (absolute) length $L\left(h\right)$ of the rectifiable curve of $\mathcal{H}_h$.
2. Let $\mathcal{H}_h$ be a hedgehog of $\mathbb{R}^2$ and let $\sigma$ denote the circular Lebesgue measure on $\mathbb{S}^1$. From the Lebesgue decomposition theorem, there is a unique pair of mutually singular measures $L_h^a$ and $L_h^s$ such that

$$L_h = L_h^a + L_h^s,$$

where $L_h^a$ is absolutely continuous with respect to $\sigma$ and $L_h^s$ mutually singular with $\sigma$. It is an easy exercise to check that these measures $L_h^a$ and $L_h^s$ are respectively the total variations $\left|l_h^a\right|$ and $\left|l_h^s\right|$ of $l_h^a$ and $l_h^s$, so that we have in particular,

$$\forall \Omega \in \mathcal{B}\left(\mathbb{S}^1\right), \; L_h^a\left(\Omega\right) = \int_\Omega |R_h| \, d\sigma.$$

**Length Function**

For every $h \in \mathcal{H}_2$, let us define a *length function* $\mathcal{L}_h : I = [0, 2\pi] \to \mathbb{R}$ as follows: for every $\theta \in I$, let $\mathcal{L}_h\left(\theta\right)$ denote the length of the rectifiable curve

$$\gamma_h : \left[0, \lambda^-\left(\theta\right)\right] \to \mathbb{R}^2, \; \lambda \longmapsto x_h\left(\theta\left(\lambda\right), t\left(\lambda\right)\right)$$
$$= p\left(\theta\left(\lambda\right)\right) u\left(\theta\left(\lambda\right)\right) + t\left(\lambda\right) u'\left(\theta\left(\lambda\right)\right),$$

where $\lambda^-\left(\theta\right) = d_h\left(\theta, p_l'\left(\theta\right)\right)$, with notations of the previous subsubsection. In other words, $\mathcal{L}_h\left(\theta\right)$ denotes the length of the subarc of $\mathcal{H}_h$ beginning at $x_h^-\left(0\right)$ and ending at $x_h^-\left(\theta\right)$. In the case where $h = h_K \in \mathcal{K}_2$, we thus have $\mathcal{L}_h\left(\theta\right) = L_h\left(u\left([0, \theta[\right)\right)$, for all $\theta \in I$. This length function $\mathcal{L}_h$ is obviously increasing and thus of bounded variation. Therefore, $\mathcal{L}_h$ admits a decomposition of the form

$$\mathcal{L}_h = j_h + r_h + s_h,$$

where $j_h$ is a jump function (with a derivative equal to $0$ except for an at most countable set of jump discontinuities), $r_h$ an absolutely continuous function, and $s_h$ a continuous singular function (with a derivative equal to $0$ almost everywhere). This decomposition is unique provided these three functions are required to be equal to $0$ at $\theta = 0$.

Let us consider the case when $h = h_K \in \mathcal{K}_2$. In this case, the length measure $L_h : \mathcal{B}\left(\mathbb{S}^1\right) \to \mathbb{R}$ of $\partial \mathcal{K}$ is (inherited from) the Lebesgue-Stieltjes measure, say $\mu_h$, associated with the increasing and left continuous function $\mathcal{L}_h$:

$$\forall \Omega \in \mathcal{B}\left(\mathbb{S}^1\right), \; L_h\left(\Omega\right) = \mu_h\left(\Omega_J\right),$$

where $\Omega_J = \{\theta \in J \,|\, u\left(\theta\right) \in \Omega\}$. Moreover, the unique decomposition of $L_h$ into discrete, absolutely continuous, and continuous singular parts (with respect to the circular Lebesgue measure $\sigma$ on $\mathbb{S}^1$) then corresponds to the decomposition $\mathcal{L}_h = j_h + r_h + s_h$ (in which $j_h$, $r_h$ and $s_h$ are required to be equal to 0 at $\theta = 0$). Let us notice that in this case

(i)  the jump function $j_h$ is given by

$$j_h\left(\theta\right) = \sum_{0 \le \alpha < \theta} \left(p_r'\left(\alpha\right) - p_l'\left(\alpha\right)\right)$$

(remember that $p_r' - p_l' \ge 0$ from Proposition 4.7.3), so that the discrete part of $L_h$ is given by

$$\forall \Omega \in \mathcal{B}\left(\mathbb{S}^1\right), \; J_h\left(\Omega\right) := \sum_{\alpha \in \Omega_J} \left(p_r'\left(\alpha\right) - p_l'\left(\alpha\right)\right);$$

(ii)  the absolutely continuous part of $L_h$ is the measure $L_h^a$, given by

$$\forall \Omega \in \mathcal{B}\left(\mathbb{S}^1\right), \; L_h^a\left(\Omega\right) = \int_\Omega R_h \, d\sigma$$

(note that the curvature function $R_h$ is then $\sigma-$almost everywhere $\ge 0$);

(iii)  the continuous singular part of $L_h$ is a positive measure on $\mathcal{B}\left(\mathbb{S}^1\right)$, so that the continuous singular function $s_h$ is increasing on $I$.


**Characterization of Support Functions of Plane Convex Bodies Among All Support Functions of Plane General Hedgehogs**

Let us prove the following characterization of support functions of plane convex bodies among support functions of plane general hedgehogs.

**Theorem 4.7.6**  *Let $h \in \mathcal{H}_2$. We have $h \in \mathcal{K}_2$ if and only if the following three conditions are satisfied:*

(*i*)  $p_r' - p_l' \ge 0$ *on $I$;*
(*ii*)  $R_h \ge 0$ $\sigma$-*almost everywhere on $\mathbb{S}^1$;*
(*iii*)  *the continuous singular part of the length function $\mathcal{L}_h$ is increasing on $I$.*

**Proof**  It follows from our above study that these three conditions are necessary. Let us check that they are also sufficient. Let us assume that these three conditions are satisfied, and denote by $L_{s_h}$ the (measure inherited from the) Lebesgue-Stieltjes measure associated with the continuous singular part of $\mathcal{L}_h$. Then, the Borel measure

$$l_h : \mathcal{B}\left(\mathbb{S}^1\right) \to \mathbb{R}, \ \Omega \longmapsto \int_\Omega R_h \, d\sigma + \sum_{\alpha \in \Omega_J} \left(p'_r(\alpha) - p'_l(\alpha)\right) + L_{s_h}(\Omega),$$

where $\Omega_J = \{\theta \in J \,|\, u(\theta) \in \Omega\}$, is positive. Moreover, as it is of the form $L_f - L_g$, where $(f, g) \in (\mathcal{K}_2)^2$, it satisfies

$$\int_{\mathbb{S}^1} u \, dm(u) = 0_{\mathbb{R}^2}.$$

Therefore, Theorem 4.7.3 ensures that there exists some $k \in \mathcal{K}_2$ such that $L_k = l_h$ and Theorem 4.7.5 that $\mathcal{H}_h$ and $\mathcal{H}_k$ must be translates, which completes the proof.    ∎

It follows from our study that the perimeter of a convex body $\mathcal{K}$ of $\mathbb{R}^2$ is given by:

$$L(\partial\mathcal{K}) = \int_0^{2\pi} \left(p + p''\right)(\theta) \, d\theta + \sum_{\theta \in J} \left(p'_r(\theta) - p'_l(\theta)\right) + L_{s_h}(J),$$

where $h = h_\mathcal{K}$ and $p(\theta) = h(u(\theta))$, $L_{s_h}$ denoting the Lebesgue-Stieltjes measure associated with the continuous singular part of $\mathcal{L}_h$.

Let us give an explicit example where

$$L(\partial\mathcal{K}) > \int_0^{2\pi} \left(p + p''\right)(\theta) \, d\theta + \sum_{\theta \in J} \left(p'_r(\theta) - p'_l(\theta)\right).$$

To this aim, let us consider the odd function $f : \mathbb{R} \to \mathbb{R}$ that satisfies

$$f(t) = \begin{cases} s(t) & \text{if} \quad 0 \le t \le 1 \\ 1 & \text{if} \quad t \ge 1, \end{cases}$$

where $s$ is the Cantor-Lebesgue function on $[0, 1]$. Now define $h : \mathbb{R}^2 \to \mathbb{R}$ by: $\forall (x, y) \in \mathbb{R}^2$,

$$h(x, y) = \begin{cases} |x| & \text{if} \quad y = 0 \\ |y| \left(1 + \displaystyle\int_1^{\frac{x}{y}} f(t) \, dt\right) & \text{if} \quad y \ne 0. \end{cases}$$

It is then easy to check that $h$ is the support function of a centered convex body $\mathcal{K}$ of $\mathbb{R}^2$ for which the required inequality is satisfied.

## Vector Length Measure

For every $h \in \mathcal{H}_2$, let us define a vector length measure $\overrightarrow{l_h} : \mathcal{B}\left(\mathbb{S}^1\right) \to \mathbb{R}^2$ whose components are (inherited from) the signed Lebesgue-Stieltjes measures associated with the components of the map $x_h^- : I \to \mathbb{R}^2$ (which are left continuous functions of bounded variation). As already noticed, the arclength parametrization $X_h : [0, L\,(h)] \to \mathcal{H}_h \subset \mathbb{R}^2$ is almost everywhere differentiable on $[0, L\,(h)]$, and there exists an increasing map $s \longmapsto \theta_s$ of $[0, L\,(h)]$ into $J$ such that, for almost every $s \in [0, L\,(h)]$, $X_h\,(s) \in \sigma_h\,(\theta_s)$ and $X'_h\,(s) = \varepsilon_h\,(s)\,u'\,(\theta_s)$, where $\varepsilon_h\,(s) \in \{-1, 1\}$. It is easy to check that

$$\forall \Omega \in \mathcal{B}\left(\mathbb{S}^1\right), \ \overrightarrow{l_h}\,(\Omega) = \int_{\Omega_h} X'_h\,(s)\,ds = \int_{\Omega_h} \varepsilon_h\,(s)\,u'\,(\theta_s)\,ds,$$

where $\Omega_h = \{s \in [0, L\,(h)] \,|\, u\,(\theta_s) \in \Omega\}$.

If the restriction of $h$ to $\mathbb{S}^1$ is of class $C^2$, then we have of course

$$\forall \Omega \in \mathcal{B}\left(\mathbb{S}^1\right), \ \overrightarrow{l_h}\,(\Omega) = \int_{\Omega_J} x'_h\,(\theta)\,d\theta = \int_{\Omega_J} R_h\,(\theta)\,u'\,(\theta)\,d\theta$$

where $\Omega_J = \{\theta \in J \,|\, u\,(\theta) \in \Omega\}$.

**Proposition 4.7.4**  *For every* $h \in \mathcal{H}_2$, *we have* $d\overrightarrow{l_h}\,(u) = u^\top\,dl_h\,(u)$, *where* $u^\top \in \mathbb{S}^1$ *is the unit vector such that* $\left(u, u^\top\right)$ *is a direct orthonormal frame of* $\mathbb{R}^2$, *which we assume equipped with the standard orientation.*

**Proof**  By linearity, it suffices to prove it for $h \in \mathcal{K}_2$. Let $l$ be the Lebesgue measure on $[0, L\,(h)]$ and let $u_h : [0, L\,(h)] \to \mathbb{S}^1$ be the measurable map defined by $u_h\,(s) = u\,(\theta_s)$. For $h \in \mathcal{K}_2$, $l_h$ is nothing but the image measure of $l$ by $u_h$, so that:

$$\forall \Omega \in \mathcal{B}\left(\mathbb{S}^1\right), \ \int_\Omega u^\top\,dl_h\,(u) = \int_{\Omega_h} u'\,(\theta_s)\,ds = \overrightarrow{l_h}\,(\Omega),$$

where $\Omega_h = \{s \in [0, L\,(h)] \,|\, u\,(\theta_s) \in \Omega\}$.                         ∎

## Geometrical Interpretation of $l_h$ and $L_h$ for $h \in \mathcal{H}_2$

The following result ensures that the absolute length measure (resp., the algebraic length measure) of $\mathcal{H}_h$ can indeed be interpreted as the length measure of $\mathcal{H}_h$ (resp., the length measure of $\mathcal{H}_h$ counted with the sign of $\varepsilon_h\,(s) = \langle X'_h\,(s)\,,\,u'\,(\theta_s)\rangle$).

**Theorem 4.7.7** *Let $h \in \mathcal{H}_2$. The algebraic length measure of $\mathcal{H}_h$ is given by*

$$\forall \Omega \in \mathcal{B}\left(\mathbb{S}^1\right), \ l_h(\Omega) = \int_{\Omega_h} \varepsilon_h(s) \, ds,$$

*where $\Omega_h = \{s \in [0, L(h)] \,|\, u(\theta_s) \in \Omega\}$ and $\varepsilon_h(s) = sgn\left(\langle X'_h(s), u'(\theta_s)\rangle\right)$.*

***Proof*** From Proposition 4.7.4, we have indeed: $\forall \Omega \in \mathcal{B}\left(\mathbb{S}^1\right)$,

$$l_h(\Omega) = \int_{\Omega} \left\langle u^\top, d\overrightarrow{l_h}(u)\right\rangle$$

$$= \int_{\Omega_h} \left\langle u'(\theta_s), X'_h(s)\right\rangle \, ds$$

$$= \int_{\Omega_h} \varepsilon_h(s) \, ds,$$

where $\langle ., .\rangle$ denotes the standard inner product on $\mathbb{R}^2$. ∎

## Algebraic Area

For every $h \in \mathcal{H}_2$, let us define the algebraic area of $\mathcal{H}_h$ as the integral

$$a(h) = \int_{\mathbb{R}^2 - \mathcal{H}_h} i_h(x) \, d\$(x),$$

where $i_h(x)$ is the winding number of $\mathcal{H}_h$ around $x \in \mathbb{R}^2 \setminus \mathcal{H}_h$ and $\$$ the Lebesgue measure on $\mathbb{R}^2$.

**Theorem 4.7.8** *For every $h \in \mathcal{H}_2$, we have*

$$a(h) = \frac{1}{2} \int_{\mathbb{S}^1} h(u) \, dl_h(u).$$

*The quadratic form $a : \mathcal{H}_2 \to \mathbb{R}$, $h \longmapsto a(h)$ satisfies: $\forall h \in \mathcal{H}_2$,*

$$a(h) = \frac{1}{2} \int_0^{2\pi} \left(p^2 - (p')^2\right)(\theta) \, d\theta,$$

*where $p(\theta) = h(u(\theta))$.*

**_Proof_** Let us define the body of $\mathcal{H}_h$ as the set

$$\mathcal{K}_h = \mathcal{H}_h \cup \left\{ x \in \mathbb{R}^2 - \mathcal{H}_h \,|\, i_h(x) \neq 0 \right\}.$$

Given $m \in \mathbb{R}^2$, let us consider $\mathcal{K}_h$ as a part of the image of $\Delta_h = [0, 1] \times [0, L(h)]$ under the map

$$\begin{aligned} X_h^m : \Delta_h &\longrightarrow \mathbb{R}^2 \\ (r, s) &\longmapsto m + r\left(X_h(s) - m\right) \end{aligned} \cdot$$

We will say that a half-line $L$ with origin $m$ in $\mathbb{R}^2$ is transverse to $\mathcal{H}_h$ if, for every $s \in [0, L(h)]$ such that $X_h(s) \in L$, the vector $X_h'(s)$ exists and is transverse to $L$. Now, almost every $x \in X_h^m(\Delta_h)$ belongs to such a half-line and $i_h(x)$ is then given by:

$$i_h(x) = \sum_{(r,s) \in E_h(x)} i_h(r, s),$$

where $E_h(x) = \left(X_h^m\right)^{-1}(x)$ and $i_h(r, s) = sgn\left[\langle X_h(s) - m, u(\theta_s)\rangle \, \varepsilon_h(s)\right]$, that is, $i_h(r, s) = sgn\left[p_m(\theta_s)\, \varepsilon_h(s)\right]$ where $p_m(\theta) = p(\theta) - \langle m, u(\theta)\rangle$. We thus have

$$\begin{aligned} a(h) &= \int\!\!\int_{\mathbb{R}^2 - \mathcal{H}_h} i_h(x_1, x_2)\, dx_1\, dx_2 \\[6pt] &= \int\!\!\int_{\Delta_h} i_h(r, s)\, \left|\det\left[\tfrac{\partial X_h^m}{\partial r}(r, s), \tfrac{\partial X_h^m}{\partial s}(r, s)\right]\right|\, dr\, ds \\[6pt] &= \int\!\!\int_{\Delta_h} \det\left[X_h(s) - m, r\, X_h'(s)\right]\, dr\, ds \\[6pt] &= \int\!\!\int_{\Delta_h} r\, p_m(\theta_s)\, \varepsilon_h(s)\, dr\, ds \\[6pt] &= \int_0^1 r\, dr \int_0^{L(h)} p_m(\theta_s)\, \varepsilon_h(s)\, ds \\[6pt] &= \tfrac{1}{2} \int_{\mathbb{S}^1} \left(h(u) - \langle m, u\rangle\right)\, dl_h(u) \\[6pt] &= \tfrac{1}{2} \int_{\mathbb{S}^1} h(u)\, dl_h(u), \end{aligned}$$

since $\int_{\mathbb{S}^1} u\, dl_h(u) = 0_{\mathbb{R}^2}$. Therefore, the map $a : \mathcal{H}_2 \to \mathbb{R}$ is a quadratic form.

As the relation

$$a(h) = \frac{1}{2} \int_0^{2\pi} \left( p^2 - (p')^2 \right) (\theta) \, d\theta$$

is well-known for $h \in \mathcal{K}_2$ (see [183, p. 188]), we can thus claim that it remains true for every $h \in \mathcal{H}_2$. ∎

**Corollary 4.7.1** *Let $a : (\mathcal{H}_2)^2 \to \mathbb{R}$, $(h, k) \longmapsto a(h, k)$ be the symmetric bilinear form obtained by polarizing $a : \mathcal{H}_2 \to \mathbb{R}$. For every $(h, k) \in (\mathcal{H}_2)^2$, $a(h, k)$ may be interpreted as an algebraic mixed area of $\mathcal{H}_h$ and $\mathcal{H}_k$, and can be given by*

$$a(h, k) = \frac{1}{2} \int_{\mathbb{S}^1} (h(u) \, dk_h(u) + k(u) \, dh_h(u)) = \frac{1}{2} \int_0^{2\pi} \left( p\,q - p'\,q' \right)(\theta) \, d\theta,$$

*where $p(\theta) = h(u(\theta))$ and $q(\theta) = k(u(\theta))$.*

**Another Way to Prove the Relation**

$$a(h) = \frac{1}{2} \int_0^{2\pi} \left( p^2 - (p')^2 \right)(\theta) \, d\theta$$

is to consider $a(h)$ as the difference of two areas. Indeed, it is easy to check that:

($i$)  the integral

$$\int_0^{2\pi} \frac{p(\theta)^2}{2} \, d\theta$$

can be seen as the area of the pedal curve of $\mathcal{H}_h$ with respect to the origin. Recall that this pedal curve, say $P(\mathcal{H}_h)$, is defined as follows: to each $x_h(\theta, t) \in \mathcal{H}_h$, we assign the foot $P(x_h(\theta, t)) = p(\theta)u(\theta)$ of the perpendicular from the origin to the support line of $\mathcal{H}_h$ at $x_h(\theta, t)$. Naturally, the area $a[P(\mathcal{H}_h)]$ of the pedal curve $P(\mathcal{H}_h)$ is defined by

$$a[P(\mathcal{H}_h)] := \int_{\mathbb{R}^2 - P(\mathcal{H}_h)} i_{P(\mathcal{H}_h)}(x) \, d\$(x) \,,$$

where $i_{P(\mathcal{H}_h)}(x)$ is the winding number of $P(\mathcal{H}_h)$ around $x \in \mathbb{R}^2 \backslash P(\mathcal{H}_h)$ and $\$$ the Lebesgue measure on $\mathbb{R}^2$;

($ii$)  the integral

$$\int_0^{2\pi} \frac{p'(\theta)^2}{2} \, d\theta$$

can be seen as the area of the image of $\Sigma_h = \{(\theta, t) \in D_h \times \mathbb{R} \mid t\,(t - p'(\theta)) < 0\}$, where $D_h = \{\theta \in J \mid p'(\theta) \text{ exists}\}$, under the map

$$T : \Sigma_h \to \mathbb{R}^2$$
$$(\theta, t) \longmapsto p(\theta)\,u(\theta) + tu'(\theta).$$

This area $a\,[T\,(\mathcal{H}_h)]$ is of course defined by

$$a\,[T\,(\mathcal{H}_h)] := \int_{T(\Sigma_h)} t_h(x)\,d\$(x),$$

where $t_h(x) = Card\,(\{(\theta, t) \in \Sigma_h \mid T(\theta, t) = x\})$, and can be given by

$$a\,[T\,(\mathcal{H}_h)] = \iint_{\Sigma_h} \left| \det\left[ \frac{\partial T}{\partial \theta}(\theta, t),\, \frac{\partial T}{\partial t}(\theta, t) \right] \right| d\theta\,dt.$$

To prove the second relation of Theorem 4.7.8, it then suffices to observe that for $\$$-almost every $x \in \mathbb{R}^2 \setminus (\mathcal{H}_h \cup P\,(\mathcal{H}_h))$, we have $i_h(x) = i_{P(\mathcal{H}_h)}(x) - t_h(x)$.

**Characterization of Support Functions of Plane Hedgehogs Among Continuous Functions**

**Theorem 4.7.9** *Functions of $\mathcal{H}_2$ are exactly functions of $C\,(\mathbb{S}^1; \mathbb{R})$ that satisfy properties $(i)$–$(iv)$ of Proposition 4.7.1.*

**Proof** We already know that functions of $\mathcal{H}_2$ satisfy these four conditions. It remains to check that if $h \in C\,(\mathbb{S}^1; \mathbb{R})$ satisfies properties $(i)$–$(iv)$, then $h \in \mathcal{H}_2$. Let $h \in C\,(\mathbb{S}^1; \mathbb{R})$ be such a function. As noticed in the third remark following Theorem 4.7.2, it defines a closed rectifiable curve $\mathcal{H}_h$ in $\mathbb{R}^2$. Let $l_h : \mathcal{B}\,(\mathbb{S}^1) \to \mathbb{R}$ be the signed Borel measure inherited from the Lebesgue-Stieltjes measure associated with the following left continuous function of bounded variation:

$$\mathcal{L}_h^\varepsilon : J = [0, 2\pi[ \,\to \mathbb{R},\, \theta \longmapsto \int_0^{\mathcal{L}_h(\theta)} \varepsilon_h(s)\,ds,$$

where the length function $\mathcal{L}_h(\theta)$ of $\mathcal{H}_h$ is defined as in the case when $h \in \mathcal{H}_2$ and where $\varepsilon_h(s) = sgn\,(\langle X_h'(s),\, u'(\theta_s)\rangle)$ (see remarks of Sect. 4.7.1); that is: $\forall \Omega \in \mathcal{B}\,(\mathbb{S}^1)$,

$$l_h(\Omega) := \mu_{\mathcal{L}_h^\varepsilon}(\Omega_J),$$

where $\mu_{\mathcal{L}_h^\varepsilon}$ is the Lebesgue-Stieltjes measure associated to $\mathcal{L}_h^\varepsilon$ on $J$ and where $\Omega_J = \{\theta \in J \,|\, u(\theta) \in \Omega\}$. This signed Borel measure $l_h : \mathcal{B}(\mathbb{S}^1) \to \mathbb{R}$ satisfies

$$\int_{\mathbb{S}^1} u \, dl_h(u) = 0_{\mathbb{R}^2}.$$

Indeed, we have

$$\int_{\mathbb{S}^1} u^\top \, dl_h(u) = \int_0^{L(h)} \varepsilon_h(s) \, u'(\theta_s) \, ds$$

$$= \int_0^{L(h)} X_h'(s) \, ds$$

$$= X_h(L(h)) - X_h(0) = 0,$$

since components of $X_h$ are absolutely continuous on $[0, L(h)]$, see the third remark following Theorem 4.7.2. Therefore, $l_h$ is the algebraic length measure of a unique (up to translations) hedgehog $\mathcal{H}_f$ of $\mathbb{R}^2$, where $f \in \mathcal{H}_2$. Now, if $\overrightarrow{l_h} : \mathcal{B}(\mathbb{S}^1) \to \mathbb{R}^2$ denotes the vector measure whose components are (inherited from) the signed Lebesgue-Stieltjes measures associated with the components of $x_h^- : I \to \mathbb{R}^2$ (which are left continuous functions of bounded variation), then $\forall \theta \in I$,

$$x_h^-(\theta) - x_h^-(0) = \overrightarrow{l_h}(\Omega_\theta)$$

$$= \int_0^{\mathcal{L}_h(\theta)} X_h'(s) \, ds = \int_0^{\mathcal{L}_h(\theta)} \varepsilon_h(s) \, u'(\theta_s) \, ds$$

$$= \int_{\Omega_\theta} u^\top \, dl_h(u) = \int_{\Omega_\theta} u^\top \, dl_f(u)$$

$$= \overrightarrow{l_f}(\Omega_\theta) = x_f^-(\theta) - x_f^-(0),$$

where $\Omega_\theta = u([0, \theta[)$. So, for all $\theta \in I$, we have $x_h^-(\theta) = x_f^-(\theta) + x$ , where $x = \left(x_h^- - x_f^-\right)(0)$, and thus $h(\theta) = f(\theta) + \langle x, u(\theta)\rangle$, so that $h \in \mathcal{H}_2$. ∎

The following result can be seen as a particular case of Theorem 4.7.5:

**Proposition 4.7.5** *If $\rho : \mathbb{R} \to \mathbb{R}$ is a $2\pi$-periodic function that is summable on $I = [0, 2\pi]$ and such that $\int_0^{2\pi} \rho(\theta)u(\theta)d\theta = 0$, then there exists a plane hedgehog $\mathcal{H}_h$ of $\mathbb{R}^2$ whose curvature function satisfies $R_h(\theta) = \rho(\theta)$ for almost every $\theta \in I$.*

***Proof*** Our proof is a mere adaptation of that of [76, Theorem 4]. Using the characterization of functions in $\mathcal{H}_2$, it consists in proving that

$$p(\theta) = \int_0^\theta \rho(\alpha) \sin(\theta - \alpha) d\alpha,$$

is a $2\pi$-periodic Lipschitzian function on $\mathbb{R}$ which admits an absolutely continuous derivative on $I$ that satisfies $\left(p + p''\right)(\theta) = \rho(\theta)$ for almost every $\theta \in I$. A first calculation shows that $p$ admits an absolutely continuous derivative on $I$, namely

$$p'(\theta) = \int_0^\theta \rho(\alpha) \cos(\theta - \alpha) d\alpha,$$

and a second one that $\left(p + p''\right)(\theta) = \rho(\theta)$ for almost every $\theta \in I$. The integral condition $\int_0^{2\pi} \rho(\theta) u(\theta) d\theta = 0$ ensures that $p$ is $2\pi$-periodic on $\mathbb{R}$.  ∎

Of course, Proposition 4.7.5 gives only an existence result: there is no uniqueness. For instance, if $\mathcal{H}_h$ in $\mathbb{R}^2$ is any polygonal hedgehog, then $p(\theta) = h(u(\theta))$ satisfies $\left(p + p''\right)(\theta) = 0$ for almost every $\theta \in I$.

### 4.7.3  Geometric Inequalities for General Plane Hedgehogs

The following theorem gives an extension to general hedgehogs of the classical Minkowski inequality (and thus of the isoperimetric inequality) for plane convex bodies:

**Theorem 4.7.10**

(i)  *If $h \in \mathcal{H}_2$ is such that $l(h) = 0$, then $a(h) \le 0$.*
(ii)  *If $(f, g) \in (\mathcal{H}_2)^2$ is such that $a(g) > 0$, then $a(f, g)^2 \ge a(f).a(g)$.*

*In particular:* $\forall h \in \mathcal{H}_2, 4\pi a(h) \le l(h)^2.$

***Proof***

(i)  If the restriction of $h$ to $\mathbb{S}^1$ is a function of class $C^2$, this is only Wirtinger's inequality. If $h = h_\mathcal{K} - h_\mathcal{L}$, where $(\mathcal{K}, \mathcal{L}) \in \mathcal{K}^2 \times \mathcal{K}^2$, then we can proceed as follows. We know there exist sequences $(\mathcal{K}_n)$ and $(\mathcal{L}_n)$ of plane convex bodies of class $C_+^2$ that converge respectively towards $\mathcal{K}$ and $\mathcal{L}$ with respect to the Hausdorff metric on $\mathcal{K}^2$ [166]. Recall that convergence of plane convex bodies with respect to the Hausdorff metric is equivalent to uniform convergence on $\mathbb{S}^1$ of the corresponding support functions [167, p. 66]. Let $(h_n)$ be the sequence defined by: $\forall n \in \mathbb{N}, h_n = f_n - \frac{1}{2\pi} l(f_n)$, where $f_n = h_{\mathcal{K}_n} - h_{\mathcal{L}_n}$. Using the assumption $l(h) - 0$, we check at once that $(h_n)$ converges uniformly towards $h$ on $\mathbb{S}^1$. Now we have $(\forall n \in \mathbb{N}, l(h_n) = 0)$, and thus $a(h_n) \le 0$

from Wirtinger's inequality. Using the bilinearity of the algebraic mixed area and the continuity of the mixed area on $\mathcal{K}^2 \times \mathcal{K}^2$, we deduce that $a\,(h) = \lim_{n \to +\infty} a\,(h_n) \leq 0$.

$(ii)$  Since $a\,(g) > 0$, we have $l\,(g) \neq 0$. Let $\tau$ be the trinomial defined on $\mathbb{R}$ by

$$\tau(t) := a(f + tg) = a(f) + 2t\,a(f, g) + t^2 a(g).$$

If $a(f, g)^2 < a(f)\,a(g)$, then $\tau(t)$ has no real root, and the assumption $a(g) > 0$ implies that $\tau\,(t) > 0$ for all $t \in \mathbb{R}$. Now, this is impossible since there exists $\lambda \in \mathbb{R}$ such that $l(f + \lambda g) = l\,(f) + \lambda l\,(g) = 0$, which implies $\tau(\lambda) \leq 0$ from $(i)$.

$\blacksquare$

# Chapter 5
# The Minkowski Problem for Hedgehogs

The classical Minkowski problem asks for necessary and sufficient conditions on a nonnegative Borel measure on the unit sphere $\mathbb{S}^n$ of $\mathbb{R}^{n+1}$ to be the surface area measure of some convex body $K$ in $\mathbb{R}^{n+1}$, unique up to translation. When restricting to the class of convex bodies of $\mathbb{R}^{n+1}$ whose surface area measures have a density with respect to spherical Lebesgue measure on $\mathbb{S}^n$, the classical Minkowski problem can be formulated as that of the existence, uniqueness and regularity of a closed convex hypersurface with preassigned curvature function (see e.g., [167, p. 397]. for more details, results and a complete bibliography).

In the last century, this Minkowski problem played an important role in the development of the theory of elliptic Monge-Ampère equations. Indeed, for $C^2_+$-hypersurfaces ($C^2$-hypersurfaces with positive Gauss curvature), this problem is equivalent to the question of solutions of certain Monge-Ampère equations of elliptic type on the unit sphere $\mathbb{S}^n$ of $\mathbb{R}^{n+1}$ [31, 134, 148]. In his book review of the Minkowski multidimensional problem, by A.V. Pogorelov, E. Calabi wrote about the Minkowski problem: *"From the geometric view point it is the Rosetta Stone, from which several other related problems can be solved"*. Some, included in Pogorelov's book, deal with the determination of closed, convex hypersurfaces by the data of other curvature functions expressed in terms of the unit normal; other generalizations deal with curvature functions expressed in terms of both the normal and the position of the point" [26]. Various analogues of the classical Minkowski problem, in which the volume and the surface area measure are replaced by a physical quantity and a measure derived from its first variations, have been investigated (see [167, 5. Applications, p. 462]).

This classical Minkowski problem admits a natural extension to hedgehogs of $\mathbb{R}^{n+1}$ to which the present section is devoted. This extended Minkowski problem is much more difficult than the classical one, even for $n = 2$ and for $C^\infty$-hedgehogs, since it essentially boils down to the question of solutions of Monge-Ampère equations of mixed type on the unit sphere $\mathbb{S}^n$ of $\mathbb{R}^{n+1}$, a class of equations

© The Author(s), under exclusive license to Springer Nature Switzerland AG 2026
Y. Martinez-Maure, *Hedgehog Theory*, Lecture Notes in Mathematics 2381,
https://doi.org/10.1007/978-3-032-04298-9_5

for which there is no global result but only local ones by Lin [84] and Zuily [199]. However, as we will see, a number of partial results have already been obtained.

## 5.1 Minkowski's Problem in Differential Geometry and Extension to Hedgehogs

In this subsection, we follow more or less [103]. If $K$ is a convex body of class $C_+^2$ in $\mathbb{R}^{n+1}$, then its surface area measure has a continuous density given by its curvature function $R$, which is the product of the principal radii of curvature of its boundary $\partial K$ (oriented by its outer normal). Its reciprocal $\kappa = 1/R$ is thus the Gauss curvature regarded as a function of the outer normal. In this context, the classical Minkowski problem can therefore be formulated as that of existence, uniqueness and regularity of a closed convex hypersurface whose curvature function is prescribed as a positive function on $\mathbb{S}^n$. In other words, given a continuous positive function $\kappa$ on $\mathbb{S}^n$, is there a closed convex hypersurface in $\mathbb{R}^{n+1}$ with Gauss curvature $\kappa$ (as a function of the outer unit normal vector)? If it does exist, is it unique? And what then can we say about its regularity and that of its support function? Using approximation by convex polyhedra, Minkowski proved the existence of a weak solution in 1903 [126]: If $\kappa$ is a continuous positive function on $\mathbb{S}^n$ satisfying the following integral condition

$$\int_{\mathbb{S}^n} \frac{u}{\kappa(u)} \, d\sigma(u) = 0,$$

where $\sigma$ is the spherical Lebesgue measure, then $\kappa$ is the Gauss curvature (in the sense of Gauss' definition) of a unique (up to translation) closed convex hypersurface $\mathcal{H}_h$. The uniqueness comes from the equality condition in a Minkowski's inequality (e.g., [167, p. 397]). The strong solution is due to Pogorelov [148] and Cheng and Yau [31] who proved independently that: if $\kappa$ is of class $C^m$, $(m \geq 3)$, then the support function of $\mathcal{H}$ is of class $C^{m+1,\alpha}$ for every $\alpha \in \,]0, 1[$.

This classical Minkowski problem has a natural extension to hedgehogs, that is to Minkowski differences $\mathcal{H} = \mathcal{K} - \mathcal{L}$ of closed convex hypersurfaces $\mathcal{K}, \mathcal{L} \in \mathbb{R}^{n+1}$. This is obviously the case if we restrict ourselves to hedgehogs whose support functions are $C^2$. Indeed, as noticed in Sect. 2.2.1, the inverse of the Gaussian curvature of such a hedgehog is well defined and continuous all over $\mathbb{S}^n$, so that the following existence question arises naturally:

($Q_1$) *Existence of a $C^2$-solution: What are necessary and sufficient conditions for a real continuous function $R$ on $\mathbb{S}^n$ to be the curvature function (that is, the inverse $\frac{1}{\kappa}$ of the Gauss curvature $\kappa$) of some $C^2$-hedgehog $\mathcal{H} = \mathcal{K} - \mathcal{L}$?*

Now let us expound the uniqueness question. As we will see later, for any $h \in C^2(\mathbb{S}^2; \mathbb{R})$, $-h$ and $h$ are the respective support functions of two hedgehogs $\mathcal{H}_{-h}$

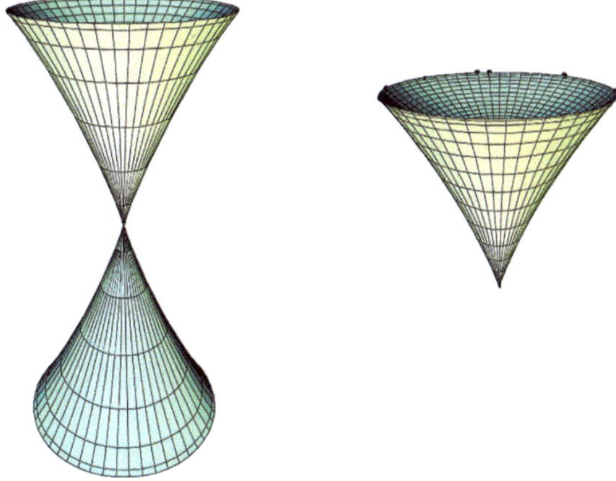

**Fig. 5.1** Noncongruent hedgehogs with the same curvature function

and $\mathcal{H}_h$ of $\mathbb{R}^3$ that have the same curvature function and are such that

$$\mathcal{H}_{-h} = s\left(\mathcal{H}_h\right),$$

where $s$ is the symmetry with respect to the origin of $\mathbb{R}^3$. Here, we have to recall that noncongruent hedgehogs of $\mathbb{R}^3$ may have the same curvature function [102]: for instance, the two smooth (but not analytic) functions $f$, $g$ defined on $\mathbb{S}^2$ by

$$f(u) := \begin{cases} \exp(-1/z^2) \text{ if } z \neq 0 \\[2mm] 0 \quad \text{ if } \quad z = 0 \end{cases} \quad \text{and} \quad g(u) := \begin{cases} sgn(z)\, f(u) \text{ if } z \neq 0 \\[2mm] 0 \quad \text{ if } \quad z = 0, \end{cases}$$

where $u = (x, y, z) \in \mathbb{S}^2$, are the support functions of two noncongruent hedgehogs $\mathcal{H}_f$ and $\mathcal{H}_g$ of $\mathbb{R}^3$ having the same curvature function $R := 1/\kappa \in C\left(\mathbb{S}^2; \mathbb{R}\right)$, (cf. Fig. 5.1).

Consequently, we state the uniqueness question as follows:

($Q_2$) ***Uniqueness of a $C^2$ -solution***: *If $R \in C\left(\mathbb{S}^n; R\right)$ is the curvature function of some hedgehog $\mathcal{H}$, what necessary and sufficient additional conditions must it satisfy in order that $\mathcal{H}$ be the unique hedgehog of which $R$ is the curvature function (up to an isometry of the space)?*

In particular, it would be very interesting to know whether there exists any pair of noncongruent analytic hedgehogs of $\mathbb{R}^3$ with the same curvature function (by 'analytic hedgehogs' we mean 'hedgehogs with an analytic support function'). We will see in Sect. 5.1.1 that this latest question presents similarities to the open question of knowing whether there exists any pair of noncongruent isometric analytic closed surfaces in $\mathbb{R}^3$.

For $n = 1$, the problem is linear and so can be solved without difficulty: Theorem 4.6 of [100] entirely solved the Christoffel-Minkowski for general plane hedgehogs.

But for $n = 2$, the problem is already very difficult: if $R \in C\left(\mathbb{S}^2; \mathbb{R}\right)$ changes sign on $\mathbb{S}^2$, the question of existence, uniqueness and regularity of a hedgehog of which $R$ is the curvature function boils down to the study of a Monge-Ampère equation of mixed type, a class of equations for which there is no global result but only local ones by Lin [84] and Zuily [199]. In the present subsection, we are mainly interested in the uniqueness question. Question $(Q_2)$ is too difficult to be solved at the present time and our main purpose will be simply to provide conditions under which two hedgehogs of $\mathbb{R}^3$ have distinct curvature functions.

**Our Results**

Let $H_3$ be the linear space of $C^2$-hedgehogs defined up to a translation in $\mathbb{R}^{n+1}$ and identified with their support function (by '$C^2$-hedgehogs' we thus mean 'hedgehogs with a $C^2$ support function'). Our first result will be the following.

**Theorem 5.1.1** *Let $\mathcal{H}$ and $\mathcal{H}'$ be $C^2$-hedgehogs that are linearly independent in $H_3$. If some linear combination of $\mathcal{H}$ and $\mathcal{H}'$ is of class $C_+^2$, then $\mathcal{H}$ and $\mathcal{H}'$ have distinct curvature functions.*

As we will recall in Sect. 5.1.2, any hedgehog can be uniquely split into the sum of its centered and projective parts. Our second result relies on this decomposition of hedgehogs into their centered and projective parts.

**Theorem 5.1.2** *Let $\mathcal{H}$ and $\mathcal{H}'$ be $C^2$-hedgehogs that are linearly independent in $H_3$ and whose centered parts are nontrivial (i.e., distinct from a point) and proportional to one and the same convex surface of class $C_+^2$. Then $\mathcal{H}$ and $\mathcal{H}'$ have distinct curvature functions.*

An immediate consequence will be that:

**Corollary 5.1.1** *If two $C^2$-hedgehogs of nonzero constant width are linearly independent in $H_3$, then their curvature functions are distinct.*

Our last result relies on the extension to hedgehogs of the notion of mixed curvature function, which we recalled in Sect. 3.1.1.

**Theorem 5.1.3** *Let $\mathcal{H}$ and $\mathcal{H}'$ be analytic (resp. projective $C^2$) hedgehogs of $\mathbb{R}^3$ that are linearly independent in $H_3$. If the mixed curvature function of $\mathcal{H}$ and $\mathcal{H}'$ does not change sign on $\mathbb{S}^2$, then $\mathcal{H}$ and $\mathcal{H}'$ have distinct curvature functions.*

In Sect. 5.1.1, we will begin by recalling what we have already seen on the Minkowski problem extended to hedgehogs. Later, we will see different ways of constructing pairs of non-congruent hedgehogs having the same curvature function.

### 5.1.1 Facts and Observations on the Minkowski Problem Extended to Hedgehogs

#### Gaussian Curvature of $C^2$-Hedgehogs

Let $H_{n+1}$ denote the linear space of $C^2$-hedgehogs defined up to a translation in the Euclidean linear space $\mathbb{R}^{n+1}$ and identified with their support functions. Analytically speaking, saying that a hedgehog $\mathcal{H}_h$ of $\mathbb{R}^{n+1}$ is defined up to a translation simply means that the first spherical harmonics of its support function is not specified.

As we saw before, elements of $H_{n+1}$ may be singular hypersurfaces. Since the parametrization $x_h$ can be regarded as the inverse of the Gauss map, the Gaussian curvature $\kappa_h$ of $\mathcal{H}_h$ at $x_h(u)$ is given by $\kappa_h(u) = 1/\det[T_u x_h]$, where $T_u x_h$ is the tangent map of $x_h$ at $u$. Therefore, singularities are the very points at which the Gaussian curvature is infinite. For every $u \in \mathbb{S}^n$, the tangent map of $x_h$ at the point $u$ is $T_u x_h = h(u)\, Id_{T_u \mathbb{S}^n} + H_h(u)$, where $H_h(u)$ is the symmetric endomorphism associated with the Hessian $\left(\nabla^2 h\right)_u$ of $h$ at $u$. Consequently, if $\lambda_1(u) \leq \ldots \leq \lambda_n(u)$ are the eigenvalues of the Hessian $\left(\nabla^2 h\right)_u$ of $h$ at $u$ then

$$R_1(u) := (\lambda_1 + h)(u) \leq \ldots \leq R_n(u) := (\lambda_n + h)(u)$$

can be interpreted as the principal radii of curvature of $\mathcal{H}_h$ at $x_h(u)$, and the so-called *curvature function* $R_h(u) := 1/\kappa_h(u) = \det[T_u x_h]$ is given for all $u \in \mathbb{S}^n$ by

$$\begin{aligned}
R_h(u) &= \det\left[h(u)\, Id_{T_u \mathbb{S}^n} + H_h(u)\right] \\
&= \det\left[H_{ij}(u) + h(u)\, \delta_{ij}\right] \\
&= (R_1 \ldots R_n)(u)
\end{aligned}$$

where $\delta_{ij}$ are the Kronecker symbols and $\left(H_{ij}(u)\right)$ the Hessian of $h$ at $u$ with respect to an orthonormal frame on $\mathbb{S}^n$. In this subsection, we are mainly interested in the case where $n = 2$.

#### The Case $n = 2$

From the above relations, the curvature function $R_h := 1/\kappa_h$ of $\mathcal{H}_h \in H_3$ is given by $R_h = (\lambda_1 + h)(\lambda_2 + h) = h^2 + h\Delta_2 h + \Delta_{22} h$, where $\Delta_2$ denotes the spherical Laplacian and $\Delta_{22}$ the Monge-Ampère operator (respectively the sum and the product of the eigenvalues $\lambda_1$, $\lambda_2$ of the Hessian of $h$). So, the equation we will be dealing with will be the following

$$h^2 + h\Delta_2 h + \Delta_{22} h = 1/\kappa.$$

Recall that the so-called 'mixed curvature function' of hedgehogs of $\mathbb{R}^3$, that is,

$$R : H_3^2 \to C\left(\mathbb{S}^2; \mathbb{R}\right)$$
$$(f, g) \mapsto R_{(f,g)} := \tfrac{1}{2}\left(R_{f+g} - R_f - R_g\right)$$

is bilinear and symmetric:

(i)   $\forall (f, g, h) \in H_3^3, \forall \lambda \in \mathbb{R}, R_{(f+\lambda g, h)} = R_{(f,h)} + \lambda R_{(g,h)}$;
(ii)   $\forall (f, g) \in H_3^2, R_{(g,f)} = R_{(f,g)}$.

For any $h \in H_3$, we have in particular $R_{-h} = R_h$. Note that $R_{(1,h)} = \tfrac{1}{2}\left(\Delta_2 h + 2h\right)$ is (up to the sign) half the sum of the principal radii of curvature of $\mathcal{H}_h \in H_3$.


**Nonexistence in the Minkowski Problem for Hedgehogs**

The point is that the curvature function $R_h := 1/\kappa_h$ of any $C^2$-hedgehog $\mathcal{H}_h$ of $\mathbb{R}^{n+1}$ is well defined and continuous all over $\mathbb{S}^n$, including at the singular points of $x_h$, so that the Minkowski problem arises naturally for hedgehogs. In this paper, we are thus interested in studying the existence and/or uniqueness of $C^2$-solutions to the Monge-Ampère equation

$$R_h = R,$$

where $R$ is a given real continuous function on $\mathbb{S}^n$.

As in the classical Minkowski problem, the following integral condition is necessary for the existence of such a solution:

$$\int_{\mathbb{S}^n} R(u)\, u d\sigma(u) = 0. \tag{5.1.1}$$

It simply expresses that any $C^2$-hedgehog $\mathcal{H}_h$ of $\mathbb{R}^{n+1}$ is a closed hypersurface.

But it is no longer sufficient: for instance, the constant function equal to $-1$ on $\mathbb{S}^2$ satisfies integral condition (5.1.1) but it cannot be the curvature function of a hedgehog since there is no compact surface with negative Gaussian curvature in $\mathbb{R}^3$.

The study of the Minkowski problem extended to hedgehogs leads to the construction of families of examples of Monge-Ampère equations of mixed type for which there is no solution. We have already proved that for every $v \in \mathbb{S}^2$, the smooth function $F_v(u) = 1 - 2\langle u, v\rangle^2$ satisfies integral condition (5.1.1) but is not a curvature function on $\mathbb{S}^2$ although it changes sign cleanly on $\mathbb{S}^2$ (see Proposition 4.3.6 as well as is proof). In other words, for every fixed $v \in \mathbb{S}^2$, the Monge-Ampère equation $h^2 + h\Delta_2 h + \Delta_{22} h = F_v$ has no $C^2$-solution on $\mathbb{S}^2$. Remind that the proof makes use of orthogonal projection techniques adapted to hedgehogs.

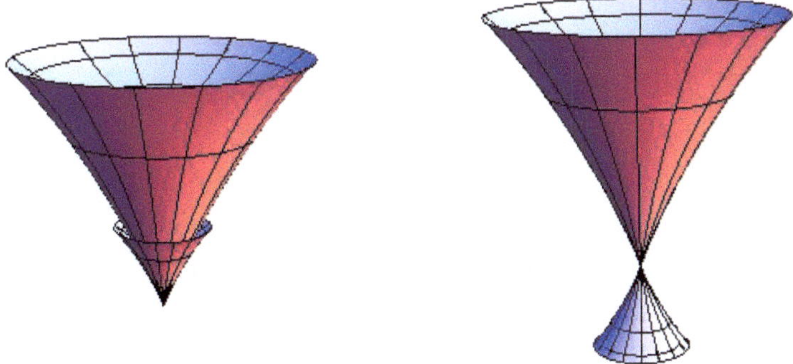

**Fig. 5.2** Noncongruent hedgehogs with the same curvature function

## Nonuniqueness in the Minkowski Problem for Hedgehogs

As recalled in the introduction, two noncongruent hedgehogs of $\mathbb{R}^3$ may have the same curvature function. By bilinearity and symmetry in the arguments of the mixed curvature function $R : H_3^2 \to C\left(\mathbb{S}^2; \mathbb{R}\right)$, if $\mathcal{H}_f$ and $\mathcal{H}_g$ are two hedgehogs of $\mathbb{R}^3$ having the same curvature function then, for all $(\lambda, \mu) \in \mathbb{R}^2$, the hedgehogs $\mathcal{H}_{\lambda f + \mu g}$ and $\mathcal{H}_{\mu f + \lambda g}$ also have the same curvature function. For instance, from the pair $\left\{\mathcal{H}_f, \mathcal{H}_g\right\}$ of noncongruent hedgehogs represented in Fig. 5.1, we deduce the pair $\left\{\mathcal{H}_{f+2g}, \mathcal{H}_{2f+g}\right\}$ of noncongruent hedgehogs (which have the same curvature function) represented in Fig. 5.2.

## On $\varepsilon_h$ Functions and the Non-uniqueness in the Minkowski Problem

In Sect. 2.9, we associate to any $C^2$-hedgehog $\mathcal{H}_h$ of $\mathbb{R}^3$ a function $\varepsilon_h : \mathbb{S}^2 \to \{-1, 0, 1\}$ whose sign at any regular point $u$ of $x_h : \mathbb{S}^2 \to \mathbb{R}^3$ tells us if the usual (i.e., relative) and the absolute transverse orientations of $\mathcal{H}_h$ coincide or not at $x_h(u)$. We proved that this function $\varepsilon_h : \mathbb{S}^2 \to \{-1, 0, 1\}$ is such that

$$\int_{\mathbb{S}^2} \varepsilon_h(u) R_h(u) u \, d\sigma(u) = 0,$$

where $\sigma$ is the spherical Lebesgue measure on $\mathbb{S}^2$ and $R_h$ the curvature function of $\mathcal{H}_h$. Now let us return to the example of the hedgehogs $\mathcal{H}_f$ and $\mathcal{H}_g$ of $\mathbb{R}^3$ that have the same curvature function: $f(u) = \exp\left(-1/z^2\right)$ and $g(u) = sgn(z) f(u)$, $u = (x, y, z) \in \mathbb{S}^2 \subset \mathbb{R}^3$ and $z \neq 0$ (see Fig. 5.1). Note that $\mathcal{H}_f$ is centrally symmetric whereas $\mathcal{H}_g$ is projective. It is interesting to notice that $\mathcal{H}_f$ and $\mathcal{H}_g$ correspond to different $\varepsilon_h$ functions. More precisely, we have $\varepsilon_f(u) = -1$ and $\varepsilon_g(u) = -sgn(z)$ for all $u = (x, y, z) \in \mathbb{S}^2$ such that $z \neq 0$. Similarly, if two hedgehogs are bounding the same centrally symmetric convex body $K$ of $\mathbb{R}^3$ but

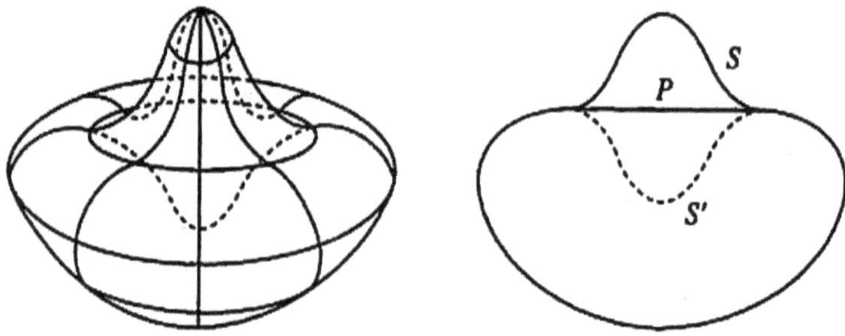

**Fig. 5.3** Noncongruent isometric surfaces of revolution [16, p. 131]

are equipped with opposite (usual) transverse orientations, then they have the same curvature function but opposite $\varepsilon_h$ functions. These examples suggest that a study of the multiplicity of solutions in the Minkowski problem for hedgehogs should probably take into account these $\varepsilon_h$ functions.

**The Analytical Case**  A natural but probably difficult question is knowing whether there exists a pair of noncongruent analytic hedgehogs of $\mathbb{R}^3$ with the same curvature function. Let us recall the similar open question of knowing whether there exists a pair of noncongruent isometric analytic closed surfaces in $\mathbb{R}^3$. Smooth closed surfaces can be isometric without being congruent: the usual way of constructing such surfaces is by gluing together smaller congruent pieces. As recalled in [16, p. 131] or [177, p. 366], we can construct a pair $\{S, S'\}$ of noncongruent isometric closed surfaces of revolution as indicated in Fig. 5.3.

We can assume that $S$ admits a parametrization of the form

$$x : \mathbb{S}^2 \to S \subset \mathbb{R}^3$$
$$u \mapsto \rho(u)\, u,$$

where $\rho$ is a smooth positive function. Then the hedgehog with support function $h = 1/\rho$ can be regarded as the dual surface of $S$ (see Sect. 2.7). This hedgehog $\mathcal{H}_h$ is a surface of revolution whose generating curve (a plane hedgehog which has a fish form) is represented in Fig. 5.4. Replacing the fish's tail by its image under the symmetry with respect to the double point (which by duality corresponds to the plane $P$) and rotating the plane hedgehog that we get around its axis of symmetry, we generate another hedgehog which has the same curvature function as $\mathcal{H}_h$ without being congruent to it.

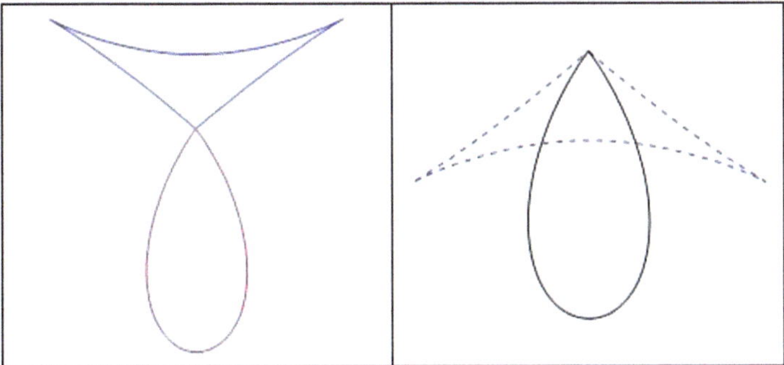

**Fig. 5.4** Generatrices of revolution hedgehogs with the same curvature


## *5.1.2   Statements of Results*

Recall that $H_3$ denotes the $\mathbb{R}$-linear space of $C^2$-hedgehogs defined up to a translation in $\mathbb{R}^3$. Our first result below will be a consequence of the classical Minkowski's uniqueness theorem.

**Theorem 5.1.1** *Let $\mathcal{H}_f$ and $\mathcal{H}_g$ be $C^2$-hedgehogs that are linearly independent in $H_3$. If some linear combination of $\mathcal{H}_f$ and $\mathcal{H}_g$ is of class $C^2_+$, then $\mathcal{H}_f$ and $\mathcal{H}_g$ have distinct curvature functions.*

Our second result makes use of the decomposition of hedgehogs into their centered and projective parts.


**Decomposition of a Hedgehog into Its Centered and Projective Parts**

Recall that a hedgehog $\mathcal{H}_h$ of $\mathbb{R}^{n+1}$ is said to be centered (resp. projective) if its support function $h$ is symmetric (resp. antisymmetric), that is, if we have:

$$\forall u \in \mathbb{S}^n, \quad h(-u) = h(u) \quad (\text{resp. } h(-u) = -h(u)).$$

For instance, the hedgehog $\mathcal{H}_f$ $\left(\text{resp. } \mathcal{H}_g\right)$ of $\mathbb{R}^3$ that is represented in Fig. 5.1a (resp. Fig. 5.1b) is centered (resp. projective). Geometrically speaking, this means that $\mathcal{H}_h$ is centrally symmetric with respect to the origin (resp. that any pair of antipodal points on the unit sphere $\mathbb{S}^n$ correspond to the same point on the hypersurface $\mathcal{H}_h = x_h(\mathbb{S}^n)$).

Now, the support function $h$ of $\mathcal{H}_h \subset \mathbb{R}^{n+1}$ can be uniquely split into the sum of its symmetric and antisymmetric parts $f$ and $g$:

$$\forall u \in \mathbb{S}^n, \quad h(u) = f(u) + g(u) \quad \text{where} \quad \begin{cases} f(u) = \tfrac{1}{2}(h(u) + h(-u)) \\[2mm] g(u) = \tfrac{1}{2}(h(u) - h(-u)) \end{cases}.$$

Consequently, any hedgehog $\mathcal{H}_h$ of $\mathbb{R}^{n+1}$ has a unique decomposition of the form $\mathcal{H}_f + \mathcal{H}_g$, where $\mathcal{H}_f$ and $\mathcal{H}_g$ are respectively centered and projective. We say that $\mathcal{H}_f$ and $\mathcal{H}_g$ are respectively the centered and the projective parts of $\mathcal{H}_h$.

**Theorem 5.1.2** *Let $\mathcal{H}_{h_1}$ and $\mathcal{H}_{h_2}$ be $C^2$-hedgehogs that are linearly independent in $H_3$ and whose centered parts are nontrivial (i.e., distinct from a point) and proportional to one and the same convex surface of class $C^2_+$. Then $\mathcal{H}_{h_1}$ and $\mathcal{H}_{h_2}$ have distinct curvature functions.*

Recall that a hedgehog $\mathcal{H}_h$ of $\mathbb{R}^{n+1}$ is said to be of constant width if its centered part has a constant support function. In other words, a hedgehog $\mathcal{H}_h$ of $\mathbb{R}^{n+1}$ is of constant width if the signed distance between the two cooriented support hyperplanes that are orthogonal to $u \in \mathbb{S}^n$ does not depend on $u$, that is, if:

$$\exists r \in \mathbb{R}, \forall u \in \mathbb{S}^n, h(u) + h(-u) = 2r.$$

A straightforward consequence of Theorem 5.1.2. is the following corollary.

**Corollary 5.1.2** *Let $\mathcal{H}_f$ and $\mathcal{H}_g$ be $C^2$-hedgehogs that are linearly independent in $H_3$. If $\mathcal{H}_f$ and $\mathcal{H}_g$ are of nonzero constant width, then their curvature functions $R_f$ and $R_g$ are distinct.*

Recall that a hedgehog $\mathcal{H}_h$ of $\mathbb{R}^{n+1}$ is said to be analytic if its support function $h$ is $C^\omega$ on $\mathbb{S}^n$.

**Theorem 5.1.3** *Let $\mathcal{H}_f$ and $\mathcal{H}_g$ be analytic (resp. projective $C^2$) hedgehogs of $\mathbb{R}^3$ that are linearly independent in $H_3$. If the mixed curvature function of $\mathcal{H}_f$ and $\mathcal{H}_g$ does not change sign on $\mathbb{S}^2$, then $\mathcal{H}_f$ and $\mathcal{H}_g$ have distinct curvature functions.*

### 5.1.3   *Proof of Results and Further Remarks*

***Proof of Theorem 5.1.1*** By assumption, there exists $(\lambda, \mu) \in \mathbb{R}^2$ such that the hedgehog $\mathcal{H}_{\lambda f + \mu g}$ is of class $C^2_+$. We can assume that $|\lambda| \neq |\mu|$.

Let assume that $R_f = R_g$. We then have:

$$R_{\lambda f + \mu g} = \lambda^2 R_f + \mu^2 R_g + 2\lambda\mu R_{(f,g)}$$

$$= \mu^2 R_f + \lambda^2 R_g + 2\mu\lambda R_{(f,g)}$$

$$= R_{\mu f + \lambda g}.$$

As the hedgehog $\mathcal{H}_{\lambda f + \mu g}$ is of class $C_+^2$, the equality $R_{\lambda f + \mu g} = R_{\mu f + \lambda g}$ implies the existence of an $\varepsilon \in \{-1, 1\}$ such that $\lambda f + \mu g = \varepsilon (\mu f + \lambda g)$ by Minkowski's uniqueness theorem. We thus have $(\lambda - \varepsilon\mu) f = \varepsilon (\lambda - \varepsilon\mu) g$ and hence $f = \varepsilon g$ since $\lambda - \varepsilon\mu \neq 0$. $\blacksquare$

Our proof of the other two theorems of the subsection requires a series of lemmas.

**Lemma 5.1.1** *Let $\mathcal{H}_f$ and $\mathcal{H}_g$ be two $C^2$-hedgehogs of $\mathbb{R}^3$. If $u \in \mathbb{S}^2$ is such that $R_g(u) > 0$, then*

$$R_{(f,g)}(u)^2 \geq R_f(u) R_g(u).$$

***Proof of Lemma 5.1.1*** Define $Q : \mathbb{R} \to \mathbb{R}$ by $Q(t) = R_{f+tg}(u)$. By bilinearity and symmetry of the mixed curvature function, we have:

$$\forall t \in \mathbb{R}, \quad Q(t) = R_f(u) + 2t R_{(f,g)}(u) + t^2 R_g(u).$$

Let us consider the 'reduced' discriminant $\Delta = R_{(f,g)}(u)^2 - R_f(u) R_g(u)$ of the quadratic trinomial $Q(t)$. On the one hand, from the assumption $R_g(u) > 0$, it follows that $Q(t) > 0$ for any large enough $t$. On the other hand, there exists some $\lambda \in \mathbb{R}$ such that $R_{(1,f+\lambda g)}(u) = R_{(1,f)}(u) + \lambda R_{(1,g)}(u) = 0$ and hence $Q(\lambda) = R_{f+\lambda g}(u) \leq 0$. Therefore $\Delta \geq 0$, which achieves the proof. $\blacksquare$

Surprisingly, as we saw in Sect. 4.4, there exist nontrivial (i.e., distinct from a point) hedgehogs of $\mathbb{R}^3$ whose curvature function is nonpositive all over $\mathbb{S}^2$, which disproves a conjectured characterization of the 2-sphere. However, the support function of such a hedgehog can be neither analytic nor antisymmetric on $\mathbb{S}^2$ (see Sect. 4.4):

**Lemma 5.1.2 ([1], and [95, Theorem 3])** *Let $\mathcal{H}_h$ be an analytic (resp. a projective $C^2$) hedgehog in $\mathbb{R}^3$. If the curvature function $R_h$ of $\mathcal{H}_h$ is nonpositive all over $\mathbb{S}^2$, then $\mathcal{H}_h$ is reduced to a single point.*

**Lemma 5.1.3** *Let $\mathcal{H}_g$ be a convex hedgehog of class $C_+^2$ in $\mathbb{R}^3$. Given a projective hedgehog $\mathcal{H}_f$ in $\mathbb{R}^3$, the mixed curvature function $R_{(f,g)}$ is equal to zero on $\mathbb{S}^2$ only if $\mathcal{H}_f$ is reduced to a single point, that is, only if $f$ is the restriction to $\mathbb{S}^2$ of a linear form on $\mathbb{R}^3$.*

**Proof of Lemma 5.1.3**  Since $\mathcal{H}_g$ is of class $C_+^2$, we have

$$R_f(u)\, R_g(u) \leq R_{(f,g)}(u)^2$$

by Lemma 5.1.1. From $R_{(f,g)} = 0$ on $\mathbb{S}^2$, we then deduce that $R_f \leq 0$ which implies the result by Lemma 5.1.2.                                                          ∎

**Proof of Theorem 5.1.2**  By assumption, $h_1$ and $h_2$ are of the form

$$\begin{cases} h_1 = f_1 + \lambda_1 k \\[2mm] h_2 = f_2 + \lambda_2 k \end{cases},$$

where $\lambda_1$, $\lambda_2$ are nonzero real numbers, $f_1$, $f_2$ the support functions of projective hedgehogs and $k$ the support function of a centered convex surface of class $C_+^2$. Assume that $R_{h_1} = R_{h_2}$. By bilinearity and symmetry of the mixed curvature function, this gives

$$R_{f_1} + \lambda_1^2 R_k + 2\lambda_1 R_{(f_1,k)} = R_{f_2} + \lambda_2^2 R_k + 2\lambda_2 R_{(f_2,k)}.$$

Splitting into symmetric and antisymmetric parts, we get

$$\begin{cases} R_{f_1} + \lambda_1^2 R_k = R_{f_2} + \lambda_2^2 R_k \\[2mm] \lambda_1 R_{(f_1,k)} = \lambda_2 R_{(f_2,k)} \end{cases}.$$

By linearity of the mixed curvature function in the first argument, the second equation is equivalent to $R_{(\lambda_1 f_1 - \lambda_2 f_2, k)} = 0$. By Lemma 5.1.3, this implies that $\mathcal{H}_{\lambda_1 f_1 - \lambda_2 f_2}$ is a point and hence that $\mathcal{H}_{\lambda_1 f_1} = \mathcal{H}_{\lambda_2 f_2}$ in $H_3$. Now, by multiplying each member of the first equation of the previous system by $\lambda_1^2$, we get

$$\lambda_1^2 R_{f_1} + \lambda_1^4 R_k = \lambda_1^2 R_{f_2} + \lambda_1^2 \lambda_2^2 R_k,$$

and hence

$$R_{\lambda_1 f_1} - R_{\lambda_1 f_2} = \lambda_1^2 \left(\lambda_1^2 - \lambda_2^2\right) R_k$$

by bilinearity of the mixed curvature function. Therefore,

$$R_{\lambda_2 f_2} - R_{\lambda_1 f_2} = \lambda_1^2 \left(\lambda_1^2 - \lambda_2^2\right) R_k,$$

that is,

$$\left(\lambda_2^2 - \lambda_1^2\right)\left(R_{f_2} - \lambda_1^2 R_k\right) = 0.$$

As $\mathcal{H}_{f_2}$ is projective (resp. $\mathcal{H}_k$ is convex of class $C_+^2$), we have (see Sect. 3.2.2):

$$\int_{\mathbb{S}^2} R_{f_2} d\sigma \leq 0 \quad \text{and} \quad \int_{\mathbb{S}^2} R_k d\sigma > 0.$$

Therefore, $R_{f_2} \neq \lambda_1^2 R_k$. From the previous equation, we thus get $\lambda_2^2 = \lambda_1^2$, that is:

$$\exists \varepsilon \in \{-1, 1\}, \lambda_2 = \varepsilon \lambda_1.$$

Now, $\lambda_1 \mathcal{H}_{f_1} = \mathcal{H}_{\lambda_1 f_1} = \mathcal{H}_{\lambda_2 f_2} = \lambda_2 \mathcal{H}_{f_2}$ and $\lambda_1, \lambda_2$ are nonzero. Therefore, we have $\mathcal{H}_{f_1} = \varepsilon \mathcal{H}_{f_2}$ in $H_3$, that is, $\mathcal{H}_{f_2} = \varepsilon \mathcal{H}_{f_1}$ and hence

$$\mathcal{H}_{h_2} = \mathcal{H}_{f_2 + \lambda_2 k} = \mathcal{H}_{f_2} + \lambda_2 \mathcal{H}_k = \varepsilon \left( \mathcal{H}_{f_1} + \lambda_1 \mathcal{H}_k \right) = \varepsilon \mathcal{H}_{h_1} \text{ in } H_3,$$

which contradicts the fact that $\mathcal{H}_{h_1}$ and $\mathcal{H}_{h_2}$ are linearly independent in $H_3$. ∎

**Lemma 5.1.4** *Let $\mathcal{H}_f$ and $\mathcal{H}_g$ be two $C^2$-hedgehogs of $\mathbb{R}^3$. If their curvature functions $R_f$ and $R_g$ are identically equal on $\mathbb{S}^2$, then either $R_{f-g}(u) \leq 0$ or $R_{f+g}(u) \leq 0$ for all $u \in \mathbb{S}^2$.*

***Proof of Lemma 5.1.4*** Assume that $R_{f-g}(u) > 0$ (resp. $R_{f+g}(u) > 0$). By Lemma 5.1.1, we then have

$$R_{(f-g, f+g)}(u)^2 \geq R_{f-g}(u) R_{f+g}(u).$$

Now the assumption $R_f = R_g$ implies

$$R_{(f-g, f+g)} = R_f - R_g = 0 \quad \text{and hence} \quad R_{f-g}(u) R_{f+g}(u) \leq 0.$$

Therefore $R_{f-g}(u) \leq 0$ (resp. $R_{f+g}(u) \leq 0$). ∎

***Proof of Theorem 5.1.3*** We prove the result by contraposition. Assume that $R_f$ and $R_g$ are identically equal on $\mathbb{S}^2$. Since $\mathcal{H}_f$ and $\mathcal{H}_g$ are analytic (resp. projective and $C^2$) hedgehogs of $\mathbb{R}^3$ that are linearly independent in $H_3$, it follows from Lemma 5.1.2 that there must exist $(u, v) \in \mathbb{S}^2 \times \mathbb{S}^2$ such that $R_{f-g}(u) > 0$ and $R_{f+g}(v) > 0$. By Lemma 5.1.4, we then deduce that $R_{f+g}(u) \leq 0$ and $R_{f-g}(v) \leq 0$. Now we have $R_{(f,g)} = \frac{1}{4}\left( R_{f+g} - R_{f-g} \right)$, so that

$$\begin{cases} R_{f-g}(u) > 0 \\ \\ R_{f+g}(u) \leq 0 \end{cases} \quad \text{and} \quad \begin{cases} R_{f+g}(v) > 0 \\ \\ R_{f-g}(v) \leq 0 \end{cases}$$

implies $R_{(f,g)}(u) < 0$ and $R_{(f,g)}(v) > 0$. ∎

## 5.2  Gauss Infinitesimal Rigidity and Volume Preservation Under Curvature-Preserving Deformations

In 1813, A. L. Cauchy proved (almost rigorously) his famous rigidity theorem: *Any convex polyhedron of* $\mathbb{R}^3$ *is rigid* (that is, no convex polyhedron of $\mathbb{R}^3$ can be continuously deformed so that its faces remain rigid) [28]. First examples of flexible polyhedra were discovered by R. Bricard in 1897 [25], but these « Bricard's flexible octahedra » are self-intersecting. The question of rigidity of embedded non-convex polyhedra remained open until 1977 when R. Connelly discovered the first example of flexible sphere-homeomorphic polyhedron [33]. In the late seventies, R. Connelly and D. Sullivan formulated the so-called « bellows conjecture » stating that whenever we perform a rigid deformation of a flexible polyhedron $P$ (that is, a continuous deformation of $P$ that changes only its dihedral angles), the volume of $P$ remains constant. The first proof of the bellows conjecture was given by I. Sabitov [161], in 1995. The second proof by R. Connelly, I. Sabitov, and A. Walz followed two years later [34].

In 2000, L. Rodriguez and H. Rosenberg gave a rigidity result for a class of polyhedral hedgehogs of $\mathbb{R}^3$ [158]. Three years later, G. Panina gave examples of flexible « virtual polytopes » (a synonym expression for « polyhedral hedgehogs ») of $\mathbb{R}^3$ which are similar to Bricard's flexible octahedra [140] and proved the following refinement of Rodriguez-Rosenberg theorem: *a virtual polytope of* $\mathbb{R}^3$ *with a convex fan is not flexible.*

In 1916, M. Dehn proved that any simplicial convex polyhedron $P$ of $\mathbb{R}^3$ is infinitesimally rigid [36]: any non-trivial first order deformation of $P$ induces a variation of its edge lengths. Gauss infinitesimal rigidity of convex polyhedra was stated and proved by A. D. Alexandrov [4]: any non-trivial first order deformation of a convex polyhedron $P$ induces a variation of its face areas. See e.g., [73] for more details.

In 1927, E. Cohn-Vossen proved that smooth closed surfaces of $\mathbb{R}^3$ with everywhere positive Gaussian curvature are rigid [32]. Smooth closed surfaces of $\mathbb{R}^3$ with everywhere positive Gaussian curvature are also infinitesimally rigid [191] (resp. Gauss infinitesimally rigid [177]), that is every isometric infinitesimal deformation of such a surface is trivial (resp. rigid with respect to the Gaussian curvature regarded as a function of the outer unit normal). See e.g., [72] for more details.

In this subsection, we consider Gauss rigidity and Gauss infinitesimal rigidity for hedgehogs of $\mathbb{R}^3$ (regarded as Minkowski differences of closed convex surfaces of $\mathbb{R}^3$ with positive Gaussian curvature). The results are essentially taken from [104]. As noticed by I. Izmestiev [72, 73], **Gauss rigidity (Gauss infinitesimal rigidity) can be interpreted as uniqueness (resp. « infinitesimal » uniqueness) in the Minkowski problem**, that is in the problem of prescribing the $n$th surface area measure of a polytope $P$ of $\mathbb{R}^{n+1}$ on the unit sphere $\mathbb{S}^n$ (resp. the Gaussian curvature of smooth strictly convex closed hypersurface of $\mathbb{R}^{n+1}$ as a function of the outer unit normal). We already studied the uniqueness part of the Minkowski problem extended to hedgehogs in the previous subsection. In particular, we saw different ways of constructing pairs of non-congruent hedgehogs that share the same curvature function (i.e., inverse of the Gaussian curvature). This will allow us to give examples of nontrivial (i.e., distinct from a point) hedgehogs that are not Gauss infinitesimally rigid.

Consider a one parameter family of $C^2$-hedgehogs $\left(\mathcal{H}_{h_t}\right)_{t\in[0,1]}$ all having the same curvature function. We do not know whether these hedgehogs are congruent in $\mathbb{R}^3$. However, we will prove a theorem of volume preservation under curvature-preserving deformations:

Under an appropriate differentiability condition of the family with respect to the parameter, we will prove that:

**All the hedgehogs of the considered family** $\left(\mathcal{H}_{h_t}\right)_{t\in[0,1]}$ **have the same algebraic volume.**

### 5.2.1  Gauss Infinitesimal Rigidity in the Context of Hedgehogs

In this subsection, we will use the Banach spaces $C_m$, $(m \in \mathbb{N})$, that were introduced by L. Nirenberg in his study of the Minkowski problem in $\mathbb{R}^3$ [134, p. 380]. The space $C_m$ is defined as follows. The unit sphere $\mathbb{S}^2$ is divided up into three pairs of regions in each of which one of the following coordinate systems is defined:

$$(X, Y) = \left(\frac{x}{z}, \frac{y}{z}\right), \ (Y, Z) = \left(\frac{y}{x}, \frac{z}{x}\right) \text{ and } (Z, X) = \left(\frac{z}{y}, \frac{x}{y}\right),$$

where $(x, y, z)$ are the standard coordinates in $\mathbb{R}^3$. Then a function $h : \mathbb{S}^2 \to \mathbb{R}$ belongs to $C_m$, $(m \in \mathbb{N})$, if in each pair of regions all its partial derivatives (with respect to the corresponding local coordinates) of order less or equal to $m$ exist and are continuous. The norm of every $h \in C_m$ is defined as the sum of the suprema of the absolute values of the partial derivatives up to order $m$ (the suprema being taken with respect to all three pairs of regions).

**Definition 5.2.1** Let $\mathcal{H}_h$ be a $C^2$-hedgehog of $\mathbb{R}^3$. A smooth deformation of $\mathcal{H}_h$ is the datum of a differentiable map $\widetilde{h} : [0, 1] \to C_2, t \mapsto h_t := h(t, .)$ such that $h_0 = h$.

**Definition 5.2.2** Let $\mathcal{H}_f$ be a $C^2$-hedgehog of $\mathbb{R}^3$. An infinitesimal isogauss deformation of $\mathcal{H}_f$ is the data of a family $\left(\mathcal{H}_{f+tg}\right)_{t\in\mathbb{R}}$ of hedgehogs of $\mathbb{R}^3$,

$$x_{f+tg} : \mathbb{S}^2 \to \mathcal{H}_{f+tg} \subset \mathbb{R}^3$$
$$u \mapsto x_f(u) + tx_g(u)$$

where $\mathcal{H}_g$ is a hedgehog of $\mathbb{R}^3$ such that the mixed curvature function $R_{(f,g)} := \frac{1}{2}\left(R_{f+g} - R_f - R_g\right)$ is identically zero on $\mathbb{S}^2$.

**Definition 5.2.3** Let $\mathcal{H}_f$ be a $C^2$-hedgehog of $\mathbb{R}^3$. If every infinitesimal isogauss deformation $\left(\mathcal{H}_{f+tg}\right)_{t\in\mathbb{R}}$ of $\mathcal{H}_f$ is trivial, that is such that $\mathcal{H}_g$ is reduced to a single point, then the hedgehog $\mathcal{H}_f$ will be said to be Gauss infinitesimally rigid.

**Remark 5.2.1** A hedgehog $\mathcal{H}_g$ is reduced to a single point if, and only if, its support function $g$ is the restriction to $\mathbb{S}^2$ of a linear form on $\mathbb{R}^3$, which amounts to saying that its curvature function $R_g$ is identically zero on $\mathbb{S}^2$ [78, Theorem 1]. Therefore, a hedgehog $\mathcal{H}_f$ is Gauss infinitesimally rigid if, and only if, we have:

$$\forall g \in C^2\left(\mathbb{S}^2; \mathbb{R}\right), \quad \left(R_{(f,g)} = 0\right) \Longrightarrow \left(R_g = 0\right).$$

**Remark 5.2.2** If a hedgehog $\mathcal{H}_f$ of $\mathbb{R}^3$ is trivial (that is, reduced to a single point), then $\mathcal{H}_f$ is not Gauss infinitesimally rigid. Indeed, for every regular $C^2$-hedgehog $\mathcal{H}_g$ of $\mathbb{R}^3$, we have $R_{(f,g)} = 0$ although $R_g$ is not identically zero on $\mathbb{S}^2$.

### 5.2.2  Gauss Infinitesimal Rigidity of Regular $C^2$-hedgehogs of $\mathbb{R}^3$

Let us recall the **proof of the Gauss infinitesimal rigidity** (with respect to the curvature function) **of regular $C^2$-hedgehogs of** $\mathbb{R}^3$ (that are closed convex surfaces of class $C_+^2$ in $\mathbb{R}^3$). It is essentially a rewriting of the proof by J. Stoker [177]: Let $\mathcal{H}_f$ be a regular $C^2$-hedgehog of $\mathbb{R}^3$. Clearly, the regularity of $\mathcal{H}_f$ is equivalent to the strict positivity of its curvature function $R_f := 1/\kappa_f$. If $\left(\mathcal{H}_{f+tg}\right)_{t\in\mathbb{R}}$ defines an isogauss deformation of $\mathcal{H}_f$, then we have (see [103, Lemma 3.1] or Lemma 5.1.1):

$$0 = R^2_{(f,g)} \geq R_f . R_g$$

and hence $R_g \leq 0$ on $\mathbb{S}^2$. By taking the origin to be an interior point of the convex body bounded by $\mathcal{H}_f$ in $\mathbb{R}^3$, we may assume without loss of generality that $f > 0$ so that $f R g \leq 0$ on $\mathbb{S}^2$. Now, by symmetry of the mixed volume of hedgehogs of $\mathbb{R}^3$, we get:

$$0 = \int_{\mathbb{S}^2} g R_{(f,g)} d\sigma = \int_{\mathbb{S}^2} f R_{(g,g)} d\sigma = \int_{\mathbb{S}^2} f R_g d\sigma,$$

where $\sigma$ is the spherical Lebesgue measure on $\mathbb{S}^2$. Therefore, $R_g$ is identically zero on $\mathbb{S}^2$ which implies that $\mathcal{H}_g$ is reduced to a single point by Remark 5.2.1.  ∎

**Relation To Minkowski Problem**

In the context of hedgehogs, there is a close connection between Gauss infinitesimal rigidity and the uniqueness question in the Minkowski problem. This is due to the following equivalence:

$$\forall (f, g) \in C^2 \left( \mathbb{S}^2; \mathbb{R} \right)^2, \quad \left( R_f = R_g \right) \Longleftrightarrow \left( R_{(f+g, f-g)} = 0 \right).$$

In [102, 103], the author gave examples of pairs of non-congruent hedgehogs of $\mathbb{R}^3$ having the same curvature function (see also the previous subsection). From each of these examples, we can deduce examples of nontrivial hedgehogs that are not Gauss infinitesimally rigid. It is for instance the case of the pair of hedgehogs of $\mathbb{R}^3$ given by:

$$f(u) := \begin{cases} 0 & \text{if } z \leq 0 \\ \exp(-1/z^2) & \text{if } z > 0 \end{cases} \quad \text{and} \quad g(u) := \begin{cases} \exp(-1/z^2) & \text{if } z < 0 \\ 0 & \text{if } z \geq 0, \end{cases}$$

where $u = (x, y, z) \in \mathbb{S}^2 \subset \mathbb{R}^3$. Indeed, we have clearly $R_{(f,g)} = 0$. Therefore, these two nontrivial hedgehogs $\mathcal{H}_f$ and $\mathcal{H}_g$ are not Gauss infinitesimally rigid. Only nonanalytic examples are known. The question of knowing whether there exists a pair of noncongruent analytic hedgehogs of $\mathbb{R}^3$ with the same curvature function remains open (by 'analytic hedgehogs', we mean 'hedgehogs with an analytic support function'). As a consequence, the question of knowing whether there exist examples of nontrivial analytic hedgehogs that are not Gauss infinitesimally rigid is also open.

### 5.2.3   *Volume Preservation Under Curvature-Preserving Deformations*

**Lemma 5.2.1** *The curvature function $R : C_2 \to C_0$, $h \mapsto R_h$ is differentiable on $C_2$, and:*

$$\forall (f, g) \in C_2 \times C_2, \qquad dR_f(g) = \lim_{\substack{t \to 0 \\ t \neq 0}} \frac{R_{f+tg} - R_f}{t} = 2R_{(f,g)}.$$

*Proof of Lemma 5.2.1* Indeed, we have:

$$\forall t \in \mathbb{R}_+^*, \qquad R_{f+tg} - R_f = R_f + 2t R_{(f,g)} + t^2 R_g - R_f$$

$$= t \left( 2R_{(f,g)} + t R_g \right),$$

and hence

$$\lim_{\substack{t \to 0 \\ t \neq 0}} \frac{R_{f+tg} - R_f}{t} = \lim_{\substack{t \to 0 \\ t \neq 0}} \left(2R_{(f,g)} + t R_g\right) = 2R_{(f,g)}.$$

Now, we have:

$$\left\| R_{f+g} - R_f - 2R_{(f,g)} \right\|_{C_0} = \left\| R_g \right\|_{C_0} = o\left(\|g\|_{C_2}\right),$$

which achieves the proof.  ∎

**Theorem 5.2.1** *Let $\mathcal{H}_h$ be a $C^2$-hedgehog of $\mathbb{R}^3$. If a smooth deformation of $\mathcal{H}_h$, say*

$$\widetilde{h} : [0, 1] \to C_2, t \mapsto h_t := h\,(t, .)\,,$$

*preserves the curvature function (that is, is such that $R_{h_t} = R_h$ for all $t \in [0, 1]$), then it also preserves the algebraic volume:*

$$\forall t \in [0, 1], \quad v\,(h_t) = v\,(h)\,.$$

***Proof of Theorem 5.2.1*** By assumption, the map $R \circ \widetilde{h} : [0, 1] \to C_0$ is constant. Since $\widetilde{h}$ is differentiable by assumption and $R$ by Lemma 5.2.1, $R \circ \widetilde{h}$ is differentiable and the chain rule gives:

$$\forall t \in [0, 1], \qquad \left(R \circ \widetilde{h}\right)'(t) = 2R_{\left(\widetilde{h}(t),\left(\frac{\partial \widetilde{h}}{\partial t}\right)(t)\right)}.$$

Therefore, the differentiation yields:

$$\forall t \in [0, 1], \qquad R_{\left(\widetilde{h}(t),\left(\frac{\partial \widetilde{h}}{\partial t}\right)(t)\right)} = 0. \tag{5.2.1}$$

Now, for every $t_0 \in [0, 1]$, we have:

$$\forall t \in [0, 1] - \{t_0\}, \qquad \frac{v\left(\widetilde{h}\,(t)\right) - v\left(\widetilde{h}\,(t_0)\right)}{t - t_0} = \frac{1}{3}\int_{\mathbb{S}^2} \frac{\widetilde{h}\,(t) - \widetilde{h}\,(t_0)}{t - t_0} R_{\widetilde{h}(t_0)} d\sigma$$

and hence:

$$\lim_{\substack{t \to t_0 \\ t \neq t_0}} \frac{v\left(\widetilde{h}\,(t)\right) - v\left(\widetilde{h}\,(t_0)\right)}{t - t_0} = \frac{1}{3}\int_{\mathbb{S}^2} \left(\frac{\partial \widetilde{h}}{\partial t}\right)(t_0)\, R_{\widetilde{h}(t_0)} d\sigma.$$

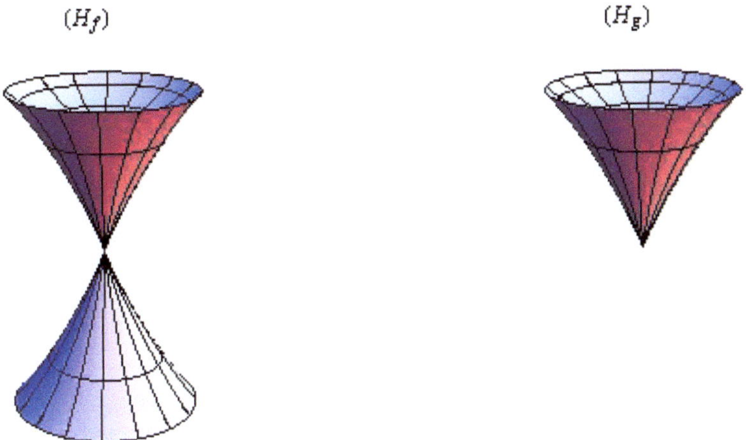


**Fig. 5.5** Same curvature function and different algebraic volumes

Besides, by symmetry of the mixed volume of hedgehogs, we have:

$$\frac{1}{3}\int_{\mathbb{S}^2}\left(\frac{\partial\widetilde{h}}{\partial t}\right)(t_0)\,R_{\widetilde{h}(t_0)}\,d\sigma = \frac{1}{3}\int_{\mathbb{S}^2}\widetilde{h}\,(t_0)\,R_{\left(\widetilde{h}(t_0),\left(\frac{\partial\widetilde{h}}{\partial t}\right)(t_0)\right)}\,d\sigma.$$

From (5.2.1), we then deduce that:

$$\forall t_0\in[0,1],\qquad \left(v\circ\widetilde{h}\right)'(t_0)=\lim_{\substack{t\to t_0\\ t\neq t_0}}\frac{v\left(\widetilde{h}\,(t)\right)-v\left(\widetilde{h}\,(t_0)\right)}{t-t_0}=0,$$

and thus all the hedgehogs of the family $\left(\mathcal{H}_{h_t}\right)$ have same (algebraic) volume.   ∎

**Remark 5.2.3** Noncongruent hedgehogs that share the same curvature function may of course have different (algebraic) volumes. It is for instance the case of the hedgehogs shown on Fig. 5.5 whose support functions $f$, $g$. are defined on $\mathbb{S}^2$ by

$$f(u):=\begin{cases}\exp(-1/z^2)\text{ if }z\neq 0\\[2mm]0\qquad\text{if}\qquad z=0\end{cases}\quad\text{and}\quad g(u):=\begin{cases}sgn\,(z)\,f(u)\text{ if }z\neq 0\\[2mm]0\qquad\text{if}\qquad z=0,\end{cases}$$

where $u=(x,y,z)\in\mathbb{S}^2\subset\mathbb{R}^3$.

# Chapter 6
# Complex Hedgehogs in $\mathbb{C}^{n+1}$ or $P^{n+1}(\mathbb{C})$

Unless explicitly states otherwise, the results of this section are essentially taken from [111]. As we saw in Chap. 2, classical hedgehogs of $\mathbb{R}^{n+1}$ can be regarded as the geometrical realizations of formal differences of convex bodies in $\mathbb{R}^{n+1}$. Like convex bodies, hedgehogs are completely determined by (and can be identified with) their support functions. Adopting a projective viewpoint, we prove in this section that any holomorphic function $h : \mathbb{C}^n \to \mathbb{C}$ can be regarded as the 'support function' of a 'complex hedgehog' $\mathcal{H}_h$ in $\mathbb{C}^{n+1}$. In the same vein, we introduce the notion of an evolute of such a hedgehog $\mathcal{H}_h$ in $\mathbb{C}^2$, and a natural (but apparently hitherto unknown) notion of complex curvature, which allows us to interpret this evolute as the locus of the centers of complex curvature. It is of course permissible to think that the development of a 'Brunn-Minkowski theory for complex hedgehogs' (replacing Euclidean volumes by symplectic ones) might be a promising way of research. We give first two results in this direction.

## 6.1   Brief Summary of the Chapter

**Complex Hedgehogs**

As we recalled it, hedgehogs of $\mathbb{R}^{n+1}$ are completely determined by (and can be identified with) their support functions, which are differences of two support functions of convex bodies of $\mathbb{R}^{n+1}$ restricted to the unit sphere $\mathbb{S}^n$. In Sect. 6.2, we adopt a projective viewpoint in order to introduce the notion of a *complex hedgehog* in the complex Euclidean space $\mathbb{C}^{n+1}$. We prove that:

*Any holomorphic function $h : \mathbb{C}^n \to \mathbb{C}$ can be regarded as 'the complex support function' of a **complex hedgehog** $\mathcal{H}_h$, which is defined by a holomorphic parametrization $x_h : \mathbb{C}^n \to \mathbb{C}^{n+1}$ in the complex Euclidean space $\mathbb{C}^{n+1}$.*

© The Author(s), under exclusive license to Springer Nature Switzerland AG 2026
Y. Martinez-Maure, *Hedgehog Theory*, Lecture Notes in Mathematics 2381,
https://doi.org/10.1007/978-3-032-04298-9_6

Of course, these complex hedgehogs can be interpreted in the metric contact geometry setting where they appear as fronts of Legendrian immersions in $\mathbb{C}^{2n+1}$ (see Sect. 6.2.2).

In passing, we introduce the notion of a *rational hedgehog* in the complex projective plane $P^2$ ($\mathbb{C}$) equipped with the usual Fubini-Study Kähler form $\omega$ (for an introduction to the Fubini-Study structure, see e.g., [27]). Such a hedgehog $\mathcal{H}_h$ is modeled on $P^1$ ($\mathbb{C}$) := $\mathbb{C} \cup \{\infty\}$ via a holomorphic map $h : \mathbb{C} \to \mathbb{C}$ that is such that $Area\,[x_h\,(\mathbb{C})] < +\infty$.

**Complex Evolutes and Complex Curvature**

In classical differential geometry of curves, the evolute of a plane curve is the locus of all its centers of curvature or, equivalently, the envelope of its normal lines. Interpreting evolutes of hedgehog curves from a projective point of view, we prove in Sect. 6.3 that:

> *There exists **a natural extension of the notion of evolute curves to complex hedgehog curves**, and **a very natural (but apparently hitherto unknown) notion of complex curvature**, which allows us to interpret any evolute of a complex hedgehog curve $\mathcal{H}_h$ as the locus of its centers of complex curvature.*

**Real and Imaginary Parts of $\mathcal{H}_h$ in $\mathbb{C}^2$ Regarded as Hedgehogs of $\mathbb{R}^3$**

In Sect. 6.4, given any complex hedgehog $\mathcal{H}_h$ in $\mathbb{R}^4$, we introduce *its real and imaginary parts as hedgehogs of* $\mathbb{R}^3$, which can be regarded globally as the images of $\mathcal{H}_h$ under the orthogonal projections onto two particular hyperplanes of $\mathbb{R}^4$, and that are determined by $\mathrm{Re}\,[h]$ and $\mathrm{Im}\,[h]$.

**Towards a Brunn-Minkowski Theory for Complex Hedgehogs**

The notion of a hedgehog curve or surface was born in 1930s from the study of the Brunn-Minkowski theory by A.D. Alexandrov, H. Geppert and some others. In the present section, we try to motivate the development of a '*theory of mixed volumes for complex hedgehogs*' (*replacing Euclidean volumes by symplectic ones*). In Sect. 6.5, we mention first two results in this direction. First, identifying complex hedgehogs with their support functions, we notice that *the complex linear space of holomorphic functions defined up to a similitude on the unit disc $\mathbb{D} \subset \mathbb{C}$ can be endowed with a scalar product which can be interpreted as a mixed symplectic area.*

Second, we give the following *sharp estimation of the (symplectic) area of $x_h$ ($\mathbb{D}$) using the energy, say $E\,(x_h)$, of the loop $x_h : \mathbb{S}^1 = \mathbb{R}/2\pi\mathbb{Z} \to \mathbb{C}^2, \theta \mapsto x_h\left(e^{i\theta}\right),$ in the case where $h : \mathbb{D} \to \mathbb{C}$ is the sum of a power series $\sum h_n z^n$ with radius of convergence $R > 1$:*

$$Area\,[x_h\,(\mathbb{D})] \leq \frac{3}{4} E\,(x_h)\,.$$

Note that this estimate is better than that well-known for an arbitrary smooth loop $\gamma : \mathbb{S}^1 \to V$ in a symplectic vector space $(V, \omega)$ (namely, $|A\,(\gamma)| \leq E\,(\gamma)$, see for instance [123, pp. 87–88]).

**Real Evolutes in Even Dimensions**

In Sect. 6.6, we return to real hedgehogs, but in $\mathbb{R}^{2n}$ endowed with a linear complex structure $J$. First of all, we introduce the notion of an *evolute of any hedgehog with a smooth support function in* $\left(\mathbb{R}^{2n}, J\right)$.

We particularly focus on the case $n = 2$. We identify $\mathbb{R}^4$ with the quaternion algebra $\mathbb{H}$ (and thus the unit sphere $\mathbb{S}^3$ with the set $\mathbb{S}^1_{\mathbb{H}}$ of unit quaternions), and, we associate to any pure unit quaternion $v$ the linear complex structure $J_v : \mathbb{R}^4 \to \mathbb{R}^4$, $x \longmapsto vx$. In other words, for any $v \in \mathbb{S}^2 \cong \mathbb{S}^1_{\mathbb{H}} \cap \mathrm{Im}\,(\mathbb{H})$, we choose to work in the Kähler vector space $\left(\mathbb{R}^4, J_v, \omega_v\right)$, where $\omega_v$ denotes the associated Kähler form (i.e. the alternating 2-form $\omega_v\,(X, Y) = \langle J_v X, Y \rangle$, where $\langle ., . \rangle$ is the standard Euclidean metric on $\mathbb{R}^4$). To any $v \in \mathbb{S}^2$, it thus corresponds a Hopf fibration and a Hopf flow leaving the Hopf fibration invariant, namely the Hopf flow $\left\{(\phi_v)_\theta\right\}_{\theta \in \mathbb{S}^1}$ given by $(\phi_v)_\theta\,(u) := (\cos \theta)\,u + (\sin \theta)\,vu$, $\left(u \in \mathbb{S}^3\right)$.

We give *a detailed study of evolutes of hedgehog hypersurfaces in these Kähler vector spaces* $\left(\mathbb{R}^4, J_v, \omega_v\right)$.

**Mixed Symplectic Area and Quaternionic Curvature Function**

In parallel, we study *the symplectic area* of images of the oriented Hopf circles under the hedgehog parametrizations $x_h : \mathbb{S}^3 \to \left(\mathbb{R}^4, J_v, \omega_v\right)$. In this setting, we introduce the notion of *mixed symplectic area* and prove what follows among other results.

**Theorem** *Let $h \in C^\infty\left(\mathbb{S}^3; \mathbb{R}\right)$, and let $v$ be a pure unit quaternion.*

*(i)* *The evolute of $\mathcal{H}_h$ in $\left(\mathbb{R}^4, J_v, \omega_v\right)$ is the hedgehog with support function*

$$\partial_v h : \mathbb{S}^3 \to \mathbb{R},\ u \longmapsto \langle \nabla h\,(-J_v\,(u))\,,\, u \rangle ,$$

*where $\langle ., . \rangle$ is the standard Euclidean metric on $\mathbb{R}^4$, and $\nabla h$ the gradient of $h$. Thus, $\partial_v h$ is such that: $\forall u \in \mathbb{S}^3$,*

$$(\partial_v h)\,(J_v\,(u)) = \langle \nabla h\,(u)\,,\, J_v\,(u) \rangle = (dh)_u\,(J_v\,(u)) ;$$

*(ii)* *For all $u \in \mathbb{S}^3$,*

$$x_{\partial_v h}\,(u) = x_h\,(u) - R_h\,(u, v)\,u,$$

*where $R_h(u, v) := -v T_u x_h(J_v\,(u))\,\overline{u}$ ; here $\overline{u}$ of course refers to the quaternion conjugate of $u$;*

*(iii)* *The map $R_h\,(., v)\ :\ u_\theta := (\cos \theta)\,u + (\sin \theta)\,vu \longmapsto R_h\,(u_\theta, v)$ can be interpreted as a quaternionic curvature function of $x_h\left(\mathbb{S}^1_{u,v}\right)$, where $\mathbb{S}^1_{u,v}$ is the unit circle of $\mathbb{C}\,(u, v) := \mathbb{R}u + \mathbb{R}J_v\,(u)$ oriented by $(u, J_v\,(u))$, in the sense that $R_h\,(., v)$ is the unique $C^\infty$-smooth quaternionic function $R\,(., v):$ $\mathbb{S}^1_{u,v} \to \mathbb{H}$ that is of the form $R\,(u_\theta, v) = -v T_{u_\theta}\,(v)$, where $T_{u_\theta}\,(v)$ is a pure*

*quaternion, and such that:*

$$\forall g \in C^{\infty}\left(\mathbb{S}^3; \mathbb{R}\right), \quad s_{u,v}(g, h) := \frac{1}{2}\int_0^{2\pi} \left\langle x_g(u_\theta), R(u_\theta, v)u_\theta\right\rangle d\theta,$$

*where $s_{u,v}(g, h)$ denotes the mixed symplectic area of $x_g\left(\mathbb{S}^1_{u,v}\right)$ and $x_h\left(\mathbb{S}^1_{u,v}\right)$.*

In other words, what is shown by $(iii)$ is that the quaternionic curvature function $R(., v)$ plays, relatively to the mixed symplectic area $s_{u,v}$, the same role as the (ordinary) curvature function of plane hedgehogs does relatively to the (ordinary) mixed area. Here, we have to recall that the mixed area of two plane hedgehogs with support functions $(g, h) \in C^{\infty}\left(\mathbb{S}^1; \mathbb{R}\right)^2$ is given by

$$a(g, h) := \frac{1}{2}\int_0^{2\pi} \left\langle x_g(u_\theta), R_h(u)u_\theta\right\rangle d\theta = \frac{1}{2}\int_0^{2\pi} g(u_\theta) R_h(u_\theta) d\theta,$$

where $u_\theta = e^{i\theta} \in \mathbb{C} \cong \mathbb{R}^2$, and where $x_g : \mathbb{S}^1 \to \mathbb{C}, \theta \mapsto g(\theta)u_\theta + g'(\theta)iu_\theta$ is the natural parametrization of $\mathcal{H}_g$, and $R_h := h + h''$ the curvature function' of $\mathcal{H}_h$.

**Relationship with the Area of Order 2**

We also show that the algebraic area of order 2 of a hedgehog $\mathcal{H}_h$ of $\mathbb{R}^4$ can be interpreted in terms of the symplectic areas of $\mathcal{H}_h$ in the Kähler vector spaces $\left(\mathbb{R}^4, J_v, \omega_v\right)$. Here, we have to recall that the algebraic area of order 2 of $\mathcal{H}_h$ is defined to be $V(h, h, 1, 1)$, where $V$ is the extension of the mixed volume (of convex bodies of $\mathbb{R}^4$) to hedgehogs of $\mathbb{R}^4$.

**Extension to $\mathbb{R}^{4n} \cong \mathbb{H}^n$**

Finally, we consider briefly evolutes of hedgehog hypersurfaces in $\mathbb{R}^{4n}$, which we identify with the hyperkähler vector space $(\mathbb{H}^n, \langle .,.\rangle, I, J, K)$, where $\langle .,.\rangle$ is the standard Euclidean metric on $\mathbb{R}^{4n}$, $(n \geq 1)$, and, the triple of complex structures $(I, J, K)$ on $\mathbb{H}^n$ is given by left multiplication by $i, j, k$ respectively.

## 6.2  Complex Hedgehogs in $\mathbb{C}^{n+1}$ or $P^{n+1}(\mathbb{C})$

### 6.2.1  Real and Complex Hedgehogs as Dual Hypersurfaces of Graphs

In order to introduce complex hedgehogs, it is convenient to recall that real hedgehogs with a smooth support function can be regarded as dual hypersurfaces of smooth graphs. In what follows, any hedgehog $\mathcal{H}_h \subset \mathbb{R}^{n+1}$ with support function $h \in C^{\infty}(\mathbb{S}^n; \mathbb{R})$ will be regarded as a hypersurface in the real projective space $P^{n+1}(\mathbb{R})$ by adding 'a hyperplane at the infinity' $H_\infty$ to $\mathbb{R}^{n+1}$: $P^{n+1}(\mathbb{R}) =$

$\mathbb{R}^{n+1} \cup H_\infty$. More precisely, we will identify $\mathbb{R}^{n+1}$ with the affine hyperplane of $P^{n+1}(\mathbb{R}) = \mathbb{R}^{n+2} - \{0\}/\mathbb{R}^*$ with equation $X_{n+2} = -1$, where $[X_1, \ldots, X_{n+2}]$ denote the homogeneous coordinates of the equivalent class of $(X_1, \ldots, X_{n+2}) \in \mathbb{R}^{n+2} \setminus \{0\}$ in $P^{n+1}(\mathbb{R})$. Then, the hedgehog hypersurface $x_h : \mathbb{S}^n \to \mathcal{H}_h \subset \mathbb{R}^{n+1} \subset P^{n+1}(\mathbb{R})$ can be regarded as the dual hypersurface of

$$\begin{aligned} \gamma_h : \mathbb{S}^n \subset \mathbb{R}^{n+1} \quad &\to P^{n+1}(\mathbb{R}) \\ u = (u_1, \ldots, u_{n+1}) &\mapsto [u_1, \ldots, u_{n+1}, h(u)]. \end{aligned}$$

Indeed, the support hyperplane with equation $\langle x, u \rangle = h(u)$ then corresponds to the point $\gamma_h(u)$ by projective duality.

It is extremely natural to follow this idea to extend the notion of hedgehog to the complex setting. We regard the complex Euclidean space $\mathbb{C}^{n+1}$ as the affine hyperplane of $P^{n+1}(\mathbb{C}) = \mathbb{C}^{n+2} - \{0\}/\mathbb{C}^*$ with equation $X_{n+2} = -1$, and we define, for any holomorphic function $h : \mathbb{C}^n \to \mathbb{C}$, the hedgehog with support function $h$ as the hypersurface of $\mathbb{C}^{n+1}$ that is the dual hypersurface of

$$\begin{aligned} \gamma_h : \mathbb{C}^n \quad &\to P^{n+1}(\mathbb{C}) \\ z = (z_1, \ldots, z_n) &\mapsto [1, z_1, \ldots, z_n, h(z)], \end{aligned}$$

that is, as the envelope of the family of hyperplanes $(H_h(z))_{z \in \mathbb{C}^n}$ with equation

$$X_1 + \sum_{k=1}^{n} z_k X_{k+1} = h(z). \tag{6.2.1}$$

In other words:

**Definition 6.2.1** Let $h : \mathbb{C}^n \to \mathbb{C}$ be a holomorphic function. The hypersurface $\mathcal{H}_h$ of the complex Euclidean space $\mathbb{C}^{n+1}$ that is parametrized by

$$\begin{aligned} x_h : \mathbb{C}^n \quad &\to \mathbb{C}^{n+1} \\ z = (z_1, \ldots, z_n) &\mapsto \left( h(z) - \sum_{k=1}^{n} z_k \frac{\partial h}{\partial z_k}(z), \frac{\partial h}{\partial z_1}(z), \ldots, \frac{\partial h}{\partial z_n}(z) \right) \end{aligned}$$

is called the hedgehog with support function $h$.

Indeed, from (6.2.1) and the contact condition $dw_0 + \sum_{j=1}^{n} z_j dw_j = 0$, where $(w_0, w_1, \ldots, w_n, z_1, \ldots, z_n) \in \mathbb{C}^{n+1} \times \mathbb{C}^n = \mathbb{C}^{2n+1}$, we deduce that for all $z \in \mathbb{C}^n$, the point $x_h(z) = (x_1(z), \ldots, x_n(z))$ is the unique solution of the system

$$\begin{cases} x_1 + \sum_{k=1}^{n} z_k x_{k+1} = h(z) & (6.2.1) \\[2em] \forall k \in \{1, \ldots, n\}, \; x_{k+1} = \dfrac{\partial h}{\partial z_k}(z), & (6.2.2) \end{cases}$$

where (6.2.2) is obtained from (6.2.1) by performing partial differentiations with respect to the complex variables $z_k$, $(1 \le k \le n)$. Thus, it appears that $\mathcal{H}_h$ is actually parametrized by

$$x_h : \mathbb{C}^n \to \mathbb{C}^{n+1},$$

$$z = (z_1, \ldots, z_n) \mapsto \left( h(z) - \sum_{k=1}^{n} z_k \frac{\partial h}{\partial z_k}(z), \frac{\partial h}{\partial z_1}(z), \ldots, \frac{\partial h}{\partial z_n}(z) \right).$$

**Example** The hedgehog of $\mathbb{C}^2$ the support function $h : \mathbb{C} \to \mathbb{C}$ of which is given by $h(z) = -z^3$ is the affine algebraic curve $\mathcal{H}_h$ of $\mathbb{C}^2$ with equation $27x^2 + 4y^3 = 0$. It is parametrized by:

$$\begin{cases} x = 2z^3 \\ y = -3z^2. \end{cases}$$

As any complex hedgehog curve $x_h : \mathbb{C} \to \mathbb{C}^2$, it is such that:

$$\forall z \in \mathbb{C}, \quad x_h'(z) = -h''(z)(z, -1) \in \mathbb{C}(z, -1).$$

Naturally, we could have introduced complex hedgehogs of $\mathbb{C}^{n+1}$ in the complex contact geometry setting, where they appear as fronts of Legendrian immersions in $\mathbb{C}^{2n+1}$ (see the next subsection).

**Remark** Of course, many other parametrizations would have been possible in order to introduce the notion of a complex hedgehog. New parametrizations can simply be obtained by performing chart changes. For instance, for any holomorphic function $g : \mathbb{C} \to \mathbb{C}$, the complex curve

$$y_g : \mathbb{C} \to \mathbb{C}, x_h : z \longmapsto \left( g'(z), g(z) - zg'(z) \right)$$

is a hedgehog, namely the hedgehog with support function $f(z) = zg(1/z)$:

$$\forall z \in \mathbb{C}^*, y_g(z) = x_f\left(\frac{1}{z}\right).$$

Therefore, this particular parametrization change only corresponds to the chart change $z \mapsto 1/z$ on the Riemann sphere $P^1(\mathbb{C}) = \mathbb{C} \cup \{\infty\}$.

### 6.2.2 Complex Hedgehogs as Fronts in $\mathbb{C}^{n+1}$ of Legendrian Immersions in $\mathbb{C}^{2n+1}$

Consider the complex Euclidean space $\mathbb{C}^{2n+1}$ endowed with the holomorphic contact form

$$\omega := dw_0 + \sum_{j=1}^{n} z_j \, dw_j \, ,$$

where $(w_0, w_1, \ldots, w_n, z_1, \ldots, z_n)$ denote the canonical complex coordinate functions on $\mathbb{C}^{2n+1}$. Recall that the projection

$$\pi : \mathbb{C}^{n+1} \times \mathbb{C}^n = \mathbb{C}^{2n+1} \qquad \rightarrow \qquad \mathbb{C}^{n+1}$$
$$(w, x) = (w_0, w_1, \ldots, w_n, z_1, \ldots, z_n) \longmapsto w = (w_0, w_1, \ldots, w_n)$$

is called the front projection.

Then, for every holomorphic function $h : \mathbb{C}^n \to \mathbb{C}$, the map

$$i_h : \mathbb{C}^n \to \mathbb{C}^{n+1} \times \mathbb{C}^n = \mathbb{C}^{2n+1}$$
$$z \longmapsto (x_h(z), z)$$

is a Legendrian immersion of $\mathbb{C}^n$ into $\left(\mathbb{C}^{2n+1}, \omega\right)$ (that is, $i_h : \mathbb{C}^n \to \mathbb{C}^{2n+1}$ is a holomorphic immersion, and $(T_z i_h)(\mathbb{C}^n) \subset Ker\left[\omega_{i_h(z)}\right]$ for all $z \in \mathbb{C}^n$) of which $\mathcal{H}_h = x_h(\mathbb{C}^n)$ is the front $(\pi \circ i_h)(\mathbb{C}^n)$ in $\mathbb{C}^{n+1}$.

Indeed, for all $z = (z_1, \ldots, z_n) \in \mathbb{C}^n$ and $i \in \{1, \ldots, n\}$, we have

$$\frac{\partial x_h}{\partial z_i}(z) = \left( -\sum_{j=1}^{n} z_j \frac{\partial^2 h}{\partial z_i \partial z_j}(z), \frac{\partial^2 h}{\partial z_i \partial z_1}(z), \ldots, \frac{\partial^2 h}{\partial z_i \partial z_n}(z) \right),$$

and hence

$$\omega_{i_h(z)}\left( \frac{\partial x_h}{\partial z_i}(z), \frac{\partial Id_{\mathbb{C}^n}}{\partial z_i}(z) \right) = -\sum_{j=1}^{n} z_j \frac{\partial^2 h}{\partial z_i \partial z_j}(z) + \sum_{j=1}^{n} z_j \frac{\partial^2 h}{\partial z_i \partial z_j}(z) = 0.$$

### 6.2.3 Rational Hedgehogs of the Complex Projective Plane $P^2(\mathbb{C})$

Here, we choose to work in the complex projective plane $P^2(\mathbb{C})$ equipped with the usual Fubini-Study Kähler form $\omega$ (see e.g., [27]). For any $(X_1, X_2, X_3) \in$

$\mathbb{C}^3 \setminus \{0\}$, $[X_1, X_2, X_3]$ will denote the homogeneous coordinates of the equivalent class of $(X_1, X_2, X_3)$ in $P^2(\mathbb{C}) = \mathbb{C}^3/\mathbb{C}^*$.

Let $h : \mathbb{C} \to \mathbb{C}$ be a holomorphic map such that the projective curve $x_h : \mathbb{C} \to P^2(\mathbb{C})$, $z \mapsto [x_h(z), -1] = [zh(z) - h'(z), h'(z), -1]$ satisfies

$$Area\,[x_h(\mathbb{C})] < +\infty.$$

Then, the hedgehog curve $x_h : \mathbb{C} \to P^2(\mathbb{C})$ extends to a rational curve

$$x_h : P^1(\mathbb{C}) \to P^2(\mathbb{C})$$
$$z \longmapsto x_h(z),$$

which we call *the rational hedgehog* $\mathcal{H}_h := x_h\left[P^1(\mathbb{C})\right]$ with support function

$$h : P^1(\mathbb{C}) \to P^1(\mathbb{C}), z \longmapsto \begin{cases} h(z) & if \quad z \in \mathbb{C} \\[2ex] \lim_{z \to \infty} h(z) & if \quad z = \infty. \end{cases}$$

Indeed Ahlfors' lemma gives a description of rational curves as entire curves of bounded area [38]:

Let $X$ be a compact complex manifold and $f : \mathbb{C} \to X$ an entire curve (i.e., a non constant holomorphic map) such that $Area\,[f(\mathbb{C})] < +\infty$. Then $f$ extends to a holomorphic map from $P^1(\mathbb{C})$ to $X$, a rational curve.

## 6.3   Evolute of a Plane Complex Hedgehog as Locus of Its Centers of Curvature

In classical differential geometry of curves, the evolute of a plane curve is the locus of all its centers of curvature or, equivalently, the envelope of its normal lines. In particular, the evolute of a plane hedgehog $\mathcal{H}_h \subset \mathbb{R}^2$ with support function $h \in C^\infty(\mathbb{S}^1; \mathbb{R})$ is the locus of all its centers of curvature $c_h(\theta) := x_h(\theta) - R_h(\theta) u(\theta)$, where $R_h(\theta) := \det\left[T_{u(\theta)} x_h\right] = (h + h'')(\theta)$ is the so-called curvature function of $\mathcal{H}_h$, and $u(\theta) := (\cos\theta, \sin\theta)$, $(\theta \in \mathbb{S}^1 = \mathbb{R}/2\pi\mathbb{Z})$. Equivalently, the evolute of $\mathcal{H}_h$ can be defined as the envelope of its 'normal lines' $\mathcal{N}_h(\theta) := \{x_h(\theta)\} + \mathbb{R}u(\theta)$, that is, the hedgehog $\mathcal{H}_{\partial h}$ with support function $(\partial h)(\theta) := h'\left(\theta - \frac{\pi}{2}\right)$. Note that in the hedgehog case, the centers of curvature $c_h(\theta)$ are well defined for all $\theta \in \mathbb{S}^1$, even if $x'_h(\theta) = R_h(\theta) u(\theta)$ is the null vector, since the curvature function $R_h(\theta) = (h + h'')(\theta)$ is well defined for all $\theta \in \mathbb{S}^1$. Likewise, the normal line to $\mathcal{H}_h$ at $x_h(\theta)$ is well defined, even if $x'_h(\theta) = 0$, as the perpendicular $\mathcal{N}_h(\theta)$ to the support line $\langle x, u(\theta) \rangle = h(\theta)$ through the point $x_h(\theta)$. For plane real hedgehogs, it is convenient to keep in mind the following commutative diagram:

$$
\begin{array}{ccc}
\gamma_h : \mathbb{S}^1 \to P^2\,(\mathbb{R}) & \xleftrightarrow[\ \ast\ ]{\text{Projective duality}} & X_h : \mathbb{S}^1 \to \mathbb{R}^2 \subset P^2\,(\mathbb{R}) \\
\theta \longmapsto [\cos\theta, \sin\theta, h\,(\theta)] & & \theta \longmapsto (x_h\,(\theta), -1) \\[2ex]
\dfrac{d}{d\theta}\ \downarrow\ \ \text{derivation} & & \partial\ \downarrow\ \ \text{evolute} \\[2ex]
\gamma_h' : \mathbb{S}^1 \to P^2\,(\mathbb{R}) & \xleftrightarrow[\ \ast\ ]{\text{Projective duality}} & (c_h, -1) : \mathbb{S}^1 \to \mathbb{R}^2 \subset P^2\,(\mathbb{R}) \\
\theta \longmapsto \left[-\sin\theta, \cos\theta, h'\,(\theta)\right] & & \theta \longmapsto (c_h\,(\theta), -1)
\end{array}
$$

where $c_h\,(\theta) = x_{\partial h}\left(\theta + \frac{\pi}{2}\right)$, $(\theta \in \mathbb{S}^1)$. The main purpose of this subsection is to extend the notion of evolute to plane complex hedgehogs, together with its interpretation as locus of the centers of curvature. To this aim, we need to **change our way of interpreting the transformation**

$$
\begin{array}{ccc}
\dfrac{d}{d\theta} : \mathbb{S}^1 \subset \mathbb{R}^2 & \to & \mathbb{S}^1 \subset \mathbb{R}^2 \\
u\,(\theta) = (\cos\theta, \sin\theta) & \longmapsto & u'\,(\theta) = (-\sin\theta, \cos\theta)
\end{array}
$$

in the above diagram since we cannot consider the complex 'normal lines' to a complex hedgehog without antiholomorphic data being involved. Our choice is to identify $\mathbb{S}^1$ with the projective line $P^1\,(\mathbb{R}) = \mathbb{R} \cup \{\infty\}$ and thus to consider the transformation

$$
\begin{array}{ccc}
P^1\,(\mathbb{R}) = \mathbb{R} \cup \{\infty\} & \to & P^1\,(\mathbb{R}) = \mathbb{R} \cup \{\infty\} \\
[\cos\theta, \sin\theta] = x & \mapsto & [-\sin\theta, \cos\theta] = \dfrac{-1}{x}.
\end{array}
$$

**In the complex hedgehogs case, it is thus the following transformation that will play the same role**:

$$
\begin{array}{ccc}
P^1\,(\mathbb{C}) = \mathbb{C} \cup \{\infty\} & \to & P^1\,(\mathbb{C}) = \mathbb{C} \cup \{\infty\} \\
[1, z] = z & \mapsto & [z, -1] = \dfrac{-1}{z}
\end{array}.
$$

In other words, we are going to consider the envelope of the family $\left(L_h'\,(z)\right)_{z \in \mathbb{C}}$ of complex lines of $\mathbb{C}^2$ given by $L_h'\,(z) := \{x_h\,(z)\} + \mathbb{C}\,(z, -1)$. For all $z \in \mathbb{C}$, $L_h'\,(z)$ can be completed into a projective line $\widehat{L_h'\,(z)}$ of $P^2\,(\mathbb{C})$ with equation

$$
zX_1 - X_2 + \left(zh\,(z) - \left(1 + z^2\right)h'\,(z)\right)X_3 = 0,
$$

where $[X_1, X_2, X_3]$ denote the homogeneous coordinates of the equivalent class of $(X_1, X_2, X_3) \in \mathbb{C}^3 \setminus \{0\}$ in $P^2(\mathbb{C})$. Now, by projective duality, this family of projective lines $\left(\widehat{L'_h(z)}\right)_{z \in \mathbb{C}}$ corresponds to the complex curve that is parametrized by

$$
\begin{aligned}
\mathbb{C} &\to P^2(\mathbb{C}) \\
z &\longmapsto \left[z, -1, zh(z) - \left(1 + z^2\right) h'(z)\right].
\end{aligned}
$$

Note that for $z \neq 0$, we have

$$
\left[z, -1, zh(z) - \left(1 + z^2\right) h'(z)\right] = [1, w, (\partial h)(w)],
$$

where $w = \frac{-1}{z}$ and $(\partial h)(w) := h\left(\frac{-1}{w}\right) + \left(w + \frac{1}{w}\right) h'\left(\frac{-1}{w}\right)$.

Therefore, we have the following commutative diagram:

<table>
<tr><td>$\gamma_h : z \mapsto [1, z, h(z)]$</td><td>Projective duality<br>$*$<br>$\leftrightarrow$</td><td>$X_h : z \mapsto [x_h(z), -1]$</td></tr>
<tr><td>$\downarrow$</td><td></td><td>$\partial \quad \downarrow \quad$ evolute</td></tr>
<tr><td>$\gamma_{\partial h} : w = \frac{-1}{z} \mapsto [1, w, (\partial h)(w)]$</td><td>Projective duality<br>$*$<br>$\leftrightarrow$</td><td>$(c_h, -1) : z \mapsto \left[x_{\partial h}\left(\frac{-1}{z}\right), -1\right]$</td></tr>
</table>

where $c_h(z) := x_{\partial h}\left(\frac{-1}{z}\right) = x_h(z) - \left(1 + z^2\right) h''(z)(1, z)$. This expression of $c_h(z)$ has to be compared with the one giving the expression of the center of curvature of a real hedgehog $\mathcal{H}_h$ at a point $x_h(\theta)$: $c_h(\theta) = x_h(\theta) - R_h(\theta) u(\theta)$, where $R_h$ is the curvature function of $\mathcal{H}_h \subset \mathbb{R}^2$. We will see below that $c_h(z) := x_{\partial h}\left(\frac{-1}{z}\right) = x_h(z) - \left(1 + z^2\right) h''(z)(1, z)$ can actually be interpreted as the center of curvature of the complex hedgehog $\mathcal{H}_h$ at the point $x_h(z)$.

**Definition 6.3.1** Let $h : \mathbb{C} \to \mathbb{C}$ be a holomorphic function. We will say that the complex hedgehog with support function $(\partial h)(z) = h\left(\frac{-1}{z}\right) + \left(z + \frac{1}{z}\right) h'\left(\frac{-1}{z}\right)$ is the **evolute** of the complex hedgehog $\mathcal{H}_h$.

Fundamental examples. If $h$ is the holomorphic function defined on the open disc $D := \{z \in \mathbb{C} \, | \, |z| < 1\}$ by $h(z) = a_1 z + a_0 + \rho\sqrt{1 + z^2}$, where $(a_0, a_1, \rho) \in \mathbb{C}^3$, then the complex hedgehog $\mathcal{H}_h = x_h(D)$ is reduced to the point $\{(a_0, a_1)\}$ if $\rho = 0$, and it lies on the complex circle $\mathcal{C}((a_0, a_1); \rho) \subset \mathbb{C}^2$ with equation

$(X_1 - a_0)^2 + (X_2 - a_1)^2 = \rho^2$ if $\rho \neq 0$. In both cases, the evolute $\mathcal{H}_{\partial h} = c_h(D)$ is reduced to the point $\{(a_0, a_1)\}$. Indeed, for all $z \in \mathbb{C}$, $x_h(z) = \left(x_h^1(z), x_h^2(z)\right) = \left(h(z) - zh'(z), h'(z)\right)$ is such that

$$\left(x_h^1(z), x_h^2(z)\right) = \left(a_0 + \rho\sqrt{1+z^2} - z\frac{z}{\sqrt{1+z^2}}, a_1 + \frac{\rho z}{\sqrt{1+z^2}}\right)$$

$$= (a_0, a_1) + \rho\frac{(1, z)}{\sqrt{1+z^2}}$$

and

$$c_h(z) = x_h(z) - \left(1+z^2\right)h''(z)(1, z) = x_h(z) - \left(1+z^2\right)\frac{\rho}{\left(1+z^2\right)^{\frac{3}{2}}}\frac{(1, z)}{\sqrt{1+z^2}}$$

$$= (a_0, a_1).$$

More generally, let us replace $h : D \to \mathbb{C}$, $z \mapsto a_1 z + a_0 + \rho\sqrt{1+z^2}$ by any holomorphic function of the form $h : \mathcal{U} \to \mathbb{C}$, $z \mapsto a_1 z + a_0 + \rho q(z)$, where $\mathcal{U}$ is a connected open subset of $\mathbb{C}\setminus\{-i, i\}$, and $q(z)$ is the support function of the complex unit circle $\mathcal{C}((0, 0); 1)$ in the neighborhood of $z$, that is:

$$q(z) = \begin{cases} \sqrt{1+z^2} & \text{if} \quad |z| < 1 \\[2em] z\sqrt{1+\left(\dfrac{1}{z}\right)^2} & \text{if} \quad |z| > 1 \\[2em] \dfrac{z+\varepsilon}{\sqrt{2}}\sqrt{1+\left(\dfrac{z-\varepsilon}{z+\varepsilon}\right)^2} & \text{if} \quad sgn\,[\mathrm{Re}\,(z)] = \varepsilon \in \{-1, 1\}. \end{cases}$$

We leave it to the reader to check that : $(i)$ the complex hedgehog $\mathcal{H}_h = x_h(\mathcal{U})$ is reduced to the point $\{(a_0, a_1)\}$ if $\rho = 0$, and it lies on the complex circle $\mathcal{C}((a_0, a_1); \rho)$ with equation $(X_1 - a_0)^2 + (X_2 - a_1)^2 = \rho^2$ if $\rho \neq 0$ ; $(ii)$ moreover, in both cases, the evolute $\mathcal{H}_{\partial h} = c_h(\mathcal{U})$ is reduced to the point $\{(a_0, a_1)\}$.

**Definition 6.3.2** Let $\mathcal{H}_f$ and $\mathcal{H}_g$ be two complex hedgehogs in $\mathbb{C}^2$, and let $z_0 \in \mathbb{C}$ be such that $x_f(z_0) = x_g(z_0)$. We will say that $\mathcal{H}_f$ and $\mathcal{H}_g$ have a contact of order $\geq 2$ at $x_f(z_0) = x_g(z_0)$, if: $\forall m \in \{0, 1, 2\}$, $f^{(m)}(z_0) = g^{(m)}(z_0)$.

Given any complex hedgehog with holomorphic support function $h : \mathcal{U} \to \mathbb{C}$, where $\mathcal{U}$ is any connected open subset of $\mathbb{C}\setminus\{-i, i\}$, a straightforward computation shows that, for any $z_0 \in \mathcal{U}$, the hedgehog with support function

$$c : \mathcal{U} \to \mathbb{C}, z \mapsto c_h(z) := c_h^2(z_0)\,z + c_h^1(z_0) + q(z_0)^3\,h''(z_0)\,q(z),$$

(which is reduced to the point $\{c_h(z_0)\}$ if $h''(z_0) = 0$, or which lies on the complex circle with equation $\left(X_1 - c_h^1(z_0)\right)^2 + \left(X_2 - c_h^2(z_0)\right)^2 = q(z_0)^6 h''(z_0)^2$ if $h''(z_0) \neq 0$), has a contact of order $\geq 2$ with $\mathcal{H}_h = x_h(\mathcal{U})$ at $x_h(z_0)$.

**Definition 6.3.3** Let $h : \mathcal{U} \to \mathbb{C}$ be a holomorphic function where $\mathcal{U}$ is a connected subset of $\mathbb{C} \setminus \{-i, i\}$. For any $z_0 \in \mathcal{U}$, we will say that $c_h(z_0)$ is **the center of curvature** of $\mathcal{H}_h = x_h(\mathcal{U})$ at $x_h(z_0)$, and, if $z_0 \in \mathcal{U}$ is a regular point of $x_h : \mathcal{U} \to \mathbb{C}^2$ (that is, if $h''(z_0) \neq 0$), we will say that the complex circle with equation

$$\left(X_1 - c_h^1(z_0)\right)^2 + \left(X_2 - c_h^2(z_0)\right)^2 = q(z_0)^6 h''(z_0)^2$$

is **the osculating complex circle** of $\mathcal{H}_h$ at $x_h(z_0)$.

Naturally, we define the complex curvature function of a hedgehog $\mathcal{H}_h = x_h(\mathcal{U})$ as follows.

**Definition 6.3.4** Let $h : \mathcal{U} \to \mathbb{C}$ be a holomorphic function where $\mathcal{U}$ is a connected subset of $\mathbb{C} \setminus \{-i, i\}$. We define the **curvature function** of $\mathcal{H}_h = x_h(\mathcal{U})$ to be the function $R_h : \mathcal{U} \to \mathbb{C}$ that is given by $R_h(z) := q(z)^3 h''(z)$ for all $z \in \mathcal{U}$.

Thus, for any $z \in \mathcal{U}$, the center of curvature of $\mathcal{H}_h = x_h(\mathcal{U})$ at $x_h(z)$ can be expressed as follows:

$$c_h(z) = x_h(z) - R_h(z)u(z),$$

where

$$u(z) := \frac{1 + z^2}{q(z)^3}(1, z) \in \mathcal{C}((0, 0); 1).$$

Of course, this expression of $c_h(z)$ has to be compared with the one giving the expression of the center of curvature of a real hedgehog $\mathcal{H}_h$ at a point $x_h(\theta)$: $c_h(\theta) = x_h(\theta) - R_h(\theta)u(\theta)$, where $R_h$ is the curvature function of $\mathcal{H}_h \subset \mathbb{R}^2$.

**Remark** With our definitions, the complex circle $\mathcal{C}((a_0, a_1); \rho) \subset \mathbb{C}^2$ with equation $(X_1 - a_0)^2 + (X_2 - a_1)^2 = \rho^2$, where $(a_0, a_1, \rho) \in \mathbb{C}^2 \times \mathbb{C}^*$, can be locally regarded as a hedgehog with radius of curvature equal to $\rho$ (possibly after a suitable chart change on the Riemann sphere).

## 6.4   Real and Imaginary Parts of $\mathcal{H}_h$ in $\mathbb{C}^2$ Regarded as Hedgehogs of $\mathbb{R}^3$

We know that if $f$ and $g$ are taken to be the real and imaginary parts respectively of a holomorphic function $h : \mathbb{C} \to \mathbb{C}$, $z = x + iy \mapsto h(z) = f(x, y) + ig(x, y)$, then $f$ and $g$ are harmonic functions satisfying the Cauchy-Riemann equations, that is,

$$\frac{\partial f}{\partial x}(x, y) = \frac{\partial g}{\partial y}(x, y) \quad \text{and} \quad \frac{\partial f}{\partial y}(x, y) = -\frac{\partial g}{\partial x}(x, y),$$

for all $(x, y) \in \mathbb{R}^2$. The aim of this subsection is to show that, in this context, $f$ and $g$ determine two hedgehogs $\mathcal{H}_F$ and $\mathcal{H}_G$ of $\mathbb{R}^3$ that can be regarded globally as the orthogonal projections of the complex hedgehog $\mathcal{H}_h$ of $\mathbb{C}^2 \cong \mathbb{R}^4$ into $e_2^{\perp}$ and $e_1^{\perp}$ respectively, where $(e_1, e_2, e_3, e_4)$ is the canonical basis of $\mathbb{R}^4$ and where $e_i^{\perp}$ denotes the 3-dimensional subspace of $\mathbb{R}^4$ that is orthogonal to $e_i$ $(1 \leq i \leq 4)$. These hedgehogs $\mathcal{H}_F$ and $\mathcal{H}_G$ of $\mathbb{R}^3$ will be modeled on the hemisphere $\mathbb{S}_+^2$ of $\mathbb{S}^2 \subset \mathbb{R} \times \mathbb{C}$ that is contained in $\mathbb{R}_+^* \times \mathbb{C}$. To any $z \in \mathbb{C}$ we associate the point $v(z) := (1, z)/\sqrt{1 + |z|^2}$ of $\mathbb{S}_+^2$. The orthogonal projection map from $\mathbb{C}^2 \cong \mathbb{R}^4$ onto $e_i^{\perp}$ will be denoted by $\pi_{e_i^{\perp}}$.

**Definition 6.4.1** Let $h : \mathbb{C} \to \mathbb{C}$ be a holomorphic function the real and imaginary parts of which are $f$ and $g$ respectively:

$$h(x + iy) = f(x, y) + ig(x, y) \quad \text{for all } (x, y) \in \mathbb{R}^2.$$

We have then

$$\pi_{e_2^{\perp}}\left[x_h(\overline{z})\right] = x_F(v(z)) \quad \text{and} \quad \pi_{e_1^{\perp}}\left[x_h(i\overline{z})\right] = x_G(v(z)),$$

where $F$ and $G$ are respectively defined by:

$$F(v(z)) = \frac{\operatorname{Re}(h(\overline{z}))}{\sqrt{1 + |z|^2}} \quad \text{and} \quad G(v(z)) = \frac{\operatorname{Im}(h(i\overline{z}))}{\sqrt{1 + |z|^2}}$$

We will of course say that the hedgehogs $\mathcal{H}_F$ and $\mathcal{H}_G$ are **the real and imaginary hedgehog parts** of $\mathcal{H}_h$.

**Proof** We first note that an easy computation making use of the Cauchy-Riemann equations gives:

$$x_h(z) = (x_1(z) + iy_1(z), x_2(z) + iy_2(z)) \in \mathbb{C}^2$$

$$\cong (x_1(z), y_1(z), x_2(z), y_2(z)) \in \mathbb{R}^4,$$

where

$$\begin{cases} x_1\,(z) = f\,(x,\,y) - x\frac{\partial f}{\partial x}\,(x,\,y) - y\frac{\partial f}{\partial y}\,(x,\,y) \\[2ex] y_1\,(z) = g\,(x,\,y) - x\frac{\partial g}{\partial x}\,(x,\,y) - y\frac{\partial g}{\partial y}\,(x,\,y) \\[2ex] x_2\,(z) = \frac{\partial f}{\partial x}\,(x,\,y) = \frac{\partial g}{\partial y}\,(x,\,y) \\[2ex] y_2\,(z) = -\frac{\partial f}{\partial y}\,(x,\,y) = \frac{\partial g}{\partial x}\,(x,\,y), \end{cases}$$

for all $z = x + iy$, $\big((x,\,y) \in \mathbb{R}^2\big)$.

Next, we consider the positively 1-homogeneous function $F\,:\,\mathbb{R}_+^* \times \mathbb{R}^2 \to \mathbb{R}$ given by

$$F\,(X,\,Y,\,Z) := Xf\left(\frac{Y}{X},\,-\frac{Z}{X}\right) \quad \text{for all } (X,\,Y,\,Z) \in \mathbb{R}_+^* \times \mathbb{R}^2.$$

A straightforward computation then shows that the Euclidean gradient of $F$ is given by

$$\nabla F\,(X,\,Y,\,Z) = \left(f\left(\tfrac{Y}{X},\,-\tfrac{Z}{X}\right) - \tfrac{Y}{X}\tfrac{\partial f}{\partial x}\left(\tfrac{Y}{X},\,-\tfrac{Z}{X}\right) + \tfrac{Z}{X}\tfrac{\partial f}{\partial y}\left(\tfrac{Y}{X},\,-\tfrac{Z}{X}\right),\right.$$

$$\tfrac{\partial f}{\partial x}\left(\tfrac{Y}{X},\,-\tfrac{Z}{X}\right),$$

$$\left.-\tfrac{\partial f}{\partial y}\left(\tfrac{Y}{X},\,-\tfrac{Z}{X}\right)\right)$$

for all $(X,\,Y,\,Z) \in \mathbb{R}_+^* \times \mathbb{R}^2$. Thus,

$$x_F\,(v\,(z)) = \nabla F\,(v\,(z))$$

$$= \left(f\,(x,\,-y) - x\frac{\partial f}{\partial x}\,(x,\,-y) - (-y)\frac{\partial f}{\partial y}\,(x,\,-y),\,\frac{\partial f}{\partial x}\,(x,\,-y),\right.$$
$$\left.-\frac{\partial f}{\partial y}\,(x,\,-y)\right)$$

$$= (x_1\,(\overline{z}),\,x_2\,(\overline{z}),\,y_2\,(\overline{z})) = \pi_{e_2^\perp}\,[x_h\,(\overline{z})],$$

for all $z = x + iy$, $\big((x,\,y) \in \mathbb{R}^2\big)$. In the same manner, we can easily check that

$$x_F\,(v\,(z)) = (y_1\,(i\overline{z}),\,x_2\,(i\overline{z}),\,y_2\,(i\overline{z})) = \pi_{e_1^\perp}\,[x_h\,(i\overline{z})] \quad \text{for all } z \in \mathbb{C}.$$

$\blacksquare$

## 6.5   Towards a Brunn-Minkowski Theory for Complex Hedgehogs

As already mentioned above, the notion of a hedgehog curve or surface was born from the study of the Brunn-Minkowski theory. It is therefore permissible to think that the development of a 'theory of mixed volumes for complex hedgehogs' (replacing Euclidean volumes by symplectic ones) might be a promising way of research. In this section, we will just mention first two observations.

### 6.5.1   Mixed Symplectic Area

Let $\mathbb{C}^2$ be the complex Euclidean space endowed with the standard Hermitian inner product $\langle ., . \rangle_{\mathbb{C}^2}$. We are interested in the symplectic area of complex hedgehogs in this Kähler manifold $\left(\mathbb{C}^2, J, \omega\right)$, where $J$ is the complex structure and $\omega$ the 2-form $\omega(X, Y) := \operatorname{Re}\left(\langle JX, Y \rangle_{\mathbb{C}^2}\right)$. Any nontrivial complex hedgehog of $\mathbb{C}^2$ modeled on the unit open disc $\mathbb{D}$ of $\mathbb{C}$ is a holomorphic curve (i.e., a nonconstant map from the complex plane to $\mathbb{C}^2$). Now, it is well-known that the Riemannian area of holomorphic curves is equal to their symplectic area, and hence that holomorphic curves have positive area (the reader that is not familiar with holomorphic curves can find details in SubSect. 1.1 of [190]). An immediate consequence is the following result, which has to be compared with classical geometric inequalities for convex bodies (see [167, p. 369 and p. 382]).

**Theorem 6.5.1**  *Let $\mathcal{H}(\mathbb{D})$ be the complex linear space of holomorphic functions $h : \mathbb{D} \to \mathbb{C}$ defined up to a similitude and consider*

$$Area\left[x_h(\mathbb{D})\right] := \int_{x_h(\mathbb{D})} \omega.$$

*Then the map $\sqrt{A} : \mathcal{H}(\mathbb{D}) \to \mathbb{R}_+, h \mapsto \sqrt{Area\left[x_h(\mathbb{D})\right]}$ is a norm associated with a scalar product $(h, k) \longmapsto A(h, k)$, which can be interpreted as a **mixed symplectic area**. In particular, for any $(h, k) \in \mathcal{H}(\mathbb{D})^2$, we have*

$$\sqrt{A(h + k)} \leq \sqrt{A(h)} + \sqrt{A(k)}$$

*and*

$$A(h, k)^2 \leq A(h)\, A(k),$$

*with equalities if, and only if, $\mathcal{H}_h$ and $\mathcal{H}_k$ are homothetic (here, "homothetic" means that there exists $(\lambda, \mu) \in \mathbb{R}^2 - \{(0, 0)\}$ such that $\lambda h + \mu k = 0$).*

### 6.5.2   A Sharp Estimation of the Area Using the Energy

Note that we have the following sharp estimate of $Area\,[x_h\,(\mathbb{D})]$, which is better than that well-known for an arbitrary smooth loop $\gamma : \mathbb{S}^1 \to V$ in a symplectic vector space $(V, \omega)$ (namely, $|A\,(\gamma)| \leq E\,(\gamma)$, see for instance [123, pp. 87-88]):

**Theorem 6.5.2** *Assume that $h : \mathbb{D} \to \mathbb{C}$ is the sum of a power series $\sum h_n z^n$ with radius of convergence $R > 1$:*

$$h\,(z) = \sum_{n=0}^{+\infty} h_n z^n \quad \text{for all } z \in \mathbb{D}.$$

*Then*

$$Area\,[x_h\,(\mathbb{D})] \leq \frac{3}{4} E\,(x_h),$$

*where $E\,(x_h)$ is the energy of the loop $x_h : \mathbb{S}^1 = \mathbb{R}/2\pi\mathbb{Z} \to \mathbb{C}^2, \theta \mapsto x_h\left(e^{i\theta}\right)$, that is:*

$$E\,(x_h) := \frac{1}{2} \int_0^{2\pi} \left| \frac{d}{d\theta} \left[ x_h\left(e^{i\theta}\right) \right] \right|^2 d\theta.$$

*Furthermore, the equality holds if, and only if, the function $h$ is of the form $h\,(z) = a_m z^m + a_1 z + a_0$, where $m \in \mathbb{N}$ and $(a_0, a_1, a_m) \in \mathbb{C}^3$.*

**Proof** Consider the Fourier expansion of $H\,(\theta) := h\left(e^{i\theta}\right)$ on $\mathbb{S}^1 = \mathbb{R}/2\pi\mathbb{Z}$:

$$H\,(\theta) := \sum_{n=0}^{+\infty} h_n e^{in\theta}.$$

An easy computation immediately gives:

$$\forall \theta \in \mathbb{S}^1, \quad x_h\,(\theta) := \sum_{n=0}^{+\infty} e^{in\theta}\left((1-n)\,h_n, (n+1)\,h_{n+1}\right).$$

Using the formula known for the action $A\,(\gamma) := (1/2) \int_0^{2\pi} \omega\left(\gamma\,(\theta), \gamma'\,(\theta)\right) d\theta$ of an arbitrary smooth loop $\gamma : \mathbb{S}^1 \to \mathbb{C}^2$ (see e.g., at the top of the page 88 in [123]), we then deduce that:

$$Area\,[x_h\,(\mathbb{D})] = -A\,(x_h) = \pi \sum_{n=0}^{+\infty} n\left((n-1)^2 |h_n|^2 + (n+1)^2 |h_{n+1}|^2\right).$$

Separating into two sums and re-indexing in the first one, we then obtain:

$$Area\left[x_h\left(\mathbb{D}\right)\right] = \pi \sum_{n=1}^{+\infty} n\left(n+1\right)\left(2n+1\right)\left|h_{n+1}\right|^2 = 6\pi \sum_{n=1}^{+\infty}\left(\sum_{k=1}^{n} k^2\right)\left|h_{n+1}\right|^2.$$

On the other hand, we have: $\forall \theta \in \mathbb{S}^1$,

$$\frac{d}{d\theta}\left[x_h\left(e^{i\theta}\right)\right] = e^{-i\theta}\left(H'\left(\theta\right) + iH''\left(\theta\right)\right)\left(e^{i\theta}, -1\right)$$

and hence

$$\left|\frac{d}{d\theta}\left[x_h\left(e^{i\theta}\right)\right]\right|^2 = 2\left|H' + iH''\right|\left(\theta\right)^2 = 2\left|H'' - iH'\right|\left(\theta\right)^2 = 2\left|h''\left(e^{i\theta}\right)\right|^2.$$

Therefore Parseval's identity yields:

$$E\left(x_h\right) := \frac{1}{2}\int_0^{2\pi}\left|\frac{d}{d\theta}\left[x_h\left(e^{i\theta}\right)\right]\right|^2 d\theta = \int_0^{2\pi}\left|h''\left(e^{i\theta}\right)\right|^2 d\theta$$

$$= 8\pi \sum_{n=1}^{+\infty}\left(\sum_{k=1}^{n} k\right)^2 \left|h_{n+1}\right|^2,$$

since

$$h''\left(z\right) = \sum_{n=1}^{+\infty} n\left(n+1\right)h_{n+1}z^{n-1} = 2\sum_{n=1}^{+\infty}\left(\sum_{k=1}^{n} k\right)h_{n+1}z^{n-1}.$$

This completes the proof.                                                                                            ∎

## 6.6   Real Hedgehogs in $\mathbb{C}^n$ and Their Evolutes

In the Euclidean plane, the evolute of a hedgehog is the locus of all its centers of curvature or, equivalently, the envelope of its normal lines. In order to find an analogue in any even higher dimension, we make use of the following trick. First, we fix a linear complex structure $J$ on $\mathbb{R}^{2n}$ (that is, an endomorphism $J$ of $\mathbb{R}^{2n}$ such that $J^2 = -Id_{\mathbb{R}^{2n}}$). Given any hedgehog with smooth support function $h$ in $\mathbb{R}^{2n}$, we then define the normal hyperplane to $\mathcal{H}_h$ at a point $x_h\left(u\right)$, say $N_h\left(u\right)$, as the affine hyperplane $\{x_h\left(u\right)\} + J\left(u^{\perp}\right)$, where $u^{\perp}$ is the $\left(2n-1\right)$-dimensional subspace of $\mathbb{R}^{2n}$ that is orthogonal to $u$. Finally, we define the evolute of $\mathcal{H}_h$ in $\left(\mathbb{R}^{2n}, J\right)$ as the envelope of the family of normal hyperplanes $(N_h(u))_{u \in \mathbb{S}^{2n-1}}$ in $\mathbb{R}^{2n}$. Let us begin by considering carefully the four dimensional case.

## 6.6.1   *Evolutes of Hedgehogs Hypersurfaces in $\mathbb{R}^4$*

In what follows, we identify $\mathbb{R}^4$ with the quaternion algebra $\mathbb{H}$ and thus the unit sphere $\mathbb{S}^3$ with the set $\mathbb{S}^1_{\mathbb{H}}$ of unit quaternions. To any pure unit quaternion $v$, we associate the linear complex structure $J_v : \mathbb{R}^4 \to \mathbb{R}^4$, $x \longmapsto vx$. We denote by $\omega_v$ the associated Kähler form (i.e. the alternating 2-form $\omega_v(X, Y) = \langle J_v X, Y \rangle$, where $\langle ., . \rangle$ is the standard Euclidean metric on $\mathbb{R}^4$). Recall that we can retrieve $\langle ., . \rangle$ from $\omega_v$: $\langle X, Y \rangle = \omega_v(X, J_v Y)$. Particularizing our definition of evolute hedgehogs to the four dimensional case, we get the following definition.

**Definition 6.6.1** Let $h \in C^\infty(\mathbb{S}^3; \mathbb{R})$. We define the **evolute of $\mathcal{H}_h$ in the Kähler vector space** $(\mathbb{R}^4, J_v, \omega_v)$ to be the envelope of the family of normal hyperplanes $(N_h^v(u))_{u \in \mathbb{S}^3}$ with equation

$$\langle x - x_h(u), J_v(u) \rangle = 0.$$

**Proposition 6.6.1** *Let $h \in C^\infty(\mathbb{S}^3; \mathbb{R})$. The evolute of $H_h$ in $(\mathbb{R}^4, J_v, \omega_v)$ is the hedgehog $\mathcal{H}_{\partial_v h}$ with support function*

$$\partial_v h : \mathbb{S}^3 \to \mathbb{R}, \, u \longmapsto \langle \nabla h(-J_v(u)), u \rangle,$$

*where $\langle ., . \rangle$ is the standard Euclidean metric on $\mathbb{R}^4$ and $\nabla h$ the gradient of $h$.*

**Proof** Since $J_v : \mathbb{R}^4 \to \mathbb{R}^4$ is an isometry such that $J_v^2 = -Id_{\mathbb{R}^4}$, the evolute of $\mathcal{H}_h$ in $(\mathbb{R}^4, J_v, \omega_v)$ can be regarded as the envelope of the family of hyperplanes $(N_h^v(-J_v(u)))_{u \in \mathbb{S}^3}$ with equation

$$\langle x - x_h(-J_v(u)), u \rangle = 0,$$

that is, as the hedgehog $\mathcal{H}_{\partial_v h}$ of $\mathbb{R}^4$ with support function

$$\partial_v h(u) = \langle x_h(-J_v(u)), u \rangle = \langle \nabla h(-J_v(u)), u \rangle.$$

∎

*By abuse of language, the hedgehog with support function $\partial_v h$ will also be called the evolute of $\mathcal{H}_h$ with respect to (the pure unit) quaternion $v$.*

**Parametrization of the Evolute of $\mathcal{H}_{\partial_v h}$ and Interpretation**

It follows immediately from definitions that $x_{\partial_v h} : \mathbb{S}^3 \to \mathbb{R}^4$ associates with each $u \in \mathbb{S}^3$ the unique solution of the system

$$\begin{cases} \langle x, J_v(u) \rangle = \langle x_h(u), J_v(u) \rangle \\ \forall X \in T_u \mathbb{S}^3, \ \langle x, J_v(X) \rangle = \langle T_u x_h(X), J_v(u) \rangle + \langle x_h(u), J_v(X) \rangle, \end{cases}$$

which is equivalent to

$$\begin{cases} \langle x - x_h(u), J_v(u) \rangle = 0 \\ \forall X \in T_u \mathbb{S}^3, \ \langle x - x_h(u), J_v(X) \rangle = \langle J_v(T_u x_h(J_v(u))), J_v(X) \rangle, \end{cases}$$

because $\langle T_u x_h(X), J_v(u) \rangle = \langle T_u x_h(J_v(u)), X \rangle = \langle J_v(T_u x_h(J_v(u))), J_v(X) \rangle$ since $T_u x_h$ is a symmetric endomorphism of $T_u \mathbb{S}^3$ and $J_v$ an isometry of $\mathbb{R}^4$. Therefore:

$$\forall u \in \mathbb{S}^3, \quad x_{\partial_v h}(u) = x_h(u) + J_v(T_u x_h(J_v(u))) = x_h(u) + v T_u x_h(J_v(u)).$$

In other words, we have the following.

**Proposition 6.6.2** *Let $h \in C^\infty(\mathbb{S}^3; \mathbb{R})$, and let $v$ be a pure unit quaternion. For all $u \in \mathbb{S}^3$,*

$$x_{\partial_v h}(u) = x_h(u) - R_h(u, v) u,$$

*where $R_h(u, v) := -v T_u x_h(J_v(u)) \overline{u}$; here $\overline{u}$ of course refers to the quaternion conjugate of $u$.*

**Comparison to the Planar Case and Interpretation**

This expression of $x_{\partial_v h}(u)$ has to be compared to the one of the center of curvature of a plane hedgehog $\mathcal{H}_h$ at a point $x_h(\theta)$:

$$c_h(\theta) := x_h(\theta) - R_h(\theta) u(\theta),$$

where $R_h := h + h''$ is the curvature function of $\mathcal{H}_h$. Identifying $\mathbb{R}^2$ with $\mathbb{C}$, and thus $T_{u_\theta} \mathbb{S}^1$ with $\mathbb{R}(i e^{i\theta})$, this last formula can be rewritten as

$$c_h\left(e^{i\theta}\right) := x_h\left(e^{i\theta}\right) - R_h\left(e^{i\theta}\right) e^{i\theta},$$

where $R_h\left(e^{i\theta}\right) := -i T_{u_0} x_h\left(i e^{i\theta}\right) e^{-i\theta}$.

We will see below that:

$R_h\left(.,v\right) : u_\theta := \left(\cos\theta\right)u + \left(\sin\theta\right)vu \longmapsto R_h\left(u_\theta,v\right)$ *can be interpreted as a* **quaternionic curvature function** *of* $x_h\left(\mathbb{S}^1_{u,v}\right)$, *where* $\mathbb{S}^1_{u,v}$ *denotes the unit circle of the vector plane* $\mathbb{C}\left(u,v\right) := \mathbb{R}u + \mathbb{R}J_v\left(u\right)$ *oriented by* $\left(u, J_v\left(u\right)\right)$.

The reason why the map $\theta \longmapsto R_h\left(e^{i\theta}\right)$ is real for $h \in C^\infty\left(\mathbb{S}^1;\mathbb{R}\right)$, whereas $u \mapsto R_h\left(u,v\right)$ is quaternionic for $h \in C^\infty\left(\mathbb{S}^3;\mathbb{R}\right)$, is because the product of two purely imaginary complex numbers is a real number, whereas the product of two purely imaginary quaternions can have both nontrivial real and imaginary parts.

**Complement to the Planar Case**

We introduced "the" evolute $\mathcal{H}_{\partial h}$ of a plane hedgehog $\mathcal{H}_h$ as the envelope of its normal lines. But in fact there are two of them if we take into account the choice of coorientation of the normal line. Of course, we could have introduced evolutes of hedgehog curves in $\mathbb{R}^2$ in the same way as we have just done for evolutes of hedgehog hypersurfaces in $\mathbb{R}^4$. Identifying $\mathbb{R}^2$ with the complex plane $\mathbb{C}$, we can associate to any $v \in \{-i, i\}$ the linear complex structure $J_v : \mathbb{C} \to \mathbb{C}, x \longmapsto vx$ and the associated Kähler form $\omega_v\left(X, Y\right) = \langle J_v X, Y\rangle$, where $\langle .,.\rangle$ is the standard Euclidean metric on $\mathbb{R}^2$, and then, define the evolute of the plane hedgehog with support function $h \in C^\infty\left(\mathbb{S}^1;\mathbb{R}\right)$ in $\left(\mathbb{R}^2, J_v\right)$ to be the envelope $\mathcal{H}_{\partial_v h}$ of the family of normal lines $\left(N^v_h\left(u\right)\right)_{u \in \mathbb{S}^1}$ with equation

$$\langle x - x_h\left(u\right), J_v\left(u\right)\rangle = 0.$$

If we do so, we can immediately check that $\mathcal{H}_{\partial_v h}$ has support function

$$\partial_v h : \mathbb{S}^1 \to \mathbb{R}, u \longmapsto \langle \nabla h\left(-J_v\left(u\right)\right), u\rangle.$$

In other words, $\left(\partial_i h\right)\left(\theta\right) = h'\left(\theta - \pi/2\right)$ and $\left(\partial_{-i} h\right)\left(\theta\right) = -h'\left(\theta + \pi/2\right)$ for all $\theta \in \mathbb{S}^1 = \mathbb{R}/2\pi\mathbb{Z}$.

Note that, in the 2 or 4-dimensional case, the evolutes $\mathcal{H}_{\partial_v h}$ and $\mathcal{H}_{\partial_{-v} h}$ are one and the same hypersurface of $\mathbb{R}^{2n}$ ($n = 1, 2$) but corresponding to opposite coorientations of the normal hyperplanes of $\mathcal{H}_h$:

$$\left(\partial_{-v} h\right)\left(u\right) = \langle \nabla h\left(-J_{-v}\left(u\right)\right), u\rangle = -\langle \nabla h\left(-J_v\left(-u\right)\right), -u\rangle = -\left(\partial_v h\right)\left(-u\right).$$

**Geometrical Interpretation of the Hodge Laplacian**

Taking the Hodge Laplacian of $h \in C^\infty\left(\mathbb{S}^1;\mathbb{R}\right)$ is tantamount to taking the evolute in $\left(\mathbb{R}^2, J_i\right)$ of the evolute of $\mathcal{H}_h$ in $\left(\mathbb{R}^2, J_{-i}\right)$, or conversely, the evolute in $\left(\mathbb{R}^2, J_{-i}\right)$ of the evolute of $\mathcal{H}_h$ in $\left(\mathbb{R}^2, J_i\right)$. Indeed, for any $h \in C^\infty\left(\mathbb{S}^1;\mathbb{R}\right)$, we have $\left(\partial_i \circ \partial_{-i}\right)(h) = \left(\partial_{-i} \circ \partial_i\right)(h) = -h'' = \Delta h$, where $\Delta$ is the Hodge Laplacian on $\mathbb{S}^1$.

This result can be extended as follows to dimension 4. Let $h \in C^\infty\left(\mathbb{S}^3; \mathbb{R}\right)$ and $u \in \mathbb{S}^3$. If $v$ is a pure unit quaternion such that $J_v(u)$ is an eigenvector of the Hessian $\left(\nabla^2 h\right)_u$ of $h$ at $u$ corresponding to the eigenvalue $\lambda$, then:

$$\partial_{-v}\left(\partial_v h\right)(u) = -\partial_v\left(\partial_v h\right)(-u) = -\partial_v^2 h\left(-u\right) = -\partial_v^2 h\left(J_v^2(u)\right)$$

$$= -\left(\nabla^2 h\right)_u\left(J_v(u), J_v(u)\right) = -\lambda.$$

Therefore, if $v_1$, $v_2$, $v_3$ are pure unit quaternions such that $J_{v_1}(u)$, $J_{v_2}(u)$, $J_{v_3}(u)$ are eigenvectors of the Riemannian Hessian $\left(\nabla^2 h\right)_u$, corresponding to eigenvalues $\lambda_1, \lambda_2, \lambda_3$, that form an orthonormal basis of $T_u\mathbb{S}^3$, then:

$$\Delta h(u) = -\sum_{i=1}^3 \lambda_i(u) = \sum_{i=1}^3 \left(\partial_{-v_i} \circ \partial_{v_i}\right)(h)(u).$$

### Decomposition of Hedgehogs into Sums of Remarkable Pedal Hypersurfaces

Let $(v, w)$ be any couple of pure unit quaternions that are orthogonal when they are regarded as vectors of $\mathbb{R}^4$. The quadruple $(1, v, w, vw)$ is then a direct orthonormal basis of $\mathbb{H} \cong \mathbb{R}^4$. For any hedgehog of $\mathbb{R}^4$ with support function $h \in C^\infty\left(\mathbb{S}^3; \mathbb{R}\right)$ and, for any $u \in \mathbb{S}^3$, we have the following decompositions

$$\begin{aligned}
x_h(u) &= h(u)\,u + \nabla h(u) \\
&= h(u)\,u + \left(\langle \nabla h(u), vu\rangle\, vu + \langle \nabla h(u), wu\rangle\, wu + \langle \nabla h(u), vwu\rangle\, vwu\right) \\
&= h(u)\,u + \partial_v h(vu)\, vu + \partial_w h(wu)\, wu + \partial_{vw} h(vwu)\, vwu \\
&= \left(h(u) + \partial_v h(vu)\, v + \partial_w h(wu)\, w + \partial_{vw} h(vwu)\, vw\right) u
\end{aligned}$$

In particular, the hedgehog $x_h : \mathbb{S}^3 \to \mathbb{H} \cong \mathbb{R}^4$ is the sum of parametrizations of 4 remarkable pedal surfaces: its own pedal surface and the pedal surfaces of its evolutes with respect to $v$, $w$, $vw$ (it being understood that, for all $u \in \mathbb{S}^3$, and, any pure unit quaternion $q$, we take the foot of the perpendicular from the origin to the support hyperplane with unit normal vector $J_q(u) := qu$).

### Evolutes and Orthogonal Projections

For every $(u, v) \in \mathbb{S}^3 \times \mathbb{S}^2$, let $\mathbb{S}^1_{u,v}$ be the oriented geodesic of $\mathbb{S}^3$ through $u$ in the direction of $J_v(u)$. This oriented circle of $\mathbb{S}^3$ can be regarded as the unit circle of the vector plane $\mathbb{C}(u, v) := \mathbb{R}u + \mathbb{R}J_v(u)$ oriented by $(u, J_v(u))$. *Restriction of support functions to $\mathbb{S}^1_{u,v}$ commutes with taking the evolutes in $\left(\mathbb{R}^4, J_v, \omega_v\right)$:*

**Proposition 6.6.3** *Let $h \in C^{\infty}\left(\mathbb{S}^3; \mathbb{R}\right)$. For all $v \in \mathbb{S}^2 \cong \mathbb{S}^1_{\mathbb{H}} \cap \mathrm{Im}\,(\mathbb{H})$,*

$$(\partial_v h)_{|\mathbb{S}^1_{u,v}} = \partial_v \left(h_{|\mathbb{S}^1_{u,v}}\right).$$

***Proof*** Define $u_\theta := (\cos\theta)\,u + (\sin\theta)\,J_v\,(u)$ for all $\theta \in \mathbb{S}^1$. We have then

$$(\partial_v h)\,(J_v\,(u_\theta)) = \langle \nabla h\,(u_\theta)\,,\,J_v\,(u_\theta)\rangle = \frac{d}{d\theta}\,[h\,(u_\theta)] = \partial_v \left(h_{|\mathbb{S}^1_{u,v}}\right)(J_v\,(u_\theta)).$$

$\blacksquare$

### Higher Order Evolutes

Of course, we can define inductively higher order evolutes. Let $\partial_v^0 h = h$ and, for any positive integer $n$, define the *$n$th* evolute of $\mathcal{H}_h$ in $\left(\mathbb{R}^4,\,J_v,\,\omega_v\right)$ to be the hedgehog with support function $\partial_v^n h := \partial_v \left(\partial_v^{n-1} h\right)$.

**Proposition 6.6.4** *Let $C^{\infty}\left(\mathbb{S}^3; \mathbb{R}\right)$. For all $n \in \mathbb{N}^*$, $v \in \mathbb{S}^2$, and $u \in \mathbb{S}^3$,*

$$\left(\partial_v^n h\right)\left(J_v^n\,(u)\right) = \frac{d^n}{d\theta^n}\,[h\,(u_\theta)]_{|\theta=0}\,,$$

*where $u_\theta := (\cos\theta)\,u + (\sin\theta)\,J_v\,(u)$.*

***Proof*** By induction, we deduce from the previous proposition that

$$\forall n \in \mathbb{N}^*, \quad \left(\partial_v^n h\right)_{|\mathbb{S}^1_{u,v}} = \partial_v^n \left(h_{|\mathbb{S}^1_{u,v}}\right),$$

and the result follows immediately. $\blacksquare$

## *6.6.2   Symplectic and Mixed Symplectic Area*

Any pure unit quaternion $v$ determines a linear complex structure $J_v : \mathbb{R}^4 \to \mathbb{R}^4$, to which it corresponds a Hopf flow induced on $\mathbb{S}^3 = \mathbb{S}^1_{\mathbb{H}}$ by the vector field $X_v\,(u) := J_v\,(u)$. We denote by $\mathbb{S}^2$ the set $\mathbb{S}^3 \cap \mathrm{Im}\,(\mathbb{H})$ of pure unit quaternions. For every $(u, v) \in \mathbb{S}^3 \times \mathbb{S}^2$, let $\mathbb{S}^1_{u,v}$ be the oriented geodesic of $\mathbb{S}^3$ through $u$ in the direction of $J_v\,(u)$. This oriented Hopf circle of $\mathbb{S}^3 \subset \left(\mathbb{R}^4,\,J_v\right)$ can be regarded as the unit circle of the vector plane $\mathbb{C}\,(u, v) := \mathbb{R}u + \mathbb{R}J_v\,(u)$ oriented by $(u, J_v\,(u))$. Conversely, any oriented vector plane $\xi$ in $\mathbb{R}^4$ determines an oriented unit circle $\mathbb{S}^1_\xi = \mathbb{S}^3 \cap \xi$ and a pure unit quaternion $v_\xi$ that is such that: $\forall u \in \mathbb{S}^1_\xi$, $T_u \mathbb{S}^1_\xi$ is oriented by the unit vector $J_{v_\xi}\,(u)$.

Now, consider the integral

$$s_\xi \left( h \right) := \int_{x_h \left( \mathbb{S}^1_\xi \right)} \alpha_{v_\xi},$$

where $\alpha_{v_\xi}$ is the 1-form given by $\left( \alpha_{v_\xi} \right)_x (dx) = \frac{1}{2} \omega_{v_\xi} (x, dx)$, which is such that $d\alpha_{v_\xi} = \omega_{v_\xi}$. This integral does not depend on the orientation of the plane $\xi$ (if we change the orientation of $\xi$, the orientation of the curve $x_h : \mathbb{S}^1_\xi \to \mathbb{R}^4$ changes as well and the 1-form $\alpha_{v_\xi}$ is changed into its opposite). Therefore, $s_\xi (h)$ can be defined for any unoriented vector plane $\xi$ in $\mathbb{R}^4$. It will be called *the symplectic area of* $x_h \left( \mathbb{S}^1_\xi \right)$ *relative to* $\xi$.

**Expression of the Symplectic Area of $x_h \left( \mathbb{S}^1_{u,v} \right)$**

Let $s_{u,v} (h)$ be this symplectic area:

$$s_{u,v} (h) := \int_{x_h \left( \mathbb{S}^1_{u,v} \right)} \alpha_v,$$

where $\alpha_v$ is the 1-form given by $(\alpha_v)_x (dx) := \frac{1}{2} \omega_v (x, dx) = \frac{1}{2} \langle x, (-J_v) (dx) \rangle$.

**Proposition 6.6.5** *For all $h \in C^\infty \left( \mathbb{S}^3; \mathbb{R} \right)$ and $(u, v) \in \mathbb{S}^3 \times \mathbb{S}^2$,*

$$s_{u,v} (h) = \frac{1}{2} \int_0^{2\pi} \langle x_h (u_\theta), R_h (u_\theta, v) u_\theta \rangle \, d\theta,$$

*where $u_\theta := (\cos \theta) u + (\sin \theta) J_v (u)$ and $R_h (u_\theta, v) := -v \left( T_{u_\theta} x_h \right) (J_v (u_\theta)) \overline{u_\theta}$, with $\overline{u_\theta}$ being the quaternion conjugate of $u_\theta$.*

**Proof** By definition

$$s_{u,v} (h) := \int_{x_h \left( \mathbb{S}^1_{u,v} \right)} \alpha_v = \frac{1}{2} \int_0^{2\pi} \left\langle x_h (u_\theta), (-J_v) \left( \frac{d}{d\theta} [x_h (u_\theta)] \right) \right\rangle d\theta.$$

Now

$$\frac{d}{d\theta} [x_h (u_\theta)] = \left( T_{u_\theta} x_h \right) (J_v (u_\theta))$$

and hence

$$(-J_v) \left( \frac{d}{d\theta} [x_h (u_\theta)] \right) = -v \left( T_{u_\theta} x_h \right) (J_v (u_\theta)) = R_h (u_\theta, v) u_\theta.$$

∎

**Proposition 6.6.6** *For all $h \in C^\infty\left(\mathbb{S}^3; \mathbb{R}\right)$ and $(u, v) \in \mathbb{S}^3 \times \mathbb{S}^2$,*

$$s_{u,v}(h) = a_{u,v}(h) + s_{u,v}^\perp(\nabla h)$$

*where $a_{u,v}(h)$ is the algebraic area of the hedgehog of $\mathbb{C}(u,v) = \mathbb{R}u + \mathbb{R}J_v(u)$ whose support function is the restriction of $h$ to $\mathbb{S}_{u,v}^1$, and where $s_{u,v}^\perp(\nabla h)$ is the symplectic area of $\nabla h\left(\mathbb{S}_{u,v}^1\right)$ in the Kähler vector space $\left(\mathbb{R}^4, J_v, \omega_v\right)$, that is,*

$$s_{u,v}^\perp(\nabla h) := \int_{\nabla h\left(\mathbb{S}_{u,v}^1\right)} \alpha_v = \frac{1}{2}\int_{\nabla h\left(\mathbb{S}_{u,v}^1\right)} \omega_v(x, dx).$$

***Proof*** It is just the fact that the symplectic area of a closed curve in the Kähler vector space $\left(\mathbb{R}^4, J_v, \omega_v\right)$ is the sum of the algebraic areas of its projections onto the planes $\mathbb{C}(u, v)$ and $\mathbb{C}(u, v)^\perp$. In the present case, we can retrieve the result as follows.

Let $\theta \in \mathbb{S}^1$. We have $x_h(u_\theta) = h(u_\theta)u_\theta + \nabla h(u_\theta)$, and

$$
\begin{aligned}
R_h(u_\theta, v)u_\theta &= (-J_v)\big(T_{u_\theta}x_h(J_v(u_\theta))\big) \\
&= (-J_v)\big(h(u_\theta)J_v(u_\theta) + \nabla_{J_v(u_\theta)}\nabla h(u_\theta)\big) \\
&= h(u_\theta)u_\theta + (-J_v)\big(\nabla_{J_v(u_\theta)}\nabla h(u_\theta)\big),
\end{aligned}
$$

where $\nabla$ is the Levi-Civita connection on $\mathbb{S}^3$. In addition,

$$
\begin{aligned}
\big\langle u_\theta, (-J_v)\left(\nabla_{J_v(u_\theta)}\nabla h(u_\theta)\right)\big\rangle &= \big\langle J_v(u_\theta), \nabla_{J_v(u_\theta)}\nabla h(u_\theta)\big\rangle \\
&= \tfrac{d}{d\theta}\left[\langle J_v(u_\theta), \nabla h(u_\theta)\rangle\right] = \tfrac{d^2}{d\theta^2}\left[h(u_\theta)\right],
\end{aligned}
$$

and

$$\big\langle \nabla h(u_\theta), (-J_v)\left(\nabla_{J_v(u_\theta)}\nabla h(u_\theta)\right)\big\rangle = \omega_v\left(\nabla h(u_\theta), \frac{d}{d\theta}\left[\nabla h(u_\theta)\right]\right)$$

since $\tfrac{d}{d\theta}\left[\nabla h(u_\theta)\right] = \nabla_{J_v(u_\theta)}\nabla h(u_\theta) - \langle \nabla h(u_\theta), J_v(u_\theta)\rangle u_\theta$. Hence

$$
\begin{aligned}
\Big\langle x_h(u_\theta), (-J_v)\left(\frac{d}{d\theta}\left[x_h(u_\theta)\right]\right)\Big\rangle &= h(u_\theta)\Big(h(u_\theta) + \frac{d^2}{d\theta^2}\left[h(u_\theta)\right]\Big) \\
&\quad + \omega_v\Big(\nabla h(u_\theta), \frac{d}{d\theta}\left[\nabla h(u_\theta)\right]\Big)
\end{aligned}
$$

The result is then an immediate consequence of the previous proposition. $\blacksquare$

## Mixed Symplectic Area

**Proposition 6.6.7 (Symmetry)** *For all* $(f, g) \in C^{\infty}(\mathbb{S}^3; \mathbb{R})^2$, *and* $(u, v) \in \mathbb{S}^3 \times \mathbb{S}^2$, *we have*

$$\int_0^{2\pi} \langle x_f(u_\theta), R_g(u_\theta, v) u_\theta \rangle d\theta = \int_0^{2\pi} \langle x_g(u_\theta), R_f(u_\theta, v) u_\theta \rangle d\theta.$$

**Proof** For all $\theta \in \mathbb{S}^1$,

$$\frac{d}{d\theta}\left[\omega_v\left(x_f(u_\theta), x_g(u_\theta)\right)\right] = \omega_v\left(T_{u_\theta}x_f(J_v(u_\theta)), x_g(u_\theta)\right)$$
$$+ \omega_v\left(x_f(u_\theta), T_{u_\theta}x_g(J_v(u_\theta))\right).$$

By integration, we deduce that

$$\int_0^{2\pi} \omega_v\left(x_f(u_\theta), T_{u_\theta}x_g(J_v(u_\theta))\right) d\theta = \int_0^{2\pi} \omega_v\left(x_g(u_\theta), T_{u_\theta}x_f(J_v(u_\theta))\right) d\theta,$$

which is the desired equality since

$$\begin{aligned}
\omega_v\left(x_h(u_\theta), T_{u_\theta}x_h(J_v(u_\theta))\right) &= \langle J_v x_h(u_\theta), T_{u_\theta}x_h(J_v(u_\theta))\rangle \\
&= \langle x_h(u_\theta), -J_v\left[T_{u_\theta}x_h(J_v(u_\theta))\right]\rangle \\
&= \langle x_h(u_\theta), R_h(u_\theta, v) u_\theta\rangle
\end{aligned}$$

for $h \in \{f, g\}$. ∎

**Definition 6.6.2** Let $(f, g) \in C^{\infty}(\mathbb{S}^3; \mathbb{R})^2$ and $(u, v) \in \mathbb{S}^3 \times \mathbb{S}^2$. We call

$$s_{u,v}(f, g) := \frac{1}{2}\left(s_{u,v}(f + g) - s_{u,v}(f) - s_{u,v}(g)\right)$$
$$= \frac{1}{2}\int_0^{2\pi} \langle x_f(u_\theta), R_g(u_\theta, v) u_\theta\rangle d\theta$$

the **mixed symplectic area** of $x_f\left(\mathbb{S}^1_{u,v}\right)$ and $x_g\left(\mathbb{S}^1_{u,v}\right)$.

A straightforward computation shows that

$$s_{u,v}(f, g) = a_{u,v}(f, g) + s^{\perp}_{u,v}(\nabla f, \nabla g),$$

where $a_{u,v}(f, g)$ is the mixed symplectic area of the hedgehogs of $\mathbb{C}(u, v)$ with support functions $f_{|\mathbb{S}^1_{u,v}}$ and $g_{|\mathbb{S}^1_{u,v}}$, and where $s^{\perp}_{u,v}(\nabla f, \nabla g)$ is the mixed symplectic area of $\nabla f\left(\mathbb{S}^1_{u,v}\right)$ and $\nabla g\left(\mathbb{S}^1_{u,v}\right)$ in $\left(\mathbb{R}^4, J_v, \omega_v\right)$, that is,

$$s^{\perp}_{u,v}\left(\nabla f, \nabla g\right) := \tfrac{1}{2}\left(s^{\perp}_{u,v}(\nabla\left(f+g\right)) - s^{\perp}_{u,v}(\nabla f) - s^{\perp}_{u,v}(\nabla g)\right)$$

$$= \tfrac{1}{2}\int_0^{2\pi} \omega_v\left(\nabla f(u_\theta), \tfrac{d}{d\theta}\left[\nabla g(u_\theta)\right]\right) d\theta$$

$$= \tfrac{1}{2}\int_0^{2\pi} \omega_v\left(\nabla g(u_\theta), \tfrac{d}{d\theta}\left[\nabla f(u_\theta)\right]\right) d\theta.$$

**Symplectic Area of $\mathcal{H}_h$**

**Definition 6.6.3** Let $h \in C^\infty\left(\mathbb{S}^3; \mathbb{R}\right)$. We define the symplectic area of $\mathcal{H}_h$ to be

$$s\left(h\right) := \frac{v_4}{v_2}\int_{G_{4,2}} s_\xi\left(h\right) d\omega_{4,2}\left(\xi\right),$$

where $v_{n+1}$ is the volume of the unit ball in $\mathbb{R}^{n+1}$, $G_{4,2}$ the Grassmann manifold of 2-dimensional subspaces of $\mathbb{R}^4$ and $\omega_{4,2}$ the normalized Haar measure on $G_{4,2}$: $\omega_{4,2}\left(G_{4,2}\right) = 1$.

Recall that the mixed volume $V : \left(\mathcal{K}^4\right)^4 \to \mathbb{R}$ extends to a symmetric 4-linear form on the vector space $\mathcal{H}^4$ of hedgehogs of $\mathbb{R}^4$. Besides, the *algebraic area of order* 2 of a hedgehog $\mathcal{H}_h$ of $\mathbb{R}^4$, denoted by $V_2\left(h\right)$, is defined to be the mixed volume $V\left(h, h, 1, 1\right)$.

**Proposition 6.6.8** *For any $h \in C^\infty\left(\mathbb{S}^3; \mathbb{R}\right)$, the symplectic area of $\mathcal{H}_h$ is equal to its algebraic area of order* 2, *that is, $s\left(h\right) := V_2\left(h\right)$.*

***Proof*** From Kubota's formula

$$V_2\left(K\right) = \frac{v_4}{v_2}\int_{G_{4,2}} V\left(p_\xi\left(K\right)\right) d\omega_{4,2}\left(\xi\right),$$

for all convex body $K$ in $\mathbb{R}^4$, where $p_\xi\left(K\right)$ is the orthogonal projection of $K$ on $\xi \in G_{4,2}$, $V\left(p_\xi\left(K\right)\right)$ its area and $V_2\left(K\right)$ the mixed volume $V\left(K, K, B, B\right)$, $B$ denoting the unit ball in $\mathbb{R}^4$ (see [167, Section 5.3]). This formula can be extended to hedgehogs by multilinearity, so that:

$$v_2\left(h\right) = \frac{v_4}{v_2}\int_{G_{4,2}} a\left(h_{|\mathbb{S}^1_\xi}\right) d\omega_{4,2}\left(\xi\right),$$

for all $h \in C^\infty\left(\mathbb{S}^3; \mathbb{R}\right)$. Note that the algebraic area of $\mathcal{H}_{h_{|\mathbb{S}^1_\xi}}$ does not depend on a choice of orientation for $\xi$. Now, we have proved above that

$$s_{u,v}(h) = a\left(h_{|\mathbb{S}^1_{u,v}}\right) + s^{\perp}_{u,v}(\nabla h)$$

for all $(u, v) \in \mathbb{S}^3 \times \mathbb{S}^2$. So, it suffices to prove that for all $u \in \mathbb{S}^3$,

$$\int_{\mathbb{S}^2} \omega_v \left(\nabla h(u), \nabla_{J_v(u)} \nabla h(u)\right) d\sigma(v) = 0,$$

where $\sigma$ is the spherical Lebesgue measure.

Now, let $(v_1, v_2, v_3) \in \mathbb{S}^2 = \mathbb{S}^1_{\mathbb{H}} \cap \mathrm{Im}(\mathbb{H})$ be such that $\left(J_{v_1}(u), J_{v_2}(u), J_{v_3}(u)\right)$ is an orthonormal basis of $T_u \mathbb{S}^3$ formed by eigenvectors of the Riemannian Hessian $\left(\nabla^2 h\right)_u$, corresponding to eigenvalues $(\lambda_1, \lambda_2, \lambda_3)$. The product of two imaginary quaternions $q_1, q_2 \in \mathrm{Im}(\mathbb{H}) = \mathbb{R}^3$ is given by $q_1 q_2 = -\langle q_1, q_2 \rangle + q_1 \times q_2$, where $\langle ., . \rangle$ is the Euclidean inner product and $\times$ the usual vector product on $\mathbb{R}^3$. Since the orthonormal basis $(v_1, v_2, v_3)$ is formed by imaginary quaternions, we thus have: $v_i v_j + v_j v_i = 0$ for all $(i, j) \in [1, 3]^2$ such that $i \neq j$. A straightforward calculation then gives, for any $v = \sum_{i=1}^{3} x_i v_i \in \mathbb{S}^2$,

$$(-J_v)\left(\nabla_{J_v(u)} \nabla h(u)\right) = (-J_v)\left(\sum_{j=1}^{3} x_j \lambda_j J_{v_j}(u)\right) = \sum_{i,j=1}^{3} x_i x_j \lambda_i \lambda_j v_i v_j u$$

$$= \sum_{1 \leq i < j \leq 3} x_i x_j \left(\lambda_i - \lambda_j\right) v_i v_j u$$

and hence

$$\int_{\mathbb{S}^2} \omega_v \left(\nabla h(u), \nabla_{J_v(u)} \nabla h(u)\right) d\sigma(v) = \int_{\mathbb{S}^2} (-J_v)\left(\nabla_{J_v(u)} \nabla h(u)\right) d\sigma(v)$$

$$= \sum_{1 \leq i < j \leq 3} \left(\underbrace{\int_{\mathbb{S}^2} x_i x_j \, d\sigma}_{=0}\right) \left(\lambda_i - \lambda_j\right) v_i v_j u.$$

$$= 0,$$

which achieves the proof.   ∎

## Quaternionic Curvature Function

Let $K$ be a convex body with class $C_+^\infty$ in $(n+1)$-Euclidean vector space $\mathbb{R}^{n+1}$. One says that $K$ has the ($C^\infty$-smooth) curvature function $R_K : \mathbb{S}^n \to \mathbb{R}$ if its surface area measure $S_n(K, .)$ has $R_K$ as density with respect to spherical area measure $\sigma$ or, equivalently, if

$$V(L, K, \ldots, K) = \tfrac{1}{n+1} \int_{\mathbb{S}^n} h_L(u) R_K(u) \, d\sigma(u)$$

$$= \tfrac{1}{n+1} \int_{\mathbb{S}^n} \langle x_{h_L}(u), R_K(u) u \rangle \, d\sigma(u),$$

for every convex body $L$ with support function $h_L : \mathbb{S}^n \to \mathbb{R}$ (see e.g. [167, p. 545]). As we saw above, the notion of curvature function naturally extends to $C^2$-hedgehogs of $\mathbb{R}^{n+1}$. The aim of this subsection is to use the notion of the mixed symplectic area of $x_g\left(\mathbb{S}^1_{u,v}\right)$ and $x_h\left(\mathbb{S}^1_{u,v}\right)$ to introduce the notion of the (quaternionic) curvature function of $x_h\left(\mathbb{S}^1_{u,v}\right)$. As already mentioned, the reason why the map $\theta \longmapsto R_h\left(e^{i\theta}\right)$ is real for $h \in C^\infty(\mathbb{S}^1; \mathbb{R})$, whereas $u \mapsto R_h(u, v)$ is quaternionic for $h \in C^\infty(\mathbb{S}^3; \mathbb{R})$, is because the product of two purely imaginary complex numbers is a real number, whereas the product of two purely imaginary quaternions can have both nontrivial real and imaginary parts.

**Proposition 6.6.9** *Let $h \in C^\infty\left(\mathbb{S}^3; \mathbb{R}\right)$, and $(u, v) \in \mathbb{S}^3 \times \mathbb{S}^2$. There exists one and only one $C^\infty$-smooth quaternionic function $R(., v) : \mathbb{S}^1_{u,v} \to \mathbb{H}$ that is of the form*

$$u_\theta := (\cos\theta) u + (\sin\theta) vu \longmapsto R(u_\theta, v) = -v T_{u_\theta}(v),$$

*where $T_{u_\theta}(v)$ denotes a pure quaternion, and such that:*

$$\forall g \in C^\infty\left(\mathbb{S}^3; \mathbb{R}\right), \quad s_{u,v}(g, h) := \frac{1}{2} \int_0^{2\pi} \langle x_g(u_\theta), R(u_\theta, v) u_\theta \rangle \, d\theta,$$

*where $u_\theta := (\cos\theta) u + (\sin\theta) vu$. Namely, the quaternionic function given by:*
*$R_h(u_\theta, v) := -v\left(T_{u_\theta} x_h\right)(vu_\theta) \overline{u_\theta}$ for all $\theta \in \mathbb{S}^1$.*

**Proof** For all $\theta \in \mathbb{S}^1$, $T_{u_\theta}\mathbb{S}^3 = \operatorname{Im}(\mathbb{H}) u_\theta$ and hence $\left(T_{u_\theta} x_h\right)(vu_\theta) \overline{u_\theta} \in \operatorname{Im}(\mathbb{H})$. Thus, $R_h(., v) : u_\theta \longmapsto R_h(u_\theta, v) = -v\left(T_{u_\theta} x_h\right)(vu_\theta) \overline{u_\theta}$ is of the required form since

$$\forall g \in C^\infty\left(\mathbb{S}^3; \mathbb{R}\right), \quad s_{u,v}(g, h) := \frac{1}{2} \int_0^{2\pi} \langle x_g(u_\theta), R_h(u_\theta, v) u_\theta \rangle \, d\theta.$$

Conversely, let $R(.,v)$ be any function satisfying the required conditions. Note that the map $u_\theta \mapsto R(u_\theta, v) u_\theta$ has then the form $u_\theta \mapsto \rho(u_\theta) u_\theta + \rho^\perp(u_\theta)$, where $\rho \in C^\infty\left(\mathbb{S}^1_{u,v}; \mathbb{R}\right)$ and $\rho^\perp \in C^\infty\left(\mathbb{S}^1_{u,v}; \mathbb{C}(u,v)^\perp\right)$. Indeed, we have

$$\langle R(u_\theta, v) u_\theta, J_v(u_\theta)\rangle = \left\langle J_v\left(-T_{u_\theta}(v) u_\theta\right), J_v(u_\theta)\right\rangle = \left\langle -T_{u_\theta}(v) u_\theta, u_\theta\right\rangle = 0$$

for all $\theta \in \mathbb{S}^1$, since $-T_{u_\theta}(v) u_\theta \in \operatorname{Im}(\mathbb{H}) u_\theta = T_{u_\theta}\mathbb{S}^3$. Besides, in the case where $R(.,v) = R_h(.,v)$, we have $R(u_\theta, v) u_\theta = R_h(u_\theta, v u_\theta) u_\theta - v\pi^\perp_{u,v}\left[\nabla_{v u_\theta}\nabla h(u_\theta)\right]$, where $R_h(u_\theta, v u_\theta)$ is the radius of curvature of $x_{h_{|\mathbb{S}^1_{u,v}}}$ : $\mathbb{S}^1_{u,v} \to \mathbb{C}(u,v)$ at $x_{h_{|\mathbb{S}^1_{u,v}}}(u_\theta)$ (or, equivalently, the tangential radius of curvature of $\mathcal{H}_h$ at $x_h(u_\theta)$ in the direction $v u_\theta$, which is given by: $R_h(u_\theta, v u_\theta) := \left\langle T_{u_\theta}x_h(v u_\theta), v u_\theta\right\rangle = h(u_\theta) + \left(\nabla^2 h\right)_{u_\theta}(v u_\theta, v u_\theta)$; see e.g. Sect. 4.3), and $\pi^\perp_{u,v}$ the orthogonal projection onto the subspace of $\mathbb{R}^4$ that is orthogonal to $\mathbb{C}(u,v)$. Indeed,

$$\begin{aligned}
R_h(u_\theta, v) u_\theta &= -v\left(T_{u_\theta}x_h\right)(v u_\theta)\\
&= -v\left(h(u_\theta) v u_\theta + \nabla_{v u_\theta}\nabla h(u_\theta)\right)\\
&= -v\left(\left(h(u_\theta) + \left(\nabla^2 h\right)_{u_\theta}(v u_\theta, v u_\theta)\right) v u_\theta + \pi^\perp_{u,v}\left[\nabla_{v u_\theta}\nabla h(u_\theta)\right]\right)
\end{aligned}$$

We already know that $\mathbb{S}^1_{u,v} \to \mathbb{R}$, $u_\theta \mapsto R_h(u_\theta, v u_\theta)$ is the unique $C^\infty$-smooth function $R: \mathbb{S}^1_{u,v} \to \mathbb{R}$ that satisfies:

$$\forall g \in C^\infty\left(\mathbb{S}^1_{u,v}; \mathbb{R}\right), \quad u_{u,v}(g, h) := \frac{1}{2}\int_0^{2\pi} g(u_\theta) R(u_\theta)\, d\theta.$$

Now, any $g \in C^\infty\left(\mathbb{S}^1_{u,v}; \mathbb{R}\right)$ can be extended to a function $g_S \in C^\infty\left(\mathbb{S}^3; \mathbb{R}\right)$ that is such that $\pi^\perp_{u,v}\left[(\nabla g_S)_{|\mathbb{S}^1_{u,v}}\right] = 0$: it suffices, for instance, to define $g_S$ by

$$\forall q \in \mathbb{S}^3, \quad g_S(q) := \begin{cases} 0 & \text{if} \quad \|p\| = 0\\[2mm] F(\|p\|) g\left(\frac{p}{\|p\|}\right) & \text{if} \quad \|p\| \neq 0, \end{cases}$$

where $p$ is the orthogonal projection of $q$ onto $\mathbb{C}(u,v)$ and,

$$F(t) := \frac{\displaystyle\int_0^t \varphi(\tau)\varphi(1-\tau)\, d\tau}{\displaystyle\int_0^1 \varphi(\tau)\varphi(1-\tau)\, d\tau},$$

where $\varphi$ is the function defined on $\mathbb{R}$ by

$$\varphi(t) := \begin{cases} 0 & \text{if } t \leq 0 \\[2ex] e^{-\frac{1}{t^2}} & \text{if } t > 0. \end{cases}$$

$(F : \mathbb{R} \to \mathbb{R}$ is $C^\infty$-smooth, and such that $F(0) = 0$, $F(1) = 1$, and: $\forall n \in \mathbb{N}^*$, $F^{(n)}(0) = F^{(n)}(1) = 0)$. For any $g \in C^\infty\left(\mathbb{S}^1_{u,v}; \mathbb{R}\right)$, such an extension $g_S$ is such that

$$s_{u,v}(g_S, h) = a_{u,v}(g, h) = \frac{1}{2}\int_0^{2\pi} g(u_\theta)\, \rho(u_\theta)\, d\theta,$$

since $s^\perp_{u,v}(\nabla g_S, \nabla h) = 0$. Therefore $\rho(u_\theta) = R_h(u_\theta, vu_\theta)$ for all $\theta \in \mathbb{S}^1$.

Now, it remains to prove that $\rho^\perp(u_\theta) = -v\pi^\perp_{u,v}\left[\nabla_{vu_\theta}\nabla h(u_\theta)\right]$ for all $\theta \in \mathbb{S}^1$. Since $\rho(u_\theta) = R_h(u_\theta, vu_\theta)$ for all $\theta \in \mathbb{S}^1$, the integral condition can be rewritten as follows:

$$\forall g \in C^\infty\left(\mathbb{S}^3; \mathbb{R}\right), \quad s^\perp_{u,v}(\nabla g, \nabla h) := \frac{1}{2}\int_0^{2\pi} \left\langle \nabla g(u_\theta)^\perp, \rho^\perp(u_\theta)\right\rangle d\theta,$$

where $\nabla g(u_\theta)^\perp := \pi^\perp_{u,v}\left[\nabla g(u_\theta)\right]$, that is,

$$\forall g \in C^\infty\left(\mathbb{S}^3; \mathbb{R}\right), \quad \int_0^{2\pi} \left\langle \nabla g(u_\theta)^\perp, \rho^\perp(u_\theta) + v\pi^\perp_{u,v}\left[\nabla_{vu_\theta}\nabla h(u_\theta)\right]\right\rangle d\theta = 0.$$

Note that $\nabla g(u_\theta)^\perp$ has the form

$$\nabla g(u_\theta)^\perp = (\langle \nabla g(u_\theta), wu_\theta\rangle + \langle \nabla g(u_\theta), vwu_\theta\rangle\, v)\, wu_\theta,$$

where $w$ is a pure unit quaternion that is $\langle ., .\rangle$ −orthogonal to $v$, so that $(v, w, vw)$ is an orthonormal basis of $\text{Im}(\mathbb{H})$. Moreover, $\rho^\perp(u_\theta) + v\pi^\perp_{u,v}\left[\nabla_{vu_\theta}\nabla h(u_\theta)\right]$ has the form $(\lambda(u_\theta) + \mu(u_\theta)\, v)\, wu_\theta$, where $\lambda$ and $\mu$ are real, since it belongs to $\mathbb{R}wu_\theta + \mathbb{R}vwu_\theta = \mathbb{C}(u, v)^\perp$. Thus, the integral condition is that the function $\mathbb{S}^1_{u,v} \to \mathbb{C}(u, v)^\perp$, $u_\theta \longmapsto \rho^\perp(u_\theta) + v\pi^\perp_{u,v}\left[\nabla_{vu_\theta}\nabla h(u_\theta)\right]$ is $L^2$-orthogonal to all the functions $\mathbb{S}^1_{u,v} \to \mathbb{C}(u, v)^\perp$, $u_\theta \longmapsto \nabla g(u_\theta)^\perp$ where $g \in C^\infty\left(\mathbb{S}^3; \mathbb{R}\right)$. Now, for any two real $C^\infty$-functions $a,b$ on $\mathbb{S}^1_{u,v}$, let us define $g : \mathbb{S}^3 \to \mathbb{H}$ by:

$$g(q(\theta, \beta, \gamma)) := [a(u_\theta)\langle q(\theta, \beta, \gamma), wu_\theta\rangle + b(u_\theta)\langle q(\theta, \beta, \gamma), vwu_\theta\rangle]\, F(\cos\beta),$$

where

$$q\left(\theta,\beta,\gamma\right)=\left(\cos\beta\right)u_\theta+\left(\sin\beta\right)\left(\left(\cos\gamma\right)wu_\theta+\left(\sin\gamma\right)vwu_\theta\right)\in\mathbb{S}^3$$
$$=\left(\cos\beta\right)\left(\cos\theta\right)u+\left(\cos\beta\right)\left(\sin\theta\right)vu$$
$$+\left(\cos\left(\gamma-\theta\right)\left(\sin\beta\right)\right)wu+\left(\sin\left(\gamma-\theta\right)\right)\left(\sin\beta\right)vwu.$$

We then obtain

$$\frac{\partial}{\partial\beta}\left[g\left(q\left(\theta,\beta,\gamma\right)\right)\right]_{|\beta=0}=\left\langle\nabla g\left(q\left(\theta,0,\gamma\right)\right),\frac{\partial q}{\partial\beta}\left(\theta,0,\gamma\right)\right\rangle$$
$$=\left\langle\nabla g\left(u_\theta\right),\left(\cos\gamma\right)wu_\theta+\left(\sin\gamma\right)vwu_\theta\right\rangle$$
$$=a\left(u_\theta\right)\cos\gamma+b\left(u_\theta\right)\sin\gamma,$$

and thus, for $\gamma=0$ and $\gamma=\pi/2$, we have respectively $a\left(u_\theta\right)=\left\langle\nabla g\left(u_\theta\right),wu_\theta\right\rangle$ and $b\left(u_\theta\right)=\left\langle\nabla g\left(u_\theta\right),vwu_\theta\right\rangle$. In other words, all the functions of the form $\mathbb{S}^1_{u,v}\to\mathbb{C}\left(u,v\right)^\perp,\ u_\theta\longmapsto\left(a\left(u_\theta\right)+b\left(u_\theta\right)v\right)wu_\theta$ can be written in the form $\mathbb{S}^1_{u,v}\to\mathbb{C}\left(u,v\right)^\perp,u_\theta\longmapsto\nabla g\left(u_\theta\right)^\perp$ where $g\in C^\infty\left(\mathbb{S}^3;\mathbb{R}\right)$.

Therefore, $\rho^\perp\left(u_\theta\right)=-v\pi^\perp_{u,v}\left[\nabla_{vu_\theta}\nabla h\left(u_\theta\right)\right]$ for all $\theta\in\mathbb{S}^1$.   ∎

**Definition 6.6.4**  For every $h\in C^\infty\left(\mathbb{S}^3;\mathbb{R}\right)$, we say that $R_h\left(.,v\right):\mathbb{S}^1_{u,v}\to\mathbb{H}$, $u_\theta\longmapsto-v\left(T_{u_\theta}x_h\right)\left(vu_\theta\right)\overline{u_\theta}$ is the **quaternionic curvature function** of $x_h\left(\mathbb{S}^1_{u,v}\right)$.


**Evolutes of Hedgehog Hypersurfaces in $\mathbb{H}^n\cong\mathbb{R}^{4n}$**

We identify $\mathbb{R}^{4n}$ with the hyperkähler vector space $\left(\mathbb{H}^n,\langle.,.\rangle,I,J,K\right)$, where $\langle.,.\rangle$ is the standard Euclidean metric on $\mathbb{R}^{4n}\cong\mathbb{H}^n$, $\left(n\geq1\right)$, and, the triple of complex structures $\left(I,J,K\right)$ on $\mathbb{H}^n$ is given by left multiplication by $i,j,k$ respectively. On this hyperkähler vector space, we have a whole $\mathbb{S}^2$ family of linear Kähler structures given by:

$$I_a:=a_1I+a_2J+a_3K\quad\text{and}\quad\omega_a\left(X,Y\right)=\left\langle I_a\left(X\right),Y\right\rangle,$$

for all $a=\left(a_1,a_2,a_3\right)\in\mathbb{S}^2\subset\mathbb{R}^3$ and, $\left(X,Y\right)\in\left(T_q\mathbb{H}^n\right)^2$. Most of the results we saw for evolutes of hedgehogs in $\mathbb{R}^4\cong\mathbb{H}$ can be extended to $\left(\mathbb{H}^n,\langle.,.\rangle,I,J,K\right)$ with a few adaptations. In particular, for all $h\in C^\infty\left(\mathbb{S}^{4n-1};\mathbb{R}\right)$, the evolute of the hedgehog $\mathcal{H}_h$ in the Kähler vector space $\left(\mathbb{R}^4,I_a,\omega_a\right)$ is defined to be the envelope of the family of normal hyperplanes $\left(N_h^a\left(u\right)\right)_{u\in\mathbb{S}^{4n-1}}$ with equation

$$\left\langle x-x_h\left(u\right),I_a\left(u\right)\right\rangle=0.$$

**Proposition 6.6.10** *Let $h \in C^{\infty}\left(\mathbb{S}^{4n-1}; \mathbb{R}\right)$. The* **evolute** *of $\mathcal{H}_h$ in $\left(\mathbb{H}^n, I_a, \omega_a\right)$ is the hedgehog $\mathcal{H}_{\partial_a h}$ with support function*

$$\partial_a h : \mathbb{S}^{4n-1} \to \mathbb{R}, \ u \longmapsto \langle \nabla h\left(-I_a\left(u\right)\right), u \rangle,$$

*where $\langle .,. \rangle$ is the standard Euclidean metric on $\mathbb{R}^{4n} \cong \mathbb{H}^n$, and $\nabla h$ the gradient of $h$. Thus, $\partial_a h$ is such that: $\forall u \in \mathbb{S}'^{n-1}$,*

$$\left(\partial_a h\right)\left(I_a\left(u\right)\right) = \langle \nabla h\left(u\right), I_a\left(u\right)\rangle = \left(dh\right)_u\left(I_a\left(u\right)\right).$$

*The proof (very similar to that of the proposition concerning evolutes of hedgehogs in $\left(\mathbb{H}, J_v, \omega_v\right)$, $\left(v \in \mathbb{S}^2 = \mathbb{S}^3 \cap \mathrm{Im}\left(\mathbb{H}\right)\right)$) is left to the reader.*

# Chapter 7
# Hedgehogs in Non-Euclidean Spaces

## 7.1 Introduction and Basics on Non-Euclidean Hedgehogs

Of course, the classical hedgehog theory is not restricted to Euclidean spaces. In [44], F. Fillastre introduced and studied '$\Gamma$-convex bodies' (or, 'Fuchsian convex bodies'), which are the closed convex sets of the Lorentz-Minkowski space $\mathbb{L}^{n+1}$ that are globally invariant under the action of some Fuchsian group $\Gamma$. In this paper, F. Fillastre gave a 'reversed Alexandrov–Fenchel inequality' and thus a 'reversed Brunn-Minkowski inequality'. This work permits to introduce 'Fuchsian hedgehogs' whose 'support functions' are differences of support functions of two $\Gamma$-convex bodies (see the remark on page 314 in [44]). In [106], the author gave a detailed study of plane Lorentzian and Fuchsian hedgehogs, including a series of Fuchsian analogues of classical geometrical inequalities (which are also reversed as compared to classical ones). These plane Lorentzian and Fuchsian hedgehogs will be introduced and studied in Sect. 7.2. For an application to marginally trapped surfaces, we will also give a short introduction to hedgehogs of $\mathbb{L}^3$ in Sect. 8.2.1. This brief introduction of Lorentzian hedgehogs of $\mathbb{L}^3$ can be easily extended to higher dimensions.

On another note, as we have already glimpsed when we spoke about projective duality in Sect. 2.7, it is possible to consider the notion of a hedgehog in real projective space $\mathbb{R}P^{n+1} = P\left(\mathbb{R}^{n+2}\right)$ (which we will denote by $\mathbb{P}^{n+1}$). As we will see, *the notion of a hedgehog is affine*: we can define the notion of a hedgehog of $\mathbb{R}^{n+1}$ regarded as an affine space over itself. Extending the Euclidean or affine space $\mathbb{R}^{n+1}$ by adding points at infinity (which we regard as corresponding to families of parallel lines, and which together make up a hyperplane at infinity), it will appear that we can define a hedgehog of $\mathbb{P}^{n+1}$ as a hedgehog of $\mathbb{R}^{n+1}$ regarded as the complement of any projective hyperplane of $\mathbb{P}^{n+1}$, and thus a hedgehog of $\mathbb{R}^{n+1}$ as a hedgehog of $\mathbb{P}^{n+1}$. Now, let us see $\mathbb{P}^{n+1}$ as an extension of the Euclidean space $\mathbb{R}^{n+1}$. Any hedgehog of $\mathbb{P}^{n+1} = \mathbb{S}^{n+1}/\{-Id, Id\}$ is contained in the complement of a projective hyperplane of $\mathbb{P}^{n+1}$, and can thus be regarded as a hedgehog of $\mathbb{S}^{n+1}$

© The Author(s), under exclusive license to Springer Nature Switzerland AG 2026     235
Y. Martinez-Maure, *Hedgehog Theory*, Lecture Notes in Mathematics 2381,
https://doi.org/10.1007/978-3-032-04298-9_7

contained in an open hemisphere of $\mathbb{S}^{n+1}$, say the open hemisphere with center $p \in \mathbb{S}^{n+1}$ which we will denote by $\mathbb{S}^{n+1}_p$. Using the gnomonic projection from $\mathbb{S}^{n+1}_p$ onto the tangent hyperplane to $\mathbb{S}^{n+1}$ at $p$, we then retrieve hedgehogs of $\mathbb{R}^{n+1}$. Here, we of course identify the tangent hyperplane $p + T_p\mathbb{S}^{n+1}$ with $\mathbb{R}^{n+1}$ via the orthogonal projection from $p + T_p\mathbb{S}^{n+1}$ onto the linear subspace orthogonal to $p$ in $\mathbb{R}^{n+2}$, which we identify with $\mathbb{R}^{n+1}$. Recall that the image of a point $m \in \mathbb{S}^{n+1}_p$ under the gnomonic projection $g_p : \mathbb{S}^{n+1}_p \to p + T_p\mathbb{S}^{n+1} \simeq \mathbb{R}^{n+1}$ is the intersection point of the linear line $\mathbb{R}m$ with $p + T_p\mathbb{S}^{n+1}$.

Similarly, regarding the hyperbolic space $\mathbb{H}^{n+1}$ as the upper sheet of the hyperboloid with equation $\langle x, x \rangle_L = -1$ in $\mathbb{R}^{n+2}$ endowed with the Lorentzian inner product given by

$$\langle x, y \rangle_L = \sum_{i=1}^{n+1} x_i y_i - x_{n+2} y_{n+2},$$

for $x = (x_1, \ldots, x_{n+2})$ and $y = (y_1, \ldots, y_{n+2})$, that is,

$$\mathbb{H}^{n+1} = \left\{ x \in \mathbb{R}^{n+2} \,|\, \langle x, x \rangle_L = -1, \, x_{n+2} > 0 \right\},$$

we can make a gnomonic projection (which preserves geodesics) from $\mathbb{H}^{n+1}$ onto the interior $\mathbb{B}^{n+1}$ of the (Euclidean) unit ball of $\mathbb{R}^{n+1}$ (identified with the affine hyperplane $\mathbb{R}^{n+1} \times \{1\}$ of $\mathbb{R}^{n+2}$). This gnomonic projection $g : \mathbb{H}^{n+1} \to \mathbb{B}^{n+1}$ sends $m \in \mathbb{H}^{n+1}$ to the intersection point of the linear line $\mathbb{R}m$ with $\mathbb{R}^{n+1} \simeq \mathbb{R}^{n+1} \times \{1\}$. Considering hedgehogs of $\mathbb{R}^{n+1}$ that are included in $\mathbb{B}^{n+1} \subset \mathbb{R}^{n+1} \times \{1\}$, and taking their images under the radial projection

$$\rho : \mathbb{B}^{n+1} \to \mathbb{H}^{n+1} \subset \mathbb{R}^{n+2}$$

$$x \mapsto x/\sqrt{|\langle x, x \rangle_L|},$$

which is the inverse $g^{-1}$ of the gnomonic projection, we can then introduce hedgehogs of $\mathbb{H}^{n+1}$. These hedgehog hypersurfaces, which are envelopes of smooth families of cooriented (totally geodesic) hyperplanes of $\mathbb{H}^{n+1}$, will be called '*g-hedgehogs*' of $\mathbb{H}^{n+1}$, where the letter g stands for indicating that these hypersurfaces are 'geodesically hedgehog hypersurfaces' (recall that hyperplanes of $\mathbb{H}^{n+1}$ are the complete **totally geodesic** hypersurfaces of $\mathbb{H}^{n+1}$) This change of names ('g-hedgehogs' instead of 'hedgehogs') aims to differentiate these g-hedgehogs from another class of hedgehogs that we will introduce in Sect. 7.4, namely '*h-hedgehogs*' of $\mathbb{H}^{n+1}$, for which geodesic hyperplanes will be replaced by horospheres (as we will see, the best analogue to Euclidean hyperplanes in $\mathbb{H}^{n+1}$ are not actually the totally geodesic hyperplanes of $\mathbb{H}^{n+1}$, but the horospheres).

## 7.2  Plane Lorentzian and Fuchsian Hedgehogs

In this subsection, we follow more or less [106]. Our main results consist in the
Fuchsian analogues of some classical geometrical inequalities that are presented
below. For the convenience of the reader, we will begin by recalling very briefly
some basic definitions and results about plane Euclidean hedgehogs. Then we will
give a short introduction to plane Lorentzian hedgehogs and first results concerning
evolutes and duality in the Lorentz-Minkowski plane $L^2$. Finally, we will present a
study of plane Fuchsian hedgehogs (convolution of Fuchsian hedgehogs, Brunn-
Minkowski and Minkowski type inequalities, reversed isoperimetric inequality,
isometric excess and area of the evolute, reversed Bonnesen inequality).

### *7.2.1  Introduction and First Results*

We begin by recalling some basic facts of plane Euclidean hedgehogs. In the
Euclidean plane $\mathbb{R}^2$, a hedgehog is the envelope of a family of cooriented lines
$L(\theta)$ parametrized by the oriented angle $\theta \in \mathbb{S}^1 = \mathbb{R}/2\pi\mathbb{Z}$ from $e_1 = (1, 0)$ to
their coorienting normal vector $u(\theta) = (\cos\theta, \sin\theta)$. These cooriented lines $L(\theta)$
have equations

$$\langle x, u(\theta)\rangle = h(\theta), \tag{7.2.1}$$

where $\langle .,.\rangle$ is the usual inner product on $\mathbb{R}^2$ and where $h \in C^1(\mathbb{S}^1; \mathbb{R})$. Partial
differentiation of (7.2.1) with respect to $\theta$ yields

$$\langle x, u'(\theta)\rangle = h'(\theta). \tag{7.2.2}$$

From (7.2.1) and (7.2.2), the parametrization of the corresponding hedgehog is

$$x_h : \mathbb{S}^1 \to \mathbb{R}^2, \theta \mapsto h(\theta) u(\theta) + h'(\theta) u'(\theta).$$

This envelope $\mathcal{H}_h := x_h(\mathbb{S}^1)$ is called the *(Euclidean) hedgehog* with support
function $h$. In this section, we are mainly interested in $C^2$-*hedgehogs*, that is,
hedgehogs with a $C^2$-support function. Recall that regular $C^2$-hedgehogs of $\mathbb{R}^2$ are
strictly convex smooth curves and that any $C^2$-hedgehog can be regarded as the
Minkowski difference of two such convex curves (see Sect. 2.2).

H. Geppert was the first to introduce hedgehogs in $\mathbb{R}^2$ and $\mathbb{R}^3$ (under the German
names *stützbare Bereiche* in $\mathbb{R}^2$ and *stützbare Flächen* in $\mathbb{R}^3$) in an attempt to
extend certain parts of the Brunn-Minkowski theory [53]. As we saw in Chap. 3,
many classical inequalities for convex curves find their counterparts in the setting
of hedgehogs. Of course, a few adaptations are necessary. In particular, lengths and

areas have to be replaced by their algebraic versions. For instance, we have seen above the following.

**Proposition 3.2.4** *For any $h \in C^3 \left(\mathbb{S}^1; \mathbb{R}\right)$, we have:*

$$0 \le l(h)^2 - 4\pi a(h) \le -4\pi a(h') = -4\pi a(\partial h), \tag{7.2.3}$$

*where $l\,(h)$ and $a\,(h)$ are respectively the signed length and area of $\mathcal{H}_h$ and where $a(h')$ is the signed area of its evolute, which is the hedgehog with support function $(\partial h)\,(\theta) = h'\left(\theta - \frac{\pi}{2}\right), \left(\theta \in \mathbb{S}^1 = \mathbb{R}/2\pi\mathbb{Z}\right)$. In each inequality of (7.2.3), the equality holds if, and only if, $\mathcal{H}_h$ is a circle or a point.*

In Sect. 7.2.6, we will prove an analogue of Proposition 3.2.4 for Fuchsian hedgehogs.


**Plane Lorentzian Hedgehogs**

In this subsection, we will undertake a similar study replacing the Euclidean plane $\mathbb{R}^2$ by the Lorentzian plane $L^2$, and the unit circle $\mathbb{S}^1$ of $\mathbb{R}^2$ by the hyperbolic line $\mathbb{H}^1$. In the Lorentzian plane $L^2$, a *spacelike hedgehog* is similarly defined to be the envelope of a family of cooriented spacelike lines $L\,(t)$ parametrized by the oriented hyperbolic angle $t \in \mathbb{H}^1 \simeq \mathbb{R}$ from $e_2 = (0, 1)$ to their coorienting normal vector $v\,(t) = (\sinh t, \cosh t)$, (see Sect. 7.2.2). These cooriented lines $L\,(t)$ have equations

$$\langle x, v\,(t)\rangle_L := h\,(t), \tag{7.2.4}$$

where $\langle x, y\rangle_L := x_1 y_1 - x_2 y_2$ is the Lorentzian inner product of the vectors $x = (x_1, x_2)$ and $y = (y_1, y_2)$ in $L^2$, and where $h \in C^1\,(\mathbb{R}; \mathbb{R})$. Note that $h\,(t)$ is the signed distance from the origin to the support line with coorienting unit normal $v\,(t)$. Partial differentiation of (7.2.4) with respect to $t$ yields

$$\left\langle x, v'\,(t)\right\rangle_L := h'\,(t). \tag{7.2.5}$$

From (7.2.4) and (7.2.5), the parametrization of the corresponding hedgehog is

$$x_h : \mathbb{H}^1 \to L^2, t \mapsto h'\,(t)\,v'\,(t) - h\,(t)\,v\,(t).$$

This envelope $\mathcal{S}_h := x_h\left(\mathbb{H}^1\right)$ is called the ***spacelike hedgehog*** of $L^2$ with support function $h \in C^1\left(\mathbb{H}^1; \mathbb{R}\right)$. As in the Euclidean case, we will generally restrict our discussion to $C^2$-hedgehogs (i.e., with a $C^2$-support function). In Sect. 7.2.2, we will give a study of their evolutes. In particular, we will prove the following.

**Theorem 7.2.1** *For any $h \in C^3 (\mathbb{R}; \mathbb{R})$, the second evolute of $\mathcal{S}_h$ is the spacelike hedgehog with support function $h''$:*

$$\mathcal{D} (\mathcal{D} (\mathcal{S}_h)) = \mathcal{S}_{h''},$$

*where $\mathcal{D} (\mathcal{E})$ denotes the evolute of the envelope $\mathcal{E} \subset L^2$ of a family of nonlightlike lines with a $C^3$-support function and no inflection point.*

In Sect. 7.2.3, we will also introduce timelike hedgehogs of $L^2$, and in Sect. 7.2.4, we will give *explicit formulas describing a natural duality relationship between spacelike hedgehogs and timelike hedgehogs.*

For a systematic study of curves and surfaces in Lorentz-Minkowski plane and space, we refer the reader to [87, Subsection 2.3 for a study of curves in Lorentz-Minkowski plane].

## Plane Fuchsian Hedgehogs

Of course, a spacelike hedgehog $\mathcal{S}_h$ of $L^2$ has no reason to be a compact curve. So, in order to develop a Brunn-Minkowski theory, we are going to replace $\mathbb{H}^1$ by its quotient by a Fuchsian group $\Gamma$. In other words: $(i)$ we identify

$$SO \, (1, 1) = \left\{ M = \begin{pmatrix} x_2 & x_1 \\ x_1 & x_2 \end{pmatrix} \in M_2 \, (\mathbb{R}) \, \middle| \, x_2^2 - x_1^2 = 1 \right\}$$

with the hyperbola

$$H = \left\{ (x_1, x_2) \in L^2 \, \middle| \, x_2^2 - x_1^2 = 1 \right\};$$

$(ii)$ we take the subgroup $\Gamma$ of $SO \, (1, 1)$ generated by $(\sinh T, \cosh T) \in \mathbb{H}^1 = \{(x_1, x_2) \in H \, | \, x_2 > 0\}$ for some $T \in \mathbb{R}_+^*$; and $(iii)$ we replace $\mathbb{H}^1$ by $\mathbb{H}^1 / \Gamma \simeq \mathbb{R} / T\mathbb{Z}$. In practice, any $h \in C^1 \left( \mathbb{H}^1 / \Gamma; \mathbb{R} \right)$ will be regarded as a $T$-periodic function $h : \mathbb{R} \longmapsto \mathbb{R}$ of class $C^1$. The $\Gamma$-*hedgehog* with support function $h \in C^1 \left( \mathbb{H}^1 / \Gamma; \mathbb{R} \right)$ is then defined to be the curve $\Gamma_h$ parametrized by

$$\gamma_h : \mathbb{R} \to L^2, t \mapsto h' \, (t) \, v' \, (t) - h \, (t) \, v \, (t) \, .$$

Note that, for any $t \in \mathbb{R}$, we have $\gamma_h \, (t + T) = g \, (T) \, [\gamma_h \, (t)]$, where $g \, (T)$ denotes the linear isometry of $L^2$ whose matrix in the canonical basis is

$$\begin{pmatrix} \cosh T & \sinh T \\ \sinh T & \cosh T \end{pmatrix} .$$

Minkowski differences of convex bodies of the Euclidean plane $\mathbb{R}^2$ not only constitute a real vector space $(\mathcal{H}^2, +, .)$, but also a commutative and associative $\mathbb{R}$-algebra. Indeed, H. Görtler [55, 56] defined the convolution product of two hedgehogs $\mathcal{H}_f$ and $\mathcal{H}_g$ of $\mathbb{R}^2$ as the hedgehog whose support function is given by

$$(f * g)(\theta) = \frac{1}{2\pi} \int_0^{2\pi} f(\theta - \alpha) g(\alpha) \, d\alpha,$$

for all $\theta \in \mathbb{S}^1 = \mathbb{R}/2\pi\mathbb{Z}$. It can then easily be verified that $(\mathcal{H}^2, +, ., *)$ is a commutative and associative algebra. The point of interest is of course that the convolution product of two Euclidean hedgehogs inherits many properties of its factors. In particular, H. Görtler noticed that the convolution product of two convex bodies of $\mathbb{R}^2$ is still a plane convex body of $\mathbb{R}^2$. The purpose of Sect. 7.2.5 will be to give *a Fuchsian analogue of Görtler's theorem.*

For every $h \in C^2(\mathbb{H}^1/\Gamma; \mathbb{R})$, the $C^1$-curve $\gamma_h : [0, T] \to L^2/\Gamma$, $t \mapsto h'(t)v'(t) - h(t)v(t)$ is rectifiable, and its length is given by

$$L(h) := \int_0^T \left\| x_h'(t) \right\|_L dt,$$

where $\|x\|_L := \sqrt{|\langle x, x \rangle|_L}$ for all $x \in L^2$. Note that $x_h' = R_h v'$, where $R_h := h'' - h$ is the so-called **curvature function** of $\Gamma_h$. Therefore

$$L(h) := \int_0^T |R_h(t)| \, dt.$$

If in this last integral we remove the absolute value to take into account the sign of the curvature function of $\Gamma_h$, we obtain the so-called *algebraic (or signed) length of* $\Gamma_h$, which is thus given by

$$l(h) := \int_0^T R_h(t) \, dt = -\int_0^T h(t) \, dt.$$

Given any $h \in C^2(\mathbb{H}^1/\Gamma; \mathbb{R})$, let $\Delta_h$ be the oriented closed curve of $L^2$ consisting of the oriented line segment joining the origin to $\gamma_h(0)$, followed by the oriented curve $\Gamma_h$, and finally the oriented line segment joining $\gamma_h(T)$ to the origin. Denote by $(\Delta_h)^-$ the curve obtained from $\Delta_h$ by taking the opposite orientation (see Fig. 7.1). Define the *algebraic (or signed) area* of the $\Gamma$-hedgehog $\Gamma_h$ to be the algebraic area bounded by $(\Delta_h)^-$, that is,

$$a(h) := \int_{L^2} i_h(x) \, d\lambda(x),$$

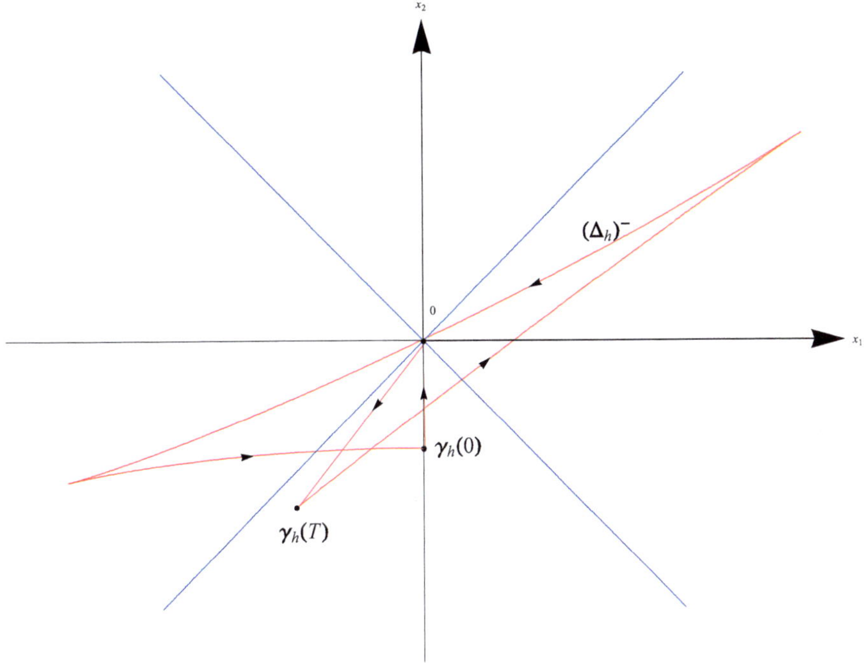


**Fig. 7.1** The oriented closed curve $(\Delta_h)^-$ when $h(t) := 1 + \cos(2\pi t)$

where $\lambda$ is the Lebesgue measure and $i_h(x)$ the winding number of $x$ with respect to $(\Delta_h)^-$ for $x \in L^2 - (\Delta_h)^-$ (we let $i_h(x) = 0$ for $x \in (\Delta_h)^-$). An easy straightforward calculation gives

$$a(h) = \frac{1}{2} \int_{(\Delta_h)^-} x_1 dx_2 - x_2 dx_1 = \frac{1}{2} \int_0^T h(t) R_h(t) dt = \frac{1}{2} \int_0^T \left( h^2 + \left( h' \right)^2 \right)(t) dt.$$

In the Fuchsian case, many geometric inequalities will be reversed. A first example is given by the following obvious result.

**Proposition 7.2.1** *The map* $\sqrt{a} : C^2 \left( \mathbb{H}^1 / \Gamma; \mathbb{R} \right) \to \mathbb{R}_+, h \longmapsto \sqrt{a(h)}$ *is a norm associated with a scalar product* $(h, k) \longmapsto a(h, k)$. *In particular, for any* $(h, k) \in C^2 \left( \mathbb{H}^1 / \Gamma; \mathbb{R} \right)^2$, *we have*

$$\sqrt{a(h + k)} \leq \sqrt{a(h)} + \sqrt{a(k)} \tag{7.2.6}$$

*and*

$$a(h, k)^2 \leq a(h) \, a(k), \tag{7.2.7}$$

*with equalities if, and only if, $\Gamma_h$ and $\Gamma_k$ are homothetic (here, "homothetic" means that there exists $(\lambda, \mu) \in \mathbb{R}^2 - \{(0, 0)\}$ such that $\lambda h + \mu k = 0$).*

Indeed, Inequality (7.2.6) (resp. (7.2.7)) has to be compared with the Brunn-Minkowski inequality (resp. Minkowski inequality) in $\mathbb{R}^2$ (e.g., see [167, Section 7]): for any pair $(H, K)$ of convex bodies of $\mathbb{R}^2$, we have

$$\sqrt{a(H + K)} \geq \sqrt{a(H)} + \sqrt{a(K)} \tag{7.2.8}$$

and

$$a(H, K)^2 \geq a(H)\, a(K), \tag{7.2.9}$$

where $a(L)$ (resp. $a(H, K)$) is the area (resp. the mixed area) of $L$ (resp. $(H, K)$). By taking $k = -1$ (that is, $\Gamma_k = \mathbb{H}^1$) in (7.2.7), we obtain the following ***reversed isoperimetric inequality***

$$a\,(h) \geq \frac{l\,(h)^2}{2T}, \tag{7.2.10}$$

with equality if, and only if, $\Gamma_h$ and $\mathbb{H}^1$ are homothetic (that is, $h$ is constant). In Sect. 7.2.6, we will prove an other reversed geometric inequality given by the following analogue of Proposition 3.2.4 for Fuchsian hedgehogs.

**Theorem 7.2.2** *Let $T \in \,]0, 2\pi]$. For any $T$-periodic function $h : \mathbb{R} \to \mathbb{R}$ of class $C^3$, we have:*

$$0 \leq 2T a(h) - l(h)^2 \leq 2T a\left(h'\right), \tag{7.2.11}$$

*where $l\,(h)$ and $a\,(h)$ are respectively the signed length and area of $\Gamma_h$ and $a(h')$ the signed area of its evolute.*

Note that $2T a(h) - l(h)^2$ provides a measure of how far $\Gamma_h$ deviates from a $\Gamma$-hedgehog given by a spacelike branch of a hyperbola. A lower bound of the isoperimetric excess $a(h) - l(h)^2/2T$ is given by the following reversed Bonnesen inequality, which we will prove in Sect. 7.2.7.

**Theorem 7.2.3 (Reversed Bonnesen Inequality)** *For any $T$-periodic function $h : \mathbb{R} \to \mathbb{R}$ of class $C^2$, we have:*

$$\frac{1}{2T}\,(R - r)^2 \leq a\,(h) - \frac{l\,(h)^2}{2T},$$

*where $l\,(h)$ and $a\,(h)$ are respectively the signed length and area of $\Gamma_h$, and where $r := \min_{0 \leq t \leq T} (-h\,(t))$ and $R := \max_{0 \leq t \leq T} (-h\,(t))$. Furthermore, the equality holds if and only if $R = r$.*

Recall that Bonnesen's sharpening of the isoperimetric inequality for a convex body $K$ with non-empty interior in $\mathbb{R}^2$ reads as follows:

$$L^2 - 4\pi A \geq \pi^2 (R - r)^2 ,$$

where $L$ and $A$ are respectively the perimeter and the area of $K$ and where $r$ and $R$ stand respectively for the inradius and the circumradius of $K$ (e.g., see [39, pp. 108–110]).

For geometric inequalities involving hedgehogs in higher dimensions in the Fuchsian case, we refer the reader to [44].


### 7.2.2  *Preliminaries*

In this subsection, the notation $x = (x_1, x_2)$ means that $(x_1, x_2)$ are the coordinates of $x \in \mathbb{R}^2$ with respect to the canonical basis of $\mathbb{R}^2$. The *Lorentzian plane* $L^2$ is the vector space $\mathbb{R}^2$ endowed with the pseudo-scalar product $\langle x, y \rangle_L := x_1 y_1 - x_2 y_2$, for any $x = (x_1, x_2)$ and $y = (y_1, y_2)$. For any $x \in L^2$, define the *norm* of $x$ by $\|x\|_L = \sqrt{|\langle x, x \rangle_L|}$ and the *sign of* $x$ by $\varepsilon(x) = sgn(\langle x, x \rangle_L)$, where $sgn$ denotes the signum function: $sgn(t)$ is $1$, $0$, or $-1$ if $t$ is positive, zero, or negative, respectively. A nonzero vector $x \in L^2$ is said to be *spacelike* if $\varepsilon(x) = 1$, *lightlike* if $\varepsilon(x) = 0$ and *timelike* if $\varepsilon(x) = -1$. Let $e_2 = (0, 1)$. A timelike vector $x = (x_1, x_2) \in L^2$ is said to be a *future vector* if $\langle x, e_2 \rangle_L < 0$, that is, if $x_2 > 0$. We will denote by $F$ the set of all future timelike vectors:

$$F = \left\{ x = (x_1, x_2) \in L^2 \, | \langle x, x \rangle_L < 0 \text{ and } x_2 > 0 \right\}.$$

The *hyperbolic line* $\mathbb{H}^1$ is the set of all unit future timelike vectors:

$$\mathbb{H}^1 := \left\{ x = (x_1, x_2) \in L^2 \, | \langle x, x \rangle_L = -1 \text{ and } x_2 > 0 \right\}.$$

In other words, $\mathbb{H}^1$ is the upper branch of the hyperbola $x_2^2 = x_1^2 + 1$. It will play in $L^2$ the same role as the one the unit circle $\mathbb{S}^1$ plays in the Euclidean plane $\mathbb{R}^2$. For any $t \in \mathbb{R}$, let $g(t)$ be the linear isometry of the Lorentzian plane whose matrix in the canonical basis of $\mathbb{R}^2$ is

$$\begin{pmatrix} \cosh t & \sinh t \\ \sinh t & \cosh t \end{pmatrix}.$$

These isometries $g(t)$ constitute the group $G$ of *hyperbolic translations* of $L^2$. Note that $G$ is an abelian subgroup of $O(1,1)$ and that $g : \mathbb{R} \to G, t \mapsto g(t)$ is a group isomorphism: $g(s+t) = g(s)g(t)$ for all $s, t \in \mathbb{R}$. The hyperbolic line $\mathbb{H}^1$ can be regarded as the orbit of $e_2$ under the action of $G$. Any $v(t) = (\sinh t, \cosh t) \in \mathbb{H}^1$ is identified with the unique $t \in \mathbb{R}$ such that $g(t)(e_2) = v(t)$. For any $x, y \in \mathbb{H}^1$, the *oriented hyperbolic angle from $x$ to $y$* is the unique $t$ such that $g(t)(x) = y$.

A (smooth) curve of $L^2$ is a differentiable map $c : I \subset \mathbb{R} \to L^2$, where $I$ is an open interval. A curve $c : I \to L^2$ is said to be *regular at $t$* if $c'(t) \neq 0$. The curve is said to be *regular* if it is regular at every $t \in I$. A curve $c : I \to L^2$ is said to be *spacelike* (resp. *lightlike*, *timelike*) at $t$ if $c'(t)$ is a spacelike (resp. null or lightlike, timelike) vector. The curve is said to be *spacelike* (resp. *timelike*) if it is spacelike (resp. timelike) at every $t \in I$.

Let $\sigma$ be the anti-isometric involutive operator of $L^2$ given by $\sigma(x_1, x_2) = (x_2, x_1)$ for all $(x_1, x_2) \in L^2$. For any nonzero vector $x$ of $L^2$, let $x^{\perp} := \varepsilon(x)\sigma(x)$. Note that, for any nonlightlike $x \in L^2 - \{(0,0)\}$, $(x, x^{\perp})$ is a positively oriented basis of $L^2$. Here, « positively oriented » means « endowed with the orientation of the canonical basis of $\mathbb{R}^2$ ».

Let $c : I \to L^2$ be a spacelike (resp. timelike) curve of class $C^2$. At any point of $c : I \to L^2$, we can define the *oriented Frenet frame* $(T(t), N(t))$ consisting of *Frenet vectors*

$$T(t) := \frac{c'(t)}{\|c'(t)\|_L} \quad \text{and} \quad N(t) := T(t)^{\perp}.$$

If $c : I \to L^2$ is parametrized by the pseudo arc length $s$ (that is, if $\|c'(s)\|_L = 1$ for all $s \in I$), then the *algebraic curvature of $c$* is defined to be the function $\kappa$ such that $T'(s) = \kappa(s)N(s)$. If it is not the case, a straightforward computation using the fact that $ds/dt = \|c'(t)\|_L$ shows that the algebraic curvature is given by

$$\kappa(t) := \frac{\langle c'(t), \sigma(c''(t)) \rangle_L}{\|c'(t)\|_L^3}.$$

If $c : I \to L^2$ is a spacelike hedgehog $x_h : \mathbb{H}^1 \to L^2$ with support function $h \in C^3(\mathbb{R}; \mathbb{R})$, then $c' = x_h' = R_h v'$, where $R_h := h'' - h$ is the so-called **curvature function** of $\mathcal{S}_h$. In this case, we hence obtain

$$T = sgn(R_h)v', \quad N = sgn(R_h)v \quad \text{and} \quad \kappa(t) = \frac{1}{|R_h|}.$$

### 7.2.3  *Evolute*

#### Evolute of a Spacelike Hedgehog $\mathcal{S}_h$ of $L^2$

In this subsubsection, $h$ will denote any $C^3$-function from $\mathbb{R}$ to $\mathbb{R}$. As in the Euclidean case, the evolute of the spacelike hedgehog $\mathcal{S}_h$ can be defined in two different but equivalent ways: as an envelope or as a locus.

#### Evolute of $\mathcal{S}_h$ as the Envelope of Its Normal Lines

For every $t \in \mathbb{R}$, the support line $L_h(t)$, with coorienting unit normal vector $v(t) := (\sinh t, \cosh t)$, has equation

$$\langle x, v(t) \rangle_L := h(t). \tag{7.2.12}$$

Let $N_h(t)$ be the line through $x$ that is orthogonal (with respect to the Lorentzian metric $\langle .,. \rangle_L$) to $L_h(t)$ in $L^2$. We will say that $N_h(t)$ is the normal line to $\mathcal{H}_h$ at $x_h(t)$. This normal line $N_h(t)$ has equation

$$\langle x, v'(t) \rangle_L := h'(t).$$

Define the **evolute** $\mathcal{D}(\mathcal{S}_h)$ of the spacelike hedgehog $\mathcal{S}_h$ of $L^2$ to be the envelope of the family $(N_h(t))_{t \in \mathbb{R}}$ of its normal lines. This evolute $\mathcal{D}(\mathcal{S}_h)$ is thus the curve of $L^2$ parametrized by

$$c_h : \mathbb{R} \to L^2, t \longmapsto c_h(t),$$

where $c_h(t)$ is the unique solution of the system

$$\begin{cases} \langle x, v'(t) \rangle_L := h'(t) \\ \langle x, v(t) \rangle_L := h''(t), \end{cases}$$

that is, $c_h(t) = h'(t) v'(t) - h''(t) v(t)$.

#### Evolute of $\mathcal{S}_h$ as the Locus of Its Centers of Curvature

The evolute $\mathcal{D}(\mathcal{S}_h)$ of the spacelike hedgehog $\mathcal{S}_h$ of $L^2$ can also be defined as the locus of all its centers of curvature. First recall that

$$x'_h(t) = R_h(t) v'(t),$$

for all $t \in \mathbb{H}^1 \simeq \mathbb{R}$. Since $\mathcal{S}_h := x_h(\mathbb{H}^1)$ is an envelope parametrized by its coorienting unit normal vector field, the center of curvature of $\mathcal{S}_h$ at $x_h(t)$ is defined to be

$$c_h(t) := x_h(t) - R_h(t)\, v(t),$$

that is

$$c_h(t) = h'(t)\, v'(t) - h''(t)\, v(t),$$

for all $t \in \mathbb{H}^1 \simeq \mathbb{R}$. Of course, if $x_h$ is regular at $t$ then

$$c_h(t) = x_h(t) - \frac{1}{\kappa(t)} N(t),$$

but the « *center of curvature* » $c_h(t)$ is well defined even if $x'_h(t) = 0$.

### Timelike Hedgehogs of $L^2$ and Their Evolutes

**Definitions** We can also define timelike hedgehogs of $L^2$. The **timelike hedgehog** with support function $h \in C^1(\mathbb{R}; \mathbb{R})$ is defined to be the envelope $\mathcal{T}_h$ of the family $\left(L'_h(t)\right)_{t \in \mathbb{R}}$ of cooriented timelike lines with equation

$$\langle x, v'(t) \rangle_L := h(t), \tag{7.2.13}$$

$v'(t) = (\cosh t, \sinh t)$ being the unit coorienting normal vector of $L'_h(t)$. Partial differentiation of (7.2.13) with respect to $t$ yields

$$\langle x, v(t) \rangle_L := h'(t). \tag{7.2.14}$$

From (7.2.13) and (7.2.14), the parametrization of the timelike hedgehog $\mathcal{T}_h$ is

$$y_h : \mathbb{R} \to L^2, \, t \mapsto h(t)\, v'(t) - h'(t)\, v(t).$$

Note that for every $t \in \mathbb{R}$, we have

$$y'_h(t) = -R_h(t)\, v(t),$$

where $R_h := h'' - h$.

The **evolute** $\mathcal{D}(\mathcal{T}_h)$ of a timelike hedgehog $\mathcal{T}_h$ of $L^2$ is defined to be the envelope of the family of its normal lines (i.e., the envelope of the family of lines with equation $\langle x, v(t) \rangle_L := h'(t)$) or, equivalently, the locus of its centers of curvature

$$d_h(t) := y_h(t) - \left(-R_h(t)\, v'(t)\right) = h''(t)\, v'(t) - h'(t)\, v(t),$$

$(t \in \mathbb{R})$.

**Relationship Between $\mathcal{S}_h = x_h(\mathbb{R})$ and $\mathcal{T}_h = y_h(\mathbb{R})$**

Let $\Sigma$ be the anti-isometric involutive operator of $L^2$ that is given by $\Sigma(x) = -\sigma(x)$ for all $x \in L^2$. Note that $\Sigma \circ v = -v'$ and $\Sigma \circ v' = -v$.

**Proposition 7.2.2** *For any $h \in C^1(\mathbb{R}; \mathbb{R})$, the spacelike hedgehog $\mathcal{S}_h$ and the timelike hedgehog $\mathcal{T}_h$ are related by*

$$\mathcal{T}_h = \Sigma(\mathcal{S}_h) \quad and \quad \mathcal{S}_h = \Sigma(\mathcal{T}_h).$$

***Proof*** Indeed, their respective parametrizations $x_h := h'v' - hv$ and $y_h := hv' - h'v$ are such that $y_h = \Sigma \circ x_h$ and $x_h = \Sigma \circ y_h$. $\blacksquare$

**Second Evolute**

**Proposition 7.2.3** *For any $h \in C^2(\mathbb{R}; \mathbb{R})$, the evolute of the spacelike hedgehog $\mathcal{S}_h$ (resp. of the timelike hedgehog $\mathcal{T}_h$) can be given by*

$$\mathcal{D}(\mathcal{S}_h) = \Sigma(\mathcal{S}_{h'}) \quad (resp. \ \mathcal{D}(\mathcal{T}_h) = \Sigma(\mathcal{T}_{h'}))$$

*and hence by*

$$\mathcal{D}(\mathcal{S}_h) = \mathcal{T}_{h'} \quad (resp. \ \mathcal{D}(\mathcal{T}_h) = \mathcal{S}_{h'})$$

*from the previous proposition.*

***Proof*** Indeed, $c_h := h'v' - h''v$ (resp. $d_h := h''v' - h'v$) satisfies $\Sigma \circ c_h = -h'v + h''v' = x_{h'}$ (resp. $\Sigma \circ d_h = -h''v + h'v' = y_{h'}$) and hence $c_h = \Sigma \circ x_{h'}$ (resp. $d_h = \Sigma \circ y_{h'}$). $\blacksquare$

See Fig. 7.2 for an illustration.

A straightforward consequence is the following.

**Corollary 7.2.1** *For any $h \in C^3(\mathbb{R}; \mathbb{R})$, the second evolute of the spacelike hedgehog $\mathcal{S}_h$ (resp. of the timelike hedgehog $\mathcal{T}_h$) is simply the spacelike (resp. timelike hedgehog) with support function $h''$:*

$$\mathcal{D}^2(\mathcal{S}_h) := \mathcal{D}(\mathcal{D}(\mathcal{S}_h)) = \mathcal{S}_{h''} \quad \left(resp. \ \mathcal{D}^2(\mathcal{T}_h) = \mathcal{D}(\mathcal{D}(\mathcal{T}_h)) = \mathcal{T}_{h''}\right).$$

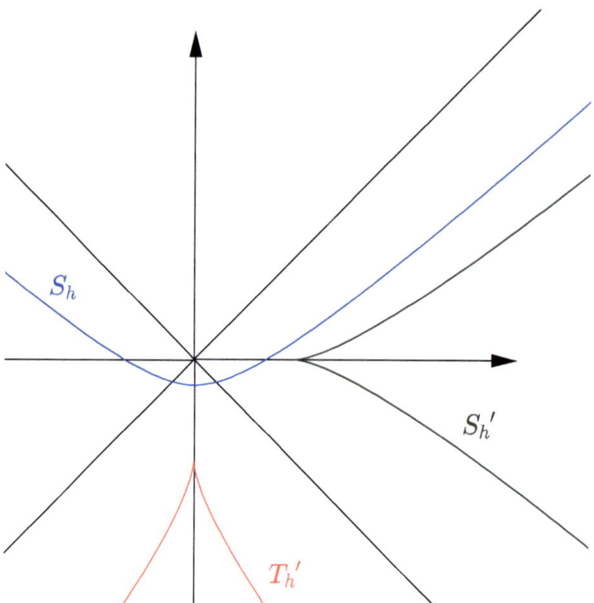



**Fig. 7.2** $S_h$ and its evolute $T_{h'}$ if $h(t) := \cosh(2t)$

### 7.2.4　Duality

Let $c : I \subset \mathbb{R} \to L^2$ be a spacelike or timelike curve of $L^2$ and let $p_c : I \to L^2$ be its pedal curve: for any $t \in I$, $p_c(t)$ is the foot of the perpendicular from the origin to the tangent line to $c$ at $c(t)$. Note that, replacing tangent lines by support lines, we can define the pedal curve of a spacelike (resp. timelike) hedgehog even if $x_h$ (resp. $y_h$) is not regular. Assume that $\|c(t)\|_L \cdot \|p_c(t)\|_L \neq 0$ for all $t \in I$. Define the *star curve of* $c$ to be the curve $c^* : I \subset \mathbb{R} \to L^2$ given by $c^* := i \circ p_C$, where

$$i(x) := \varepsilon(x) \frac{x}{\|x\|_L^2} \quad \text{for all } x \in L^2 \text{ such that } \|x\|_L \neq 0,$$

(recall that $\varepsilon(x) := sgn(\langle x, x \rangle_L)$). If $c : I \to L^2$ is the restriction of a spacelike hedgehog $x_h$ (resp. a timelike hedgehog $y_h$) to some open interval $I$, then $p_c = -hv$ (resp. $p_c = hv'$) and, assuming that $h \cdot \|x_h\|_L \neq 0$ (resp. $h \cdot \|y_h\|_L \neq 0$), we can define its *star curve* in the same way.

**Proposition 7.2.4** *Let $I$ be an open interval of $\mathbb{R}$. If $c : I \to L^2$ is the restriction to $I$ of a spacelike hedgehog $x_h$ (resp. a timelike hedgehog $y_h$) such that $h \cdot \|x_h\|_L \neq 0$ (resp. $h \cdot \|y_h\|_L \neq 0$) on $I$, then $(c^*)^* = c$.*

***Proof*** If $c = x_h$ (resp. $c = y_h$), then $p_c = -hv$ (resp. $p_c = hv'$). Thus,

$$x_h^* = \frac{v}{h} \quad \left(\text{resp. } y_h^* = \frac{v'}{h}\right).$$

Differentiation gives

$$\left(x_h^*\right)' = \frac{y_h}{h^2} \quad \left(\text{resp. } \left(y_h^*\right)' = -\frac{x_h}{h^2}\right).$$

Now

$$x_h^* = \frac{h'y_h - hx_h}{h\left(h^2 - (h')^2\right)} \quad \left(\text{resp. } y_h^* = \frac{hy_h - h'x_h}{h\left(h^2 - (h')^2\right)}\right).$$

Therefore

$$p_{x_h^*} = \frac{x_h}{(h')^2 - h^2} \quad \left(\text{resp. } p_{y_h^*} = \frac{y_h}{h^2 - (h')^2}\right),$$

and hence

$$\left(x_h^*\right)^* = x_h \quad \left(\text{resp. } \left(y_h^*\right)^* = y_h\right).$$

$\blacksquare$

**Definition 7.2.1**  For any $h \in C^1(\mathbb{R}; \mathbb{R})$ such that $h.\|x_h\|_L \neq 0$ (resp. $h.\|y_h\|_L \neq 0$), we will say that $\mathcal{S}_h^* := x_h^*(\mathbb{R})$ (resp. $\mathcal{T}_h^* := y_h^*(\mathbb{R})$) is the **dual curve** of the spacelike (resp. timelike) hedgehog $\mathcal{S}_h$ (resp. $\mathcal{T}_h$).

### 7.2.5  *Convolution*

Differences of convex bodies of the Euclidean plane $\mathbb{R}^2$ not only constitute a real vector space $\left(\mathcal{H}^2, +, .\right)$, but also a commutative and associative $\mathbb{R}$-algebra. As observed by H. Görtler [55, 56], we can define the convolution product of two hedgehogs $\mathcal{H}_f$ and $\mathcal{H}_g$ of $\mathbb{R}^2$ as the hedgehog whose support function is given by

$$(f * g)(\theta) = \frac{1}{2\pi} \int_0^{2\pi} f(\theta - \alpha) g(\alpha) \, d\alpha,$$

for all $\theta \in \mathbb{S}^1 = \mathbb{R}/2\pi\mathbb{Z}$. It can be readily checked that $\left(\mathcal{H}^2, +, ., *\right)$ is then a commutative and associative algebra. The point of interest is, of course, that the convolution product of two Euclidean hedgehogs inherits many properties of its factors. In particular, H. Görtler observed that the convolution product of two plane convex bodies remains a plane convex body. The purpose of the present section is to give a similar result for Fuchsian hedgehogs.

Let $h \in C^2 \left( \mathbb{H}^1; \mathbb{R} \right)$. Recall that, for all $t \in \mathbb{H}^1 \simeq \mathbb{R}$, we have:

$$x'_h (t) = R_h (t) \, v' (t),$$

where $R_h := h'' - h$. Therefore, the spacelike hedgehog $\mathcal{S}_h = x_h \left( \mathbb{H}^1 \right)$ is a regular curve if, and only if, its curvature function $R_h$ is everywhere nonzero. In that case, $\mathcal{S}_h$ will be said to be *convex*.

**Definition**  Let $h \in C^2 \left( \mathbb{H}^1; \mathbb{R} \right)$. The spacelike hedgehog $\mathcal{S}_h$ is said to be convex if its curvature function $R_h := h'' - h$ is everywhere nonzero on $\mathbb{H}^1$. It is said to be future convex (resp. past convex) if its curvature function is everywhere positive (resp. negative) on $\mathbb{H}^1$.

**Definition**  Let $h \in C^2 \left( \mathbb{H}^1 / \Gamma; \mathbb{R} \right)$. The $\Gamma$-hedgehog $\Gamma_h$ is said to be a $\Gamma$-hedgehog of class $C^2_+$ of $F = \left\{ x = (x_1, x_2) \in L^2 \, | \, \langle x, x \rangle_L < 0 \text{ and } x_2 > 0 \right\}$ if $h < 0$ and $R_h > 0$.

**Remark**  A $\Gamma$-hedgehog of class $C^2_+$ of $F$ can indifferently be regarded as a convex curve of $F$ or as a convex closed curve of $F / \Gamma$.

**Definition**  Let $\Gamma_f$ and $\Gamma_g$ be $\Gamma$-hedgehogs whose respective support functions $f$ and $g$ are in $C^1 \left( \mathbb{H}^1 / \Gamma; \mathbb{R} \right)$. The convolution of $\Gamma_f$ and $\Gamma_g$ is the $\Gamma$-hedgehog $\Gamma_{f*g}$ whose support function is defined by

$$(f * g)(t) = - \int_0^T f(t - s) \, g(s) \, ds \quad \text{for all } t \in [0, T].$$

The operation of convolution of $\Gamma$-hedgehogs is of course commutative, associative and distributive over addition. Here is an analogous result of Görtler's theorem.

**Proposition 7.2.5**  *Let $\Gamma_f$ and $\Gamma_g$ be $\Gamma$-hedgehogs whose respective support functions $f$ and $g$ are in $C^2 \left( \mathbb{H}^1 / \Gamma; \mathbb{R} \right)$. If $\Gamma_g$ is a $\Gamma$-hedgehog of class $C^2_+$ of $F$ and if $f$ is negative, then $\Gamma_{f*g}$ is a $\Gamma$-hedgehog of class $C^2_+$ of $F$.*

**Proof**  If $f < 0$, $g < 0$ and $R_g > 0$, then $(f * g) < 0$ and $R_{f*g} > 0$. Indeed, the first inequality is trivial and a simple computation shows that

$$R_{f*g}(t) = (f * g)''(t) - (f * g)(t) = \left( f * g'' \right)(t) - (f * g)(t)$$
$$= \left( f * \left( g'' - g \right) \right)(t) = \left( f * R_g \right)(t) = - \int_0^T f(t - s) \, R_g(s) \, ds$$

is positive for all $t \in [0, T]$ since $f < 0$ and $R_g > 0$.  ∎

In particular:

**Corollary 7.2.2** *If $\Gamma_f$ and $\Gamma_g$ are $\Gamma$-hedgehogs of class $C_+^2$ of $F$, then $\Gamma_{f*g}$ is also a $\Gamma$-hedgehog of class $C_+^2$ of $F$.*

### 7.2.6   Isometric Excess and Area of the Evolute

The following theorem is analogous to Proposition 3.2.4 for Fuchsian hedgehogs.

**Theorem 7.2.2** *Let $T \in \,]0, 2\pi]$. For any $T$-periodic function $h : \mathbb{R} \to \mathbb{R}$ of class $C^3$, we have:*

$$0 \leq 2T a(h) - l(h)^2 \leq 2T a\left(h'\right),$$

*where $l\,(h)$ and $a\,(h)$ are respectively the signed length and area of $\Gamma_h$, and $a(h')$ the signed area of its evolute.*

**Proof** The first inequality is simply the isoperimetric inequality (7.2.10). Let us prove the second one. First note that:

$$a\,(h) - a\left(h'\right) = \frac{1}{2}\left(\int_0^T h^2 dt - \int_0^T \left(h''\right)^2 dt\right).$$

Let $a_n\,(f)$ and $b_n\,(f)$ denote the Fourier coefficients of a $T$-periodic differentiable function $f : \mathbb{R} \to \mathbb{R}$:

$$a_0\,(f) := \frac{1}{T}\int_0^T f\,(t)\,dt,$$

$$a_n\,(f) := \frac{2}{T}\int_0^T f\,(t)\cos n\omega t\,dt$$

$$\text{and } b_n\,(f) := \frac{2}{T}\int_0^T f\,(t)\sin n\omega t\,dt$$

where $\omega := 2\pi/T$ and $n \in \mathbb{N}^*$. Recall that $a_n\left(h''\right) = -(n\omega)^2 a_n\,(h)$ and $b_n\left(h''\right) = -(n\omega)^2 b_n\,(h)$ for all $n \in \mathbb{N}^*$. By applying the Parseval equality, we thus obtain

$$\frac{1}{2}\int_0^T \left(h^2 - \left(h''\right)^2\right)(t)\,dt = \frac{T}{2}a_0\,(h)^2 + \frac{T}{4}\sum_{n=1}^{+\infty}\left(1 - (n\omega)^4\right)\left(a_n\,(h)^2 + b_n\,(h)^2\right).$$

Since the sum in the right-hand side is obviously nonpositive, we deduce that

$$\frac{1}{2}\int_0^T \left(h^2 - (h'')^2\right)(t)\,dt \le \frac{T}{2}a_0\,(h)^2 = \frac{l\,(h)^2}{2T}.$$

Therefore

$$a\,(h) - \frac{l\,(h)^2}{2T} \le a\,(h'),$$

which achieves the proof.                                                      ∎

**Remarks**

1. For $T \in\, ]0, 2\pi[$, the equality $2T a(h) - l(h)^2 = 2T a\left(h'\right)$ holds if, and only if, $h$ is constant.
2. For $T = 2\pi$, the equality $2T a(h) - l(h)^2 = 2T a\left(h'\right)$ may hold for nonconstant $h \in C^3\left(\mathbb{H}^1/\Gamma; \mathbb{R}\right)$. Consider for instance $h\,(t) := \cos t$.
3. The assumption $T \in\, ]0, 2\pi]$ is necessary even if we restrict to $\Gamma$-hedgehogs that are $\Gamma$-hedgehogs of class $C_+^2$ of $F$. Consider for instance $h\,(t) := -2 + \cos\left(\frac{2\pi}{7}t\right)$, which is such that $h < 0$ and $R_h > 0$.

### 7.2.7  Reversed Bonnesen Inequality

Let $K$ be a convex body with non-empty interior in $\mathbb{R}^2$. In the 1920s, T. Bonnesen gave various sharpenings of the classical isoperimetric inequality

$$A \le \frac{L^2}{4\pi},$$

where $L$ and $A$ denote respectively the perimeter and the area of $K$. In particular, he proved the inequality

$$L^2 - 4\pi A \ge \pi^2\,(R - r)^2,\tag{7.2.15}$$

where $r$ and $R$ are respectively the inradius and the circumradius of $K$ (i.e., the radii of the largest inscribed and the smallest circumscribed circles of the boundary of $K$, respectively). He further proved that the equality holds in (7.2.15) if and only if $R = r$, i.e., if $K$ is a disc. The proof by Bonnesen is reproduced in [39, pp. 108–110]. For a survey of Bonnesen-type inequalities in Euclidean spaces, we refer the reader to [136]. Let us prove the following reversed Bonnesen inequality for Fuchsian hedgehogs.

**Theorem 7.2.3** *For any $T$-periodic function $h : \mathbb{R} \to \mathbb{R}$ of class $C^2$, we have:*

$$\frac{1}{2T} (R - r)^2 \leq a(h) - \frac{l(h)^2}{2T},$$

*where $l(h)$ and $a(h)$ are respectively the signed length and area of $\Gamma_h$ and where $r := \min_{0 \leq t \leq T} (-h(t))$ and $R := \max_{0 \leq t \leq T} (-h(t))$. Furthermore, the equality holds if, and only, if $R = r$.*

**Proof** Since $-h : \mathbb{R} \to \mathbb{R}$ is continuous, there exists $(t_0, t_1) \in [0, T]^2$ such that $r = -h(t_0)$ and $R = -h(t_1)$. Thus we have

$$(R - r)^2 = (h(t_1) - h(t_0))^2 = \left( \int_{t_0}^{t_1} h'(t)\, dt \right)^2.$$

By the Cauchy-Schwarz inequality, we deduce that

$$(R - r)^2 \leq |t_1 - t_0| \cdot \int_{\min(t_0,t_1)}^{\max(t_0,t_1)} h'(t)^2\, dt \leq T \int_0^T h'(t)^2\, dt.$$

Now

$$\int_0^T h'(t)^2\, dt = 2a(h) - \int_0^T h(t)^2\, dt,$$

and again by the Cauchy-Schwarz inequality

$$l(h)^2 = \left( \int_0^T h(t)\, dt \right)^2 \leq T \int_0^T h(t)^2\, dt.$$

Therefore

$$(R - r)^2 \leq 2Ta(h) - l(h)^2,$$

which achieves the proof of the reversed Bonnesen inequality.

Finally, considering equality cases at each step of the reasoning, we immediately see that the equality holds if, and only if, $h$ is constant, which completes the proof. $\blacksquare$

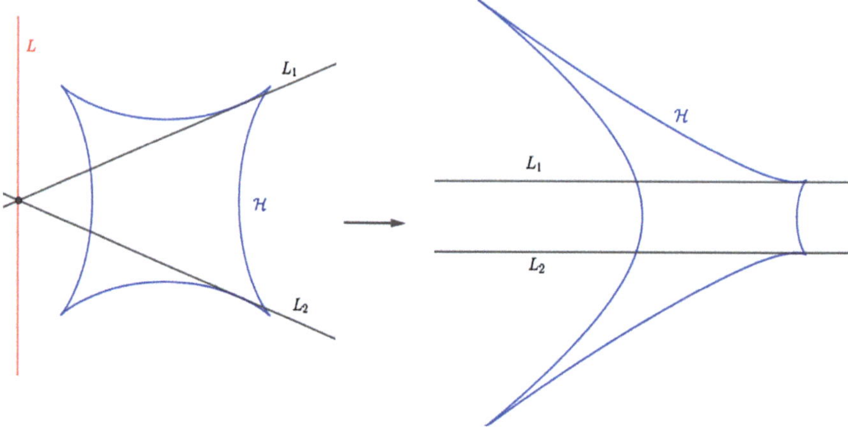

**Fig. 7.3**  Sending $L$ to infinity does not affect the hedgehog nature of $C$

## 7.3  $C^2$-hedgehogs in Real Affine or Projective Space

### 7.3.1  The Two Dimensional Case

Let us begin with the two dimensional case. The following remark reveals the projective nature of the notion of a hedgehog.

**Remark 7.3.1**  Consider any $C^2$-hedgehog curve, say $C$, in the Euclidean plane $\mathbb{R}^2$, and any line, say $L$, that does not meet $C$ in $\mathbb{R}^2$. We know by Theorem 2.8.1 that there are exactly two cooriented support lines of $C$ through any point of $L$. Therefore, if we send the line $L$ to infinity, the curve $C$ remains a hedgehog curve since it still has exactly two cooriented support lines that are parallel to any given line direction (see Fig. 7.3).

In fact, like convexity, the notion of a hedgehog is affine (i.e. invariant under any affine transformation: see Fig. 7.5). We can indeed introduce the notion of a hedgehog of $\mathbb{R}^2$ regarded as an affine space over itself (without any metric structure) as follows. First, we define the notion of a cooriented line of $\mathbb{R}^2$: a cooriented line of the affine plane $\mathbb{R}^2$ is a straight line of $\mathbb{R}^2$ together with a transverse orientation given by a crossing transverse direction indicated by an arrow (or by the half-plane where the arrow is pointing to): see Fig. 7.4. The orientation of a line (that is, of a contact element) of $\mathbb{R}^2$ is thus the choice of one of the two half-planes into which it divides the (tangent) plane.

Fixing an orientation of $\mathbb{R}^2$, we may identify the set, say $S$, of cooriented lines through the origin of $\mathbb{R}^2$ with the set of oriented lines through the origin of $\mathbb{R}^2$, in

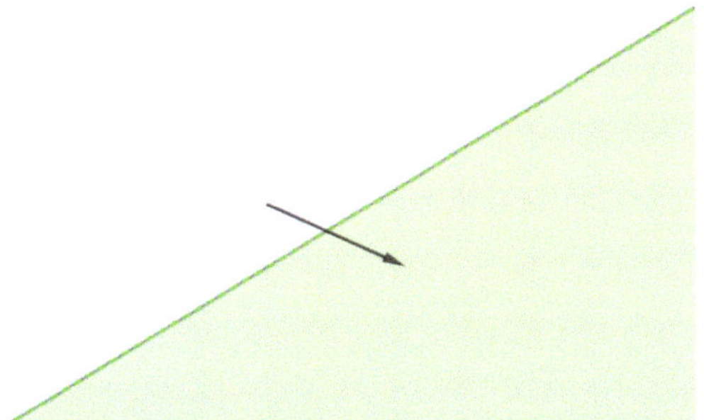

**Fig. 7.4** A cooriented line of $\mathbb{R}^2$

order that, for each line, (a coorienting vector, an orienting vector) gives the positive orientation of $\mathbb{R}^2$. Then $S$ is naturally equipped with the structure of a compact smooth 1-manifold, which is diffeomorphic to $\mathbb{S}^1$. So, we can state the following.

**Definition 7.3.1** If $\mathcal{H}$ in $\mathbb{R}^2$, regarded as an affine space over itself, is the envelope of a $C^\infty$-smooth family of cooriented lines $(L_l)_{l \in S}$, where $L_l$ is an affine line that is parallel to $l$ and whose coorientation corresponds to that of $l$, then we will say that $\mathcal{H}$ is a $C^\infty$-hedgehog of the affine plane $\mathbb{R}^2$.

Another way of presenting $C^\infty$-hedgehogs of the affine plane $\mathbb{R}^2$ is of course by making use of contact geometry. Consider $\mathbb{R}^2$ as an affine space over itself. Fix an orientation of the vector space $\mathbb{R}^2$ by choosing a basis $\mathcal{B} = (e_1, e_2)$ of $\mathbb{R}^2$, and let $(x_1, x_2)$ denote the coordinates of a point $x$ in $\mathcal{B}$. For all $\theta \in \mathbb{S}^1 = \mathbb{R}/2\pi\mathbb{Z}$, identify the cooriented line that is spanned and oriented by the vector $u'(\theta) = -(\sin\theta)\, e_1 + (\cos\theta)\, e_2$ with the coorienting vector $u(\theta) = (\cos\theta)\, e_1 + (\sin\theta)\, e_2$ of the line $\mathbb{R}u'(\theta)$. Then, the set $S$ of all of cooriented lines through the origin of $\mathbb{R}^2$ is identified with $\left\{ u(\theta) = (\cos\theta)\, e_1 + (\sin\theta)\, e_2 \,\middle|\, \theta \in \mathbb{S}^1 = \mathbb{R}/2\pi\mathbb{Z} \right\}$. Now equip $\mathbb{R}^2 \times S$ with the contact form $\omega = u_1 dx_1 + u_2 dx_2$, where $(u_1, u_2)$ are the coordinates of $u \in S \subset \mathbb{R}^2$. Consider a $C^\infty$-immersion of the form

$$ i : \mathbb{S}^1 \to \mathbb{R}^2 \times S, \theta \mapsto (x(\theta), u(\theta)). $$

This immersion is Legendrian if, and only if, we have $x_1'(\theta) \cos\theta + x_2'(\theta) \sin\theta = 0$ for all $\theta \in \mathbb{S}^1$. Let $h(\theta) = x_1(\theta) \cos\theta + x_2(\theta) \sin\theta$ for all $\theta \in \mathbb{S}^1$. If $i : \mathbb{S}^1 \to \mathbb{R}^2 \times S$ is Legendrian, then $h'(\theta) = -x_1(\theta) \sin\theta + x_2(\theta) \cos\theta$, and from

$$ \begin{cases} h(\theta) = x_1(\theta) \cos\theta + x_2(\theta) \sin\theta \\ h'(\theta) = -x_1(\theta) \sin\theta + x_2(\theta) \cos\theta \end{cases} $$

we deduce that

$$ \begin{aligned} x(\theta) &= x_1(\theta)\, e_1 + x_2(\theta)\, e_2 \\ &= \left( h(\theta) \cos\theta - h'(\theta) \sin\theta \right) e_1 + \left( h(\theta) \sin\theta + h'(\theta) \cos\theta \right) e_2 \\ &= h(\theta)\, u(\theta) + h'(\theta)\, u'(\theta) \end{aligned} $$

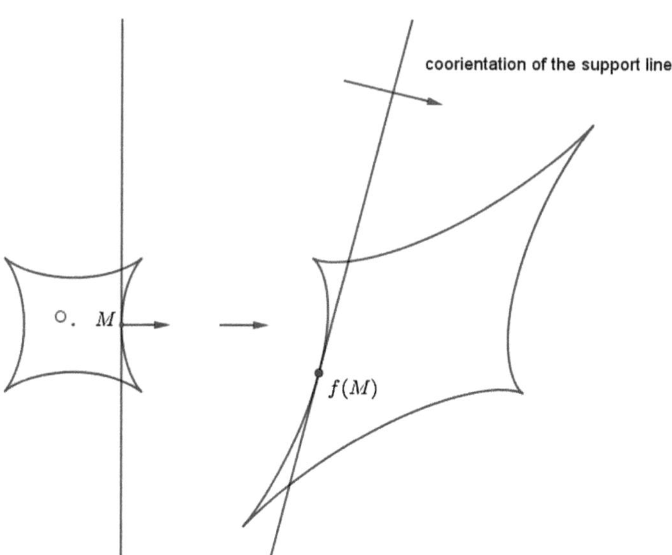


**Fig. 7.5** The hedgehog $\mathcal{H}_h$ with $h(\theta) = \cos(2\theta)$, and its image under the affine transformation $f : \mathbb{R}^2 \to \mathbb{R}^2$, $(x, y) \mapsto \left(7 - 2x - \frac{y}{2}, -x - 2y\right)$

Thus, for any Legendrian immersion of the form $i : \mathbb{S}^1 \to \mathbb{R}^2 \times S$, $\theta \mapsto (x(\theta), u(\theta))$, there exists $h \in C^\infty(\mathbb{S}^1; \mathbb{R})$, such that $x : \mathbb{S}^1 \to \mathbb{R}^2$ is given by $x(\theta) = x_h(\theta)$, where $x_h(\theta) = h(\theta)u(\theta) + h'(\theta)u'(\theta)$ for all $\theta$. Then we will say that the Legendrian front $x_h : \mathbb{S}^1 \to \mathbb{R}^2$ parametrizes an affine $C^\infty$-hedgehog $\mathcal{H}_h = x_h(\mathbb{S}^1)$ of $\mathbb{R}^2$. For all integer $k \geq 2$, we may of course define affine $C^k$-hedgehogs of $\mathbb{R}^2$ in the same way.

Note however that given such an affine hedgehog $\mathcal{H}_h$ of $\mathbb{R}^2$, we can endow $\mathbb{R}^2$ with a scalar product $\langle ., . \rangle$ so that $\mathcal{H}_h$ be a Euclidean hedgehog of $\left(\mathbb{R}^2, \langle ., . \rangle\right)$.

The curve shown on the right side of Fig. 7.5 is an example of a hedgehog of the affine plane $\mathbb{R}^2$. Extending the Euclidean or affine plane $\mathbb{R}^2$ by adding points at infinity (which may be regarded as corresponding to families of parallel lines, and which together make up a line at infinity), we can consider such a curve $\mathcal{H}$ in the real projective plane $\mathbb{P}^2$. Now we know from the previous remark that for any straight line $L$ not meeting $\mathcal{H}$, the curve $\mathcal{H}$ can still be viewed as a hedgehog of the affine plane $\mathbb{R}^2$ regarded as the complement of $L$ in $\mathbb{P}^2$. We will therefore define the notion of a hedgehog of the projective plane $\mathbb{P}^2$ as follows.

**Definition 7.3.2** If $\mathcal{H}$ in the real projective plane $\mathbb{P}^2$, is such that, for any projective line $L$ that does not meet $\mathcal{H}$, $\mathcal{H}$ is a hedgehog of the real affine plane deduced from $\mathbb{P}^2$ by removing the line $L$, then we will say that $\mathcal{H}$ is a hedgehog of $\mathbb{P}^2$.

**Remarks**

1. It is worth noting that, by Remark 7.3.1, $\mathcal{H}$ in $\mathbb{P}^2$ is a hedgehog of $\mathbb{P}^2$ provided there exists some line $L$ of $\mathbb{P}^2$ that does not meet $\mathcal{H}$, and such that $\mathcal{H}$ is an affine hedgehog of $\mathbb{P}^2 \backslash L$.
2. The above definition is coherent with the one given in Sect. 2.7 for a convex body of $\mathbb{P} = \mathbb{P}\left(\mathbb{R}^{n+2}\right)$.

Note that if $L$ is a projective line that does not meet $\mathcal{H}$, then any projective line distinct from $L$, say $L'$, can be cooriented by the choice of a component of the complement of $L \cup L'$ in $\mathbb{P}^2$. Now, if we consider $\mathbb{P}^2$ as the extension of the Euclidean plane $\mathbb{R}^2$ obtained by adding points at infinity which correspond to families of parallel lines, then we can see $\mathbb{P}^2$ as a closed hemisphere of $\mathbb{S}^2$ of which antipodal points of the boundary great circle are identified to form the projective line $L = \mathbb{P}^1$. In that case, any projective line $L'$, distinct from $L$, can be cooriented by the choice of one of the two open lunes that are the components of the complement of $L \cup L'$ in $\mathbb{P}^2$.


### 7.3.2  $C^2$-hedgehogs in Real Affine or Projective Space

The above remarks and definitions can be extended to higher dimensions. As in the plane, the following remark reveals the projective nature of the notion of a hedgehog. Given any $C^2$-hedgehog $\mathcal{H} = \mathcal{H}_h$ in the Euclidean space $\mathbb{R}^{n+1}$, and any hyperplane $H$ that does not meet $\mathcal{H}$ in $\mathbb{R}^{n+1}$, there are exactly two cooriented support hyperplanes of $\mathcal{H}$ through any $(n-1)$-plane $P$ of $H$. To see this, it suffices to apply Theorem 2.8.1 to the plane hedgehog whose support function is the restriction of $h$ to $\mathbb{S}^n \cap \overrightarrow{P}^\perp$, where $\overrightarrow{P}^\perp$ is the linear 2-plane that is orthogonal to the direction $\overrightarrow{P}$ of $P$. Therefore, if we send the hyperplane $H$ to infinity, the hypersurface $\mathcal{H}$ remains a hedgehog since it still has exactly two cooriented hyperplanes that are parallel to any given hyperplane direction.

Of course, we can still introduce the notion of a hedgehog of $\mathbb{R}^{n+1}$ regarded as an affine space over itself (without any metric structure). We still define a cooriented hyperplane of the affine space $\mathbb{R}^{n+1}$ as an affine hyperplane of $\mathbb{R}^{n+1}$ together with a transverse orientation given by a crossing transverse direction indicated by an arrow (or by the half-space where the arrow is pointing to).


## 7.4  From h-Convexity to h-Hedgehogs in $\mathbb{H}^{n+1}$

### 7.4.1  Introduction and Definitions

We will make use of the Poincaré ball model of $\mathbb{H}^{n+1}$, which is the open unit ball $\mathbb{B}^{n+1}$ in $\mathbb{R}^{n+1}$ equipped with the Poincaré metric given by:

$$ds^2 = \frac{4}{\left(1 - \|x\|^2\right)^2} \sum_{i=1}^{n+1} dx_i^2,$$

where $x = (x_1, \ldots, x_{n+1})$, and $\|x\|^2 = \sum_{i=1}^{n+1} x_i^2$. A **horoball** of $\mathbb{H}^{n+1}$ is the limit of a sequence of increasing balls sharing a tangent hyperplane in a given point, as the corresponding sequence of their radii go towards the infinity. A **horosphere** is the boundary of a horoball. As already mentioned in the introduction, the best analogue to Euclidean hyperplanes in $\mathbb{H}^{n+1}$ are not actually the totally geodesic hyperplanes of $\mathbb{H}^{n+1}$: indeed, horospheres can also be regarded as 'hyperplanes', and they are in some sense closer to Euclidean hyperplanes, even though they are not totally geodesic hypersurfaces. In the Poincaré model, horospheres are represented by Euclidean spheres internally tangent to $\partial_\infty \mathbb{H}^{n+1}$ (the ideal boundary sphere at infinity of $\mathbb{H}^{n+1}$, also denoted by $\mathbb{S}^n_\infty$), minus the point of tangency. A horosphere can be regarded as a sphere of infinite radius whose center at infinity is none other than the ideal point where the corresponding Euclidean sphere is tangent to the sphere at infinity: it can also be regarded as a hypersurface of $\mathbb{H}^{n+1}$ whose normal geodesics all converge asymptotically to the same ideal point. Note that if such a 'limit sphere' is a horosphere in $\mathbb{H}^{n+1}$, it would be a hyperplane in the Euclidean space $\mathbb{R}^{n+1}$. It is also important to recall that the intrinsic geometry of horospheres of $\mathbb{H}^{n+1}$ is Euclidean, whereas totally geodesic hyperplanes of $\mathbb{H}^{n+1}$ are isometric to $\mathbb{H}^n$. In our definition of h-hedgehogs, horospheres will play in $\mathbb{H}^{n+1}$ the role assigned to cooriented hyperplanes in $\mathbb{R}^{n+1}$, and the corresponding ideal points the one of the unit normal vectors (unless explicitly stated otherwise).

Before going any further, we must also recall that there are likewise two natural notions of convexity in hyperbolic space: geodesical convexity and horospherical convexity, which is stronger. A body $K$ of $\mathbb{H}^{n+1}$ is said to be **g-convex** (geodesically convex), or convex, if for every pair of points in $K$, the geodesic segment joining them is completely contained in $K$. Given any pair of points in $\mathbb{H}^{n+1}$, there is a whole family of horocycle segments joining them. Here, a horocycle segment is a segment of a (Euclidean) circle internally tangent to $\mathbb{S}^n_\infty$. A body $K$ of $\mathbb{H}^{n+1}$ is said to be **h-convex** (horospherically convex) if for every pair of points in $K$, every horocycle segment joining them is completely contained in $K$. Equivalently, a body $K$ is said to be h-convex (horospherically convex) if every point $p$ of its boundary $\partial K$ has a **support horoball**, that is, a horoball $B$ such that $K \subset B$ and $p$ belongs to the horosphere that bounds $B$. If $\partial K$ is a smooth hypersurface, this implies that all its principal curvatures are greater than or equal to 1. Of course, any h-convex body in $\mathbb{H}^{n+1}$ is a also a g-convex body, but the converse is not true: for instance, g-convex polygons of $\mathbb{H}^2$ are not h-convex bodies.

In Euclidean space $\mathbb{R}^{n+1}$, convex bodies can be regarded as intersections of their support halfspaces, and the support function of a given convex body $K$ is determined by the signed distance from the origin to its support hyperplanes (cooriented towards the exterior). Similarly, h-convex bodies can be regarded as intersections of their support horoballs, and the support function of a given h-convex body $K$ is determined by the signed distance from a fixed origin to the horospheres (cooriented towards the corresponding ideal points) that bounds its support horoballs.

In [117], D. Rochera and the author introduced h-hedgehogs as envelopes of smooth families of horospheres in such a way that regular h-hedgehogs of $\mathbb{H}^{n+1}$ be h-convex hypersurfaces (that is, boundaries of h-convex bodies). Our aim was essentially to define h-hedgehogs of $\mathbb{H}^2$ in order to study those that are of constant width. But a large part of our introduction of h-hedgehogs extends naturally to higher dimensions.

In a preprint and two papers dating back to the 1980s [40–42], C. L. Epstein introduced a hyperbolic analogue for the Gauss map, and he associated to any $C^\infty$-smooth function $\rho$ on $\mathbb{S}^n_\infty \equiv \mathbb{S}^n$, the envelope (which can be regarded in some sense as 'parametrized by its hyperbolic Gauss map') $\Sigma_\rho$ of the smooth family of horospheres $(H\,(u;\rho\,(u)))_{u\in\mathbb{S}^n_\infty}$, where $H\,(u;\rho\,(u))$ is the horosphere with center (ideal point) $u$ such that $\rho\,(u)$ is the signed hyperbolic distance from $0$ in $\mathbb{B}^{n+1}$ to $H\,(u;\rho\,(u))$ (cooriented by $u$, so that $\rho\,(u) < 0$ if $0$ lies in the 'interior' to the horosphere, and nonnegative otherwise). In [40], Epstein proved the hypersurface $\Sigma_\rho$ determined by the smooth function $\rho$ admits the following parametrization

$$x_\rho : \mathbb{S}^n = \mathbb{S}^n_\infty \rightarrow \Sigma_\rho \subset \mathbb{B}^{n+1}$$

$$u \mapsto \frac{\left(\|\nabla\rho\,(u)\|^2 - 1 + e^{2\rho(u)}\right) u + 2\nabla\rho\,(u)}{\|\nabla\rho\,(u)\|^2 + \left(e^{\rho(u)} + 1\right)^2},$$

where $\nabla\rho\,(u)$ is the gradient of $\rho$ at $u$ on $\mathbb{S}^n$ with respect the round metric, and $\|\nabla\rho\,(u)\|^2$ is its scalar square.

For a smoothly immersed, oriented hypersurface $\Sigma$ in the Poincaré ball model of $\mathbb{H}^{n+1}$, Epstein introduced two hyperbolic Gauss maps as follows: for each $p \in \Sigma$, there exists an ordered pair $(G_-\,(p)\,,G_+\,(p))$ of distinct points of $\mathbb{S}^n_\infty$ such that the geodesic line starting from $G_-\,(p)$ towards $G_+\,(p)$ is the oriented normal line to $\Sigma$ at $p$; the maps $G_-$ and $G_+$ from $\Sigma$ to $\mathbb{S}^2_\infty$ are respectively called *the negative and positive hyperbolic Gauss maps* of $\Sigma$. Recall that in the Poincaré ball model of $\mathbb{H}^{n+1}$: $(i)$ the geodesic lines are all arcs of Euclidean circles in $\mathbb{B}^{n+1}$ that are orthogonal to $\mathbb{S}^n_\infty$, plus all diameters of the ball; $(ii)$ every pair of points of $\mathbb{S}^n_\infty$ uniquely determines a geodesic line and vice versa. Note that for $\varepsilon \in \{-, +\}$ there is a uniquely determined horosphere centered at $G_\varepsilon\,(p) \in \mathbb{S}^2_\infty$ that is tangent to $\Sigma$, and that a reversal of the orientation of $\Sigma$ changes each of the hyperbolic Gauss maps into the other.

If an envelope $\Sigma_\rho = x_\rho\,(\mathbb{S}^n)$ determined by a smooth function $\rho : \mathbb{S}^n \rightarrow \mathbb{R}$ fails to be a smooth hypersurface, then for any $u \in \mathbb{S}^n$ we will say that the geodesic line oriented towards $u$ that is orthogonal to $H\,(u;\rho\,(u))$ at $X_\rho\,(u)$ is the oriented normal line to $\Sigma_\rho$ at $x_\rho\,(u)$. This allows us to define the negative and positive hyperbolic Gauss maps, $G_-$ and $G_+$, of such an envelope $\Sigma_\rho$. We will distinguish between a pair of oriented normal lines of $\Sigma_\rho = x_\rho\,(\mathbb{S}^n)$ that does not meet in $\mathbb{B}^{n+1}$, and a pair of distinct oriented normal lines that are coming from the same ideal point $v \in \mathbb{S}^n_\infty$.

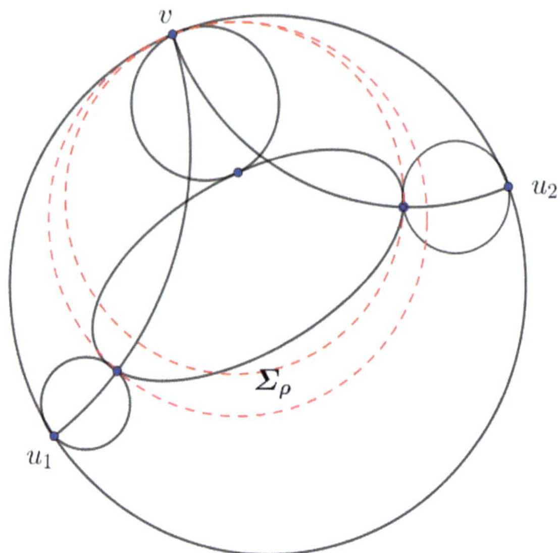

**Fig. 7.6** Smooth $\Sigma_\rho$ with parallel oriented normal lines

The first ones will be said pair of *nonintersecting normal lines* and the second ones will be said to be *parallel*. Even if an envelope $\Sigma_\rho = x_\rho\left(\mathbb{S}^n\right)$ is a smooth convex hypersurface, it may have parallel normal lines if it fails to be a h-convex hypersurface (see Fig. 7.6). In order to qualify an envelope $\Sigma_\rho = x_\rho\left(\mathbb{S}^n\right)$ as a h-hedgehog, we naturally want that it has no pair of distinct oriented normal lines that are parallel. In fact, we will require that the map sending every $u = G_+\left(x_\rho\left(u\right)\right)$ to the other ideal endpoint $v\left(u\right)$ of the oriented normal line to $\Sigma_\rho$ at $x_\rho\left(u\right)$ be a $C^\infty$-diffeomorphism $v = G_- \circ G_+^{-1} : \mathbb{S}^n_\infty \to \mathbb{S}^n_\infty$, which is a stronger condition.

**Definition 7.4.1** Let $h \in C^\infty\left(\mathbb{S}^n; \mathbb{R}\right)$, and let $\Sigma_h$ be the hypersurface of the Poincaré ball model of $\mathbb{H}^{n+1}$ that is the envelope of the smooth family of horospheres $\left(H\left(u; h\left(u\right)\right)\right)_{u \in \mathbb{S}^n_\infty}$, where $H\left(u; h\left(u\right)\right)$ is the horosphere with center (ideal point) $u$ such that $h\left(u\right)$ is the signed hyperbolic distance from $0$ in $\mathbb{B}^{n+1}$ to $H\left(u; h\left(u\right)\right)$ (cooriented by $u$, so that $h\left(u\right) < 0$ if $0$ lies in the 'interior' to the horosphere, and nonnegative otherwise). We will say that $\Sigma_h = x_h\left(\mathbb{S}^n\right)$ is *the h-hedgehog $\mathcal{H}_h$ of $\mathbb{H}^{n+1}$ with support function $h$* if the map sending every $u = G_+\left(x_\rho\left(u\right)\right)$ to the other ideal endpoint $v\left(u\right)$ of the oriented normal line to $\mathcal{H}_h$ at $x_h\left(u\right)$ is a $C^\infty$-diffeomorphism $v : \mathbb{S}^n_\infty \to \mathbb{S}^n_\infty$.

Note that a reversal of the orientation of normal lines of a h-hedgehog $\mathcal{H}_h$ amounts to change $\mathcal{H}_h$ into another h-hedgehog $\mathcal{H}_{\tilde{h}}$ such that $x_{\tilde{h}}\left(\mathbb{S}^n_\infty\right) = x_h\left(\mathbb{S}^n_\infty\right)$ since the initial family of horospheres $\left(H\left(u, h\left(u\right)\right)\right)_{u \in \mathbb{S}^n_\infty}$ is then replaced by the family of horospheres

$$\left(H\left(v; \tilde{h}\left(v\right)\right)\right)_{v \in \mathbb{S}^n_\infty} = \left(\mathcal{H}\left(x_h\left(u\right); v\left(u\right)\right)\right)_{u \in \mathbb{S}^n_\infty},$$

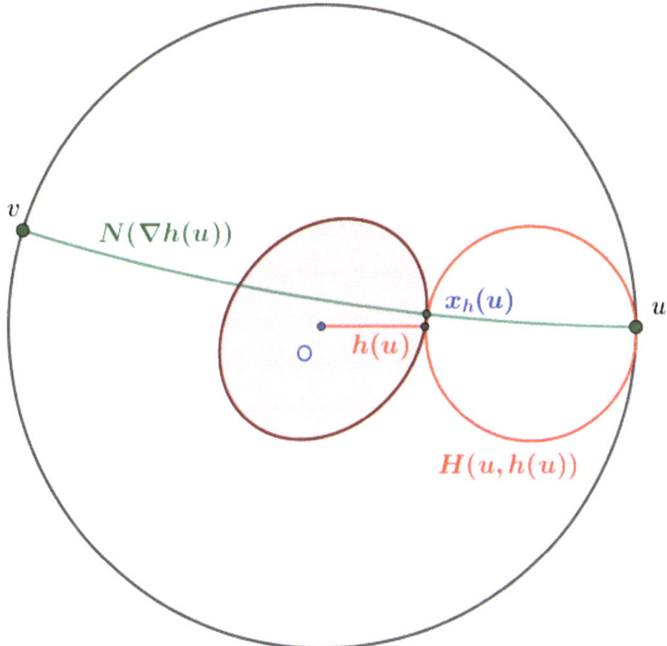



**Fig. 7.7** Determination of $\mathcal{H}_h$ by $h$

where $\mathcal{H}\left(x_h\left(u\right); v(u)\right)$ is the horosphere with center (ideal point)

$$v := v\left(u\right) = \frac{\left(\|\nabla h\left(u\right)\|^2 - 1\right)u + 2\nabla h\left(u\right)}{\|\nabla h\left(u\right)\|^2 + 1}$$

passing through $x_h\left(u\right)$ (see [41, Proposition 3.2]). Here, $v : \mathbb{S}^n_\infty \rightarrow \mathbb{S}^n_\infty$ is the diffeomorphism sending any $u \in \mathbb{S}^n_\infty$ to the other ideal endpoint of the oriented normal line to $\mathcal{H}_h$ at $x_h\left(u\right)$. If $\mathcal{H}_h$ is a Euclidean hedgehog of $\mathbb{R}^{n+1}$, then for all $u \in \mathbb{S}^n$, $x_h\left(u\right)$ lies at the intersection of the support hyperplane $H_h\left(u\right)$, with equation $\langle x, u \rangle = h\left(u\right)$, with the normal line $\mathcal{N}_{\nabla h}\left(u\right) := \{\nabla h\left(u\right)\} + \mathbb{R}u$. Similarly, if $\mathcal{H}_h$ is a h-hedgehog of $\mathbb{H}^{n+1}$, then for all $u \in \mathbb{S}^n_\infty$, $x_h\left(u\right)$ lies at the intersection of the support horosphere $H\left(u; h\left(u\right)\right)$ with the geodesic normal line $N\left(\nabla h\left(u\right)\right)$ joining $u$ and $v = v\left(u\right)$ (see Fig. 7.7).

It is an easy exercise to deduce from above formulas that the support function $\widetilde{h}$ is in fact determined by

$$\widetilde{h}\left(v\right) = \ln\left(1 + \|\nabla h\left(u\right)\|^2\right) - h(u), \text{ where } v = v\left(u\right),$$

and that we have

$$\nabla \widetilde{h}\left(v\right) = \frac{2\left\|\nabla h\left(u\right)\right\|^{2} u + \left(1 - \left\|\nabla h\left(u\right)\right\|^{2}\right)\nabla h\left(u\right)}{\left(1 + \left\|\nabla h\left(u\right)\right\|^{2}\right)} \quad \text{and thus}$$

$$\left\|\nabla \widetilde{h}\left(v\right)\right\| = \left\|\nabla h\left(u\right)\right\|.$$

From the expression of $v : \mathbb{S}^{n}_{\infty} \rightarrow \mathbb{S}^{n}_{\infty}$, we can deduce, in passing, that $\nabla h\left(u\right)$ can also be expressed as follows

$$\nabla h\left(u\right) = \frac{v\left(u\right) - \left\langle u, v\left(u\right)\right\rangle u}{1 - \left\langle u, v\left(u\right)\right\rangle},$$

where $u$ and $v\left(u\right)$ are considered as being in $\mathbb{R}^{n+1}$, and $\left\langle .,.\right\rangle$ is the standard Euclidean inner product on $\mathbb{R}^{n+1}$.

### 7.4.2   On the h-Width of h-Hedgehogs in $\mathbb{H}^{n+1}$

Given a g-convex or h-convex body $K$ in $\mathbb{H}^{n+1}$, there have been many different attempts to define the width function of $K$, among which the Santaló width function [163], the Fillmore width function [45], and the Leichtweiss width function [83] (the first two width functions were introduced by their authors only in $\mathbb{H}^{2}$, but they have a natural extension to higher dimensions). There are two more recent attempts in $\mathbb{H}^{2}$ [67, 75]. In the last one, the reader can find a comparison of these different widths in the hyperbolic plane.

The notion of h-width of a h-hedgehog in $\mathbb{H}^{n+1}$, which we are going to introduce below by inspiring us of our approach to define the width function of a Euclidean hedgehog, is closely related to the Santaló width function (except that our hypersurfaces are hedgehogs, and that they come with their hyperbolic Gauss map).

In the Euclidean space $\mathbb{R}^{n+1}$, we saw that a reversal of the orientation of normal lines of a hedgehog $\mathcal{H}_h$ amounts to change $\mathcal{H}_h$ into the hedgehog $\mathcal{H}_{\widetilde{h}}$, where $\widetilde{h}\left(u\right) = -h\left(u\right)$ for all $u \in \mathbb{S}^{n}$, and that $x_{\widetilde{h}}\left(\mathbb{S}^{n}_{\infty}\right) = x_{h}\left(\mathbb{S}^{n}_{\infty}\right)$. We defined for all $u \in \mathbb{S}^{n}$, the signed width in direction $u$ of such a Euclidean hedgehog $\mathcal{H}_h$ in $\mathbb{R}^{n+1}$ to be the signed distance $w_h\left(u\right) = h\left(u\right) + h\left(-u\right) = \left(h - \widetilde{h}\right)\left(u\right)$ between the two cooriented support hyperplanes orthogonal to $u$ (see Chap. 4). In the hyperbolic space $\mathbb{H}^{n+1}$, we will define the signed h-width of a h-hedgehog $\mathcal{H}_h$ in direction $u \in \mathbb{S}^{n} = \mathbb{S}^{n}_{\infty}$ in the same way. The **h-width** of $\mathcal{H}_h$ in direction $u$ is defined to be the signed hyperbolic distance $w_h\left(u\right)$ between the two parallel horospheres $H_h\left(u\right) := H\left(u; h\left(u\right)\right)$ (whose normal lines are oriented towards $u$) and $H_{\widetilde{h}}\left(u\right) := H\left(x_{\widetilde{h}}\left(u\right); u\right)$ (the horosphere with center $u$ passing through $x_{\widetilde{h}}\left(u\right)$ whose normal lines are coming from $u$), that is, $w_h\left(u\right) = \left(h - \widetilde{h}\right)\left(u\right)$ since the geodesic line passing through 0 and going towards $u$ is orthogonal to both horospheres (see Fig. 7.8). Therefore, a h-hedgehog $\mathcal{H}_h$ is of constant h-width $2r$,

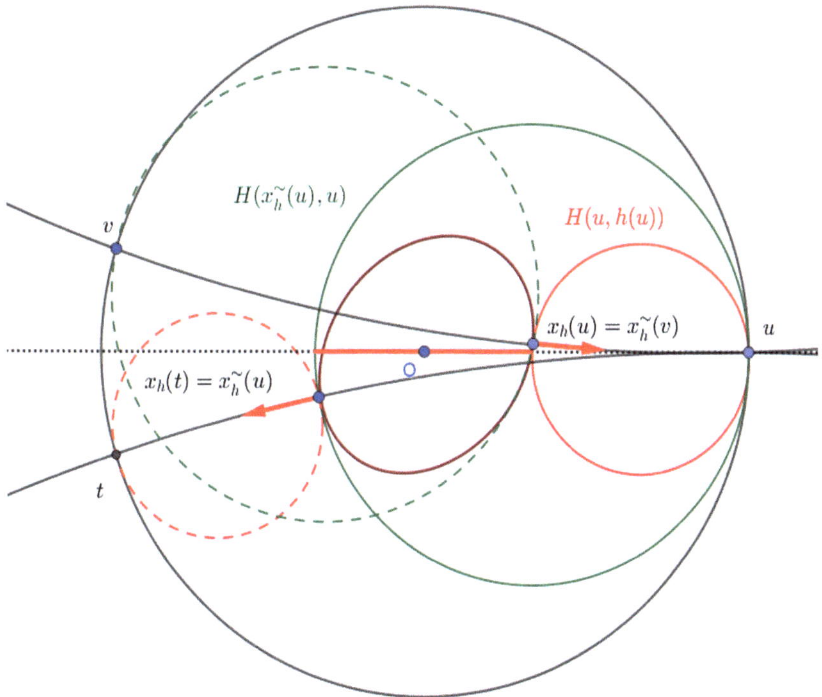


**Fig. 7.8** The h-width of a regular h-hedgehog

$(r \in \mathbb{R})$ if, and only if, it satisfies

$$h(u) - \widetilde{h}(u) = 2r \quad \text{for all } u \in \mathbb{S}_\infty^n,$$

or equivalently

$$h(v(u)) + h(u) - \ln\left(1 + \|\nabla h(u)\|^2\right) = 2r \quad \text{for all } u \in \mathbb{S}_\infty^n.$$

It is worth to note following C. L. Epstein that the parallel hypersurface at hyperbolic distance $r$ from a h-hedgehog $\mathcal{H}_h$ is none other than the h-hedgehog $\mathcal{H}_{h+r}$, $(r \in \mathbb{R})$. Moreover, for all $u \in \mathbb{S}_\infty^n$, $r \mapsto x_{h+r}(u)$ is the unit speed parametrization of a hyperbolic geodesic normal to the family of parallel horo-spheres $(H(u; (h+r)(u)))_{r \in \mathbb{R}}$.

# Chapter 8
# Marginally Trapped Hedgehogs

Unless explicitly states otherwise, the results of this section are essentially taken from [109].[1] In this section, our first aim will be to show using fundamental examples that a huge class of marginally trapped surfaces arises naturally from a 'lightlike co-contact structure', exactly in the same way as Legendrian fronts from a contact one (by projection of a Legendrian submanifold to the base of a Legendrian fibration). Furthermore, we will observe that there is an adjunction relationship between these two notions. We especially focus on marginally trapped hedgehogs and study their relationships with Laguerre geometry and Brunn-Minkowski theory.

## 8.1 Introduction and Statement of Main Results

*Trapped surfaces* were introduced in general relativity by R. Penrose [146] to study singularities of spacetimes. They appeared in a natural way earlier in the work of Blaschke, in the context of conformal and Laguerre geometry [19]. These surfaces play an extremely important role in general relativity where they are of central importance in the study of black holes, those regions of spacetime where everything is trapped, and nothing can escape, even light. A closed embedded spacelike 2-surface of a 4-dimensional spacetime is said to be *trapped* if its mean curvature vector is everywhere timelike. The limiting case of *marginally trapped surfaces* (i.e., surfaces whose mean curvature vector is everywhere lightlike) plays the role of apparent horizons of black holes. Mathematically, marginally trapped surfaces are regarded as spacetime analogues of minimal surfaces in Riemannian geometry. Even though they received considerable attention both from mathematicians and

---

[1] This article was published in Advances in Applied Math. 101, Y. Martinez-Maure, New insights on marginally trapped surfaces. The hedgehog theory point of view, pp. 320–353. Copyright Elsevier 2001.

© The Author(s), under exclusive license to Springer Nature Switzerland AG 2026

Y. Martinez-Maure, *Hedgehog Theory*, Lecture Notes in Mathematics 2381,

https://doi.org/10.1007/978-3-032-04298-9_8

physicists, these surfaces are still not very well understood. For a recent survey on marginally trapped surfaces, see the book by B.Y. Chen [30].

As we have said, our main objective here will be to show by the study of a series of fundamental examples that:

*A huge class of singular marginally trapped surfaces arises naturally from a 'lightlike co-contact structure', exactly in the same way as Legendrian fronts arise from a contact one (by projection of a Legendrian submanifold to the base of a Legendrian fibration), and there is an adjunction relationship between these two notions.*

In addition, a huge class of singular marginally trapped surfaces correspond by adjunction to hedgehogs (envelopes parametrized by their Gauss map) and can thus benefit directly from contributions of hedgehog theory, which can be seen as an extension of the Brunn-Minkowski one. This correspondence is naturally promising in terms of new geometric inequalities, and we know how important geometric inequalities are  in gravitation. We will give examples of geometric inequalities involving hedgehogs and marginally trapped surfaces in Sect. 8.1.2.

In order to explain precisely what we mean here by a *'lightlike co-contact structure'*, let us begin by the presentation of a fundamental example in the 4-dimensional Lorentz-Minkowski space  $\mathbb{L}^4$. This example will be detailed in Sect. 8.1.3.


### *8.1.1  Marginally Trapped Hedgehogs or Co-hedgehogs in $\mathbb{L}^4$*

**Characterization and Definitions in $\mathbb{L}^4$**

For simplicity, we will restrict our presentation to surfaces in $\mathbb{L}^4$ but our results extend, without much change, to higher dimensions. To any $h \in C^\infty\left(\mathbb{S}^2; \mathbb{R}\right)$ corresponds the envelope $\mathcal{H}_h$ of the family $(P_h(u))_{u \in \mathbb{S}^2}$ of cooriented planes of $\mathbb{R}^3$ with equation $(E)$  $\langle x, u \rangle = h(u)$, where $\langle ., . \rangle$ is the standard scalar product on $\mathbb{R}^3$. We say that $\mathcal{H}_h$ is the hedgehog with support function $h$. From $(E)$ and the contact condition $\langle dx, u \rangle = 0$, we deduce

$$\begin{cases} \langle x, u \rangle = h(u) \\ \langle x, du \rangle = dh_u, \end{cases}$$

for all $u \in \mathbb{S}^2$. Thus, it appears that $\mathcal{H}_h$ can be parametrized by $x_h : \mathbb{S}^2 \to \mathbb{R}^3$, $u \mapsto h(u)u + (\nabla h)(u)$, where $(\nabla h)(u)$ stands for the gradient of $h$ at $u$. The parametrization $x_h$ can be interpreted as the inverse of its Gauss map (if $u$ is a regular point of $x_h$, then $u$ is normal to $\mathcal{H}_h$ at $x_h(u)$). Note that $(\mathcal{H}_{h+t})_{t \in \mathbb{R}}$ is a family of parallel hedgehogs in $\mathbb{R}^3$: for all $(u, t) \in \mathbb{S}^2 \times \mathbb{R}$, $x_{h+t}(u) = x_h(u) + tu$.

Now, we claim that:

*To the differential dh of h corresponds naturally a 'marginally trapped hedgehog' of the Lorentz Minkowski 4-space $\mathbb{L}^4 = \left(\mathbb{R}^4, \langle ., . \rangle_L\right)$ via a 'lightlike co-contact condition'.*

Here, the pseudo-scalar product $\langle ., . \rangle_L$ is defined by

$$\left\langle (x, t), \left(x', t'\right) \right\rangle_L := x_1 x_1' + x_2 x_2' + x_3 x_3' - tt',$$

for all $(x, t) := \left(\left(x_1, x_2, x_3\right), t\right)$ and $\left(x', t'\right) := \left(\left(x_1', x_2', x_3'\right), t'\right)$ in $\mathbb{R}^4 = \mathbb{R}^3 \times \mathbb{R}$. Indeed, $dh$ determines the family $(L_{h+t}(u))_{(u,t)\in\mathbb{S}^2\times\mathbb{R}}$ of oriented null lines of $\mathbb{L}^4$ defined by:

$$\forall (u, t) \in \mathbb{S}^2 \times \mathbb{R}, \, L_{h+t}(u) = \{l_{h+t}(u)\} + \mathbb{R}(u, -1),$$

where $l_{h+t}(u) := (\nabla h(u), h(u) + t)$; and this family of null lines determines a 'marginally trapped hedgehog', which is unique up to translations parallel to the time axis, via a 'lightlike co-contact condition':

**Theorem 8.1.1 (Determination of Marginally Trapped Hedgehogs of $\mathbb{L}^4$ by 1-Jet and Co-contact Condition)** *For all $t \in \mathbb{R}$,*

(i) *there is a unique map $x \colon \mathbb{S}^2 \to \mathbb{L}^4$ of class $C^\infty$ satisfying $x(u) \in L_{h+t}(u)$ for all $u \in \mathbb{S}^2$ together with the **'lightlike co-contact condition'***

$$\langle \delta(dx), u_L \rangle_L = 0,$$

*where $\delta$ is the Hodge codifferential on $\mathbb{S}^2$ (i.e., the formal adjoint to the exterior differentiation $d$) and $u_L := \frac{1}{\sqrt{2}}(u, -1)$, namely*

$$x = x_{l_{h+t}} \colon \mathbb{S}^2 \to \mathbb{L}^4, \, u \mapsto l_{h+t}(u) + \delta(\partial h)(u) u_L,$$

*where $\partial h(u) := dh/\sqrt{2}$;*

(ii) *this map $x \colon \mathbb{S}^2 \to \mathbb{L}^4$ is such that $x^*g = \frac{1}{4}(R_1 - R_2)^2 g_S$, where $g_S$ is the standard metric on $\mathbb{S}^2$, $g$ the first fundamental form on $x(\mathbb{S}^2)$, $x^*g$ the pullback of $g$ along $x$, and $R_1(u), R_2(u)$ are the principal radii of curvature of $\mathcal{H}_h$ at $x_h(u)$;*

(iii) *for all $u \in \mathbb{S}^2$, $\Delta x(u) \in \mathbb{R}u_L$, where $\Delta$ is the Hodge Laplacian on $\mathbb{S}^2$, so that the mean curvature vector of $x(\mathbb{S}^2)$ at $x(u)$, say $\overrightarrow{H}_x(u)$, is parallel to the lightlike vector $\overrightarrow{u_L}$ whenever $x_h(u)$ is not an umbilical point of $\mathcal{H}_h$, that is:*

$$\forall u \in \mathbb{S}^2 \setminus \left\{u \in \mathbb{S}^2 \mid R_1(u) = R_2(u)\right\}, \quad \overrightarrow{H}_x(u) \in \mathbb{R}\overrightarrow{u_L}$$

Of course, the Hodge-Laplacian of a $C^\infty$–map $x : \mathbb{S}^2 \to \mathbb{L}^4$, $u \mapsto (x_i\,(u))_{i=1}^4$ is understood to be the vector function $\Delta X : \mathbb{S}^2 \to \mathbb{L}^4$, $u \mapsto (\Delta x_i\,(u))_{i=1}^4$, where $\Delta x_i$ is the Hodge-Laplacian of the coordinate function $x_i$ in the intrinsic metric on $\mathbb{S}^2$, $(1 \le i \le 4)$.

**Definition 8.1.1** For all $t \in \mathbb{R}$, *we call* $\mathcal{H}_{l_{h+t}} := x_{l_{h+t}}\left(\mathbb{S}^2\right)$ *the (marginally trapped) hedgehog with support* 1-*jet* $l_{h+t} : \mathbb{S}^2 \to J^1\left(\mathbb{S}^2\right)$, $u \mapsto (\nabla h\,(u)\,,\,h\,(u) + t)$.

Note that $\sqrt{2}\,\partial$ stands for the Hodge-Dirac operator $D = d + \delta$ on $\mathbb{S}^2$ and that the datum of $\partial h := dh/\sqrt{2}$ is equivalent to the one of the family of 1-jets $(l_{h+t})_{t\in\mathbb{R}}$, where $l_{h+t} := (\nabla h,\, h + t)$. All the marginally trapped hedgehogs of the family $\left(\mathcal{H}_{l_{h+t}}\right)_{t\in\mathbb{R}}$ are equal up to translations parallel to the time axis $\mathbb{R}\,\overrightarrow{\partial_t}$ in $\mathbb{L}^4$, where $\overrightarrow{\partial_t} := \left(\overrightarrow{0_{\mathbb{R}^3}},\,1\right)$, since: for all $(u,t) \in \mathbb{S}^2 \times \mathbb{R}$, $x_{l_{h+t}}\,(u) = x_{l_h}\,(u) + t\,\overrightarrow{\partial_t}$. Therefore, this family can be regarded as one and only one marginally trapped hedgehog defined up to a translation parallel to the time axis $\mathbb{R}\,\overrightarrow{\partial_t}$ in $\mathbb{L}^4$. This hedgehog will be denoted by $\mathcal{H}_{\partial h}$, where $\partial h := dh/\sqrt{2}$. The factor $1/\sqrt{2}$ is chosen so that the hedgehog with support function $\partial^2 h := \partial\,(\partial h)$ coincides with the mean evolute of $\mathcal{H}_h$ (see Sect. 8.1.2 for details and a precise definition of the mean evolute).

These remarks make natural the following definition:

**Definition 8.1.2** We call $\mathcal{H}_{\partial h}$ the **marginally trapped hedgehog** (or, the **co-hedgehog**) of $\mathbb{L}^4$ with support differential $\partial h := dh/\sqrt{2}$. Anyone of the marginally trapped hedgehogs $\mathcal{H}_{l_{h+t}}$, $(t \in \mathbb{R})$, will be regarded as a representative of $\mathcal{H}_{\partial h}$ in $\mathbb{L}^4$. We will say that $\mathcal{H}_{\partial h}$ is the **co-evolute of** $\mathcal{H}_h$ **in** $\mathbb{L}^4$.

Thus, once the lightlike 'co-contact condition' $\langle \delta\,(dx)\,,\,u_L \rangle_L = 0$ is fixed, the (co)hedgehog $\mathcal{H}_{\partial h}$ is defined in $\mathbb{L}^4$ as the 'co-envelope' of the family of oriented null lines determined by its support differential $\partial h$ in the same way as, once the contact condition $\langle dx,\,u \rangle = 0$ is fixed, the hedgehog $\mathcal{H}_h$ is defined in $\mathbb{R}^3$ as the envelope of the family of cooriented planes determined by its support function $h$. Of course, it is important to note that $\mathcal{H}_h$ and $\mathcal{H}_{\partial h}$ are not assumed to be embedded surfaces of respectively $\mathbb{R}^3$ and $\mathbb{L}^4$: $\mathcal{H}_h$ and $\mathcal{H}_{\partial h}$ may possibly be singular and self-intersecting. Note that regularity assumptions on the support function $h$ can be weakened in many cases. For instance, to define the co-envelope $\mathcal{H}_{\partial h}$ in $\mathbb{L}^4$ we only need to assume that $h$ is of class $C^2$ on $\mathbb{S}^2$.

In everyday language, hedgehogs are spiny mammals. Note that Langevin, Levitt and Rosenberg chose to call $\mathcal{H}_h$ a hérisson [82], which is the French name for hedgehog, to illustrate the fact that, in each direction $u \in \mathbb{S}^2$, there is one and only one 'normal' $N_h\,(u) := x_h\,(u) + \mathbb{R}u$ pointing out from $\mathcal{H}_h$ (as, in each direction, a unique spine is pointing out from the skin of the spiny mammal). Similarly, it is worth pointing out, that, in each null direction $u_L := \frac{1}{\sqrt{2}}\,(u,\,-1)) \in \left(\mathbb{S}^2 \times \{-1\}\right)/\sqrt{2} \subset \mathbb{S}^3$, there is one and only one null line $L_h\,(u) := x_{\partial h}\,(u) + \mathbb{R}u_L$ that is normal to the co-envelope $\mathcal{H}_{\partial h}$ in $\mathbb{L}^4$.

## Geometrical Interpretation of Co-hedgehogs of $\mathbb{L}^4$

First, let us remind some basic facts about the curvature of hedgehogs of $\mathbb{R}^3$. Since the parametrization $x_h : \mathbb{S}^2 \to \mathcal{H}_h \subset \mathbb{R}^3$ can be regarded as the inverse of the Gauss map, the Gauss curvature $\kappa_h$ of $\mathcal{H}_h$ at $x_h(u)$ is given by $\kappa_h(u) = 1/\det[T_u x_h]$, where $T_u x_h$ is the tangent map of $x_h$ at $u$. Therefore, singularities of $\mathcal{H}_h$ are the points at which its Gauss curvature becomes infinite. For all $u \in \mathbb{S}^2$, the tangent map of $x_h$ at the point $u$ is $T_u x_h = h(u)\, Id_{T_u \mathbb{S}^2} + H_h^u$, where $H_h^u$ is the symmetric endomorphism associated with the Hessian of $h$ at $u$. Consequently, if $\lambda_1(u)$ and $\lambda_2(u)$ are the eigenvalues of the Hessian of $h$ at $u$ then $R_1(u) := (\lambda_1 + h)(u)$ and $R_2(u) := (\lambda_2 + h)(u)$ can be interpreted as the principal radii of curvature of $\mathcal{H}_h$ at $x_h(u)$, and the so-called *curvature function* $R_h := 1/\kappa_h$ is given by

$$R_h(u) = \det\left[H_{ij}(u) + h(u)\,\delta_{ij}\right] = (R_1 R_2)(u),$$

where $\delta_{ij}$ are the Kronecker symbols and $\left(H_{ij}(u)\right)$ the Hessian of $h$ at $u$ with respect to an orthonormal frame on $\mathbb{S}^2$. In computations, it is often more convenient to replace $h$ by its positively $1-$homogeneous extension to $\mathbb{R}^3 \setminus \{0\}$, which is given by,

$$\varphi(x) := \|x\|\, h\left(\frac{x}{\|x\|}\right),$$

for $x \in \mathbb{R}^3 \setminus \{0\}$, where $\|.\|$ is the Euclidean norm on $\mathbb{R}^3$. A straightforward computation gives:

$(i)$   $x_h$ is the restriction of the gradient of $\varphi$ to the unit sphere $\mathbb{S}^2$ ;

$(ii)$  For all $u \in \mathbb{S}^2$, the tangent map $T_u x_h$ identifies with the symmetric endomorphism associated with the Hessian of $\varphi$ at $u$.

In order to bring out our geometrical interpretation of the co-evolute $\mathcal{H}_{\partial h}$, we now give another expression for $x_{l_h}$. For all $u \in \mathbb{S}^2$, we have

$$x_{l_h}(u) := l_h(u) + \delta(\partial h)(u)\, u_L = (\nabla h(u), h(u)) + \frac{\Delta h(u)}{2}(u, -1)$$

$$= \left(\nabla h(u) + \frac{\Delta h(u)}{2}u,\, h(u) - \frac{\Delta h(u)}{2}\right)$$

$$= \left(x_h(u) - R_{(1,h)}(u)\, u,\, R_{(1,h)}(u)\right) = \left(c_h(u),\, R_{(1,h)}(u)\right),$$

where $c_h(u)$ is the midpoint of the focal segment (which connects the two centers of principal curvatures of $\mathcal{H}_h$ at $x_h(u)$), and $R_{(1,h)} := h - \frac{\Delta h}{2} = \frac{1}{2}(R_1 + R_2)$ is the mean radius of curvature of $\mathcal{H}_h$ (for the interpretation of $R_{(1,h)}$ as a mixed curvature function see Proposition 3.1.1). Therefore, for all $u \in \mathbb{S}^2$, $x_{l_h}(u)$ can be interpreted geometrically as the (possibly reduced to a point) **middle sphere** of $\mathcal{H}_h$ at $x_h\,(u)$ that is cooriented by the vector $u$, which is normal to it, by identifying $\mathbb{L}^4$ with the so-called **Laguerre space**, say $\Sigma$, of cooriented spheres and (non-cooriented) point spheres of $\mathbb{R}^3$ via the bijection

$$\mathbb{L}^4 \to \Sigma$$

$$(a, r) \mapsto S\,(a; r)\,,$$

where $S\,(a; r)$ denotes the sphere of radius $|r|$ centered at $a$ that is cooriented by its outward (resp. inward) pointing normal if $r > 0$ (resp. $r < 0$) holds, and the (non-cooriented) point sphere $\{a\}$ if $r = 0$ holds. Thus:

**Proposition 8.1.1** *The co-evolute $\mathcal{H}_{\partial h}$ of $\mathcal{H}_h$ can be regarded as the locus of all the (possibly reduced to a point) cooriented middle spheres of $\mathcal{H}_h \subset \mathbb{R}^3$ in $\mathbb{L}^4 \cong \Sigma$.*

### Relationship with Laguerre Geometry

Laguerre geometry in $\mathbb{R}^3$ is based on oriented planes, cycles (i.e., oriented spheres and points regarded as unoriented spheres of radius zero) and oriented contact between them. The orientation is determined by a unit normal vector field or, equivalently, by a signed radius in the case of a sphere. An oriented sphere or an oriented hyperplane of $\mathbb{R}^3$ is said to be in oriented contact with another oriented sphere or hyperplane if they are tangent and moreover if their unit normals coincide at the point of tangency. An unoriented point sphere is said to be in oriented contact with an oriented sphere or hyperplane if it is contained in it. An affine Laguerre transformation of $\mathbb{L}^4 \cong \Sigma$ is an affine transformation $\mathcal{A}\,(x) := \mathcal{L}(x) + \mathcal{C}$ of $\mathbb{L}^4$, where $L$ is a linear transformation that preserves the pseudo-scalar product $\langle ., . \rangle_L$ (i.e., $\mathcal{L} \in O\,(3, 1)$), and thus the tangential distance between spheres, and $\mathcal{C}$ is a vector of $\mathbb{L}^4$.

Classical references on sphere geometries of Laguerre and Lie are Blaschke's book [19] and, for a modern account, T. Cecil's book [29]. In these two references, it is shown that Laguerre geometry in $\mathbb{R}^3$ can be built as a subgeometry of Lie sphere geometry. The subgroup of 'Laguerre transformations' then consists of those Lie sphere transformations that map planes to planes. Each of these Laguerre transformations corresponds to an affine Laguerre transformation of $\mathbb{L}^4$ (as we have defined them above).

Laguerre geometry of surfaces studies properties and invariants of surfaces of $\mathbb{R}^3$ under the Laguerre transformation group. It has been extensively developed by Blaschke and its school [19]. Let us consider the case of hedgehogs of $\mathbb{R}^3$. Let $\Omega$ be an open domain of $\mathbb{S}^2$ such that $x_h : \Omega \to \mathbb{R}^3$ is an umbilic-free (piece of) hedgehog $\mathcal{H}_h$ with support function $h \in C^\infty\,(\Omega; \mathbb{R})$. The geometric invariants of the (co)hedgehog $\mathcal{H}_{\partial h} \cong x_{l_h}\,(\Omega)$ of $\mathbb{L}^4$ are exactly the Laguerre invariants of the original hedgehog $\mathcal{H}_h$. The conformal immersion $x_{l_h} : \Omega \to \mathbb{L}^4$, which we consider as a natural parametrization $x_{\partial h} : \Omega \to \mathbb{L}^4$ of the co-evolute $\mathcal{H}_{\partial h}$ of $\mathcal{H}_h$, assigns to each $u \in \Omega$ the point of $\mathbb{L}^4$ corresponding to the oriented middle sphere or unoriented middle point $S\left(c_h(u); R_{(1,h)}(u)\right) \in \mathbb{L}^4 \cong \Sigma$ that is in oriented contact with the oriented support plane of $\mathcal{H}_h$ at $x_h(u)$. It will be called the **Laguerre Gauss map** of $x_h : \Omega \to \mathbb{R}^3$.

Note that the set of marginally trapped hedgehogs $\mathcal{H}_{\partial h} \cong x_{l_h}\left(\mathbb{S}^2\right)$ of $\mathbb{L}^4$, identified with their support differentials $\partial h := dh/\sqrt{2}$, constitute a real linear space, say $\partial\mathcal{H}$. The surface area of marginally trapped hedgehogs of $\mathbb{L}^4$ defines a particularly interesting Laguerre invariant functional on $\partial\mathcal{H}$, the so-called **Laguerre functional**:

$$L : \quad \partial\mathcal{H} \to \mathbb{R}$$
$$\partial h \mapsto L(\partial h) := \tfrac{1}{4}\int_{\mathbb{S}^2}(R_1 - R_2)^2 \, d\sigma = \int_{\mathbb{S}^2}\left(R_{(1,h)}^2 - R_h\right)d\sigma,$$

where $\sigma$ is the spherical Lebesgue measure on $\mathbb{S}^2$, $R_{(1,h)}$ the mean radius of curvature, and $R_h$ the curvature function of $\mathcal{H}_h$. For any $h \in C^\infty(\Omega; \mathbb{R})$, $L(\partial h)$ will be called the **Laguerre area** of $\mathcal{H}_h$ and will sometimes be denoted by $L(x_h)$.

The surface area element of $\mathcal{H}_{\partial h}$ can be interpreted in $\mathbb{R}^3$ by considering the mean surface of $\mathcal{H}_h$, say $\mathcal{M}_h$, which we define as the (possibly singular and self-intersecting) surface that is parametrized by $c_h : \mathbb{S}^2 \to \mathbb{R}^3$, $u \mapsto c_h(u)$. Indeed, if we denote by $d\mu_h(u)$ the corresponding surface area element of this mean surface $\mathcal{M}_h$, then it is pure routine to check that:

$\left(R_{(1,h)}^2 - R_h\right)(u)\,d\sigma(u)$ *can be regarded as the orthogonal projection of* $d\mu_h(u)$ *into the plane that is parallel to the support plane* $P_h(u)$ *of* $\mathcal{H}_h$ *and passes through* $c_h(u) \in \mathcal{M}_h$.

The following theorem is a first step towards a 'Brunn-Minkowski theory' for marginally trapped hedgehogs, which will be the topic of our next subsubsection.

**Theorem 8.1.2** *The map* $\sqrt{L} : \partial\mathcal{H} \to \mathbb{R}_+$, $\partial h \longmapsto \sqrt{L(\partial h)}$ *is a norm associated with the scalar product*

$$L : \quad (\partial\mathcal{H})^2 \mapsto \mathbb{R},$$
$$(\partial f, \partial g) \longmapsto L(\partial f, \partial g) := \int_{\mathbb{S}^2}\left(R_{(1,f)}R_{(1,g)} - R_{(f,g)}\right)d\sigma,$$

*where* $R_{(.,.)}$ *denotes the mixed curvature function of hedgehogs (see Sect. 3.1.1 for a detailed definition and fundamental properties of* $R_{(.,.)}$*).*

*In particular, for all* $(\partial f, \partial g) \in (\partial\mathcal{H})^2$, *we have*

$$\sqrt{L(\partial f + \partial g)} \leq \sqrt{L(\partial f)} + \sqrt{L(\partial g)},$$

*and*

$$L(\partial f, \partial g)^2 \leq L(\partial f)\,L(\partial g),$$

*with equalities if, and only if, $\mathcal{H}_{\partial f}$ and $\mathcal{H}_{\partial g}$ are homothetic (here, "homothetic" means that there exists $(\lambda, \mu) \in \mathbb{R}^2 - \{(0, 0)\}$ such that $\lambda \partial f + \mu \partial g = 0$).*

The scalar product $L(\partial f, \partial g)$ will of course be called the **mixed Laguerre area** of $\mathcal{H}_{\partial f}$ and $\mathcal{H}_{\partial g}$, and will sometimes be denoted by $L(x_f, x_g)$.

We will return to the Laguerre geometry of hedgehogs in the framework of contact geometry.

### 8.1.2  Relationship with the Brunn-Minkowski Theory

Classical hedgehog theory is an extension of the Brunn-Minkowski theory. The relationship between hedgehogs and marginally trapped surfaces is thus very promising in terms of new geometric inequalities. We already mentioned two fundamental geometric inequalities for marginally trapped hedgehogs in our Theorem 8.1.2. Here, it is worth pointing out how important geometric inequalities are in gravitation since they provide information on the relationship between physically relevant magnitudes in a robust way. We will see that marginally trapped hedgehogs arise naturally in a 3-dimensional equivalent of a classical geometric inequality for hedgehogs of the Euclidean plane.

#### Classical Isoperimetric Inequalities Involving Evolutes for Hedgehog Curves Are Involving (Co)evolutes for Hedgehog Surfaces

In [101] the author already introduced $\mathcal{H}_{\partial h}$ as a hedgehog with support differential $\partial h := dh/\sqrt{2}$ but without precisely grasping the essence of the co-contact structure. Let us recall how marginally trapped hedgehogs appeared in order to play the role of evolutes in a natural 3-dimensional equivalent of a known upper bound of the isoperimetric deficit of plane hedgehog curves in terms of signed area of their evolute.

As we have seen previously, many classical inequalities for convex curves find their counterparts for hedgehogs. Of course, adaptations are necessary. In particular, lengths and areas have to be replaced by algebraic versions. For instance, Theorem 8.1.3 below extends the isoperimetric inequality $L^2 - 4\pi A \geq 0$, which holds for any planar convex body $K$ with perimeter $L$, and area $A$, to any hedgehog $\mathcal{H}_h$ of the Euclidean plane $\mathbb{R}^2$, and gives an **upper bound of the isoperimetric deficit in terms of signed area of the evolute** (see Sect. 3.2.2.).

**Theorem 8.1.3** *For any $h \in C^4(\mathbb{S}^1; \mathbb{R})$, let $\mathcal{H}_h$ be the hedgehog of $\mathbb{R}^2$ with support function $h$. We have:*

$$0 \leq l(h)^2 - 4\pi a(h) \leq -4\pi a(\partial h) = -4\pi a\left(h, \partial^2 h\right), \tag{I}$$

*where $l(h)$ is the length of $\mathcal{H}_h$, $a(h)$ its area, $\partial h(\theta) := h'\left(\theta - \frac{\pi}{2}\right)$ the support function of its evolute $\mathcal{H}_{\partial h}$, $\partial^2 h := -h''$ the one of its second evolute $\mathcal{H}_{\partial^2 h}$, and $a\left(h, \partial^2 h\right)$ the mixed area of $\mathcal{H}_h$ and $\mathcal{H}_{\partial^2 h}$. In each inequality of (I), equality holds if and only if $\mathcal{H}_h$ is a circle or a point.*

The isoperimetric inequality $L^2 - 4\pi A \geq 0$ admits the following 3-dimensional equivalent. Given a convex body $K$ in the Euclidean 3-space $\mathbb{R}^3$, with surface area $S$ and integral of the mean curvature $M$, the first Minkowski inequality for mixed volumes ensures that

$$M^2 - 4\pi S \geq 0,$$

where equality holds if and only if $K$ is a ball (see e.g., [165] , and Proposition 3.2.5 for the hedgehog case). In [101] the author proved the following 3-dimensional equivalent of Theorem 8.1.3:

**Theorem 8.1.4 ([101, Theorem 1: Upper bound of the 'deficit' $M^2 - 4\pi S$ in terms of signed area of the co-evolute surface])** *For any $h \in C^4\left(\mathbb{S}^2; \mathbb{R}\right)$, let $\mathcal{H}_h$ denote the hedgehog of $\mathbb{R}^3$ with support function $h$, (i.e., the envelope of the family of cooriented planes with equation $\langle x, u \rangle = h(u)$ in $\mathbb{R}^3$). We have:*

$$0 \leq m(h)^2 - 4\pi s(h) \leq -4\pi s(\partial h) = -4\pi s\left(h, \partial^2 h\right), \tag{II}$$

*where $m(h)$ is the integral of the mean curvature of $\mathcal{H}_h$, $s(h)$ its signed surface area, $s(\partial h)$ the signed surface area of $\mathcal{H}_{\partial h}$, and $s\left(h, \partial^2 h\right)$ the mixed surface area of $\mathcal{H}_h$ and of its mean evolute $\mathcal{H}_{\partial^2 h}$. In each inequality of (II), the equality holds if and only if $\mathcal{H}_h$ is a sphere or a point.*

Here, *the signed surface area $s(\partial h)$ is the opposite of the Laguerre area $L(\partial h)$*, which we have introduced in the previous subsubsection. In other words, $s(\partial h) = -L(\partial h)$. Besides, $\sqrt{2}\,\partial$ stands for the Hodge-Dirac operator $D = d + \delta$ on $\mathbb{S}^2$, where $d$ is the exterior differentiation and $\delta = -*d*$ the codifferential, so that $\partial h = dh/\sqrt{2}$ and $\partial^2 h = \frac{1}{2}\delta(dh) = \frac{1}{2}\Delta h(u)$. Recall that the **mean evolute** of $\mathcal{H}_h$ is the envelope of the family of planes parallel to the support planes to $\mathcal{H}_h$ and passing through the midpoints of the focal segments (which connect the two centers of principal curvature of $\mathcal{H}_h$ in $\mathbb{R}^3$). Now in the planar case, $\partial$ can be interpreted as the Hodge-Dirac operator $D = d + \delta$ on $\mathbb{S}^1$. Indeed, $dh_\theta = h'(\theta)\,d\theta$ can be interpreted as the support function $\partial h(\theta)$ of the evolute $\mathcal{H}_{\partial h}$ in the sense that $\mathcal{H}_{\partial h}$ is the envelope of the family of cooriented lines with equation $\langle x, u'(\theta)\rangle = h'(\theta)$, and $D^2 h = -h''$. Thus, in this equivalence between these two isometric inequalities (for hedgehog curves and surfaces), we have the following correspondences:

<table>
<tr><td>

Hedgehog curve $\mathcal{H}_h$ in $\mathbb{R}^2$

$\downarrow \partial$

Evolute curve $\mathcal{H}_{\partial h}$ in $\mathbb{R}^2$

$\downarrow \partial$

Second evolute curve $\mathcal{H}_{\partial^2 h}$ in $\mathbb{R}^2$

</td><td>

Hedgehog surface $\mathcal{H}_h$ in $\mathbb{R}^3$

$\downarrow \partial$

Co-evolute surface of $\mathcal{H}_h$    in $\mathbb{L}^4$
(Marginally trapped hedgehog $\mathcal{H}_{\partial h}$)

$\downarrow \partial$

Mean evolute surface $\mathcal{H}_{\partial^2 h}$ in $\mathbb{R}^3$

</td></tr>
</table>

where $\sqrt{n}\,\partial$ denote the Hodge-Dirac operator $D = d + \delta$ on $\mathbb{S}^n$, with respectively $n = 1$ and 2 for curves and surfaces.

**Remark in the 3-Dimensional Case** For any $h \in C^4(\mathbb{S}^2; \mathbb{R})$, we have the following commutative diagram:

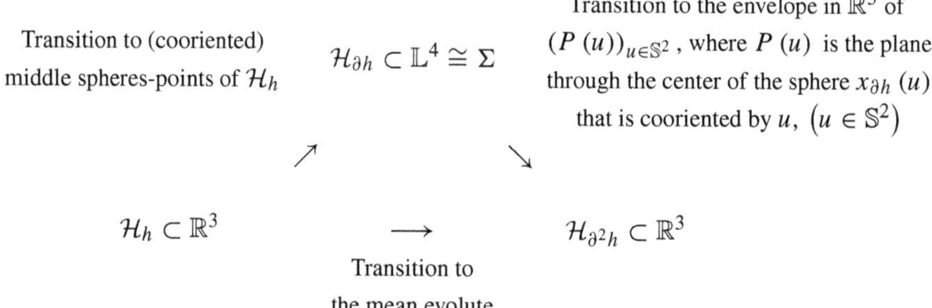


**Passage over into the Spacetime and Reincarnation in $\mathbb{R}^3$** The marginally trapped hedgehog $\mathcal{H}_{\partial h}$ can be thought as the 'Hodge-Dirac evolute' of $\mathcal{H}_h$. This evolute therefore lives in $\mathbb{L}^4$ (and not in $\mathbb{R}^3$) whereas the mean evolute of $\mathcal{H}_h$ (which can be thought as the 'second Hodge-Dirac evolute' $\mathcal{H}_{\partial^2 h}$ of $\mathcal{H}_h$, or equivalently, as the 'first Hodge-Dirac evolute' of $\mathcal{H}_{\partial h}$) does live in $\mathbb{R}^3$. For any $h \in C^\infty(\mathbb{S}^2; \mathbb{R})$, we can thus consider the sequence $(\mathcal{H}_{\partial^n h})_{n \in \mathbb{N}}$ of 'hedgehogs', which is such that:

(i)  for any even $n = 2k$, $(k \in \mathbb{N}^*)$, $\mathcal{H}_{\partial^{2k} h}$ is the (ordinary) hedgehog of $\mathbb{R}^3$ that is the mean evolute of $\mathcal{H}_{\partial^{2(k-1)} h} \subset \mathbb{R}^3$;

(ii)  for any odd $n = 2k+1$, $(k \in \mathbb{N})$, $\mathcal{H}_{\partial^{2k+1} h}$ is the (marginally trapped) hedgehog of $\mathbb{L}^4$ that is the co-evolute of $\mathcal{H}_{\partial^{2k} h} \subset \mathbb{R}^3$.

**The Support Vector Field of $\mathcal{H}_{\partial h}$**   For any $h \in C^4\left(\mathbb{S}^2; \mathbb{R}\right)$, let $\overrightarrow{\partial h}$ denote the vector field that corresponds to the 1-form $\partial h := dh/\sqrt{2}$ by the canonical musical duality:

$$\forall \overrightarrow{X} \in T\mathbb{S}^2, \qquad (\partial h)\left(\overrightarrow{X}\right) = g\left(\overrightarrow{\partial h}, X\right) = \overrightarrow{\partial h}.\overrightarrow{X},$$

where $g : \left(\overrightarrow{X}, \overrightarrow{Y}\right) \mapsto g\left(\overrightarrow{X}, \overrightarrow{Y}\right) = \overrightarrow{X}.\overrightarrow{Y}$ denotes the standard metric on $\mathbb{S}^2$.

**Similarity of Formulas for Hedgehogs and (Co)hedgehogs**

**Theorem 8.1.5**  *For any $h \in C^4\left(\mathbb{S}^2; \mathbb{R}\right)$, the signed surface area $s(\partial h)$ of $\mathcal{H}_{\partial h}$ can be expressed in the following forms:*

$$s(\partial h) = \int_{\mathbb{S}^2} \left(\left(\overrightarrow{\partial h}\right)^2 - \left(\partial^2 h\right)^2\right) d\sigma = \int_{\mathbb{S}^2} \overrightarrow{\partial h}.\overrightarrow{\partial R_{(1,h)}}d\sigma,$$

*where $\sigma$ is the spherical Lebesgue measure on $\mathbb{S}^2$, $\partial^2 h$ the support function of the mean evolute of $\mathcal{H}_h$, and $R_{(1,h)}$ the mean radius of curvature of $\mathcal{H}_h$.*

It is worth noting the similarity with the known formulas for the signed surface area of a hedgehog $\mathcal{H}_h$ of $\mathbb{R}^3$:

$$s\left(h\right) = \int_{\mathbb{S}^2} \left(h^2 - \left(\overrightarrow{\partial h}\right)^2\right) d\sigma = \int_{\mathbb{S}^2} h.R_{(1,h)}d\sigma,$$

where $h \in C^2\left(\mathbb{S}^2; \mathbb{R}\right)$.

Let us recall the proof of these two last inequalities. From the definition

$$s\left(h\right) = \int_{\mathbb{S}^2} R_h d\sigma,$$

we obtain $s\left(h\right) = \int_{\mathbb{S}^2} h.R_{(1,h)}d\sigma$ by the symmetry of the mixed volume, and then

$$s\left(h\right) = \int_{\mathbb{S}^2} \left(h^2 - \left(\overrightarrow{\partial h}\right)^2\right) d\sigma$$

by integrating by parts. In the same vein as Theorem 8.1.5, we can notice that the signed surface area of the mean evolute $\mathcal{H}_{\partial^2 h}$ of $\mathcal{H}_h$ is given by:

$$s\left(\partial^2 h\right) = \int_{\mathbb{S}^2} \left(\left(\partial^2 h\right)^2 - \overline{\partial\left(\partial^2 h\right)^2}\right) d\sigma = \int_{\mathbb{S}^2} \left(\partial^2 h\right).\partial^2 R_{(1,h)}d\sigma.$$

From the above formulas, we deduce that the mixed surface area of two hedgehogs $\mathcal{H}_f$ and $\mathcal{H}_g$ of $\mathbb{R}^3$ is given by:

$$s\left(f,g\right) = \int_{\mathbb{S}^2}\left(fg - \overrightarrow{\partial f}.\overrightarrow{\partial g}\right)d\sigma = \int_{\mathbb{S}^2}f.R_{(1,g)}d\sigma = \int_{\mathbb{S}^2}g.R_{(1,f)}d\sigma.$$

Now, the integration by parts formula can be written as

$$\int_{\mathbb{S}^2}\overrightarrow{\partial f}.\overrightarrow{\partial g}d\sigma = \int_{\mathbb{S}^2}f.\partial^2 g d\sigma$$

for all $(f,g) \in C^4\left(\mathbb{S}^1;\mathbb{R}\right)^2$. Therefore

$$\int_{\mathbb{S}^2}\overrightarrow{\partial f}.\overrightarrow{\partial R_{(1,g)}}d\sigma = \int_{\mathbb{S}^2}f.\partial^2 R_{(1,g)}d\sigma = \int_{\mathbb{S}^2}f.R_{(1,\partial^2 g)}d\sigma = s\left(f,\partial^2 g\right),$$

and so, by the symmetry of the mixed surface area of hedgehogs,

$$\int_{\mathbb{S}^2}\overrightarrow{\partial f}.\overrightarrow{\partial R_{(1,g)}}d\sigma = s\left(\partial^2 g, f\right) = \int_{\mathbb{S}^2}\partial^2 g.R_{(1,f)}d\sigma,$$

and again by the integration by parts formula

$$\int_{\mathbb{S}^2}\overrightarrow{\partial f}.\overrightarrow{\partial R_{(1,g)}}d\sigma = \int_{\mathbb{S}^2}\overrightarrow{\partial g}.\overrightarrow{\partial R_{(1,f)}}d\sigma.$$

Thus, we have deduced the following corollary to Theorem 8.1.5.

**Corollary 8.1.1** *For any* $(f,g) \in C^4\left(\mathbb{S}^2;\mathbb{R}\right)^2$, *let* $s(\partial f,\partial g)$ *denote the mixed surface area of* $\mathcal{H}_{\partial f}$ *and* $\mathcal{H}_{\partial g}$, *that is* $s(\partial f,\partial g) := -L(\partial f,\partial g)$. *This mixed area* $s(\partial f,\partial g)$ *can be expressed in the following forms:*

$$s(\partial f,\partial g) = \int_{\mathbb{S}^2}\overrightarrow{\partial f}.\overrightarrow{\partial R_{(1,g)}}d\sigma = \int_{\mathbb{S}^2}\overrightarrow{\partial g}.\overrightarrow{\partial R_{(1,f)}}d\sigma = s\left(f,\partial^2 g\right) = s\left(\partial^2 f, g\right),$$

*where* $\sigma$ *is the spherical Lebesgue measure on* $\mathbb{S}^2$.

*In particular, the mixed surface area of the marginally trapped hedgehogs* $\mathcal{H}_{\partial f}$ *and* $\mathcal{H}_{\partial g}$ *is equal to the mixed surface area of the (ordinary) hedgehogs* $\mathcal{H}_f$ *and* $\mathcal{H}_{\partial^2 g}$, *which is the mean evolute of* $\mathcal{H}_g$.

### 8.1.3   Synthetic and Comparative Co-presentation of Hedgehogs and (Co)hedgehogs

In this subsubsection, we will now compare carefully the definitions of hedgehogs in $\mathbb{R}^3$ and of (co)hedgehogs (i.e., marginally trapped hedgehogs) in $\mathbb{L}^4$, and then for the convenience of the reader, we will summarize schematically our comparison in tables.

As recalled previously, the datum of any $h \in C^\infty(\mathbb{S}^2; \mathbb{R})$ determines the hedgehog $\mathcal{H}_h$ in $\mathbb{R}^3$ as the envelope of the family $\mathcal{P} = (P_h(u))_{u \in \mathbb{S}^2}$ of cooriented planes with equation $\langle x, u \rangle = h(u)$, $(u \in \mathbb{S}^2)$. Analytically, taking the envelope of $\mathcal{P}$ amounts to solving the following system of equations

$$\begin{cases} \langle x, u \rangle = h(u) \\ \langle x, du \rangle = dh_u, \end{cases}$$

for all $u \in \mathbb{S}^2$, where the second equation is deduced from the first via the contact condition $\langle dx, u \rangle = 0$ (and thus by performing a partial differentiation with respect to $u$).

Now, as we saw previously, the datum of $\partial h := dh/\sqrt{2}$ determines the (co)hedgehog (i.e., marginally trapped hedgehog) $\mathcal{H}_{\partial h}$ in $\mathbb{L}^4$ as the co-envelope of the family $\mathcal{L} = (L_{h+t}(u))_{(u,t) \in \mathbb{S}^2 \times \mathbb{R}}$ of oriented null lines defined by $L_{h+t}(u) := \{(\nabla h(u), h(u) + t)\} + \mathbb{R}u_L$. Analytically, taking the co-envelope of $\mathcal{L}$ amounts to solving the following system of equations

$$(S) \quad \begin{cases} \begin{cases} \langle x, u_L \rangle_L = (h(u) + t)/\sqrt{2} \\[2mm] \langle x, du_L \rangle_L = \partial h(u) \\[2mm] \langle x, \Delta u_L \rangle_L = \delta(dh)(u), \end{cases} \end{cases}$$

for all $(u, t) \in \mathbb{S}^2 \times \mathbb{R}$, where the third equation is deduced from the two first ones via the co-contact condition $\langle \Delta x, u_L \rangle_L = 0$ (or, equivalently, by performing a partial codifferentiation with respect to $u$ in the second equation). Here, the Hodge-Laplace operator of a vector function $X : \mathbb{S}^2 \to \mathbb{L}^4$, $u \mapsto (X_i(u))_{i=1}^4$ is of course understood to be the vector function $\Delta X : \mathbb{S}^2 \to \mathbb{L}^4$, $u \mapsto (\Delta X_i(u))_{i=1}^4$, where $\Delta X_i$ is the Hodge-Laplace operator of the coordinate function $X_i$ in the intrinsic metric on $\mathbb{S}^2$, $(1 \le i \le 4)$. Note that the system formed by the two first equations of $(S)$ simply traduces the fact that $x \in L_{h+t}(u)$. The second equation can be deduced from the first by using the condition $\langle dx, u_L \rangle_L = 0$ (and thus by performing a partial differentiation with respect to $u$).

For any $h \in C^\infty(\mathbb{S}^2; \mathbb{R})$, we therefore have the following comparison table between both definitions:

| The hedgehog $\mathcal{H}_h$ can be defined in $\mathbb{R}^3$ as the envelope of the family $(P_h(u))_{u \in \mathbb{S}^2}$ of cooriented planes with equation $\langle x, u \rangle = h(u)$. | The (co)hedgehog $\mathcal{H}_{\partial h}$ can be defined in $\mathbb{L}^4$ as the co-envelope of the family $(L_{h+t}())_{(u,t) \in \mathbb{S}^2 \times \mathbb{R}}$ of oriented null lines defined by : $L_{h+t}(u) := \{(\nabla h(u), h(u)+t)\} + \mathbb{R} u_L.$ |
|---|---|
| (1)    $\langle x, u \rangle = h(u)$ <br><br> [via the contact condition <br><br> $\Downarrow$    (d)    $\langle dx, u \rangle = 0$ <br><br> or a partial differentiation with respect to $u$] <br> (2)    $\langle x, du \rangle = dh_u$ | (1) $\begin{cases} (1.a) \quad \langle x, u_L \rangle_L = (h(u)+t)/\sqrt{2} \\ \quad\Downarrow \;\; \text{(via } \langle dx, u_L \rangle_L = 0 \text{ or a partial} \\ \qquad \text{differentiation with respect to } u) \\ (1.b) \quad \langle x, du_L \rangle_L = \partial h(u) \end{cases}$ <br> (via the co-contact condition <br> $\Downarrow$    $(\delta)$    $\langle \delta(dx), u_L \rangle_L = 0,$ <br> or a codifferentiation with respect to $u$) <br><br> (2)    $\langle x, \Delta u_L \rangle = \delta(dh)(u)$ |
| For every $u \in \mathbb{S}^2$, the system <br><br> $\begin{cases} (1) & \langle x, u \rangle = h(u) \\ (2) & \langle x, du \rangle = dh_u, \end{cases}$ <br><br> implies $x = h(u)u + \nabla h(u)$ | For every $(u, t) \in \mathbb{S}^2 \times \mathbb{R}$, the system <br><br> $\begin{cases} (1) = (1.a) \text{ and } (1.b): & x \in L_{h+t}(u) \\ (2) & \langle x, \Delta u_L \rangle = \delta(dh)(u), \end{cases}$ <br><br> implies $x = (\nabla h(u), h(u)+t) + \delta(\partial h)(u) u_L$ |
| In the absence of singularities, <br><br> $x_h : \mathbb{S}^2 \to \mathcal{H}_h, u \mapsto h(u)u + \nabla h(u)$ <br><br> can be interpreted as the inverse of the Gauss map of $\mathcal{H}_h$. | In the absence of umbilical points, <br><br> $x_{\partial h} : \Omega \to \mathcal{H}_{\partial h}, u \mapsto l_h(u) + \delta(\partial h)(u) u_L$ <br><br> can be interpreted as the Laguerre Gauss map of $x_h : \Omega \to \mathbb{R}^3$. |

The cooriented support plane $P_h(u)$ is determined by $h(u)u$ and thus by the value of the support function $h$ at $u$, and $x_h(u)$ is then determined on $P_h(u)$ by the contact condition which imposes $x_h(u) - h(u)u = \nabla h(u)$. Analogously, the oriented support null line $L_h(u)$ is determined by $l_h(u)$ and thus by the value of the support differential $\partial h$ at $u$, and $x_{\partial h}(u)$ is then determined on $L_h(u)$ by the co-contact condition which imposes $x_{\partial h}(u) - l_h(u) = \delta(\partial h)(u)\,u_L$:

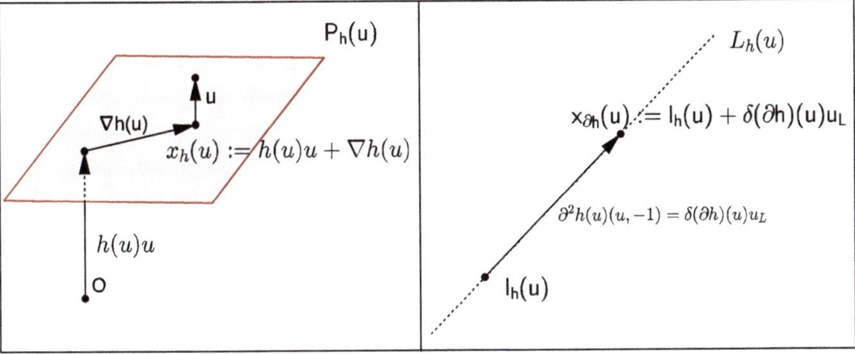


## The Role of $(d)$ and $(\delta)$ in the Determination of $\mathcal{H}_h$ and $\mathcal{H}_{\partial h}$

| The contact condition $(d)$ allows us to write | The co-contact condition $(\delta)$ allows us to write $\langle x, \Delta u_L\rangle_L = \Delta\left(\langle x, u_L\rangle_L\right)$ |
|---|---|
| $\langle x, du\rangle = d\left(\langle x, u\rangle\right).$ | provided that $\langle dx, u_L\rangle = 0.$ |

## What Is Meant by the Contact and Co-contact Conditions $(d)$ and $(\delta)$

Recall first that, for every $C^2$-map $x : \mathbb{S}^2 \to \mathbb{L}^4$ and every $u \in \mathbb{S}^2$, $\Delta x(u) = \delta(dx)(u)$ can be interpreted as a second order derivative that measures, both in direction and magnitude, how $x(u)$ deviates from the average of $x$ over an infinitesimal sphere centered at $u$ in $\mathbb{S}^2$.

| Condition $(d)$   $\langle dx, u\rangle = 0$ on $x : \mathbb{S}^2 \to \mathbb{R}^3$ | Condition $(\delta)$   $\langle \delta(dx), u_L\rangle_L = 0$ on $x : \mathbb{S}^2 \to \mathbb{L}^4$ |
|---|---|
| For all $u \in \mathbb{S}^2$, $u$ is orthogonal to the tangent space $T_{x(u)}x\left(\mathbb{S}^2\right) = (T_u x)\left(T_u\mathbb{S}^2\right)$ in the Euclidean space $\left(\mathbb{R}^3, \langle .,.\rangle\right)$. | For all $u \in \mathbb{S}^2$, $u_L$ is orthogonal in $\left(\mathbb{L}^4, \langle .,.\rangle_L\right)$ to the above mentioned measure of deviation (in direction and magnitude) $\Delta x(u) = \delta(dx)(u)$. |

## 8.2  Marginally Trapped Hedgehogs in Other Spaces and Generalizations

### 8.2.1  *Marginally Trapped Hedgehogs or Co-hedgehogs in* $\mathbf{M}^4$

**Characterization and Definitions in $\mathbf{M}^4$**

As already mentioned, the hedgehog theory is not restricted to Euclidean spaces. For instance, in our preliminary study of Sect. 8.1.1, we can replace $\left(\mathbb{R}^3, \langle ., .\rangle\right)$ by the Lorentzian-Minkowski 3-space $\mathbb{L}^3 = \left(\mathbb{R}^3, \langle ., .\rangle_L\right)$, where

$$\langle x, x'\rangle_L = x_1 x_1' + x_2 x_2' - x_3 x_3',$$

for all $x := \left(x_1, x_2, x_3\right)$ and $x' := \left(x_1', x_2', x_3'\right)$ in $\mathbb{R}^3$, and the unit sphere $\mathbb{S}^2$ of $\mathbb{R}^3$ by the 'unit sphere'

$$\mathbf{H}^2 := \left\{ x = (x_1, x_2, x_3) \in \mathbb{L}^3 \,|\, \langle x, x\rangle_L = -1 \right\},$$

which is a two-sheeted hyperboloid with constant Gaussian curvature $-1$ with respect to the induced metric. To any $h \in C^\infty\left(\mathbf{H}^2; \mathbb{R}\right)$ corresponds the envelope $\mathcal{H}_h$ of the family $\left(P_h\left(v\right)\right)_{v \in \mathbb{S}^2}$ of cooriented spacelike planes of $\mathbb{L}^3$ with equation $(E)$    $\langle x, v\rangle_L = h\left(v\right)$. We say that $\mathcal{H}_h$ is the hedgehog of $\mathbb{L}^3$ with **support function** $h$. From $(E)$ and the contact condition $\langle dx, v\rangle_L = 0$, we deduce

$$\begin{cases} \langle x, v\rangle_L = h\left(v\right) \\ \langle x, dv\rangle_L = dh_v, \end{cases}$$

for all $v \in \mathbf{H}^2$. Thus, it appears that $\mathcal{H}_h$ can be parametrized by $x_h : \mathbf{H}^2 \to \mathbb{L}^3$, $v \mapsto (\nabla h)\left(v\right) - h(v)v$, where $(\nabla h)\left(v\right)$ stands for the gradient of $h$ at $v$. The parametrization $x_h$ can be interpreted as the inverse of its Gauss map (if $v$ is a regular point of $x_h$, then $v$ is normal to $\mathcal{H}_h$ at $x_h\left(v\right)$). Note that $(\mathcal{H}_{h+t})_{t \in \mathbb{R}}$ is a family of parallel hedgehogs in $\mathbb{R}^3$: for all $(v, t) \in \mathbf{H}^2 \times \mathbb{R}$, $x_{h+t}\left(v\right) = x_h\left(v\right) - tv$.

Now, we claim that:

> *To the differential dh of h corresponds naturally a 'marginally trapped hedgehog' of the Lorentz Minkowski 4-space* $\mathbf{M}^4 = \left(\mathbb{R}^4, \langle ., .\rangle_M\right)$ *via a 'lightlike co-contact condition'.*

Here, the pseudo-scalar product $\langle ., .\rangle_M$ is defined by

$$\langle (x, t), \left(x', t'\right)\rangle_M := x_1 x_1' + x_2 x_2' - x_3 x_3' + tt',$$

for all $(x, t) := ((x_1, x_2, x_3), t)$ and $(x', t') := ((x'_1, x'_2, x'_3), t')$ in $\mathbb{R}^4 = \mathbb{R}^3 \times \mathbb{R}$. Indeed, $dh$ determines the family $(L_{h+t}(v))_{(v,t) \in \mathbb{H}^2 \times \mathbb{R}}$ of oriented null lines of $\mathbf{M}^4$ defined by:

$$\forall (v, t) \in \mathbb{H}^2 \times \mathbb{R}, \ L_{h+t}(v) = l_{h+t}(v) + \mathbb{R}(v, -1),$$

where $l_{h+t}(v) := (\nabla h(v), -h(v) - t)$; and this family of null lines determines a 'marginally trapped hedgehog', which is unique up to translations parallel to the time axis, via a 'lightlike co-contact condition':

**Theorem 8.2.1 (Determination of Marginally Trapped Hedgehogs of $\mathbf{M}^4$ by 1-Jet and Co-contact Condition)**   *For all $t \in \mathbb{R}$,*

*(i)   there is a unique map $x : \mathbf{H}^2 \to \mathbf{M}^4$ of class $C^\infty$ satisfying $x(v) \in L_{h+t}(v)$ for all $v \in \mathbf{H}^2$ together with the **'lightlike co-contact condition'***

$$\langle \delta(dx), v_M \rangle_M = 0,$$

*where $\delta$ is the Hodge codifferential on $\mathbf{H}^2$ (i.e., the formal adjoint to the exterior differentiation $d$) and $v_M := \frac{1}{\sqrt{2}}(v, -1)$, namely*

$$x = x_{l_{h+t}} : \mathbf{H}^2 \to \mathbf{M}^4, \ v \mapsto l_{h+t}(v) + \delta(\partial h)(v) v_M,$$

*where $\partial h(v) := dh/\sqrt{2}$;*

*(ii)   this map $x : \mathbf{H}^2 \to \mathbf{M}^4$ is such that $x^*g - \frac{1}{4}(R_1 - R_2)^2 g_H$, where $g_H$ is the standard metric on $\mathbf{H}^2$, $g$ the first fundamental form on $x(\mathbf{H}^2)$, $x^*g$ the pullback of $g$ along $x$ and $R_1(v), R_2(v)$ the principal radii of curvature of $\mathcal{H}_h$ at $x_h(v)$;*

*(iii)   for all $v \in \mathbf{H}^2$, $\Delta x(v) \in \mathbb{R}v_M$, where $\Delta$ is the Hodge Laplacian on $\mathbf{H}^2$, so that the mean curvature vector of $x(\mathbf{H}^2)$ at $x(v)$, say $\overrightarrow{H}_x(v)$, is parallel to the lightlike vector $\overrightarrow{v_M}$ whenever $x_h(v)$ is not an umbilical point of $\mathcal{H}_h$, that is:*

$$\forall v \in \mathbf{H}^2 \setminus \left\{ v \in \mathbf{H}^2 \mid R_1(v) = R_2(v) \right\}, \quad \overrightarrow{H}_x(v) \in \mathbb{R}\overrightarrow{v_M}$$

Here, *the principal radii of curvature of $\mathcal{H}_h$ at $x_h(v)$, $R_1(v)$ and $R_2(v)$, are defined as the eigenvalues of $x_h : T_v \mathbf{H}^2 \to T_{x_h(v)} \mathcal{H}_h \subset T_v \mathbf{H}^2$.*

**Definition 8.2.1**   For all $t \in \mathbb{R}$, we call $\mathcal{H}_{l_{h+t}} := x_{l_{h+t}}(\mathbf{H}^2)$ the (marginally trapped) hedgehog with support 1-jet $j^1_{h+t} : \mathbf{H}^2 \to J^1(\mathbf{H}^2)$, $v \mapsto (\nabla h(v), h(v) + t)$.

Note that $\sqrt{2}\,\partial$ stands for the Hodge-Dirac operator $D = d + \delta$ on $\mathbf{H}^2$ and the datum of $\partial h := dh/\sqrt{2}$ is equivalent to the one of the family of 1-jets $\left( j^1_{h+t} \right)_{t \in \mathbb{R}}$, where $j^1_{h+t} := (\nabla h, h + t)$. All the marginally trapped hedgehogs of the family

$\left(\mathcal{H}_{l_{h+t}}\right)_{t\in\mathbb{R}}$ are equal up to translations parallel to the time axis $\mathbb{R}\,\overrightarrow{\partial_t}$ in $\mathbf{M}^4$, where $\overrightarrow{\partial_t} := \left(\overrightarrow{0_{\mathbb{L}^3}}, 1\right)$, since: for all $(v, t) \in \mathbf{H}^2 \times \mathbb{R}$, $x_{l_{h+t}}(v) = x_{l_h}(v) - t\,\overrightarrow{\partial_t}$. Therefore, this family can be regarded as one and only one marginally trapped hedgehog defined up to a translation parallel to the time axis $\mathbb{R}\,\overrightarrow{\partial_t}$ in $\mathbf{M}^4$. This hedgehog will be denoted by $\mathcal{H}_{\partial h}$, where $\partial h := dh/\sqrt{2}$ (the factor $1/\sqrt{2}$ is chosen for the same reasons as in the case of marginally trapped hedgehogs of $\mathbb{L}^4$).

These remarks make natural the following definition:

**Definition 8.2.2**   We call $\mathcal{H}_{\partial h}$ the **marginally trapped hedgehog** (or, **co-hedgehog**) of $\mathbf{M}^4$ with support differential $\partial h := dh/\sqrt{2}$. Anyone of the marginally trapped hedgehogs $\mathcal{H}_{l_{h+t}}$, $(t \in \mathbb{R})$, will be regarded as a representative of $\mathcal{H}_{\partial h}$ in $\mathbf{M}^4$. We will say that $\mathcal{H}_{\partial h}$ is the co-evolute of $\mathcal{H}_h$ in $\mathbf{M}^4$.

Thus, once the lightlike 'co-contact condition' $\langle \delta(dx), v_M \rangle_M = 0$ is fixed, the hedgehog $\mathcal{H}_{\partial h}$ is defined in $\mathbf{M}^4$ as the 'co-envelope' of the family of oriented null lines determined by its support differential $\partial h$ in the same way as, once the contact condition $\langle dx, v \rangle_L = 0$ is fixed, the hedgehog $\mathcal{H}_h$ is defined in $\mathbb{L}^3$ as the envelope of the family of cooriented spacelike planes determined by its support function $h$. Of course, it is important to note that $\mathcal{H}_h$ and $\mathcal{H}_{\partial h}$ are not assumed to be embedded surfaces of respectively $\mathbb{L}^3$ and $\mathbf{M}^4$: $\mathcal{H}_h$ and $\mathcal{H}_{\partial h}$ may possibly be singular and self-intersecting.

**Geometrical Interpretation of Co-hedgehogs of $\mathbb{M}^4$**   First, let us remind some basic facts about the curvature of hedgehogs of $\mathbb{L}^3$. Since the parametrization $x_h : \mathbf{H}^2 \to \mathcal{H}_h \subset \mathbb{L}^3$ can be regarded as the inverse of the Gauss map, the Gauss curvature $\kappa_h$ of $\mathcal{H}_h$ at $x_h(v)$ is given by $\kappa_h(v) = 1/\det[T_v x_h]$, where $T_v x_h$ is the tangent map of $x_h$ at $v$. Therefore, singularities of $\mathcal{H}_h$ are the points at which its Gauss curvature becomes infinite. For all $v \in \mathbf{H}^2$, the tangent map of $x_h$ at the point $u$ is $T_u x_h = H_h(v) - h(v)\,Id_{T_v \mathbf{H}^2}$, where $H_h(v)$ is the symmetric endomorphism associated with the Hessian of $h$ at $v$. Consequently, if $\lambda_1(v)$ and $\lambda_2(v)$ are the eigenvalues of the Hessian of $h$ at $v$ then $R_1(v) := (\lambda_1 - h)(u)$ and $R_2(v) := (\lambda_2 - h)(v)$ can be interpreted as the principal radii of curvature of $\mathcal{H}_h$ at $x_h(v)$, and the so-called ***curvature function*** $R_h := 1/\kappa_h$ is given by

$$R_h(v) = \det\left[H_{ij}(u) - h(u)\,\delta_{ij}\right] = (R_1 R_2)(v),$$

where $\delta_{ij}$ are the Kronecker symbols and $\left(H_{ij}(v)\right)$ the Hessian of $h$ at $v$ with respect to an orthonormal frame on $\mathbf{H}^2$. In computations, it is often more convenient to replace $h$ by its positively $1-$homogeneous extension to the interior of the light cone $\mathcal{U} = \left\{x \in \mathbb{L}^3 \,|\, \langle x, x \rangle_L < 0\right\}$, that is, by

$$\varphi(x) := \|x\|_L\, h\left(\frac{x}{\|x\|_L}\right),$$

for $x \in \mathcal{U}$, where $\|x\|_L = \sqrt{-\langle x, x\rangle_L}$. A straightforward computation gives:

*(i)*  $x_h$ is the restriction of the Lorentzian gradient $\nabla_L \varphi := \left(\frac{\partial \varphi}{\partial x_1}, \frac{\partial \varphi}{\partial x2}, -\frac{\partial \varphi}{\partial x_3}\right)$ of $\varphi$
to the unit sphere $\mathbf{H}^2$ ;

*(ii)*  For all $v \in \mathbf{H}^2$, the tangent map $T_v x_h$ identifies with the symmetric endomorphism associated with the Hessian of $\varphi$ at $v$.

In order to bring out our geometrical interpretation of the co-evolute $\mathcal{H}_{\partial h}$, we now give another expression for $x_{l_h}$. For all $v \in \mathbf{H}^2$, we have

$$x_{l_h}(v) := l_h(v) + \delta\,(\partial h)(v)\,v_M = (\nabla h(v), -h(v)) + \frac{\Delta h(v)}{2}(v, -1)$$

$$= \left(\nabla h(v) + \frac{\Delta h(v)}{2}v, -h(v) - \frac{\Delta h(v)}{2}\right)$$

$$= \left(x_h(v) - R_{(-1,h)}(v)\,v, R_{(-1,h)}(v)\right) = \left(c_h(v), R_{(-1,h)}(v)\right),$$

where $\Delta$ is the Hodge Laplacian on $\mathbf{H}^2$, $c_h(v)$ is the midpoint of the segment connecting the two centers of principal curvatures of $\mathcal{H}_h$ at $x_h(v)$ and, $R_{(-1,h)} := -\left(h + \frac{\Delta h}{2}\right) = \frac{1}{2}(R_1 + R_2)$ is the mean radius of principal curvature of $\mathcal{H}_h$. Thus, for all $v \in \mathbf{H}^2$, $x_{l_h}(v)$ can be interpreted geometrically as the **middle pseudosphere** (or possibly non-cooriented light cone) of $\mathcal{H}_h$ at $x_h(v)$ that is cooriented by the vector $v$, which is normal to it, by identifying $\mathbf{M}^4$ with the space $\Pi$ of cooriented pseudospheres and (non-cooriented) light cones of $\mathbb{L}^3$ via the bijection

$$\mathbf{M}^4 = \mathbb{L}^3 \times \mathbb{R} \qquad \rightarrow \Pi$$
$$(a, r) = ((a_1, a_2, a_3), r) \mapsto H^2(a; r),$$

where $H^2(a; r)$ is the two-sheeted hyperboloid (pseudosphere) with equation $\langle x - a, x - a\rangle_L = -r^2$ that is cooriented so that its upper sheet $H^2_+(a; r) := H^2(a; r) \cap \{x_3 > a_3\}$ is cooriented by its future (resp. pass) pointing normal if $r > 0$ (resp. $r < 0$) holds, and the (non-cooriented) light cone with apex at $a$ if $r = 0$ holds. Thus:

**Proposition 8.2.1** *The co-evolute $\mathcal{H}_{\partial h}$ of $\mathcal{H}_h$ can be regarded as the locus of all the middle cooriented pseudospheres (or non-cooriented light cones) of $\mathcal{H}_h \subset \mathbb{L}^3$ in* $\mathbf{M}^4 \cong \Pi$.

### 8.2.2  *Towards Other Spaces and Generalizations*

As already mentioned, most of our results extend, without much change, to higher dimensions. Furthermore, the hedgehog theory is of course not restricted to the only two examples we have considered above. For instance, we could have adapted our presentation to Fuchsian hedgehogs, which were introduced by François Fillastre [44]. We could also have considered multihedgehogs or $N$-hedgehogs, provided of course that we pay proper attention to the fact that an $N$-hedgehog may have parabolic points (i.e., points where their Gauss curvature vanishes, and thus, points at which their curvature function is not defined) for $N \geq 2$. Recall that an $N$-hedgehog of $\mathbb{R}^3$ is any envelope of a family of cooriented planes of $\mathbb{R}^3$ such that the number of cooriented support planes with a given coorienting unit normal vector is finite and constant equal to $N$ (at least for an open dense set of directions). Thus 1-hedgehogs of $\mathbb{R}^3$ are simply hedgehogs. But in fact, all these geometrical objects of different spaces can be regarded as wavefronts of a special class of Legendrian submanifolds (i.e., images of Legendrian maps) of a given metric contact manifold. Before going on, we first recall some basic definitions and facts on contact, symplectic and almost-hermitian structures, and next present and study hedgehogs and marginally trapped hedgehogs in this setting.

**Contact Manifolds and Metric Contact Manifolds**

A *contact structure* on an oriented $(2n + 1)$-dimensional $C^\infty$-manifold $M$ is the datum of a smooth field $V$ of tangent hyperplanes on $M$, called *contact hyperplanes*, satisfying the following condition of maximal non-integrability: any (and hence every) 1-form $\alpha$ defining $V$ (i.e., such that $V = Ker(\alpha)$) satisfies $\alpha \wedge (d\alpha)^n \neq 0$ everywhere on $M$. Any 1-form $\alpha$ defining such a maximally non-integrable hyperplane field $V$ on $M$ is called a *contact form* on $M$. Given such a contact structure (or a contact form $\alpha$ defining it), the pair $(M, V)$ (or the pair $(M, \alpha)$ if we want to fix the contact form defining $V$) is then called a *contact manifold*. On $(M, \alpha)$, the *Reeb vector field* $\xi_\alpha$ associated to the contact form $\alpha$ is defined to be the unique smooth vector field satisfying

$$\alpha\left(\xi_\alpha\right) = 1 \qquad \text{and} \qquad \xi_\alpha \in Ker\left(d\alpha\right).$$

A submanifold $L$ of a contact manifold $(M, V)$ is said to be *integral* if $T_m L \subset V_m$ for all $m \in L$. A *Legendrian submanifold* of $(M, V)$ is an integral submanifold of $(M, V)$ with maximal dimension $n = (\dim M - 1)/2$. A fibration of a contact manifold is said to be *Legendrian* if all its fibers are Legendrian submanifolds.

Let $i : L \to E$ be an immersed Legendrian submanifold $L$ in the total space of a Legendrian fibration $\pi : E \to B$. The restriction of $\pi$ to $L$, that is $x = \pi \circ i : L \to B$ is called a *Legendrian map* and its image $x(L)$ in $B$ is called its *Legendrian front* or *wavefront*.

**Example** Unit tangent bundles of Riemannian manifolds are among the most classical examples of contact manifolds. Let us recall briefly how this is done. Let $(M, g)$ be a Riemannian manifold and let

$$UTM = \{u \in TM \mid g(u, u) = 1\}$$

be its unit tangent bundle with canonical projection $\pi : UTM \to M$ ; the metric $g$ induces a contact form $\alpha$ (and thus a contact structure $V$) on $UTM$ as follows: for any $u \in UTM$ and $v \in T_u(UTM)$, we let

$$\alpha_u(v) = g(u, T_u\pi(v)),$$

where $T_u\pi(v) = \pi_*(v)$ is the pushforward along $\pi$ of the vector $v$. Moreover, $\pi : UTM \to M$ is an example of a Legendrian fibration.

In particular, if we let

$$\alpha_{(x,u)} := \langle u, dx \rangle = \sum_{i=0}^{n+1} u_i dx_i$$

for all $(x, u) \in U\mathbb{R}^{n+1} = \mathbb{R}^{n+1} \times \mathbb{S}^n$, where $(x_1, \cdots, x_{n+1}; u_1, \cdots, u_{n+1})$ are the canonical coordinate functions on $U\mathbb{R}^{n+1} = \mathbb{R}^{n+1} \times \mathbb{S}^n \subset \mathbb{R}^{2n+2}$, we obtain a contact manifold $\left(U\mathbb{R}^{n+1}; \alpha\right)$.

A *contactomorphism* from a contact manifold $(M_1, V_1)$ to a contact manifold $(M_2, V_2)$ is a diffeomorphism $f : M_1 \to M_2$ that preserves the contact structure, i.e., such that $Tf(V_1) = V_2$, where $Tf : TM_1 \to TM_2$ denotes the tangent map of $f$. If $V_i = Ker(\alpha_i)$, $(i = 1, 2)$, this is equivalent to the existence of a nowhere zero function $\lambda : M_1 \to M_2$ such that $f^*\alpha_2 = \lambda\alpha_1$.

**Example** Another example of a contact manifold is defined as follows: on the manifold $T\mathbb{S}^n \times \mathbb{R}$, where the tangent bundle $T\mathbb{S}^n$ is identified with

$$\left\{(u, p) \in \left(\mathbb{R}^{n+1}\right)^2 \mid \|u\| = 1 \text{ and } \langle u, p \rangle = 0\right\}$$

( $\|.\|$ and $\langle .,. \rangle$ denoting respectively the Euclidean norm and scalar product in $\mathbb{R}^{n+1}$), we define a contact form $\beta$ by putting $\beta_{(u,p,z)} := dz - pdu$ for all $(u, p, z) \in T\mathbb{S}^n \times \mathbb{R}$. Moreover

$$
\begin{aligned}
f : \quad & U\mathbb{R}^{n+1} = \mathbb{R}^{n+1} \times \mathbb{S}^n \to T\mathbb{S}^n \times \mathbb{R} \\
& (x, u) \mapsto (u, x - \langle x, u \rangle u, \langle x, u \rangle)
\end{aligned}
$$

is a diffeomorphism such that $f^*\beta = \alpha$, and hence a contactomorphism from $\left(U\mathbb{R}^{n+1}, \alpha\right)$ to $(T\mathbb{S}^n \times \mathbb{R}, \beta)$.

A *metric contact manifold* is defined to be a tuple $(M, g, \alpha, J)$, where $(M, g)$ is a Riemannian manifold, $\alpha$ a smooth 1-form on $M$ and $J$ a section of the endomorphism bundle $End\,(TM)$ which satisfy the three following conditions:

*(i)* $\alpha\,(\xi_\alpha) = 1$, where $\xi_\alpha$ is the metric dual of $\alpha$;
*(ii)* $d\alpha\,(X, Y) = g\,(JX, Y)$ for any vector fields $X, Y$ on $M$;
*(iii)* $J^2 X = -X + \alpha\,(X)\,\xi_\alpha$ for any vector field $X$ on $M$.

Then $(M, Ker\,(\alpha))$ is a contact manifold (i.e., $\alpha \wedge (d\alpha)^n \neq 0$ on $M$), $\xi_\alpha$ is the Reeb vector field associated to $\alpha$, $J\xi_\alpha = 0$ and $g$ is determined by $\alpha$ and $J$ through the equality $g\,(X, Y) = \alpha\,(X)\,\alpha\,(Y) + d\alpha\,(X, JY)$, (see e.g., [176]).

**Example** In the case of hedgehogs of $\mathbb{R}^{n+1}$, we will consider the metric contact manifold $\left(U\mathbb{R}^{n+1}, g, \alpha, J\right)$, where $g$ is the Riemannian product metric on $U\mathbb{R}^{n+1} = \mathbb{R}^{n+1} \times \mathbb{S}^n$ and $J : TU\mathbb{R}^{n+1} \to TU\mathbb{R}^{n+1}$, $(X, Q) \mapsto (Q, \langle X, q \rangle q - X)$.

**Hedgehogs as Legendrian Fronts**
Let us consider first the case where $(M, g) = \left(\mathbb{R}^{n+1}, g_{can}\right)$, where $g_{can} = \langle ., . \rangle$ is the canonical Euclidean metric. Let $\mathcal{H}_h$ be a hedgehog of $\mathbb{R}^{n+1}$ with support function $h \in C^\infty\,(\mathbb{S}^n; \mathbb{R})$. Let us recall that its natural parametrization $x_h : \mathbb{S}^n \to \mathbb{R}^{n+1}$, $u \mapsto x_h\,(u) = h(u)u + (\nabla h)\,(u)$ can be interpreted as the inverse of its Gauss map. Thus it appears that

$$
\begin{aligned}
i_h : \quad & \mathbb{S}^n \to U\mathbb{R}^{n+1} = \mathbb{R}^{n+1} \times \mathbb{S}^n \\
& u \mapsto (x_h\,(u)\,, u)
\end{aligned}
$$

is the immersion of a Legendrian submanifold in $U\mathbb{R}^{n+1}$ of which $\mathcal{H}_h$ is the Legendrian front in $\mathbb{R}^{n+1}$ and $x_h = \pi \circ i_h$ the corresponding Legendrian map. Recall that on $U\mathbb{R}^{n+1}$, the contact form and the associated Reeb vector field are respectively given by

$$
\alpha_{(x,u)} := \langle u, dx \rangle = \sum_{i=0}^{n+1} u_i dx_i \qquad \text{and} \qquad \xi\,(x, u) := \left(u; 0_{T_u\mathbb{S}^n}\right),
$$

for all $(x, u) \in U\mathbb{R}^{n+1}$, where $(x_1, \cdots, x_{n+1}; u_1, \cdots, u_{n+1})$ are the canonical coordinate functions on $U\mathbb{R}^{n+1} = \mathbb{R}^{n+1} \times \mathbb{S}^n \subset \mathbb{R}^{2n+2}$.

Thus, hedgehogs of $\mathbb{R}^{n+1}$ are the Legendrian fronts of those Legendrian submanifolds of $\left(U\mathbb{R}^{n+1}, \alpha\right)$ whose Legendrian maps can be interpreted as the inverse of the Gauss map of their image (i.e., of the Legendrian front).

$$i_h$$
$$\mathbb{S}^n \xrightarrow{i_h} i_h\left(\mathbb{S}^n\right) \subset \left(U\mathbb{R}^{n+1}, Ker\left(\alpha\right)\right)$$

$$x_h \searrow \qquad \downarrow \pi$$

$$\mathcal{H}_h \subset \mathbb{R}^{n+1}.$$

***This can of course be adapted to hedgehogs of other spaces***. We can, for instance, replace:

- $\left(\mathbb{R}^{n+1}, \langle ., . \rangle\right)$ by the Lorentzian-Minkowski $(n+1)$-space $\mathbb{L}^{n+1} = \left(\mathbb{R}^{n+1}, \langle ., . \rangle_L\right)$, where

$$\langle x, x' \rangle_L = \sum_{k=0}^{n} x_k x'_k - tt',$$

for all $x := \left(\left(x_1, \ldots, x_n\right), t\right)$ and $x := \left(\left(x'_1, \ldots, x'_n\right), t'\right)$ in $\mathbb{L}^{n+1} = \mathbb{R}^n \times \mathbb{R}$;
- $\mathbb{S}^n$ by $\mathbf{H}^n := \left\{ x = \left(\left(x_1, \ldots, x_n\right), t\right) \in \mathbb{L}^{n+1} \,|\, \langle x, x \rangle_L = -1 \right\}$;
- $U\mathbb{R}^{n+1} = \mathbb{R}^{n+1} \times \mathbb{S}^n$ by $T_{-1}\mathbf{H}^n := \mathbb{L}^{n+1} \times \mathbf{H}^n$;
- $x_h : \mathbb{S}^n \to \mathbb{R}$ by $x_h : \mathbf{H}^n \to \mathbb{L}^{n+1}$, $v \mapsto x_h(v) = (\nabla h)(v) - h(v)v$;
- $i_h : \mathbb{S}^n \to U\mathbb{R}^{n+1}$ by $i_h : \mathbf{H}^n \to T_{-1}\mathbf{H}^n$, $v \mapsto \left(x_h(v), v\right)$;
- $\alpha_{(x,u)}$ by $\alpha_{(x,v)} := \langle v, dx \rangle_L$ and $\xi_\alpha(x, u)$ by $\xi_\alpha(x, v) := \left(v; 0_{T_v \mathbf{H}^n}\right)$.

### Symplectic and Almost-Hermitian Structures

A ***symplectic structure*** on a $2n$-dimensional $C^\infty$-manifold $M$ is a closed differentiable 2-form $\omega$ that is nondegenerate (i.e., such that: $\omega(u, v) = 0$ for all $u \in T_m M$ implies $v = 0_{T_m M}$). The pair $(M, \omega)$ is then called a ***symplectic manifold***.

A submanifold $L$ of a symplectic manifold $(M, \omega)$ is said to be ***Lagrangian*** if the restriction of $\omega$ to $L$ is equal to 0 and $\dim L = (\dim M)/2$. A fibration $\pi : E \to B$ of a symplectic manifold $E$ is called a ***Lagrangian fibration*** if all the fibers are Lagrangian submanifolds. Let $i : L \to E$ be an immersed Lagrangian submanifold $L$ in the total space of a Lagrangian fibration $\pi : E \to B$. The restriction of $\pi$ to $L$, that is $x = \pi \circ i : L \to B$ is called a ***Lagrangian map***.

Given any contact manifold $(M, \alpha)$, there is a canonical symplectic structure on $M \times \mathbb{R}$, which is given by $\omega = d\left(e^t \alpha\right)$. We say that $\left(M \times \mathbb{R}, d\left(e^t \alpha\right)\right)$ is the ***symplectization*** of the contact manifold $(M, \alpha)$.

**Example** The symplectization of the contact manifold $\left(U\mathbb{R}^3, \alpha\right)$, where $\alpha_{(y,u)} := \langle u, dy \rangle$ for all $(y, u) \in \mathbb{R}^3 \times \mathbb{S}^2$, can be regarded as the symplectic manifold $\left(\mathbb{L}^4 \times \mathbb{S}^2, \omega\right)$, where $\omega_{(y,t,u)} = d\left(e^t \alpha_{(y,u)}\right)$, and $\pi_L : \mathbb{L}^4 \times \mathbb{S}^2 \to \mathbb{R}^3$, $(y, t, u) \mapsto y$

is a Lagrangian fibration. Moreover, for any $h \in C^\infty\left(\mathbb{S}^2; \mathbb{R}\right)$,

$$I_h : \quad \mathbb{S}^2 \times \mathbb{R} \to \mathbb{L}^4 \times \mathbb{S}^2$$
$$(u, t) \mapsto (x_{h-t}(u), t, u)$$

appears to be the immersion of a Lagrangian submanifold $\mathcal{L}_h := I_h\left(\mathbb{S}^2 \times \mathbb{R}\right)$ in $\mathbb{L}^4 \times \mathbb{S}^2$. Note that this Lagrangian submanifold $\mathcal{L}_h$ of $\mathbb{L}^4 \times \mathbb{S}^2$ is obtained by lifting to $\mathbb{L}^4 \times \mathbb{S}^2$ the family $(i_{h-t})_{t \in \mathbb{R}}$ of Legendrian immersions $i_{h-t} : \mathbb{S}^2 \to U\mathbb{R}^3$, $u \mapsto (x_{h-t}(u), u)$ whose Legendrian fronts $x_{h-t}\left(\mathbb{S}^2\right)$ form the family of parallel hedgehogs $(\mathcal{H}_{h-t})_{t \in \mathbb{R}}$ in $\mathbb{R}^3$.

A ***symplectomorphism*** from a symplectic manifold $(M_1, \omega_1)$ to a symplectic manifold $(M_2, \omega_2)$ is a diffeomorphism $f : M_1 \to M_2$ that preserves the symplectic structure (i.e., such that $f^* \omega_2 = \omega_1$).

**Example** It is easy to check that

$$f : \quad \left(\mathbb{L}^4 \times \mathbb{S}^2, d\left(e^t \alpha\right)\right) \to \left(T\left(\mathbb{S}^2\right) \times \mathbb{R}^2, d\left(e^t \beta\right)\right)$$
$$(y, t, u) \mapsto (u, y - \langle y, u \rangle u, \langle y, u \rangle, t)$$

is a symplectomorphism between the symplectizations of $\left(U\mathbb{R}^3, \alpha\right)$ and $\left(T\left(\mathbb{S}^2\right) \times \mathbb{R}, \beta\right)$.

An ***almost complex manifold*** is defined to be a tuple $(M, J)$, where $(M, g)$ is a Riemannian manifold and $J$ an ***almost complex structure*** on $TM$, that is, a vector bundle endomorphism $J : TM \to TM$ such that $J^2 = -1$. An ***almost Hermitian manifold*** is defined to be a tuple $(M, g, J)$, where $(M, J)$ is an almost complex manifold and $g$ an ***almost hermitian metric*** on $(M, J)$, that is, a Riemannian metric on $M$ satisfying:

$$\forall m \in M, \forall (X, Y) \in (T_m M)^2, \quad g(X, JY) = -g(JX, Y).$$

To any metric contact manifold $(M, g, \alpha, J)$, we can also associate a manifold $\widehat{M} = M \times \mathbb{R}$ which carries an almost-hermitian structure $\left(\widehat{g}, \widehat{J}\right)$ extending the one we have on the contact distribution $V = Ker(\alpha)$: indeed, $\left(\widehat{C}, \widehat{g}, \widehat{J}\right)$ is almost Hermitian if we let $\widehat{g} = g + dt^2$, $\widehat{J}_{|V} = J_{|V}$, $\widehat{J}\xi_\alpha = -\partial_t$ and $\widehat{J}\partial_t = \xi_\alpha$, where $t$ is the coordinate on the $\mathbb{R}$ factor (see e.g., [176]).

**Marginally Trapped Hedgehogs of an Associated Lagrangian Submanifold**
Given any $h \in C^\infty\left(\mathbb{S}^2; \mathbb{R}\right)$, let us consider the marginally trapped hedgehog $\mathcal{H}_{\partial h}$ of $\mathbb{L}^4$ with support differential $\partial h := dh/\sqrt{2}$. As we have seen above, $\left(\mathbb{L}^4 \times \mathbb{S}^2, \omega := d\left(e^t \alpha\right)\right)$ can be regarded as the symplectization of the contact manifold $\left(U\mathbb{R}^3, \alpha\right)$, where $\alpha := \langle u, dy \rangle$, $\pi_L : \mathbb{L}^4 \times \mathbb{S}^2 \to \mathbb{R}^3$, $(y, t, u) \mapsto y$ is a Lagrangian fibration and $I_h : \mathbb{S}^2 \times \mathbb{R} \to \mathbb{L}^4 \times \mathbb{S}^2$, $(u, t) \mapsto (x_{h-t}(u), t, u)$ is the immersion of a Lagrangian submanifold $\mathcal{L}_h := I_h\left(\mathbb{S}^2 \times \mathbb{R}\right)$ in $\mathbb{L}^4 \times \mathbb{S}^2$. If we endow the symplectic manifold $\left(\mathbb{L}^4 \times \mathbb{S}^2, \omega\right)$ with the Lorentzian metric $g_L := g - dt^2$,

where $g$ is the Riemannian product metric on $U\mathbb{R}^3 = \mathbb{R}^3 \times \mathbb{S}^2$, then

$$i_{\partial h} : \quad \mathbb{S}^2 \to \left(\mathbb{L}^4 \times \mathbb{S}^2, g_L\right)$$

$$u \mapsto (x_{\partial h}(u), u) = \left(\nabla h(u) + \partial^2 h(u)\, u, h(u) - \partial^2 h(u), u\right)$$

is the parametrization of a marginally trapped surface $\mathcal{I}_{\partial h} := i_{\partial h}\left(\mathbb{S}^2\right)$ included in the Lagrangian submanifold $\mathcal{L}_h$ of $\mathbb{L}^4 \times \mathbb{S}^2$: indeed, $\mathcal{I}_{\partial h}$ is spacelike (i.e., its induced metric is Riemannian) and its mean curvature vector $\mathcal{H}_{i_{\partial h}}$ is lightlike at each point. More precisely, it is a routine to check that:

(i) $i_{\partial h}^* g = \left(1 + \frac{1}{4}(R_1 - R_2)^2\right) g_{\mathbb{S}}$, where $g_S$ is the standard metric on $\mathbb{S}^2$, $g$ the first fundamental form on $\mathcal{I}_{\partial h}$, $i_{\partial h}^* g$ the pullback of $g$ along $i_{\partial h}$ and $R_1(u)$, $R_2(u)$ are the principal radii of curvature of $\mathcal{H}_h$ at $x_h(u)$;

(ii) for all $u \in \mathbb{S}^2$, $\mathcal{H}_{i_{\partial h}}(i_{\partial h}(u)) \in \mathbb{R}(\xi_\alpha - \partial_t)$, where $\xi_\alpha := \left(u, 0, 0_{T_u \mathbb{S}^2}\right)$, $\partial_t := \left(0, 1, 0_{T_u \mathbb{S}^2}\right)$ and hence $\xi_\alpha - \partial_t := \left(u, -1, 0_{T_u \mathbb{S}^2}\right) \in T_{i_{\partial h}(u)}\mathcal{I}_{\partial h} \subset \mathbb{L}^4 \times T_u \mathbb{S}^2$.

Thus, $\mathcal{H}_{\partial h}$ is the image of $\mathcal{I}_{\partial h}$ under the projection $\pi_L : \mathbb{L}^4 \times \mathbb{S}^2 \to \mathbb{L}^4$, and we have the following commutative diagram:

$$
\begin{array}{ccc}
 & i_{\partial h} & \\
\mathbb{S}^2 & \to & \mathcal{I}_{\partial h} \text{ marginally trapped in } \mathcal{L}_h \subset \left(\mathbb{L}^4 \times \mathbb{S}^2, g_L\right) \\
 & & \\
x_{\partial h} \searrow & & \downarrow \pi_L \\
 & & \\
 & & \mathcal{H}_{\partial h} \text{ marginally trapped in } \mathbb{L}^4.
\end{array}
$$

In other words, any marginally trapped hedgehog $x_{\partial h} : \mathbb{S}^2 \to \mathcal{H}_{\partial h} \subset \mathbb{L}^4$ lifts to a marginally trapped surface $\mathcal{I}_{\partial h}$ of the Lagrangian submanifold $\mathcal{L}_h$ of $\left(\mathbb{L}^4 \times \mathbb{S}^2, \omega, g_L\right)$ that is deduced from the family of parallel hedgehogs $\left(\mathcal{H}_{h-t}\right)_{t \in \mathbb{R}}$.

Now, if $M$ is another 6-dimensional $C^\infty$-manifold endowed with a symplectic structure $\omega_M$ and a Lorentzian metric $g_M$, and if $f : \mathbb{L}^4 \times \mathbb{S}^2 \to M$ is a diffeomorphism preserving both the symplectic structure and the Lorentzian metric, then $f(\mathcal{I}_{\partial h})$ is a marginally trapped surface of the Lagrangian submanifold $f(\mathcal{L}_h)$ of $(M, \omega_M, g_M)$. For instance, if we use the diffeomorphism

$$f : \quad \mathbb{L}^4 \times \mathbb{S}^2 \to T\left(\mathbb{S}^2\right) \times \mathbb{R}^2$$

$$(y, t, u) \mapsto (u, y - \langle y, u \rangle\, u, \langle y, u \rangle, t),$$

to transport the Lorentzian metric $g_L$ to a Lorentzian metric $g_M$ on $T\left(\mathbb{S}^2\right) \times \mathbb{R}^2$, then we see that

$$
\begin{aligned}
j_{\partial h}: \quad & \mathbb{S}^2 \to \left(T\left(\mathbb{S}^2\right) \times \mathbb{R}^2, d\left(e^t \beta\right), g_M\right) \\
& u \mapsto \left(u, \nabla h\left(u\right), \partial^2 h\left(u\right), h\left(u\right) - \partial^2 h\left(u\right)\right)
\end{aligned}
$$

defines a marginally trapped surface $j_{\partial h}\left(\mathbb{S}^2\right)$ of the Lagrangian submanifold of $\left(T\left(\mathbb{S}^2\right) \times \mathbb{R}^2, d\left(e^t \beta\right)\right)$ that is parametrized by:

$$
\begin{aligned}
J_h: \quad & \mathbb{S}^2 \times \mathbb{R} \to T\left(\mathbb{S}^2\right) \times \mathbb{R}^2 \\
& (u, t) \mapsto \left(j_{h-t}^1\left(u\right), t\right) := \left(u, \nabla h\left(u\right), h\left(u\right) - t, t\right).
\end{aligned}
$$

Let us conclude this section with a remark concerning the definition of the marginally trapped hedgehog $x_{\partial h} : \mathbb{S}^2 \to \mathcal{H}_{\partial h} \subset \mathbb{L}^4$, where $\mathbb{L}^4$ is identified with $\Sigma$. For any $u \in \mathbb{S}^2$, $x = x_{\partial h}\left(u\right)$ is the (possibly reduced to a point) cooriented sphere of $\mathbb{R}^3$ that satisfies the following system of conditions:

$$
\begin{cases}
(1) \ x \in L_h\left(u\right) \\
(2) \ \langle x, \Delta u_L \rangle = \delta\left(dh\right)\left(u\right),
\end{cases}
$$

(see Sect. 8.1.1). The first condition simply ensures that the 'contact element' $i_h\left(u\right) := \left(x_h\left(u\right), u\right)$ is in oriented contact with the sphere $x_{\partial h}\left(u\right)$ (i.e., $x_h\left(u\right)$ belongs to the sphere $x_{\partial h}\left(u\right)$ and $u$ is the unit normal to $x_{\partial h}\left(u\right)$ at $x_h\left(u\right)$). The second condition (which is the co-contact one) then ensures that $x_{\partial h}\left(u\right)$ is more precisely the oriented middle sphere (or unoriented middle point) $S\left(c_h\left(u\right); R_{(1,h)}\left(u\right)\right)$.

## 8.3   Proof of the Main Results and Further Remarks

We first prove Theorem 8.1.1 and then Theorem 8.2.1, which is its analogue in $\mathbf{M}^4$.

***Proof of Theorem 8.1.1*** In this proof, $\nabla_S$ (resp. $\nabla$) stands for the gradient on $\mathbb{S}^2$ (resp. $\mathbb{R}^3$), $\Delta_S$ (resp. $\Delta$) for the Laplace-Beltrami operator on $\mathbb{S}^2$ (resp. $\mathbb{R}^3$) and the Laplace-Beltrami operator of a vector function $X : \mathbb{S}^2 \to \mathbb{L}^4$, $u \mapsto \left(X_i\left(u\right)\right)_{i=1}^4$ is understood to be the vector function $\Delta_S X : \mathbb{S}^2 \to \mathbb{L}^4$, $u \mapsto \left(\Delta_S X_i\left(u\right)\right)_{i=1}^4$, where $\Delta_S X_i$ is the Laplace-Beltrami operator of the coordinate function $X_i$ in the intrinsic metric on $\mathbb{S}^2$, $(1 \leq i \leq 4)$.

(i) For all $t \in \mathbb{R}$, let $x : \mathbb{S}^2 \to \mathbb{L}^4$ be any $C^\infty$ map such that

$$
x\left(u\right) \in L_{h+t}\left(u\right) := \{l_{h+t}\left(u\right)\} + \mathbb{R} u_L
$$

for all $u \in \mathbb{S}^2$. Then there exists some $\lambda \in C^\infty\left(\mathbb{S}^2; \mathbb{R}\right)$ such that $x$ is of the form

$$
\begin{aligned}
X_\lambda: \quad &\mathbb{S}^2 \to \mathbb{L}^4 \\
&u \mapsto \left(\nabla_S h\left(u\right), h\left(u\right) - t\right) + \lambda\left(u\right)\left(u, -1\right) \\
&= \left(x_h\left(u\right) + f\left(u\right) u, t - f\left(u\right)\right),
\end{aligned}
$$

where $f := \lambda - h$. For all $u \in \mathbb{S}^2$, we have:

$$
\Delta_S X_\lambda\left(u\right) = \left(\Delta_S x_h\left(u\right) + \Delta_S\left(f\left(u\right) u\right), -\left(\Delta_S f\right)\left(u\right)\right).
$$

Now, $\Delta_S x_h = \Delta_S\left(\nabla\varphi\right) = \Delta\left(\nabla\varphi\right) = \nabla\left(\Delta\varphi\right)$, where $\varphi$ is the positively 1-homogeneous extension of $h$ to $\mathbb{R}^3\setminus\{0\}$, that is,

$$
\varphi\left(x\right) := \|x\|\, h\left(\frac{x}{\|x\|}\right),
$$

for $x \in \mathbb{R}^3\setminus\{0\}$, where $\|.\|$ is the Euclidean norm on $\mathbb{R}^3$. Indeed, $\nabla\varphi$ is positively 0-homogeneous on $\mathbb{R}^3\setminus\{0\}$ and equal to $x_h$ on $\mathbb{S}^2$. Thus

$$
\langle\Delta_S x_h\left(u\right), u\rangle = -\left(\Delta\varphi\right)\left(u\right) = -\left(\Delta_S h + 2h\right)\left(u\right),
$$

for $u \in \mathbb{S}^2$, since $\Delta\varphi$ is positively $-1$-homogeneous on $\mathbb{R}^3\setminus\{0\}$. Besides, we have:

$$
\begin{aligned}
\Delta_S\left(f\left(u\right) u\right) &= \left(\Delta_S f\right)\left(u\right) u + 2\left(\langle\left(\nabla_S f\right)\left(u\right), \left(\nabla_S x_i\right)\left(u\right)\rangle\right)_{i=1}^3 \\
&\quad + f\left(u\right)\left(\Delta_S id_{\mathbb{S}^2}\right)\left(u\right) \\
&= \left(\Delta_S f - 2f\right)\left(u\right) u + 2\left(\nabla_S f\right)\left(u\right),
\end{aligned}
$$

for all $u \in \mathbb{S}^2$. Therefore, we have:

$$
\begin{aligned}
\langle\Delta_S X_\lambda\left(u\right), u_L\rangle_L = 0 &\Leftrightarrow -\left(\Delta_S h + 2h\right)\left(u\right) + \left(\Delta_S f - 2f\right)\left(u\right) \\
&\qquad - \left(\Delta_S f\right)\left(u\right) = 0 \\
&\Leftrightarrow -\left(\Delta_S h + 2h\right)\left(u\right) + 2\left(h - \lambda\right)\left(u\right) = 0 \\
&\Leftrightarrow \lambda = -\tfrac{1}{2}\left(\Delta_S h\right)\left(u\right),
\end{aligned}
$$

for all $u \in \mathbb{S}^2$. Thus

$$
x\left(u\right) = X_\lambda\left(u\right) = l_{h+t}\left(u\right) - \frac{\left(\Delta_S h\right)\left(u\right)}{2}\left(u, -1\right) = l_{h+t}\left(u\right) + \delta\left(\partial h\right)\left(u\right) u_L,
$$

for all $u \in \mathbb{S}^2$.

(ii)  We know that: $x\left(u\right) - \left(x_h\left(u\right), t\right) - R_{(1,h)}\left(u\right)\left(u, 1\right)$ for all $u \in \mathbb{S}^2$, where $R_{(1,h)} := h - \frac{\Delta h}{2} = \frac{1}{2}\left(R_1 + R_2\right)$ is the mean radius of principal curvature of

$\mathcal{H}_h$. From this we deduce that

$$(T_u x)(v) = ((T_u x_h)(v), 0) - \left(d R_{(1,h)}\right)_u (v)(u, -1) - R_{(1,h)}(u)(v, 0);$$

and thus that

$$\langle (T_u x)(v), (T_u x)(v) \rangle_L = \big\langle (T_u x_h)(v) - R_{(1,h)}(u)\, v,$$

$$(T_u x_h)(v) - R_{(1,h)}(u)\, v \big\rangle,$$

for all $v \in T_u \mathbb{S}^2$. Now, considering an orthonormal basis $(e_1, e_2)$ of $T_u \mathbb{S}^2$ made of eigenvectors of $T_u x_h$ (i.e., $(T_u x_h)(e_i) = R_i(u)\, e_i$ for $i \in \{1, 2\}$), we conclude that: $\forall v \in T_u \mathbb{S}^2$,

$$\left(x^* g\right)_u (v, v) := \langle (T_u x)(v), (T_u x)(v) \rangle_L = \frac{1}{4}(R_1 - R_2)(u)^2\, g_S(v, v).$$

(*iii*) We know that: $\forall u \in \mathbb{S}^2$, $x(u) = (x_h(u), t) - f(u)(u, -1)$, where $f = R_{(1,h)}$. From this, we deduce that: $\forall u \in \mathbb{S}^2$,

$$(\Delta_S x)(u) = (\Delta_S x_h(u) - \Delta_S(f(u)\, u), (\Delta_S f)(u)).$$

Now: $\forall u \in \mathbb{S}^2$, $\Delta_S(f(u)\, u) = (\Delta_S f - 2f)(u)\, u + 2(\nabla_S f)(u)$, (see above the proof of (*i*)). Besides, $\Delta_S x_h = \Delta_S(\nabla \varphi) = \Delta(\nabla \varphi) = \nabla(\Delta \varphi)$, where $\varphi$ is the positively 1-homogeneous extension of $h$ to $\mathbb{R}^3 \setminus \{0\}$ (see ib.), so that $\Delta_S x_h(u) = \nabla_S(\Delta \varphi)(u) - (\Delta \varphi)(u)\, u = 2(\nabla_S f(u) - f(u)\, u)$ for all $u \in \mathbb{S}^2$, since $\Delta \varphi$ is positively $-1$-homogeneous on $\mathbb{R}^3 \setminus \{0\}$, and equal to $2f$ on $\mathbb{S}^2$. Therefore: $\forall u \in \mathbb{S}^2$,

$$(\Delta_S x)(u) = -(\Delta_S f)(u)(u, -1) = -\left(\Delta_S R_{(1,h)}\right)(u)(u, -1) \in \mathbb{R} u_L.$$

We know that under a conformal change of metric $\tilde{g} = e^{2\phi} g$ on a surface $M^2$, $\left(\psi \in C^\infty(M^2; \mathbb{R})\right)$, the Laplace-Beltrami operator $\Delta_g$ transforms according to the formula

$$\Delta_{\tilde{g}} f = e^{-2\phi} \Delta_g f \quad \text{for all } f \in C^\infty\left(M^2; \mathbb{R}\right).$$

Above, we have demonstrated that $x^* g = \frac{1}{4}(R_1 - R_2)(u)^2\, g_S = \left(R^2_{(1,h)} - R_h\right) g_S$, where $R_h := R_1 R_2$ is the curvature function of $\mathcal{H}_h$, and that $(\Delta_S x)(u) = -\left(\Delta_S R_{(1,h)}\right)(u)(u, -1)$ for all $u \in \mathbb{S}^2$. Therefore

$$\left(\Delta_{x^* g} x\right)(u) = -\frac{\Delta_S R_{(1,h)}}{R^2_{(1,h)} - R_h}(u)(u, -1)$$

outside umbilical points of $\mathcal{H}_h$ (i.e., for all $u \in \mathbb{S}^2$ such that $R_1(u) \neq R_2(u)$). We also know that the mean vector field of any surface $X : \Omega \subset \mathbb{S}^2 \to \mathbb{L}^4$, $u \mapsto (X_i(u))_{i=1}^4$ is given by $\overrightarrow{H}_X = \frac{1}{2}(\Delta X) = \frac{1}{2}(\Delta X_i)_{i=1}^4$, where $\Delta X_i$ is the Laplace-Beltrami operator of the coordinate function in the intrinsic metric of the surface. Therefore: $\forall u \in \mathbb{S}^2 \setminus \{u \in \mathbb{S}^2 \mid R_1(u) = R_2(u)\}$,

$$\overrightarrow{H}_x(u) = -\frac{\Delta_S R_{(1,h)}}{2\left(R_{(1,h)}^2 - R_h\right)}(u)(u,-1) = \frac{\partial^2 R_{(1,h)}}{R_{(1,h)}^2 - R_h}(u)(u,-1) \in \mathbb{R}\overrightarrow{u_L}.$$

$\blacksquare$

***Proof of Theorem 8.2.1*** The steps of the proof are the same as those in the proof of Theorem 8.1.1. There are just some slight changes in formulas. For the convenience of the reader, we resume below the different steps of the proof.

In this proof, $\nabla_H$ (resp. $\nabla_L$) stands for the gradient on $\mathbf{H}^2$ (resp. $\mathbb{L}^3$), $\Delta_H$ for the Laplace-Beltrami operator on $\mathbf{H}^2$, and $\square$ for the d'Alembertian (or wave operator) on $\mathbb{L}^3$. Let $(x_1, x_2, x_3)$ be the standard coordinates on $\mathbb{L}^3$. For all differentiable function $\psi : \mathbb{L}^3 \to \mathbb{R}$ and all $x \in \mathbb{L}^3$, $\nabla_L \psi(x)$ thus denotes the vector with entries $\left(\frac{\partial \psi}{\partial x_1}(x), \frac{\partial \psi}{\partial x_2}(x), -\frac{\partial \psi}{\partial x_3}(x)\right)$ in $\mathbb{L}^3$. For its part, the d'Alembertian has the form $\square := \frac{\partial^2}{\partial x_1^2} + \frac{\partial^2}{\partial x_2^2} - \frac{\partial^2}{\partial x_3^2}$. Besides, the Laplace-Beltrami operator of a vector function $X : \mathbf{H}^2 \to \mathbf{M}^4$, $u \mapsto (X_i(u))_{i=1}^4$ is understood to be the vector function $\Delta_H X : \mathbf{H}^2 \to \mathbf{M}^4$, $u \mapsto (\Delta_H X_i(u))_{i=1}^4$, where $\Delta_H X_i$ is the Laplace-Beltrami operator of the coordinate function $X_i$ in the intrinsic metric on $\mathbf{H}^2$, $(1 < i \leq 4)$.

*(i)* For all $t \in \mathbb{R}$, let $x : \mathbf{H}^2 \to \mathbf{M}^4$ be any $C^\infty$ map such that

$$x(v) \in L_{h+t}(v) := \{l_{h+t}(v)\} + \mathbb{R}v_L,$$

for all $v \in \mathbf{H}^2$. Then there exists some $\lambda \in C^\infty(\mathbb{S}^2; \mathbb{R})$ such that $x$ is of the form

$$\begin{aligned}
X_\lambda : \quad & \mathbf{H}^2 \to \mathbf{M}^4 \\
& v \mapsto (\nabla_H h(v), -h(v) - t) + \lambda(v)(v, -1) \\
& = (x_h(v) + f(v)v, -f(v) - t),
\end{aligned}$$

where $f := h + \lambda$. For all $v \in \mathbf{H}^2$, we have:

$$\Delta_H X_\lambda(v) = (\Delta_H x_h(v) + \Delta_H(f(v)v), -(\Delta_H f)(v)).$$

Now, $\Delta_H x_h = \Delta_H(\nabla_L \varphi) = \square(\nabla_L \varphi) = \nabla(\square \varphi)$, where $\varphi$ is the positively 1-homogeneous extension of $h$ to $\mathcal{U} = \{x \in \mathbb{L}^3 \mid \langle x, x \rangle_L < 0\}$, that is,

$$\varphi(v) := \|x\|_L \, h\left(\frac{x}{\|x\|_L}\right),$$

for $x \in \mathcal{U}$, where $\|x\|_L := \sqrt{-\langle x, x \rangle_L}$. Indeed, $\nabla \varphi$ is positively $0$-homogeneous on $\mathcal{U}$ and equal to $x_h$ on $\mathbf{H}^2$. Thus

$$\langle \Delta_H x_h \left( v \right), v \rangle_L = - \left( \Box \varphi \right) \left( v \right) = \left( -\Delta_H h + 2h \right) \left( v \right),$$

for $v \in \mathbf{H}^2$, since $\Box \varphi$ is positively $-1$-homogeneous on $\mathcal{U}$. Besides, we have:

$$\Delta_H \left( f \left( v \right) v \right) = \left( \Delta_H f \right) \left( u \right) u + 2 \left( \langle \left( \nabla_H f \right) \left( v \right), \left( \nabla_H x_i \right) \left( v \right) \rangle_L \right)_{i=1}^3$$
$$+ f \left( v \right) \left( \Delta_H i d_{\mathbf{H}^2} \right) \left( v \right)$$
$$= \left( \Delta_H f + 2f \right) \left( v \right) v + 2 \left( \nabla_H f \right) \left( v \right),$$

for all $v \in \mathbf{H}^2$. Therefore, we have:

$$\langle \Delta_H X_\lambda \left( v \right), v_M \rangle_M = 0 \Leftrightarrow \left( -\Delta_H h + 2h \right) \left( v \right)$$
$$- \left( \Delta_H f + 2f \right) \left( u \right) + \left( \Delta_H f \right) \left( v \right) = 0$$
$$\Leftrightarrow - \left( -\Delta_H h + 2h \right) \left( u \right) - 2 \left( h + \lambda \right) \left( v \right) = 0$$
$$\Leftrightarrow \lambda = -\tfrac{1}{2} \left( \Delta_H h \right) \left( v \right),$$

for all $v \in \mathbf{H}^2$. Thus

$$x \left( v \right) = X_\lambda \left( v \right) = l_{h+t} \left( v \right) - \frac{\left( \Delta_H h \right) \left( v \right)}{2} \left( v, -1 \right) = l_{h+t} \left( v \right) + \delta \left( \partial h \right) \left( v \right),$$

for all $v \in \mathbf{H}^2$.

*(ii)* We know that $x \left( v \right) = \left( x_h \left( v \right), t \right) - R_{\left( -1, h \right)} \left( v \right) \left( v, -1 \right)$ for all $v \in \mathbf{H}^2$, where $R_{\left( -1, h \right)} := -h + \frac{\Delta_H h}{2} = \tfrac{1}{2} \left( R_1 + R_2 \right)$ is the mean radius of principal curvature of $\mathcal{H}_h$. From this we deduce that

$$\left( T_v x \right) \left( w \right) = \left( \left( T_v x_h \right) \left( w \right), 0 \right) - \left( d R_{\left( -1, h \right)} \right)_v \left( w \right) \left( v, -1 \right) - R_{\left( -1, h \right)} \left( v \right) \left( w, 0 \right);$$

and thus that

$$\langle \left( T_v x \right) \left( w \right), \left( T_v x \right) \left( w \right) \rangle_M = \big\langle \left( T_u x_h \right) \left( v \right) - R_{\left( 1, h \right)} \left( u \right) v,$$
$$\left( T_u x_h \right) \left( v \right) - R_{\left( 1, h \right)} \left( u \right) v \big\rangle,$$

for all $w \in T_v \mathbf{H}^2$. Now, considering an orthonormal basis $\left( e_1, e_2 \right)$ of $T_v \mathbf{H}^2$ made of eigenvectors of $T_v x_h$ (i.e., $\left( T_v x_h \right) \left( e_i \right) = R_i \left( v \right) e_i$ for $i \in \{1, 2\}$), we conclude that: $\forall w \in T_v \mathbf{H}^2$,

$$\left( x^* g \right)_v \left( w, w \right) := \langle \left( T_v x \right) \left( w \right), \left( T_v x \right) \left( w \right) \rangle_L = \frac{1}{4} \left( R_1 - R_2 \right) \left( u \right)^2 g_H \left( v, v \right).$$

*(iii)* We know that: $\forall v \in \mathbf{H}^2$, $x(v) = (x_h(v), -t) - f(v)(v, -1)$, where $f = R_{(-1,h)}$. From this, we deduce that: $\forall v \in \mathbf{H}^2$,

$$(\Delta_H x)(v) = (\Delta_H x_h(v) - \Delta_H(f(v)v), (\Delta_H f)(v)).$$

Now: $\forall v \in \mathbf{H}^2$, $\Delta_H(f(v)v) = (\Delta_H f + 2f)(v)v + 2(\nabla_H f)(v)$, (see above the proof of *(i)*). Besides, $\Delta_H x_h = \Delta_H(\nabla_L \varphi) = \Box(\nabla_L \varphi) = \nabla(\Box\varphi)$, where $\varphi$ is the positively 1-homogeneous extension of $h$ to $\mathcal{U} = \{x \in \mathbb{L}^3 \,|\, \langle x, x\rangle_L < 0\}$ (see ib.), so that $\Delta_H x_h(v) = \nabla_H(\Box\varphi)(v) + (\Box\varphi)(v)v = 2(\nabla_H f(v) + f(v)v)$ for all $v \in \mathbf{H}^2$, since $\Box\varphi$ is positively $-1$-homogeneous on $\mathcal{U}$, and equal to $2f$ on $\mathbf{H}^2$. Therefore: $\forall v \in \mathbf{H}^2$,

$$(\Delta_H x)(v) = -(\Delta_H f)(v)(v, -1) = -\left(\Delta_H R_{(-1,h)}\right)(v)(v, -1) \in \mathbb{R}\overrightarrow{v_M}.$$

Above, we have demonstrated that $x^* g = \tfrac{1}{4}(R_1 - R_2)(v)^2 g_H = \left(R_{(-1,h)}^2 - R_h\right)g_H$, where $R_h := R_1 R_2$ is the curvature function of $\mathcal{H}_h$, and that $(\Delta_H x)(v) = -\left(\Delta_H R_{(-1,h)}\right)(v)(v, -1)$ for all $v \in \mathbf{H}^2$. Therefore

$$\left(\Delta_{x^* g} x\right)(v) = -\frac{\Delta_H R_{(-1,h)}}{R_{(-1,h)}^2 - R_h}(v)(v, -1)$$

outside umbilical points of $\mathcal{H}_h$ (i.e., for all $v \in \mathbf{H}^2$ such that $R_1(v) \neq R_2(v)$). We also know that the mean vector field of any surface $X : \Omega \subset \mathbf{H}^2 \to \mathbf{M}^4$, $v \mapsto (X_i(v))_{i=1}^4$ is given by $\overrightarrow{H}_X = \tfrac{1}{2}(\Delta_H X) = \tfrac{1}{2}(\Delta_H X_i)_{i=1}^4$, where $\Delta_H X_i$ is the Laplace-Beltrami operator of the coordinate function in the intrinsic metric of the surface. Therefore: $\forall v \in \mathbf{H}^2 \setminus \{v \in \mathbf{H}^2 \,|\, R_1(v) = R_2(v)\}$,

$$\overrightarrow{H}_X(v) = -\frac{\Delta_H R_{(-1,h)}}{2\left(R_{(-1,h)}^2 - R_h\right)}(v)(v, -1) = \frac{\partial^2 R_{(-1,h)}}{R_{(-1,h)}^2 - R_h}(v)(v, -1) \in \mathbb{R}\overrightarrow{v_M}.$$

$\blacksquare$

Let us now turn to the proofs of Theorems 8.1.2 and 8.1.5.

**Proof of Theorem 8.1.2** Since the mixed curvature function $R : C^\infty(\mathbb{S}^2, \mathbb{R})^2 \to \mathbb{R}$, $(f, g) \mapsto R_{(f,g)}$ is bilinear and symmetric, it follows that the same holds true for the mixed Laguerre area

$$L : \quad (\partial\mathcal{H})^2 \mapsto \mathbb{R}$$
$$(\partial f, \partial g) \mapsto \int_{\mathbb{S}^2} \left(R_{(1,f)} R_{(1,g)} - R_{(f,g)}\right) d\sigma.$$

Furthermore, for all $\partial h \in \partial \mathcal{H}$, we have:

$$(s\,(\partial h) = 0) \Longleftrightarrow (\mathcal{H}_h \text{ is totally umbilical}) \Longleftrightarrow (\partial h = 0_{\partial \mathcal{H}}).$$

This completes the proof. $\qquad\qquad\qquad\qquad\qquad\qquad\qquad\qquad\qquad\qquad\blacksquare$

***Proof of Theorem 8.1.5*** For any $h \in C^4\left(\mathbb{S}^2, \mathbb{R}\right)$, we know from $(ii)$ of Theorem 8.1.1 (which remains true under our present smoothness assumption) that $x^*g = \frac{1}{4}(R_1 - R_2)^2 g_S$, where $g_S$ is the standard metric on $\mathbb{S}^2$, $g$ the first fundamental form on $x\left(\mathbb{S}^2\right)$, $x^*g$ the pullback of $g$ along $x$, and $R_1\,(u)$, $R_2\,(u)$ are the principal radii of curvature of $\mathcal{H}_h$ at $x_h\,(u)$ for all $u \in \mathbb{S}^2$. Therefore

$$s\,(\partial h) = -L\,(\partial h) = -\frac{1}{4}\int_{\mathbb{S}^2} (R_1 - R_2)^2\, d\sigma.$$

Now $R_1 = \lambda_1 + h$ and $R_2 = \lambda_2 + h$, where $\lambda_1\,(u)$, $\lambda_2\,(u)$ denote the eigenvalues of the Hessian of $h$ at $u$ for all $u \in \mathbb{S}^2$. Hence:

$$s\,(\partial h) = -\frac{1}{4}\int_{\mathbb{S}^2} (\lambda_1 - \lambda_2)^2\, d\sigma = \frac{1}{4}\int_{\mathbb{S}^2} \left(4\,(\Delta_{22}h) - (\Delta_S h)^2\right) d\sigma,$$

where $\Delta_2$ and $\Delta_{22}$ are respectively the Laplace-Beltrami operator and the Monge-Ampère operator on $\mathbb{S}^2$ (that is, respectively the sum and the product of the eigenvalues $\lambda_1, \lambda_2$ of the Hessian of $h$). From the equality

$$\int_{\mathbb{S}^2} R_h d\sigma = \int_{\mathbb{S}^2} h.R_{(1,h)} d\sigma,$$

which is a direct consequence of the symmetry of the mixed volume, we obtain

$$\int_{\mathbb{S}^2} (\Delta_{22}h)\, d\sigma = -\frac{1}{2}\int_{\mathbb{S}^2} h\,(\Delta_S h)\, d\sigma,$$

after development and simplification, and then

$$\int_{\mathbb{S}^2} (\Delta_{22}h)\, d\sigma = \frac{1}{2}\int_{\mathbb{S}^2} (\nabla h)^2\, d\sigma = \int_{\mathbb{S}^2} \left(\overrightarrow{\partial h}\right)^2 d\sigma$$

by integrating by parts. It follows that:

$$s\,(\partial h) = \int_{\mathbb{S}^2} \left(\left(\overrightarrow{\partial h}\right)^2 - \left(\partial^2 h\right)^2\right) d\sigma$$

Note that the integration by parts formula for functions $f, g \in C^2\left(\mathbb{S}^2; \mathbb{R}\right)$ can be written as

$$\int_{\mathbb{S}^2} \overrightarrow{\partial f}.\overrightarrow{\partial g}\, d\sigma = \int_{\mathbb{S}^2} f\left(\partial^2 g\right) d\sigma.$$

Therefore

$$\int_{\mathbb{S}^2}\left(\left(\overrightarrow{\partial h}\right)^2 - \left(\partial^2 h\right)^2\right) d\sigma = \int_{\mathbb{S}^2}\left(\left(\overrightarrow{\partial h}\right)^2 - \overrightarrow{\partial h}.\overrightarrow{\partial\left(\partial^2 h\right)}\right) d\sigma$$

$$= \int_{\mathbb{S}^2} \overrightarrow{\partial h}.\overrightarrow{\partial\left(h - \partial^2 h\right)}\, d\sigma$$

$$= \int_{\mathbb{S}^2} \overrightarrow{\partial h}.\overrightarrow{\partial R_{(1,h)}}\, d\sigma,$$

since $R_{(1,h)} = h + \frac{(\Delta_S h)}{2} = h - \frac{\Delta h}{2} = h - \partial^2 h$. $\blacksquare$

**Curvature Function of Marginally Trapped Hedgehogs**

In analogy to the cases of ordinary hedgehogs and convex bodies of $\mathbb{R}^3$, we will say that a marginally trapped hedgehog $\mathcal{H}_{\partial h}$ of $\mathbb{L}^4$ has the **curvature function** $R_{\overrightarrow{\partial h}}$ : $\mathbb{S}^2 \to \mathbb{R}$ if its signed surface area measure

$$s_{\partial h} : \quad \mathcal{B}\left(\mathbb{S}^2\right) \to \mathbb{R}$$
$$\omega \longmapsto -\int_\omega \left(R^2_{(1,h)} - R_h\right) d\sigma,$$

where $\mathcal{B}\left(\mathbb{S}^2\right)$ is the Borel algebra on $\mathbb{S}^2$, has $R_{\partial h}$ as a density with respect to spherical Lebesgue measure $\sigma$. In other words, $\mathcal{H}_{\partial h}$ has the curvature function $R_{\overrightarrow{\partial h}} = -\left(R^2_{(1,h)} - R_h\right) = -\frac{1}{4}(R_1 - R_2)^2$, where $R_1$ and $R_2$ are the principal radii of curvature of $\mathcal{H}_h$.

**Proposition 8.3.1** *Let* $h \in C^\infty\left(\mathbb{S}^2; \mathbb{R}\right)$. *The curvature function of the marginally trapped hedgehog* $\mathcal{H}_{\partial h}$ *of* $\mathbb{L}^4$ *can be expressed in the form*

$$R_{\overrightarrow{\partial h}} = \overrightarrow{\partial h}^2 + \overrightarrow{\partial h}.\left(\nabla \circ \operatorname{div}\right)\left(\overrightarrow{\partial h}\right) + \frac{1}{2}\left(\left(\operatorname{div}\overrightarrow{\partial h}\right)^2 - \left(\operatorname{div}\circ\nabla\right)\left(\overrightarrow{\partial h}^2\right)\right),$$

*where* div *stands for the divergence operator on* $\mathbb{S}^2$.

This expression of $R_{\overrightarrow{\partial h}}$ has to be compared with the following one, which gives an expression for the curvature function of an ordinary hedgehog $\mathcal{H}_h$ of $\mathbb{R}^3$:

$$R_h = h^2 + h\left(\operatorname{div}\circ\nabla\right)(h) + \frac{1}{2}\left(\left(\operatorname{tr} H_h\right)^2 - \operatorname{tr}\left(H_h^2\right)\right),$$

where $tr$ stands for the trace operator and $H_h$ is the Hessian of $h$. Of course, as for ordinary hedgehogs of $\mathbb{R}^3$, we can define a ***mixed curvature function*** for marginally trapped hedgehogs of $\mathbb{L}^4$:

$$R : (\partial \mathcal{H})^2 \to C^\infty \left(\mathbb{S}^2; \mathbb{R}\right), \left(\overrightarrow{\partial f}, \overrightarrow{\partial g}\right) \longmapsto R_{\left(\overrightarrow{\partial f}, \overrightarrow{\partial g}\right)} := \left(R_{(f,g)} - R_{(1,f)} R_{(1,g)}\right).$$

***Proof of Proposition 8.3.1*** We know that

$$R_{\overrightarrow{\partial h}} = -\frac{1}{4}(\lambda_1 - \lambda_2)^2 = \frac{1}{4}\left((tr\ H_h)^2 - 2tr\left(H_h^2\right)\right),$$

where $\lambda_1$ and $\lambda_2$ are the eigenvalues of $H_h$. Now, in the present case, the classical Bochner-Lichnerowicz-Weitzenböck formula can be rewritten in the form

$$\Delta_S \left(\overrightarrow{\partial h}^2\right) = 2\overrightarrow{\partial h}.(\nabla \circ \mathrm{div})\left(\overrightarrow{\partial h}\right) + tr\left(H_h^2\right) + 2\overrightarrow{\partial h}^2,$$

where $\Delta_S = \mathrm{div} \circ \nabla$ is the Laplace-Beltrami operator on $\mathbb{S}^2$. Combining the above formulas yields the desired result. ∎

Now let us mention briefly how the curvature function of marginally trapped hedgehogs of $\mathbb{L}^4$ is also involved in the volume of focal surfaces of ordinary hedgehogs of $\mathbb{R}^3$. Let $h \in C^\infty\left(\mathbb{S}^2; \mathbb{R}\right)$. The focal surface, say $\mathcal{F}_h$, of $\mathcal{H}_h$ is defined as the locus of the centers of principal curvature of $\mathcal{H}_h$ (or, which is equivalent, as the envelope of its normal lines). In [101], we defined the volume of $\mathcal{F}_h$ by:

$$v(\nabla h) := -\int_{\mathbb{R}^3} i_{\mathcal{F}_h}(x)\,dx,$$

where $i_{\mathcal{F}_{\nabla h}}(x) := 1 - \frac{1}{2}N_{\nabla h}(x)$, denoting by $N_{\nabla h}(x)$ the number of oriented normal lines to $\mathcal{H}_h$ through $x$. Note that $\mathcal{F}_h$ and hence its volume only depend on $\nabla h$. In fact, from [101, Theorems 2 and 3], we have:

$$v(\nabla h) = \frac{4}{3}\int_{\mathbb{S}^2}\left(-R_{\overrightarrow{\partial h}}\right)^{\frac{3}{2}}d\sigma$$

and the geometric inequality

$$4L(\partial h)^3 \le 9\pi\, v(\nabla h)^2.$$

## Minkowski Problem for Marginally Trapped Hedgehogs

When restricting to the class of convex bodies of $\mathbb{R}^{n+1}$ whose surface area measures have a density with respect to spherical Lebesgue measure, the classical Minkowski problem can be formulated as that of the existence, uniqueness and regularity of convex bodies of this class whose curvature function is prescribed (see e.g., [167, Section 8.2] for more details, results and a complete bibliography). This Minkowski problem admits a natural extension to hedgehogs of $\mathbb{R}^{n+1}$, which we considered in [103]. The extension to hedgehogs is much more difficult since it involves the study of a Monge-Ampère equation of mixed type (instead of a Monge-Ampère equation of elliptic type in the convex case). Since we have defined the curvature function $R_{\overrightarrow{\partial h}}$ of any marginally trapped hedgehog $\mathcal{H}_{\partial h}$ of $\mathbb{L}^4$ by analogy to the cases of hedgehogs and convex bodies of $\mathbb{R}^3$, it is now natural to consider the analogue of the Minkowski problem for marginally trapped hedgehogs (modulo some slight changes in the statement, as in the Minkowski problem for ordinary hedgehogs, due for instance to the fact that for any $h \in C^\infty\left(\mathbb{S}^2; \mathbb{R}\right)$, $\partial\left(-h\right)$ and $\partial h$ are the respective support differentials of two marginally trapped hedgehogs $\mathcal{H}_{\partial(-h)}$ and $\mathcal{H}_{\partial h}$ of $\mathbb{L}^4$ that have the same curvature function $R_{\overrightarrow{\partial(-h)}} = R_{\overrightarrow{\partial h}}$ and are such that

$$\mathcal{H}_{\partial(-h)} = s\left(\mathcal{H}_{\partial h}\right),$$

where $s$ is the symmetry with respect to the origin in $\mathbb{L}^4$).

This problem is certainly very difficult since to solve it properly, it would be necessary in particular to know if any convex closed and sufficiently smooth surface of $\mathbb{R}^3$ admits at least two umbilical points, that is, to have a complete solution to the Carathéodory conjecture.

# Chapter 9
# Focal of Hedgehogs in $\mathbb{R}^{n+1}$ and Concurrent Normals Conjecture

Unless explicitly states otherwise, the results of this section are essentially taken from [114].[1]

## 9.1 Basics on Focal of Hedgehogs in $\mathbb{R}^{n+1}$

### 9.1.1 Focal and Normal Lines of a $C^\infty$-hedgehog of $\mathbb{R}^{n+1}$

Let $\mathcal{H}_h$ be a $C^\infty$-hedgehog of $\mathbb{R}^{n+1}$. For all $u \in \mathbb{S}^n$, the normal line to $\mathcal{H}_h$ at $x_h(u)$ is defined to be the line passing through $x_h(u)$ and oriented by $u$; this normal line $\mathcal{N}_{\nabla h}(u) := \{\nabla h(u)\} + \mathbb{R}u$ is the perpendicular to the support hyperplane to $\mathcal{H}_h$ at $x_h(u)$. The focal set (or evolute) of $\mathcal{H}_h$ is the locus of its centers of principal curvatures, or equivalently, the envelope of its normal lines. This set is also the union of all the singular points of all the parallel hedgehogs $\mathcal{H}_{h+r}$, $(r \in \mathbb{R})$. Since it only depends on the gradient of $h$, we will denote it by $\mathcal{F}_{\nabla h}$. The focal (or evolute) $\mathcal{F}_{\nabla h}$ of $\mathcal{H}_h$ is the singular hypersurface of $\mathbb{R}^{n+1}$ formed by all the principal centers of curvature of $\mathcal{H}_h$ and it consists of $n$ sheets $\mathcal{F}_{\nabla h}^1, \ldots, \mathcal{F}_{\nabla h}^n$ corresponding respectively to the principal radii of curvature $R_h^1, \ldots, R_h^n$ of $\mathcal{H}_h$, which we label so that $R_h^1 \leq R_h^2 \leq \ldots \leq R_h^n$. For all $k \in [1, n]$, the sheet $\mathcal{F}_{\nabla h}^k$ can be parametrized by

$$c_{\nabla h}^k : \mathbb{S}^n \to \mathbb{R}^{n+1}, u \mapsto c_{\nabla h}^k(u) = x_h(u) - R_h^k(u)u = \nabla h(u) - \lambda_{\nabla h}^k(u)u,$$

---

[1] This article was published in Mathematika 68. Y. Martinez-Maure, On the concurrent normals conjecture for convex bodies, pp. 620–650. Copyright Wiley 2022.

© The Author(s), under exclusive license to Springer Nature Switzerland AG 2026
Y. Martinez-Maure, *Hedgehog Theory*, Lecture Notes in Mathematics 2381,
https://doi.org/10.1007/978-3-032-04298-9_9

since $x_h(u) = \nabla h(u) + h(u)u$ and $R_h^i(u) = \left(\lambda_{\nabla h}^k + h\right)(u)$, where $\nabla h(u)$ is the gradient of $h$ at $u$, and $\lambda_{\nabla h}^1(u), \ldots, \lambda_{\nabla h}^n(u)$ are the respective eigenvalues of the Hessian of $\mathcal{H}_h$ at $u$.

In most of the papers on concurrent normals to a convex body $K$ with a smooth boundary $\partial K$ in $\mathbb{R}^{n+1}$, the evolute of $\partial K$ is also regarded as the complement of the set of points $x \in \mathbb{R}^{n+1}$ such that the square of the distance function from $x$ induces a Morse function on $\partial K$:

$$d_x : \quad \partial K \to \mathbb{R}$$
$$y \mapsto \|x - y\|^2,$$

where $\|.\| : \mathbb{R}^{n+1} \to \mathbb{R}_+$ is the Euclidean norm.

In this chapter, we will adopt another point of view. For any $x \in \mathbb{R}^{n+1}$, we will consider the support function of $\mathcal{H}_h$ with respect to $x$, that is $h_x : \mathbb{S}^n \to \mathbb{R}$, $u \mapsto h(u) - \langle x, u \rangle$, and we will regard the evolute $\mathcal{F}_{\nabla h}$ as the complement of the set of points $x \in \mathbb{R}^{n+1}$ such that $h_x : \mathbb{S}^n \to \mathbb{R}$ is a Morse function. In other words, we will regard the evolute of $\mathcal{H}_h$ as the subset $\mathcal{F}_{\nabla h}$ of $\mathbb{R}^{n+1}$ on which the number and nature of the critical points of $h_x$ change. We will make use of singularity theory and Morse theory viewpoints in order to describe and summarize these changes. In particular, we will make essential use of the following proposition.

**Proposition 9.1.1** *Let $\mathcal{H}_h$ be a $C^\infty$-hedgehog of $\mathbb{R}^{n+1}$, and let $x \in \mathbb{R}^{n+1}\backslash\mathcal{F}_{\nabla h}$. For all $u \in \mathbb{S}^n$, the normal line $\mathcal{N}_{\nabla h}(u)$ to $\mathcal{H}_h$ at $x_h(u)$ is passing through $x$ if, and only if, $\nabla(h_x)(u) = 0$. Thus, the number of normal lines to $\mathcal{H}_h$ passing through $x$ is given by*

$$N_{\nabla h}(x) = \#\left\{u \in \mathbb{S}^n \,|\, \nabla(h_x)(u) = 0\right\}$$

***Proof of Proposition 9.1.1*** The point $\{x\}$ can be regarded as the hedgehog with support function $l_x : \mathbb{S}^n \to \mathbb{R}$, $u \mapsto \langle x, u \rangle$, and for all $u \in \mathbb{S}^n$, we have:

$$x = l_x(u)u + \nabla(l_x)(u) = \langle x, u \rangle u + \pi_{u^\perp}(x),$$

where $\pi_{u^\perp}(x)$ is the orthogonal projection of $x$ onto the linear hyperplane that is orthogonal to $u$. Therefore, the hedgehog $\mathcal{H}_{h_x}$ is the Minkowski difference $\mathcal{H}_h - \{x\}$, and for all $u \in \mathbb{S}^n$, we have:

$$x_{h_x}(u) = h_x(u)u + \nabla(h_x)(u) = x_h(u) - x = h_x(u)u + \left(\nabla h(u) - \pi_{u^\perp}(x)\right),$$

so that $\nabla(h_x)(u) = \nabla h(u) - \pi_{u^\perp}(x)$. Now, for all $u \in \mathbb{S}^n$, we thus have:

$$(x \in \mathcal{N}_{\nabla h}(u)) \Leftrightarrow (\exists r \in \mathbb{R}, x_h(u) + ru = x)$$
$$\Leftrightarrow \left(\exists r \in \mathbb{R}, (h_x(u) + r)u + \nabla h(u) - \pi_{u^\perp}(x) = 0\right)$$
$$\Leftrightarrow \left(\nabla h(u) - \pi_{u^\perp}(x) = 0\right) \Leftrightarrow (\nabla(h_x)(u) = 0).$$

∎

### *9.1.2  Singularity and Morse Theories Viewpoints*

In this subsubsection, we essentially follow [151, 9.1]. Given $h \in C^\infty (\mathbb{S}^n; \mathbb{R})$, we consider the following family of functions

$$
\begin{aligned}
F_h : \quad & U\mathbb{R}^{n+1} := \mathbb{R}^{n+1} \times \mathbb{S}^n \to \mathbb{R} \\
& (x, u) \mapsto h_x (u) = h (u) - \langle x, u \rangle .
\end{aligned}
$$

The critical set (or catastrophe manifold) of $F_h$ is the set

$$
C_{\nabla h} = \left\{ (x, u) \in U\mathbb{R}^{n+1} \left| \frac{\partial F_h}{\partial u} (x, u) = 0 \quad i.e. \quad \nabla (h_x) (u) = 0 \right. \right\} .
$$

The catastrophe map $\chi := \chi_{\nabla h}$ is the restriction to the critical set $C_{\nabla h}$ of the projection $\pi : U\mathbb{R}^{n+1} \to \mathbb{R}^{n+1}$, $(x, u) \mapsto x$. The singularity set $\mathcal{S}_{\nabla h}$ is the set of points of $C_{\nabla h}$ at which the catastrophe map $\chi$ has rank less than $n + 1$:

$$
\begin{aligned}
\mathcal{S}_{\nabla h} &:= \left\{ (x, u) \in C_{\nabla h} \left| rk \left[ T_{(x,u)} \chi \right] < N + 1 \right. \right\} \\
&= \left\{ (x, u) \in C_{\nabla h} \left| \exists k \in [1, n] , x = c^k_{\nabla h} (u) \right. \right\} \\
&= \left\{ (x, u) \in C_{\nabla h} \, | \exists k \in [1, n], \, 0 = \quad c^k_{\nabla h} (u) - x = c^k_{\nabla (h_x)} (u) \right. \\
&\qquad\qquad\qquad\qquad\qquad\qquad = \underbrace{\nabla (h_x) (u)}_{=0} - \lambda^k_{\nabla (h_x)} (u) u \Big\} \\
&= \left\{ (x, u) \in C_{\nabla h} \left| \exists k \in [1, n] , \lambda^k_{\nabla (h_x)} (u) = 0 \right. \right\} \\
&= \left\{ (x, u) \in C_{\nabla h} \left| \left( \det \circ \nabla^2 \right) (h_x) (u) = 0 \right. \right\} ,
\end{aligned}
$$

where $\nabla^2$ is the Hessian operator and thus $M = \det \circ \nabla^2$ the Monge-Ampère one. It is thus the set of points $(x, u) \in C_{\nabla h}$ at which $F_h : U\mathbb{R}^{n+1} \to \mathbb{R}$ has a degenerate critical point. The bifurcation set of $F_h : U\mathbb{R}^{n+1} \to \mathbb{R}$ is defined to be the image of the singularity set $\mathcal{S}_{\nabla h}$ under the catastrophe map $\chi$. It is none other than the focal $\mathcal{F}_{\nabla h}$ of the hedgehog $\mathcal{H}_h$ in $\mathbb{R}^{n+1}$ (and of any hedgehog that is parallel to $\mathcal{H}_h$, that is of the form $\mathcal{H}_{h+\lambda}$, where $\lambda \in \mathbb{R}$):

$$
\chi (\mathcal{S}_{\nabla h}) = \mathcal{F}_{\nabla h} = \bigcup_{k=1}^{n} \mathcal{F}^k_{\nabla h} ,
$$

where
$$\mathcal{F}_{\nabla h}^k := c_{\nabla h}^k \left(\mathbb{S}^n\right) = \left\{ x \in \mathbb{R}^{n+1} \,\middle|\, \exists u \in \mathbb{S}^n, \, x = c_{\nabla h}^k (u) \right\}$$
$$= \left\{ x \in \mathcal{F}_{\nabla h} \,\middle|\, \exists u \in \mathbb{S}^n, \, \nabla (h_x) (u) = 0 \text{ and } \lambda_{\nabla (h_x)}^k (u) = 0 \right\}.$$

Note that

$$\mathbb{R}^{n+1} \setminus \mathcal{F}_{\nabla h} = \left\{ x \in \mathbb{R}^{n+1} \,\middle|\, h_x : \mathbb{S}^n \to \mathbb{R} \text{ is a Morse function} \right\},$$

while the bifurcation set $\mathcal{F}_{\nabla h} = \chi \left(\mathcal{S}_{\nabla h}\right)$ is the subset of $\mathbb{R}^{n+1}$ on which the number and nature of the critical points of $h_x : \mathbb{S}^n \to \mathbb{R}$ change (for by structural stability of Morse functions such a change can only occur passing through a degenerate critical point). This can be of course be checked by a direct computation.

**Index of a Point with Respect to the Focal**  Of course, every $x$ belonging to the unbounded connected component of $\mathbb{R}^{n+1} \setminus \mathcal{F}_{\nabla h}$ lies on exactly two normal lines to $\mathcal{H}_h$. We have proved in Proposition 9.1.1 that, for every $x \in \mathbb{R}^{n+1} \setminus \mathcal{F}_{\nabla h}$, the number of normal lines to $\mathcal{H}_h$ passing through $x$ is given by:

$$N_{\nabla h} (x) = \# \left\{ u \in \mathbb{S}^n \,|\, \nabla (h_x) (u) = 0 \right\}.$$

This number $N_{\nabla h} (x)$ is constant and even on any connected component of $\mathbb{R}^{n+1} \setminus \mathcal{F}_{\nabla h}$ [66, Theorem 4], and it suddenly changes by two units every time $x$ transversally crosses $\mathcal{F}_{\nabla h}$ at a simple regular point, so that $N_{\nabla h} (x)$ is even for any $x \in \mathbb{R}^{n+1} \setminus \mathcal{F}_{\nabla h}$. It is thus natural to define, for every $x \in \mathbb{R}^{n+1} \setminus \mathcal{F}_{\nabla h}$, the index $i_{\mathcal{F}_{\nabla h}} (x)$ of $x$ with respect to $\mathcal{F}_{\nabla h}$ by putting:

$$i_{\mathcal{F}_{\nabla h}} (x) := 1 - \frac{1}{2} N_{\nabla h} (x),$$

This index induces a transverse orientation of the regular part of $\mathcal{F}_{\nabla h}$ (and thus of any of its $n$ sheets): $\mathcal{F}_{\nabla h}$ is transversely oriented so that the number of normal lines to $\mathcal{H}_h$ passing through $x$ increases by two units when $x$ transversally crosses $\mathcal{F}_{\nabla h}$ at a simple regular point in the direction of the transverse orientation.

Let $x \in \mathbb{R}^{n+1} \setminus \mathcal{F}_{\nabla h}$ so that $h_x : \mathbb{S}^n \to \mathbb{R}$ is a Morse function. Denote by $C_m (x)$ and $C_M (x)$ the respective numbers of critical points of index $0$ and $n$ of this Morse function (that is, the respective numbers of its local minima and maxima), and, for every $i \in [1, n-1]$, denote by $S_i (x)$ the number of its critical points of index $i$. We know that the number of normal lines to $\mathcal{H}_h$ passing through $x$ is then given by:

$$N_{\nabla h} (x) = \# \left( \left\{ u \in \mathbb{S}^n \,|\, \nabla (h_x) (u) = 0 \right\} \right)$$
$$= C_m (x) + \sum_{i=1}^{n-1} S_i (x) + C_M (x) \tag{9.1.1}$$

Furthermore, by virtue of the Morse-Euler relationship, we have:

$$C_m(x) + \sum_{i=1}^{n-1} (-1)^i S_i(x) + (-1)^n C_M(x) = \chi(\mathbb{S}^n), \qquad (9.1.2)$$

where $\chi(\mathbb{S}^n)$ is the Euler characteristic of $\mathbb{S}^n$ (i.e., $1 + (-1)^n$).

From (9.1.1) and (9.1.2), we can immediately deduce that:

$$i_{\mathcal{F}_{\nabla h}}(x) := \begin{cases} -\sum_{l=1}^{p} S_{2l-1}(x) & \text{if} \quad n = 2p \in 2\mathbb{Z}^* \\[2mm] 1 - \left(C_m + \sum_{l=1}^{p} S_{2l}\right)(x), \ or \ equivalently, \\[2mm] 1 - \left(C_M + \sum_{l=0}^{p-1} S_{2l+1}\right)(x) & if \quad n - 1 = 2p \in 2\mathbb{Z}^*, \end{cases} \qquad (9.1.3)$$

**Index Decomposition**  For any $k \in [1, n]$, assume that $x$ moves and transversally crosses the $k$th sheet $\mathcal{F}_{\nabla h}^k$ of $\mathcal{F}_{\nabla h}$ at a simple regular point of $\mathcal{F}_{\nabla h}^k$ in the direction of the transverse orientation. We know this crossing of $\mathcal{F}_{\nabla h}^k$ results in a two-units increase in the number of normal lines to $\mathcal{H}_h$ passing through $x$ (i.e., in the number of critical points of $h_x : \mathbb{S}^n \to \mathbb{R}$). From the Morse-Euler relationship, the two new critical points of $h_x$ cannot be of the same index (since the two indices must have different parities). Since, moreover, $\mathcal{F}_{\nabla h}^k$ is given by

$$\mathcal{F}_{\nabla h}^k = \left\{ x \in \mathcal{F}_{\nabla h} \,\middle|\, \exists u \in \mathbb{S}^n, \, \nabla(h_x)(u) = 0 \text{ and } \lambda_{\nabla(h_x)}^k(u) = 0 \right\},$$

it appears that the two critical points of $h_x : \mathbb{S}^n \to \mathbb{R}$ that arise at the moment of the crossing of $\mathcal{F}_{\nabla h}^k$ have adjacent indices $k - 1$ and $k$, the sign of the function $\lambda_{\nabla(h_x)}^k :$ $\mathbb{S}^n \to \mathbb{R}$ being different at these two points. Therefore, this crossing of $\mathcal{F}_{\nabla h}^k$ has the effect of transforming the vector $v(x) = (C_m(x), S_1(x), \ldots, S_{n-1}(x), C_M(x))$ into the vector $v(x) + e_{k-1} + e_k$, where $(e_0, \ldots, e_n)$ is the canonical basis of $\mathbb{R}^{n+1}$.

Now, let us distinguish two cases according to the parity of $n - 1$. In both cases, and for every $x \in \mathbb{R}^{n+1} \setminus \mathcal{F}_{\nabla h}$, we will split the index $i_{\mathcal{F}_{\nabla h}}(x)$ of $x$ with respect to the focal $\mathcal{F}_{\nabla h}$ into the sum over $k \in [1, n]$ of the (appropriately defined) index $i_{\mathcal{F}_{\nabla h}^k}(x)$ of $x$ with respect to the $k$th sheet $\mathcal{F}_{\nabla h}^k$ of $\mathcal{F}_{\nabla h}$.

In the case that $n = 2p \in 2\mathbb{Z}^*$, define the indices $i_{\mathcal{F}_{\nabla h}^k}(x)$, $(k \in [1, 2p])$, by

$$
\begin{cases}
i_{\mathcal{F}_{\nabla h}^1}(x) = 1 - C_m(x) \\[2ex]
i_{\mathcal{F}_{\nabla h}^2}(x) = (C_m - S_1)(x) - 1 \\[1ex]
\quad\vdots \\[1ex]
i_{\mathcal{F}_{\nabla h}^p}(x) = (-1)^{p-1}\left(1 - \left(C_m - S_1 + \ldots + (-1)^{p-1} S_{p-1}(x)\right)\right) \\[2ex]
i_{\mathcal{F}_{\nabla h}^{p+1}}(x) = (-1)^{p-1}\left(1 - \left(C_M - S_{2p-1} + \ldots + (-1)^{p-1} S_{p+1}(x)\right)\right) \\[1ex]
\quad\vdots \\[1ex]
i_{\mathcal{F}_{\nabla h}^{2p-1}}(x) = \left(C_M - S_{2p-1}\right)(x) - 1 \\[2ex]
i_{\mathcal{F}_{\nabla h}^{2p}}(x) = 1 - C_M(x)
\end{cases}
$$

for all $x \in \mathbb{R}^{2p+1} \setminus \mathcal{F}_{\nabla h}$, so that when $x$ moves and transversally crosses the $k$th sheet $\mathcal{F}_{\nabla h}^k$ of $\mathcal{F}_{\nabla h}$ at a simple regular point of $\mathcal{F}_{\nabla h}^k$ in the direction of the transverse orientation, then the index $i_{\mathcal{F}_{\nabla h}^k}(x)$ decreases by one unit. Thus, for every $x \in \mathbb{R}^{2p+1} \setminus \mathcal{F}_{\nabla h}$, $i_{\mathcal{F}_{\nabla h}^k}(x)$ can be interpreted as the index of $x$ with respect to $\mathcal{F}_{\nabla h}^k$ equipped with its transverse orientation.

Note that, for every $x \in \mathbb{R}^{2p+1} \setminus \mathcal{F}_{\nabla h}$, the $2p$ strong Morse inequalities, which give lower bounds for the number of critical points of each index of the Morse function $h_x : \mathbb{S}^{2p} \to \mathbb{R}$ in terms of the Betti numbers of $\mathbb{S}^{2p}$, can simply be rewritten as: $\forall k \in [1, 2p]$, $i_{\mathcal{F}_{\nabla h}^k}(x) \leq 0$. In other words, these $2p$ indices are nonpositive, and, for every $k \in [1, 2p]$, the index $i_{\mathcal{F}_{\nabla h}^k}(x)$ negatively measures how far from equality we are in the $k$th strong Morse inequality when considering the Morse function $h_x : \mathbb{S}^{2p} \to \mathbb{R}$.

Similarly, if $n - 1 = 2p \in 2\mathbb{Z}^*$, define the indices $i_{\mathcal{F}_{\nabla h}^k}(x)$, $(k \in [1, 2p + 1])$, by

$$
\begin{cases}
i_{\mathcal{F}_{\nabla h}^1}(x) = 1 - C_m(x) \\[2ex]
i_{\mathcal{F}_{\nabla h}^2}(x) = (C_m - S_1)(x) - 1 \\[1ex]
\quad\vdots \\[1ex]
i_{\mathcal{F}_{\nabla h}^{p+1}}(x) = (-1)^p \left(1 - \left(C_m + \sum_{l=1}^{p}(-1)^l S_l(x)\right)\right) \\[1ex]
\quad\vdots \\[1ex]
i_{\mathcal{F}_{\nabla h}^{2p}}(x) = \left(C_M - S_{2p-1}\right)(x) - 1 \\[2ex]
i_{\mathcal{F}_{\nabla h}^{2p+1}}(x) = 1 - C_M(x).
\end{cases}
$$

for all $x \in \mathbb{R}^{2p+2} \setminus \mathcal{F}_{\nabla h}$, so that when $x$ moves and transversally crosses the $k$th sheet $\mathcal{F}_{\nabla h}^k$ of $\mathcal{F}_{\nabla h}$ at a simple regular point of $\mathcal{F}_{\nabla h}^k$ in the direction of the transverse orientation, then the index $i_{\mathcal{F}_{\nabla h}^k}(x)$ decreases by one unit. Thus, for every $x \in \mathbb{R}^{2p+2} \setminus \mathcal{F}_{\nabla h}$, $i_{\mathcal{F}_{\nabla h}^k}(x)$ can be interpreted as the index of $x$ with respect to $\mathcal{F}_{\nabla h}^k$ equipped with its transverse orientation.

Again, for every $x \in \mathbb{R}^{2p+2} \setminus \mathcal{F}_{\nabla h}$, the $2p + 1$ strong Morse inequalities, which give lower bounds for the number of critical points of each index of the Morse function $h_x : \mathbb{S}^{2p+1} \to \mathbb{R}$ in terms of the Betti numbers of $\mathbb{S}^{2p+1}$, can simply be rewritten as: $\forall k \in [1, 2p + 1]$, $i_{\mathcal{F}_{\nabla h}^k}(x) \leq 0$. These $2p + 1$ indices are nonpositive, and, for every $k \in [1, 2p + 1]$, the index $i_{\mathcal{F}_{\nabla h}^k}(x)$ negatively measures how far from equality we are in the $k$th strong Morse inequality when considering the Morse function $h_x : \mathbb{S}^{2p+1} \to \mathbb{R}$.

In both cases ($n$ even or odd), using the Morse-Euler relationship we can verify that:

**Proposition 9.1.2** *For every $x \in \mathbb{R}^{n+1} \setminus \mathcal{F}_{\nabla h}$, we have*

$$i_{\mathcal{F}_{\nabla h}}(x) = \sum_{k=1}^{n} i_{\mathcal{F}_{\nabla h}^k}(x),$$

*and, for every $k \in [1, n]$, $i_{\mathcal{F}_{\nabla h}^k}(x)$ can be interpreted as the index of $x$ with respect to the $k$th sheet $\mathcal{F}_{\nabla h}^k$ of $\mathcal{F}_{\nabla h}$ equipped with its transverse orientation; the index $i_{\mathcal{F}_{\nabla h}^k}(x)$ negatively measures how far from equality we are in the $k$th strong Morse inequality when considering the Morse function $h_x : \mathbb{S}^n \to \mathbb{R}$, ($k \in [1, n]$).*

We define *the **interior*** and the ***body*** *of the focal* $\mathcal{F}_{\nabla h}$ by setting, respectively,

$$Int\left(\mathcal{F}_{\nabla h}\right) := \left\{ x \in \mathbb{R}^{n+1} \setminus \mathcal{F}_{\nabla h} \,\middle|\, i_{\mathcal{F}_{\nabla h}}(x) < 0 \right\},$$

and

$$\mathcal{K}\left(\mathcal{F}_{\nabla h}\right) := \mathcal{F}_{\nabla h} \cup Int\left(\mathcal{F}_{\nabla h}\right).$$

Of course, for every $k \in [1, n]$, we can also define the **interior** and the **body** of the $k$th sheet $\mathcal{F}_{\nabla h}^k$ of the focal by setting, respectively,

$$Int\left(\mathcal{F}_{\nabla h}^k\right) := \left\{ x \in \mathbb{R}^{n+1} \setminus \mathcal{F}_{\nabla h}^k \,\middle|\, i_{\mathcal{F}_{\nabla h}^k}(x) < 0 \right\},$$

where $i_{\mathcal{F}_{\nabla h}^k}(x)$ is defined as indicated above, and

$$\mathcal{K}\left(\mathcal{F}_{\nabla h}^k\right) := \mathcal{F}_{\nabla h}^k \cup Int\left(\mathcal{F}_{\nabla h}^k\right).$$

It has been proved that arbitrarily close to the center of the minimal spherical shell of a convex body $K \subset \mathbb{R}^{n+1}$ with support function $h \in C^\infty (\mathbb{S}^n; \mathbb{R})$, there exist points $x \in \mathbb{R}^{n+1} \backslash \mathcal{F}_{\nabla h}$ such that $C_m (x) \geq 2$ and $C_M (x) \geq 2$ [66, Lemma 3], so that $i_{\mathcal{F}_{\nabla h}^1} (x) = 1 - C_m (x) \leq -1$ and $i_{\mathcal{F}_{\nabla h}^n} (x) = 1 - C_M (x) \leq -1$. In other words, we have:

$$Int \left( \mathcal{F}_{\nabla h}^1 \right) \cap Int \left( \mathcal{F}_{\nabla h}^n \right) \neq \varnothing.$$

### 9.1.3  Singular Locus of the Focal $\mathcal{F}_{\nabla h}$ of $\mathcal{H}_h \subset \mathbb{R}^{n+1}$

Here $\mathcal{H}_h$ is a hedgehog of $\mathbb{R}^{n+1}$ such that $h \in C^\infty (\mathbb{S}^n; \mathbb{R})$. When two principal radii of curvature $R_h^k$, $R_h^l$ of $\mathcal{H}_h$ coincide at a point $u \in \mathbb{S}^n$, then the corresponding sheets $\mathcal{F}_{\nabla h}^k$, $\mathcal{F}_{\nabla h}^l$ of $\mathcal{F}_{\nabla h}$ intersect at $c_{\nabla h}^k (u) = c_{\nabla h}^l (u)$, and this point is a singular point of both sheets $\mathcal{F}_{\nabla h}^k$, $\mathcal{F}_{\nabla h}^l$. When all the principal radii of curvature are pairwise distinct at $u \in \mathbb{S}^n$, the focal $\mathcal{F}_{\nabla h}$ is locally the union of $n$ disjoint patches of hypersurfaces, parametrized by the maps $c_{\nabla h}^k : \mathbb{S}^n \to \mathbb{R}^{n+1}$, $u \mapsto x_h (u) - R_h^k (u) u$, $(k \in [1, n])$. Let us examine the regularity of these patches at their point corresponding to $u$. In the case that the principal radii of curvature are pairwise distinct at $u \in \mathbb{S}^n$, there exists an orthonormal basis of $T_u \mathbb{S}^n$ consisting of eigenvectors $v_1, \ldots, v_n$ of the tangent map $T_u x_h$ associated respectively with $R_h^1 (u), \ldots, R_h^n (u)$: $\forall k \in [1, n]$, $(T_u x_h) (v_k) = R_h^k (u) v_k$. For every $k \in [1, n]$, we have then:

$$\frac{\partial c_{\nabla h}^k}{\partial v_k} (u) = (T_u x_h) (v_k) - \left( \frac{\partial R_h^k}{\partial v_k} (u) u + R_h^k (u) v_k \right) = - \frac{\partial R_h^k}{\partial v_k} (u) u,$$

and for every $l \in [1, n] \backslash \{k\}$,

$$\frac{\partial c_{\nabla h}^k}{\partial v_l} (u) = (T_u x_h)(v_l) - \left( \frac{\partial R_h^k}{\partial v_l} (u) u + R_h^k (u) v_l \right)$$
$$= \left( R_h^l - R_h^k \right)(u) v_l - \frac{\partial R_h^k}{\partial v_l} (u) u.$$

Denote by $\mathcal{B}$ the orthonormal system $(u, v_1, \ldots, \widehat{v_k}, \ldots, v_n)$, where the hat over the term $v_k$ means that it must be omitted. A straightforward computation shows that

$$\det_{\mathcal{B}} \left[ \left( \frac{\partial c_{\nabla h}^k}{\partial v_k}, \frac{\partial c_{\nabla h}^k}{\partial v_1}, \ldots, \frac{\widehat{\partial c_{\nabla h}^k}}{\partial v_k}, \ldots, \frac{\partial c_{\nabla h}^k}{\partial v_n} \right) (u) \right],$$

where the hat means again that the corresponding term must be omitted, is equal to

$$-\frac{\partial R_h^k}{\partial v_k}(u) \prod_{\substack{1 \leq l \leq n \\ l \neq k}} \left(R_h^l - R_h^k\right)(u).$$

Thus:

**Proposition 9.1.3** *For every $k \in [1, n]$ and every $u \in \mathbb{S}^n$, $c_{\nabla h}^k(u)$ is a singular point of $\mathcal{F}_{\nabla h}^k$ if, and only if, one of the following two conditions is satisfied:*

- $R_h^k(u)$ *is an eigenvalue of multiplicity at least 2 of the tangent map $T_u x_h$;*
- $R_h^k(u)$ *is a simple eigenvalue of $T_u x_h$, and $\dfrac{\partial R_h^k}{\partial v_k}(u) = 0$, $v_k$ being a unit eigenvector of $T_u x_h$ associated with $R_h^k(u)$.*

For every $k \in [1, n]$, the singular locus of the $k$th sheet $\mathcal{F}_{\nabla h}^k$ of $\mathcal{F}_{\nabla h}$ is thus given by:

$$\mathcal{S}ing\left(\mathcal{F}_{\nabla h}^k\right) = c_{\nabla h}^k(S_k),$$

where $S_k$ is the set of points $u \in \mathbb{S}^n$ that satisfy one of the two conditions of the above proposition; and the singular locus of the focal $\mathcal{F}_{\nabla h}$ is of course defined by:

$$\mathcal{S}ing\left(\mathcal{F}_{\nabla h}\right) := \bigcup_{k=1}^{n} \mathcal{S}ing\left(\mathcal{F}_{\nabla h}^k\right).$$

**The Three-Dimensional Case**

**Corollary 9.1.1** *Let $\mathcal{H}_h$ be a hedgehog of $\mathbb{R}^3$ such that $h \in C^\infty\left(\mathbb{S}^2; \mathbb{R}\right)$, and let $u \in \mathbb{S}^2$ be such that*

$$R_h^1(u) = 0, \quad R_h^2(u) \neq 0 \quad and \quad \frac{\partial R_h^1}{\partial v_1}(u) \neq 0$$

$$\left(resp. \quad R_h^2(u) = 0, \quad R_h^1(u) \neq 0 \quad and \quad \frac{\partial R_h^2}{\partial v_2}(u) \neq 0\right),$$

*where $(v_1, v_2)$ is an orthonormal basis of $T_u \mathbb{S}^2$ made of eigenvectors $v_1, v_2$ of $T_u x_h$ associated respectively with $R_h^1(u)$, $R_h^2(u)$. Then, $x_h(u)$ is equal to $c_{\nabla h}^1(u)$ (resp. $c_{\nabla h}^2(u)$), which is a regular point of $\mathcal{F}_{\nabla h}^1$ (resp. $\mathcal{F}_{\nabla h}^2$), and a regular point of a cuspidal edge of $\mathcal{H}_h$.*

***Proof of Corollary 9.1.1*** The first part of the corollary is a straightforward consequence of the proposition. Now, let $(i, j) = (1, 2)$ (resp. $(i, j) = (2, 1)$) so that:

$$R_h^i(u) = 0, \quad R_h^j(u) \neq 0 \quad \text{and} \quad \frac{\partial R_h^i}{\partial v_i}(u) \neq 0.$$

Since $R_h(u) = 0$ and $\nabla R_h(u) \neq 0$, the level set $R_h$ can be parametrized as a regular smooth curve $\Gamma$ in a neighborhood of $u$ on $\mathbb{S}^2$. Let $\gamma : I \to \mathbb{S}^2, t \mapsto \gamma(t)$ be this regular parametrization of $\Gamma$, and let $t_0 \in I$ be such that $\gamma(t_0) = u$ and $\gamma'(t_0) = \lambda_1 v_1 + \lambda_2 v_2$, where $(\lambda_1, \lambda_2) \in \mathbb{R}^2 \setminus \{(0, 0)\}$. Since $R_h \circ \gamma$ is identically equal to zero on $I$, we have $(R_h \circ \gamma)'(t_0) = \langle \nabla R_h(u), \gamma'(t_0) \rangle = 0$. On the other hand, $\frac{\partial R_h^i}{\partial v_i}(u) := \langle \nabla R_h(u), v_i \rangle \neq 0$. Therefore $\gamma'(t_0)$ is not colinear to $v_i$, and thus $\lambda_j \neq 0$. As a result $(x_h \circ \gamma)'(t_0) = (T_u x_h)(\gamma'(t_0)) = \lambda_i R_h^i(u) v_i + \lambda_j R_h^j(u) v_j \neq 0$, that is $u = \gamma(t_0)$ is a regular point of the cuspidal edge $x_h(\Gamma)$.   ∎

Note in passing that any cuspidal edge of $\mathcal{H}_h$ is locally separating a hyperbolic region from an elliptic one.

### 9.1.4   *Volume of the Focal $\mathcal{F}_{\nabla h}$ of $\mathcal{H}_h$ in $\mathbb{R}^{n+1}$*

Here again $\mathcal{H}_h$ is a hedgehog of $\mathbb{R}^{n+1}$ such that $h \in C^\infty(\mathbb{S}^n; \mathbb{R})$. We can define the (absolute) volume of its focal $\mathcal{F}_{\nabla h}$, and, for any $k \in [1, n]$ the one of the $k$th sheet $\mathcal{F}_{\nabla h}^k$ of $\mathcal{F}_{\nabla h}$, to be respectively

$$v(\mathcal{F}_{\nabla h}) := -\int_{\mathbb{R}^{n+1}} i_{\mathcal{F}_{\nabla h}}(x)\, dx, \quad \text{and} \quad v\left(\mathcal{F}_{\nabla h}^k\right) := -\int_{\mathbb{R}^{n+1}} i_{\mathcal{F}_{\nabla h}^k}(x)\, dx,$$

where the integrals are with respect to the Lebesgue measure on $\mathbb{R}^{n+1}$. By the index decomposition, we thus have:

$$v(\mathcal{F}_{\nabla h}) = \sum_{k=1}^{n} v\left(\mathcal{F}_{\nabla h}^k\right).$$

In the case where $\mathcal{H}_h \subset \mathbb{R}^3$, the volume of $\mathcal{F}_{\nabla h}$ had already been introduced by the author in [101], where the following is proved.

**Theorem ([101])** *Let $\mathcal{H}_h$ be a hedgehog of $\mathbb{R}^3$ such that $h \in C^\infty(\mathbb{S}^2; \mathbb{R})$. The volume of its focal $\mathcal{F}_{\nabla h}$ is given by*

$$v(\mathcal{F}_{\nabla h}) = \frac{1}{6} \int_{\mathbb{S}^2} |R_1 - R_2|^3 \, d\sigma,$$

*where $\sigma$ is the spherical Lebesgue measure on $\mathbb{S}^2$, and $R_1(u)$, $R_2(u)$ are the principal radii of curvature of $\mathcal{H}_h$ at $x_h(u)$. Besides, the map*

$$v : \mathcal{H} := \left\{ \nabla h \,\middle|\, h \in C^4\left(\mathbb{S}^2; \mathbb{R}\right) \right\} \to \mathbb{R}_+, \quad \nabla h \longmapsto v\left(\mathcal{F}_{\nabla h}\right)^{\frac{1}{3}}$$

*is a norm on the real vector space of families of parallel hedgehogs with support function of class $C^4$ in $\mathbb{R}^3$.*

## 9.2 Introduction to the Concurrent Normal Conjecture

It is conjectured that every convex body in $n$-dimensional Euclidean space $\mathbb{R}^n$ has an interior point lying on normals through $2n$ distinct boundary points. Using the existence of a minimal spherical shell for any convex body and, a combination of Morse theory and approximation, E. Heil has proved this concurrent normals conjecture for $n = 2$ and $n = 3$ in [64–66]. For $n = 4$, J. Pardon put forward a proof of the conjecture under a smoothness assumption on the boundary [144]. For $n \geq 5$, it is only known that any convex body in $\mathbb{R}^n$ has an interior point lying on normals through six distinct boundary points. However Zamfirescu has shown that, in the sense of Baire category based on the Hausdorff distance between convex sets, most interior points of most convex bodies lie on infinitely many normals [196]. For $n \in \{3, 4\}$, we will prove in this section that any normal through a boundary point to any convex body $K$ (with a smooth enough support function) in $\mathbb{R}^n$ passes arbitrarily close to the set of interior points of $K \cup L$ lying on normals through at least 6 distinct points of $\partial K$, where $L$ is the body bounded by the smallest convex parallel hypersurface to $\partial K$ whose unit normal points in the opposite direction. Motivated by this work, Grebennikov and Panina gave a proof of almost the same fact for any $n \geq 3$ via bifurcation theory [57].

### 9.2.1 Our Setting and Tools

In this section, we assume for the sake of simplicity of the presentation that the support function is $C^\infty$ but our results remain true provided that the support function is smooth enough (say at least of class $C^4$). Actually, our main argument essentially relies on the Morse lemma, which remains true for $C^2$-functions [138], and on elementary properties of the focal of the boundary.

In most of the papers on concurrent normals to a convex body $K$ with a smooth boundary $\partial K$ in $\mathbb{R}^{n+1}$, the focal (or evolute) of $\partial K$ is regarded as the complement of the set of points $x \in \mathbb{R}^{n+1}$ such that the square of the distance function from $x$ induces a Morse function on $\partial K$. In this section, we will adopt another point of view. For any $x \in \mathbb{R}^{n+1}$, we will consider the support function of $\partial K$ with respect to $x$, that is $h_x : \mathbb{S}^n \to \mathbb{R}$, $u \mapsto h(u) - \langle x, u \rangle$, where $h : \mathbb{S}^n \to \mathbb{R}$ is the support function of $K$, and we will regard the evolute of $\partial K$ as the complement of set of points $x \in \mathbb{R}^{n+1}$ such that $h_x : \mathbb{S}^n \to \mathbb{R}$ is a Morse function.

We will also make intensive use of hedgehogs of $\mathbb{R}^{n+1}$, which are the (possibly singular and self-intersecting) hypersurfaces of $\mathbb{R}^{n+1}$ that are parametrized by their Gauss map and parallel to some $C^2$ convex hypersurface in $\mathbb{R}^{n+1}$. For $h \in C^\infty(\mathbb{S}^n; \mathbb{R})$, and $u \in \mathbb{S}^n$, the normal line to the hedgehog $\mathcal{H}_h$ at $x_h(u)$ is defined to be the line passing through $x_h(u)$ and oriented by $u$; this normal line $\mathcal{N}_{\nabla h}(u) := \{\nabla h(u)\} + \mathbb{R}u$ is the perpendicular to the support hyperplane to $\mathcal{H}_h$ at $x_h(u)$. Recall that the focal set (or evolute) $\mathcal{F}_{\nabla h}$ of $\mathcal{H}_h$ is defined as the locus of its centers of principal curvatures, or equivalently, as the envelope of its normal lines $(\mathcal{N}_{\nabla h}(u))_{u \in \mathbb{S}^n}$.

### 9.2.2  Our Main Statements

For $n + 1 = 3$, we will prove the following result which will turn out to be a refinement of Heil's theorem.

**Theorem 9.2.1**  *Let $\mathcal{H}_h$ be a hedgehog of $\mathbb{R}^3$ such that $h \in C^\infty(\mathbb{S}^2; \mathbb{R})$, and let $\mathcal{F}_{\nabla h}$ denote its focal surface. If $u \in \mathbb{S}^2$ is such that the normal line $\mathcal{N}_{\nabla h}(u)$ to $\mathcal{H}_h$ at $x_h(u)$ does not meet the singular locus of $\mathcal{F}_{\nabla h}$, then there exists some $r \in \mathbb{R}$ such that $x_{h-r}(u) = x_h(u) - ru \in \mathcal{N}_{\nabla h}(u)$ is an (at least) double hyperbolic point of $\mathcal{H}_{h-r}$.*

Here, "$x_{h-r}(u) = x_h(u) - ru \in \mathcal{N}_{\nabla h}(u)$ is an (at least) double hyperbolic point of $\mathcal{H}_{h-r}$" means that there exists $v \in \mathbb{S}^2 \setminus \{u\}$, such that:

$$x_{h-r}(u) = x_{h-r}(v), \quad R_{h-r}(u) < 0, \quad \text{and } R_{h-r}(v) < 0$$

where $R_{h-r}$ is the curvature function of $\mathcal{H}_{h-r}$ (that is, the inverse $1/\kappa_{h-r}$ of the Gauss curvature $\kappa_{h-r}$ of $\mathcal{H}_{h-r}$).

We will deduce the following reformulation of Heil's theorem without making use of the notion of a minimal spherical shell.

**Corollary 9.2.1 (Heil's Theorem [64, 65])**  *Let $\mathcal{H}_h$ be a hedgehog of $\mathbb{R}^3$ such that $h \in C^\infty(\mathbb{S}^2; \mathbb{R})$. Either there exists a point of $\mathbb{R}^3$ lying on infinitely many normals to $\mathcal{H}_h$ or there exists an open set formed by points of $\mathbb{R}^3 \setminus \mathcal{F}_{\nabla h}$ lying on at least 6 normals to $\mathcal{H}_h$.*

We will in fact prove the following stronger result.

**Corollary 9.2.2**  *Let $\mathcal{H}_h$ be a hedgehog of $\mathbb{R}^3$ such that $h \in C^\infty(\mathbb{S}^2; \mathbb{R})$. If there does not exist a point of $\mathbb{R}^3$ lying on infinitely many normals to $\mathcal{H}_h$, then infinitely many normals to $\mathcal{H}_h$ do not meet the singular locus of $\mathcal{F}_{\nabla h}$, and any one of them* *meets the closure of*

$$\left\{ x \in \mathbb{R}^3 \backslash \mathcal{F}_{\nabla h} \,|\, N_{\nabla h}(x) \geq 6 \right\},$$

*where $N_{\nabla h}(x)$ denotes the number of normal lines to $\mathcal{H}_h$ passing through $x$.*

The focal (or evolute) $\mathcal{F}_{\nabla h}$ of $\mathcal{H}_h$ is the singular hypersurface of $\mathbb{R}^{n+1}$ formed by all the principal centers of curvature of $\mathcal{H}_h$ and it consists of $n$ sheets $\mathcal{F}_{\nabla h}^1, \ldots, \mathcal{F}_{\nabla h}^n$ corresponding respectively to the principal radii of curvature $R_h^1, \ldots, R_h^n$ of $\mathcal{H}_h$, which we label so that $R_h^1 \leq R_h^2 \leq \ldots \leq R_h^n$. In Sect. 9.1, we introduced for each $k \in [1, n]$ the index of a point $x \in \mathbb{R}^{n+1}\backslash \mathcal{F}_{\nabla h}$ with respect to the $k$th sheet $\mathcal{F}_{\nabla h}^k$ of $\mathcal{F}_{\nabla h}$, and the interior $Int\left(\mathcal{F}_{\nabla h}^k\right)$ of this sheet. Under the assumptions of Corollary 9.2.2, we will in fact prove that any normal to $\mathcal{H}_h$ that does not meet the singular locus of $\mathcal{F}_{\nabla h}$ meets the closure of the nonempty interiors of both sheets of the focal.

We will obtain the following result as a corollary of our Theorem 9.4.1 stated and proved in Sect. 9.4.

**Theorem 9.2.2** *Let $\mathcal{H}_h$ be a hedgehog of $\mathbb{R}^4$ such that $h \in C^\infty\left(\mathbb{S}^3; \mathbb{R}\right)$, let $\mathcal{F}_{\nabla h}$ denote its focal surface, and let $\mathcal{F}_{\nabla h}^2$ be its second sheet. For all $x \in \mathbb{R}^4\backslash \mathcal{F}_{\nabla h}$, we have:*

$$\left( x \in Int\left(\mathcal{F}_{\nabla h}^2\right) \right) \Longleftrightarrow \left( \begin{array}{l} h_x : \quad \mathbb{S}^3 \rightarrow \mathbb{R} \text{ admits a smooth} \\ \text{level surface with nonzero genus} \end{array} \right).$$

*The following theorem is an adaptation of Theorem 9.2.1 to dimension 4.*

**Theorem 9.2.3** *Let $\mathcal{H}_h$ be a hedgehog of $\mathbb{R}^4$ such that $h \in C^\infty\left(\mathbb{S}^3; \mathbb{R}\right)$, and let $\mathcal{F}_{\nabla h}$ denote its focal surface. If $u \in \mathbb{S}^3$ is such that the normal line $\mathcal{N}_{\nabla h}(u)$ to $\mathcal{H}_h$ at $x_h(u)$ does not meet the singular locus of $\mathcal{F}_{\nabla h}$, then there exists $(r_1, r_2) \in \mathbb{R}^2$ such that, for each $i \in [1, 2]$, $x_{h-r_i}(u) = x_h(u) - r_i u$ is an (at least) type $i$ double hyperbolic point of $\mathcal{H}_{h-r_i}$.*

**Corollary 9.2.3** *Let $\mathcal{H}_h$ be a hedgehog of $\mathbb{R}^4$ such that $h \in C^\infty\left(\mathbb{S}^3; \mathbb{R}\right)$. If there does not exist a point of $\mathbb{R}^4$ lying on infinitely many normals to $\mathcal{H}_h$, then infinitely many normals to $\mathcal{H}_h$ do not meet the singular locus of $\mathcal{F}_{\nabla h}$ and any one of them meets the closure of*

$$\left\{ x \in \mathbb{R}^4\backslash \mathcal{F}_{\nabla h} \,|\, N_{\nabla h}(x) \geq 6 \right\},$$

*where $N_{\nabla h}(x)$ denotes the number of normal lines to $\mathcal{H}_h$ passing through $x$; more precisely, any one of these normals meets the closures of*

$$Int\left(\mathcal{F}_{\nabla h}^1\right) \cap Int\left(\mathcal{F}_{\nabla h}^2\right) \qquad and \qquad Int\left(\mathcal{F}_{\nabla h}^2\right) \cap Int\left(\mathcal{F}_{\nabla h}^3\right).$$

Finally, we will prove that it is not true that for any convex body $K$ of $\mathbb{R}^4$, there are at least 8 normal lines passing through the center of the minimal spherical shell of $K$ (Theorem 9.5.6).


## 9.3　The Three-Dimensional Case

### 9.3.1　$i_h$-Index and Usual Transverse Orientation

Recall that any hedgehog $\mathcal{H}_h$ of $\mathbb{R}^{n+1}$ is a (possibly singular and self-intersecting) parametrized hypersurface $x_h : \mathbb{S}^n \rightarrow \mathcal{H}_h \subset \mathbb{R}^{n+1}$ that is equipped with the transverse orientation defined as follows: at each regular point $x_h(u)$ of $\mathcal{H}_h$, the usual transverse orientation of $\mathcal{H}_h$ is given by the normal vector $sgn\,[R_h(u)]\,u$, where $sgn$ is the sign function and $R_h := 1/\kappa_h$ the curvature function of $\mathcal{H}_h$ ($\kappa_h$ denoting the Gauss curvature of $\mathcal{H}_h$). As already mentioned, the Kronecker index $i_h(x)$ of a point $x \in \mathbb{R}^{n+1}\setminus\mathcal{H}_h$ with respect to $\mathcal{H}_h$ can be defined as the degree of the map

$$\mathcal{U}_{(h,x)} : \mathbb{S}^n \rightarrow \mathbb{S}^n,\ u \longmapsto \frac{x_h(u) - x}{\|x_h(u) - x\|},$$

and interpreted as the algebraic intersection number of an oriented half-line with origin $x$ with the hypersurface $\mathcal{H}_h$ equipped with its usual transverse orientation (number independent of the oriented half-line for an open dense set of directions). The usual transverse orientation and the Kronecker index are thus mutually associated. It is worth noting that if we let $\tilde{h}(u) = -h(-u)$ for all $u \in \mathbb{S}^n$, where $h \in C^2(\mathbb{S}^n; \mathbb{R})$, then the hedgehogs $\mathcal{H}_h = x_h(\mathbb{S}^n)$ and $\mathcal{H}_{\tilde{h}} = x_{\tilde{h}}(\mathbb{S}^n)$ are identical as hypersurfaces of $\mathbb{R}^{n+1}$ except that they have opposite transverse orientations when $n + 1$ is odd. Indeed

$$x_{\tilde{h}}(-u) = x_h(u)\ \text{ for all } u \in \mathbb{S}^n,$$

but

$$sgn\left[R_{\tilde{h}}(-u)\right](-u) = (-1)^{n+1}\,sgn\left[R_h(u)\right]u,$$

and thus

$$i_{\tilde{h}}(x) = (-1)^{n+1}\,i_h(x)\ \text{ for all } x \in \mathbb{R}^{n+1}\setminus\mathcal{H}_h.$$

For $n + 1 = 3$, we already proved the following result in Sect. 2.8:

**Theorem 2.8.2** *Let $\mathcal{H}_h$ be a hedgehog of $\mathbb{R}^3$ such that $h \in C^2\left(\mathbb{S}^2; \mathbb{R}\right)$. Then, for all $x \in \mathbb{R}^3 \backslash \mathcal{H}_h$, we have:*

$$i_h(x) = r_h^+(x) - r_h^-(x),$$

*where $r_h^-(x)$ (resp. $r_h^+(x)$) denotes the number of connected components of $\mathbb{S}^2 \backslash h_x^{-1}(\{0\})$ on which $h_x$ is negative (resp. positive).*

Since the usual transverse orientation does depend on the orientation of normal lines to $x_h\left(\mathbb{S}^2\right) = x_{\widetilde{h}}\left(\mathbb{S}^2\right)$ in $\mathbb{R}^3$, we also call it 'the relative transverse orientation' of $\mathcal{H}_h$ in $\mathbb{R}^3$, so as to distinguish it from 'the absolute transverse orientation', which corresponds to the $j_h$-index.

## 9.3.2   $j_h$-index and Absolute Transverse Orientation

In Sect. 2.9, we introduced the following notion of index of a point $x \in \mathbb{R}^3 \backslash \mathcal{H}_h$ with respect to $\mathcal{H}_h$:

$$j_h(x) := 1 - c_h(x),$$

where $c_h(x)$ is the number of connected components of $(h_x)^{-1}(\{0\})$ on $\mathbb{S}^2$, that is the number of closed spherical curves formed by points $u \in \mathbb{S}^2$ such that $x$ belongs to the support hyperplane of $\mathcal{H}_h$ at $x_h(u)$. Moreover, we saw that this $j_h$-index corresponds to the transverse orientation of $\mathcal{H}_h$ that is such that whenever $x_h(u)$ is a simple regular point of $\mathcal{H}_h$, then the normal line to $\mathcal{H}_h$ at $x_h(u)$, is oriented in the direction that $j_h$ decreases by one unit. Since this transverse orientation of $\mathcal{H}_h$ does not depend on the choice of the orientation of normal lines to $x_h\left(\mathbb{S}^2\right) = x_{\widetilde{h}}\left(\mathbb{S}^2\right)$, we called it 'the absolute transverse orientation' of $\mathcal{H}_h$. The reading of Sect. 2.9.1 will be essential for understanding the rest of this subsection. We remind that the absolute transverse orientation cannot change on elliptic regions (i.e., on regions on which the Gauss curvature of $\mathcal{H}_h$ remains positive) since it is then simply given by the direction of convexity. For our present study, the crucial point is that a hedgehog of $\mathbb{R}^3$ may admit reversals of its absolute transverse orientation along certain of its self-intersection curves formed by double hyperbolic points  However, and this is the key point, such a reversal will not necessarily occur on any curve of hyperbolic double points of $\mathcal{H}_h$ but only on certain of them: it depends on the global geometry of the hedgehog.

**Absolute Body of $\mathcal{H}_h$ in $\mathbb{R}^3$**

We call **absolute body** of $\mathcal{H}_h$ in $\mathbb{R}^3$, and we denote by $\mathcal{K}_h$, the set

$$\mathcal{K}_h := \mathcal{H}_h \cup Int\,(\mathcal{H}_h),$$

where $Int\,(\mathcal{H}_h) := \left\{ x \in \mathbb{R}^3 \backslash \mathcal{H}_h \,|\, j_h\,(x) \neq 0 \right\}.$

**Comparison to the Body of $\mathcal{F}_{\nabla h}$ in $\mathbb{R}^3$**

**Proposition 9.3.1** *Let $\mathcal{H}_h$ be a hedgehog of $\mathbb{R}^3$ such that $h \in C^\infty\left(\mathbb{S}^2; \mathbb{R}\right)$. We have $i_{\mathcal{F}_{\nabla h}} \leq j_h$ on $\mathbb{R}^3 \backslash (\mathcal{H}_h \cup \mathcal{F}_{\nabla h})$.*

**Proof** Let $x \in \mathbb{R}^3 \backslash (\mathcal{H}_h \cup \mathcal{F}_{\nabla h})$. Denote by $A$ a connected component of $\mathbb{S}^2 \backslash (h_x)^{-1}\,(\{0\})$, and by $b_A$ the number of connected component of its boundary in $\mathbb{S}^2$. Now, denote by $C_A\,(x)$ (resp. $S_A\,(x)$) the number of local extrema (resp. saddle points) of $h_x$ in $A$. We know that:

$$C_A\,(x) - S_A\,(x) = 2 - b_A,$$

since the Euler characteristic of $A$ is given by $\chi\,(A) = 2 - b_A$. Thus the number $N_A\,(x) = C_A\,(x) + S_A\,(x)$ of critical points of $h_x$ in $A$ is such that $N_A\,(x) = b_A + 2\,(C_A\,(x) - 1) \geq b_A$. Therefore $N_{\nabla h}\,(x) = \sum_A N_A\,(x) \geq \sum_A b_A = 2c_h\,(x)$, where the sums are taken over the set of connected components of $\mathbb{S}^2 \backslash (h_x)^{-1}\,(\{0\})$. Thus:

$$i_{\mathcal{F}_{\nabla h}}\,(x) = 1 - \frac{N_{\nabla h}\,(x)}{2} \leq 1 - c_h\,(x) = j_h\,(x).$$

$\blacksquare$

**Corollary 9.2.4** *If the mean of $h \in C^\infty\left(\mathbb{S}^2; \mathbb{R}\right)$ on $\mathbb{S}^2$ is equal to $0$, then the absolute body $\mathcal{K}_h$ of $\mathcal{H}_h$ is included in the body $\mathcal{K}\,(\mathcal{F}_{\nabla h})$ of its focal (since in this case $(h_x)^{-1}\,(\{0\})$ is nonempty for all $x \in \mathbb{R}^3$).*

### 9.3.3  Proofs of Theorem 9.2.1 and Corollary 9.2.2

**Proof of Theorem 9.2.1** We have assumed that $\mathcal{N}_{\nabla h}\,(u)$ does not meet the singular locus of $\mathcal{F}_{\nabla h}$, and we want to prove that there exists some $r \in \mathbb{R}$ such that $x_{h-r}\,(u)$ is an at least double hyperbolic point of $\mathcal{H}_{h-r}$. We thus consider the family $(\mathcal{H}_{h-r})_{r \in \mathbb{R}}$ of the hedgehogs that are parallel to $\mathcal{H}_h$. Note that the point $x_{h-r}\,(u) = \nabla h\,(u) + (h\,(u) - r)\,u$ describes the entire normal line $\mathcal{N}_{\nabla h}\,(u)$ when $r$ describes the entire real line. Our proof relies on the comparative evolution of the absolute and relative transverse orientations of $\mathcal{H}_{h-r}$ at $x_{h-r}(u)$ when $r$ describes the entire real line.

Define a function $\varepsilon_h^u : \mathbb{R} \to \{-1, 0, 1\}$ as follows: put $\varepsilon_h^u(r) = 1$ if both transverse orientations of $\mathcal{H}_{h-r}$ are well defined and identical at $x_{h-r}(u)$; put $\varepsilon_h^u(r) = -1$ if both transverse orientations of $\mathcal{H}_{h+r}$ are well defined and opposite at $x_{h-r}(u)$; finally put $\varepsilon_h^u(r) = 0$ if one the two transverse orientations cannot be defined at $x_{h-r}(u)$. If $x_{h-r}(u)$ is a simple elliptic point of $\mathcal{H}_{h-r}$, the relative transverse orientation of $\mathcal{H}_{h-r}$ at $x_{h-r}(u)$ is given by $u$, whereas the absolute one is given by the direction of convexity at $x_{h-r}(u)$, that is by $\left( R_{h-r}^1 + R_{h-r}^2 \right)(u)\, u$, with $R_{h-r}^1(u)$, $R_{h-r}^2(u)$ being the principal radii of curvature of $\mathcal{H}_{h-r}$ at $x_{h-r}(u)$. In other words, we have in this case:

$$\varepsilon_h^u(r) = sgn\left[ \left( R_{h-r}^1 + R_{h-r}^2 \right)(u) \right] = sgn\left[ \left( R_h^1 + R_h^2 \right)(u) - 2r \right].$$

In fact this definition makes sense for any $r \in \mathbb{R}$ such that $R_{h-r}(u) > 0$ (i.e., such that $x_{h-r}(u)$ is an elliptic point of $\mathcal{H}_{h-r}$). Note we have thus

$$\varepsilon_h^u(r) = \begin{cases} 1 & \text{if} \quad r < R_h^1(u) \\ -1 & \text{if} \quad r > R_h^2(u) \end{cases}$$

(since $R_{h-r}(u) = \left( R_{h-r}^1 R_{h-r}^2 \right)(u) = \left( R_h^1(u) - r \right)\left( R_h^2(u) - r \right) > 0$ if $r \notin \left[ R_h^1(u), R_h^2(u) \right]$), so that $\varepsilon_h^u : \mathbb{R} \to \{-1, 0, 1\}$ has a sign change on $\left[ R_h^1(u), R_h^2(u) \right]$.

If $r \in \left\{ R_h^1(u), R_h^2(u) \right\}$, then $R_{h-r}(u) = 0$ so that $x_{h-r}$ is a singular point of $\mathcal{H}_{h-r}$, and thus a point of $\mathcal{F}_{\nabla h}$ (namely, $c_{\nabla h}^1(u)$ or $c_{\nabla h}^2(u)$). Since $\mathcal{N}_{\nabla h}(u)$ does not contain any singular point of $\mathcal{F}_{\nabla h}$, $x_{h-r}(u)$ is then a regular point of a cuspidal edge $x_{h-r}(\Gamma)$ (see Sect. 9.1.3), which is separating a hyperbolic region of $\mathcal{H}_{h-r}$ from an elliptic one, and a sign change of $\varepsilon_h^u$ cannot occur at such a point. In fact, we still could define $\varepsilon_h^u(r)$ by $\varepsilon_h^u(r) = sgn\left[ \left( R_{h-r}^1 + R_{h-r}^2 \right)(u) \right]$ in order to be consistent with the changes of the $i_{h-r}$ and $j_{h-r}$ indices at the points of the adjacent elliptic region.

Finally, let us come to the case where $r \in \left] R_h^1(u), R_h^2(u) \right[$. We then have $R_{h-r}^1(u) = R_h^1(u) - r < 0$ and $R_{h-r}^2(u) = R_h^2(u) - r > 0$ so that $x_{h-r}(u)$ is a hyperbolic point of $\mathcal{H}_{h-r}$: $R_{h-r}(u) = R_{h-r}^1(u) R_{h-r}^2(u) < 0$. If $x_{h-r}(u)$ is moreover a single point of $\mathcal{H}_{h-r}$, then for any $\rho$ that is close enough to $r$ in $\left] R_h^1(u), R_h^2(u) \right[$, $x_{h-\rho}(u)$ is a hyperbolic point of $\mathcal{H}_{h-\rho}$, and the configuration of $h_{x_{h-\rho}(u)}(\mathbb{R}_-)$, $h_{x_{h-\rho}(u)}(\{0\})$ and $h_{x_{h-\rho}(u)}(\mathbb{R}_+)$ on $\mathbb{S}^2$ is qualitatively the same as the one of $h_{x_{h-r}(u)}(\mathbb{R}_-)$, $h_{x_{h-r}(u)}(\{0\})$ and $h_{x_{h-r}(u)}(\mathbb{R}_+)$, so that there is no sign change of $\varepsilon_h^u$ at $r$ in such a case (see the relationships between $i_h(x)$, $j_h(x)$ and $r_h^-(x)$, $r_h^+(x)$, $c_h(x)$ that we have recalled above). Therefore, if a sign change of $\varepsilon_h^u$ occurs at $r \in \left] R_h^1(u), R_h^2(u) \right[$, then the hyperbolic point $x_{h-r}(u)$ of $\mathcal{H}_{h-r}$ is a multiple point of $\mathcal{H}_{h-r}$. Now, a sign change of $\varepsilon_h^u$ at $r \in \left] R_h^1(u), R_h^2(u) \right[$ cannot be due to the fact that $x = x_{h-r}(u)$ is also an elliptic point $x_{h-r}(v)$ or a singular point $x_{h-r}(v)$ of $\mathcal{H}_{h-r}$, which must be a regular point of a cuspidal edge of $\mathcal{H}_{h-r}$ since by assumption $\mathcal{N}_{\nabla h}(u)$ does not meet $\mathcal{S}ing(\mathcal{F}_{\nabla h})$. Indeed, if $x = x_{h-r}(u)$ is such a point $x_{h-r}(v)$, then there exists a neighborhood $V$ of $v$ in $\mathbb{S}^2$ such that the absolute transverse orientation of $x_{h-r}(V)$ is fully determined by the direction of convexity on its elliptic part, so that no change of this absolute transverse orientation can be due to the crossing with the image under $x_{h-r}$ of a neighborhood of $u$ in $\mathbb{S}^2$. Thus, in the present case, if a sign change of $\varepsilon_h^u$ occurs at $r$ then $x_{h-r}(u)$ is an (at least double) hyperbolic point of $\mathcal{H}_{h-r}$, which achieves the proof. ∎

Our proof of Corollary 9.2.2 will make use of the following remark.

**Remark** Let $\mathcal{H}_h$ be a hedgehog such that $h \in C^\infty(\mathbb{S}^2; \mathbb{R})$. For any $u \in \mathbb{S}^2$ such that $x = x_h(u) \in \mathbb{R}^3 \backslash \mathcal{F}_{\nabla h}$, denote by $\mathcal{F}(\nabla h_x)$ the foliation of $\mathbb{S}^2$ by the level lines of $h_x : \mathbb{S}^2 \to \mathbb{R}$. For any such $u \in \mathbb{S}^2$, we have:

$$(u \text{ is a saddle point of } \mathcal{F}(\nabla h_x)) \iff (x = x_h(u) \text{ is a hyperbolic point of } \mathcal{H}_h).$$

***Proof of Corollary 9.2.2*** Assume first that there exists some $u \in \mathbb{S}^2$ such that $\mathcal{N}_{\nabla h}(u) \cap \mathcal{S}ing(\mathcal{F}_{\nabla h}) = \varnothing$. We then know from Theorem 2.9.1 that there exists some $r \in \mathbb{R}$ such that $x_{h-r}(u) = x_h(u) - ru \in \mathcal{N}_{\nabla h}(u)$ is an (at least) double hyperbolic point of $\mathcal{H}_{h-r}$: that is, there exists $v \in \mathbb{S}^2 \backslash \{u\}$, such that:

$$x_{h-r}(u) = x_{h-r}(v), \quad R_{h-r}(u) < 0, \quad \text{and } R_{h-r}(v) < 0.$$

If this point $x = x_{h-r}(u)$ belongs to $\mathbb{R}^3 \backslash \mathcal{F}_{\nabla h}$, then

$$i_{\mathcal{F}_{\nabla h}}(x) = -S(x) \le -2,$$

where $S(x)$ is the number of saddle points of $\mathcal{F}(\nabla h_x)$ (see (9.1.3) and the remark just above), and thus $N_{\nabla h}(x) = 2\left(1 - i_{\mathcal{F}_{\nabla h}}(x)\right) \ge 6$. In that case, knowing that $N_{\nabla h}$ remains constant on each component of $\mathbb{R}^3 \backslash \mathcal{F}_{\nabla h}$, there thus exists an open set formed by points of $\mathbb{R}^3 \backslash \mathcal{F}_{\nabla h}$ lying on at least 6 normal lines to $\mathcal{H}_h$. To reach the same conclusion in the case where $x = x_{h-r}(u) \in \mathcal{F}_{\nabla h}$, it suffices to prove that, arbitrarily close to $x$, there exist points $y \in \mathbb{R}^3 \backslash \mathcal{F}_{\nabla h}$ such that $i_{\mathcal{F}_{\nabla h}}(y) = -S(y) \le -2$, and thus $N_{\nabla h}(y) = 2\left(1 - i_{\mathcal{F}_{\nabla h}}(y)\right) \ge 6$. Since $u, v \in \mathbb{S}^2$ are such that $u \ne v$, $R_{h-r}(u) < 0$ and $R_{h-r}(v) < 0$, there exist $U \times I, V \times J$ neighborhoods of respectively $(u, r)$ and $(v, r)$ in $\mathbb{S}^2 \times \mathbb{R}$, such that $U \cap V \ne \varnothing$ and $R_{h-\rho}(\omega) < 0$ for all $(\omega, \varrho)$ in $(U \times I)$ or $(V \times J)$. Besides, there exists a neighborhood $\mathcal{Y}$ of $x = x_{h-r}(u) = x_{h-r}(v)$ in $\mathbb{R}^3$ such that any point $y \in \mathcal{Y} \cap \left(\mathbb{R}^3 \backslash \mathcal{F}_{\nabla h}\right)$ can be written in the form $x_{h-\rho}(\omega)$ with $(\omega, \varrho) \in U \times I$ (resp. $(\omega, \varrho) \in V \times J$). Therefore, any $y \in \mathcal{Y} \cap \left(\mathbb{R}^3 \backslash \mathcal{F}_{\nabla h}\right)$ is such that $i_{\mathcal{F}_{\nabla h}}(y) = -S(y) \le -2$, and thus $N_{\nabla h}(y) = 2\left(1 - i_{\mathcal{F}_{\nabla h}}(y)\right) \ge 6$. In deed, the foliation of $\mathbb{S}^2$ by the level lines of $h_y : \mathbb{S}^2 \to \mathbb{R}$ has at least two distinct saddle points: one in $U$ and the other in $V$. Thus, in that case also, there exists an open set formed by points of $\mathbb{R}^3 \backslash \mathcal{H}_h$ lying on at least 6 normal lines to $\mathcal{H}_h$.

Therefore, it only remains to consider the case where all the normal lines to $\mathcal{H}_h$ meet the singular locus $Sing\,(\mathcal{F}_{\nabla h})$ of its focal $\mathcal{F}_{\nabla h}$, that is: $\forall u \in \mathbb{S}^2,\ \mathcal{N}_{\nabla h}\,(u) \cap Sing\,(\mathcal{F}_{\nabla h}) \neq \varnothing$. Now in this last case, there must exist a point of $\mathbb{R}^3$ lying on infinitely many normal lines to $\mathcal{H}_h$. $\blacksquare$

### 9.3.4  Critical Saddle Points

It is worth to stress that $x = x_{h-r}\,(u) = x_{h-r}\,(v)$, the double hyperbolic point of $\mathcal{H}_{h-r}$ that Theorem 2.9.1 states to exist, is not an arbitrary double hyperbolic point of $\mathcal{H}_{h-r}$ but a double hyperbolic point of $\mathcal{H}_{h-r}$ in the vicinity of which the absolute transverse orientation of $\mathcal{H}_{h-r}$ is reversed. This property can be read on the foliation $\mathcal{F}\,(\nabla h_x)$ of $\mathbb{S}^2$ by the level lines of $h_x : \mathbb{S}^2 \to \mathbb{R}$. Saying that $x = x_{h-r}\,(u) = x_{h-r}\,(v)$ is a double hyperbolic point of $\mathcal{H}_{h-r}$ is equivalent to saying that $u$ and $v$ are saddle points of $\mathcal{F}\,(h_x)$ lying on $(h_x)^{-1}\,(\{r\})$. Now, saying that $x$ is a double hyperbolic point of $\mathcal{H}_{h-r}$ in the vicinity of which the absolute transverse orientation of $\mathcal{H}_{h-r}$ is reversed is equivalent to saying that these saddle points $u, v$ of $\mathcal{F}\,(\nabla h_x)$ lying on $(h - r)_x^{-1}\,\{0\} = (h_x)^{-1}\,(\{r\})$ are 'critical saddle points' in the following sense.

**Definition 9.3.1**  Let $\mathcal{H}_h$ be a hedgehog of $\mathbb{R}^3$ such that $h \in C^\infty\,(\mathbb{S}^2; \mathbb{R})$, and let $u \in \mathbb{S}^2$ be such that $x = x_h\,(u)$ is a hyperbolic point of $\mathcal{H}_h$. Note that $u \in \mathbb{S}^2$ is such that any $v \in \mathbb{S}^2$ that is sufficiently close to $u$ is a saddle point of $\mathcal{F}\,(\nabla h_y)$, where $y = x_h\,(v)$.

We say that such a point $v$ disconnects $(h_y)^{-1}\,(\mathbb{R}_-)$ (resp. $(h_y)^{-1}\,(\mathbb{R}_+)$) if $(h_y)^{-1}\,(\mathbb{R}_-)$ and $(h_y)^{-1}\,(\mathbb{R}_-) \backslash \{v\}$ (resp. $(h_y)^{-1}\,(\mathbb{R}_+)$ and $(h_y)^{-1}\,(\mathbb{R}_+) \backslash \{v\}$) does not have the same number of connected components.

Now, the saddle point $u$ of $\mathcal{F}\,(\nabla h_x)$ is said to be critical if the disconnecting or nondisconnecting character of $v$ with respect to $(h_y)^{-1}\,(\mathbb{R}_-)$ and $(h_y)^{-1}\,(\mathbb{R}_+)$ changes in every neighborhood of $u$ in the region of $(R_h)^{-1}\,(\mathbb{R}_-^*)$ containing it.

By considering carefully the following three types of pairs of saddle points of $\mathcal{F}\,(\nabla h_x)$ shown in Fig. 9.1, we check that $u$ and $v$ are critical saddle points of $\mathcal{F}\,(\nabla h_x)$ in the first case $(a)$, and only in this case.

Recall that the critical and singular sets, $\mathcal{C}_{\nabla h}$ and $\mathcal{S}_{\nabla h}$, can respectively be defined by

$$
\begin{aligned}
\mathcal{C}_{\nabla h} &:= \left\{ (x, u) \in U\mathbb{R}^3 \,|\, \nabla\,(h_x)\,(u) = 0 \right\} \\
&= \left\{ (x, u) \in U\mathbb{R}^3 \,|\, x \in \mathcal{N}_{\nabla h}\,(u) \right\},
\end{aligned}
$$

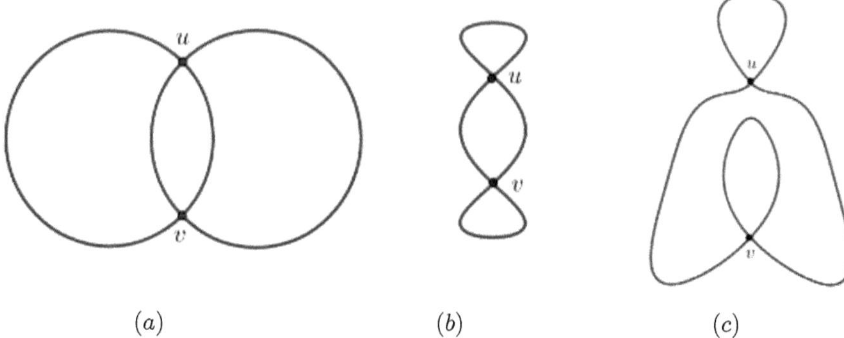

|     |     |     |
|:---:|:---:|:---:|
| $(a)$ | $(b)$ | $(c)$ |

**Fig. 9.1** The three types of pairs of saddle points

where $\mathcal{N}_{\nabla h}(u) := \{\nabla h(u)\} + \mathbb{R}u$ is the normal line to $\mathcal{H}_h$ at $x_h(u)$, and

$$\mathcal{S}_{\nabla h} := \left\{ (x, u) \in \mathcal{C}_{\nabla h} \,\middle|\, \exists k \in [1, 2], x = c_{\nabla h}^k(u) \right\}$$
$$= \left\{ (x, u) \in \mathcal{C}_{\nabla h} \,\middle|\, (\det \circ \nabla^2)(h_x)(u) = 0 \right\}.$$

The following subset of $\mathcal{C}_{\nabla h} \backslash \mathcal{S}_{\nabla h}$ played an important role in our proof of Theorem 9.2.1:

$$\Sigma_{\nabla h}^c =: \left\{ (x, u) \in \mathcal{C}_{\nabla h} \,\middle|\, u \text{ is a critical saddle point of } \mathcal{F}(\nabla h_x) \right\}.$$

More precisely, the key point is the fact that under the assumption $\mathcal{N}_{\nabla h}(u) \cap Sing(\mathcal{F}_{\nabla h}) = \varnothing$, the focal segment $\left[ c_{\nabla h}^1(u), c_{\nabla h}^2(u) \right]$ contains some interior point $x$ such that $(x, u) \in \Sigma_{\nabla h}^c$. Indeed, for every $x \in \mathcal{N}_{\nabla h}(u)$, $(x, u) \in \Sigma_{\nabla h}^c$ if, and only if, there exist some $r \in \mathbb{R}$ such that $x = x_{h-r}(u)$ is a hyperbolic point of $\mathcal{H}_{h-r}$ and a neighborhood $\mathcal{U}$ of $u$ on $\mathbb{S}^2$ such that the absolute transverse orientation of $x_{h-r}(\mathcal{U})$ is reversed in the vicinity of $x_{h-r}(u)$. From this remark we can deduce the following refinement of Corollary 9.2.2.

**Proposition 9.3.2** *Let $\mathcal{H}_h$ be a hedgehog of $\mathbb{R}^3$ such that $h \in C^\infty\left(\mathbb{S}^2; \mathbb{R}\right)$. If there does not exist a point of $\mathbb{R}^3$ lying on infinitely many normal lines to $\mathcal{H}_h$, then there exist infinitely many normal lines to $\mathcal{H}_h$ that do no meet the singular locus of $\mathcal{F}_{\nabla h}$ and any one of these lines meets the closure of $Int\left(\mathcal{F}_{\nabla h}^1\right) \cap Int\left(\mathcal{F}_{\nabla h}^2\right).$*

**Proof** Let $u$ be any point of $\mathbb{S}^2$ such that $\mathcal{N}_{\nabla h}(u) \cap Sing(\mathcal{F}_{\nabla h}) = \varnothing$. For such a point $u$, we know from our proof of Theorem 9.2.1 that there exists some $r \in \mathbb{R}$ such that $u$ is a critical saddle point of $\mathcal{F}(\nabla h_x)$, where $x = x_{h-r}(u)$. By the very definition of a critical saddle point, this shows that $C_m(x) \geq 2$ and $C_M(x) \geq 2$. If $x \in \mathbb{R}^3 \backslash \mathcal{F}_{\nabla h}$, this exactly means that $x \in Int\left(\mathcal{F}_{\nabla h}^1\right) \cap Int\left(\mathcal{F}_{\nabla h}^2\right)$. Now if $x \in \mathcal{F}_{\nabla h}$, we know that arbitrarily close to $x$, there are points $y \in \mathbb{R}^3 \backslash \mathcal{F}_{\nabla h}$, and that if $y \in \mathbb{R}^3 \backslash \mathcal{F}_{\nabla h}$ is closed enough to $x$, then $C_m(y) \geq 2$ and $C_M(y) \geq 2$ (the arguments are the same as those used by E. Heil in [65]), so that $y \in Int\left(\mathcal{F}^1_{\nabla h}\right) \cap Int\left(\mathcal{F}^2_{\nabla h}\right)$. Therefore, in any case, $x \in \overline{Int\left(\mathcal{F}^1_{\nabla h}\right) \cap Int\left(\mathcal{F}^2_{\nabla h}\right)}$. $\blacksquare$

In higher dimension, our idea is to adopt the same type of approach.

## 9.4   The Four-Dimensional Case

In order to follow a similar approach in the four dimensional case, we must overcome a series of difficulties.

First of all, when considering a hedgehog $\mathcal{H}_h$ in $\mathbb{R}^4$, we have to deal with two types of 'hyperbolic points', that is, of points $x = x_h(u)$ in which the principal radii of curvature are nonzero and not all of the same sign: a hyperbolic point $x = x_h(u)$ is of type 1 if $R_h(u) < 0$, and then it corresponds to a type 1 saddle point $u$ of $h_x$ (i.e. to a critical point $u$ of index 1 of $h_x$); and a hyperbolic point $x = x_h(u)$ is of type 2 if $R_h(u) > 0$, and then it corresponds to a type 2 saddle point $u$ of $h_x$ (i.e. to a critical point $u$ of index 2 of $h_x$). We will often say simply '$i$-saddle of $\mathcal{F}(\nabla h_x)$' instead of 'type $i$ saddle point $u$ of $h_x$', $(i \in [1, 2])$.

Besides, in the four dimensional case the usual transverse orientation of $x_h\left(\mathbb{S}^3\right) = x_{\tilde{h}}\left(\mathbb{S}^3\right)$ is no longer relative but absolute due to the even parity of the dimension of $\mathbb{R}^4$: it does not depend on the choice between $h$ and $\tilde{h}$ as the support function (i.e. it does not depend on the choice of the orientation of the normal lines to the hypersurface). On the other hand, if the index $r_h$ defined by

$$r_h(x) = r_h^+(x) - r_h^-(x) \quad \text{for all } x \in \mathbb{R}^{n+1} \backslash \mathcal{H}_h,$$

where $r_h^-(x)$ (resp. $r_h^+(x)$) is the number of connected components of $\mathbb{S}^n \backslash h_x^{-1}(\{0\})$ on which $h_x : \mathbb{S}^n \to \mathbb{R}$ is negative (resp. positive), is none other than the usual index $i_h$ for $n + 1 = 3$ (see above and Sect. 2.8.1), it provides us with a new index and a new transverse orientation which is relative (to the choice of the orientation of the normal lines to $x_h\left(\mathbb{S}^3\right) = x_{\tilde{h}}\left(\mathbb{S}^3\right)$) for $n + 1 = 4$. We will call it 'the relative transverse orientation'.

### 9.4.1   $r_h$-Index and Transverse Orientation

Let $\mathcal{H}_h$ be a hedgehog of $\mathbb{R}^4$ such that $h \in C^\infty\left(\mathbb{S}^3; \mathbb{R}\right)$. Above, we defined the index $r_h : \mathbb{R}^4 \backslash \mathcal{H}_h \to \mathbb{N}$ as follows:

$$r_h(x) = r_h^+(x) - r_h^-(x) \quad \text{for all } x \in \mathbb{R}^4 \backslash \mathcal{H}_h,$$

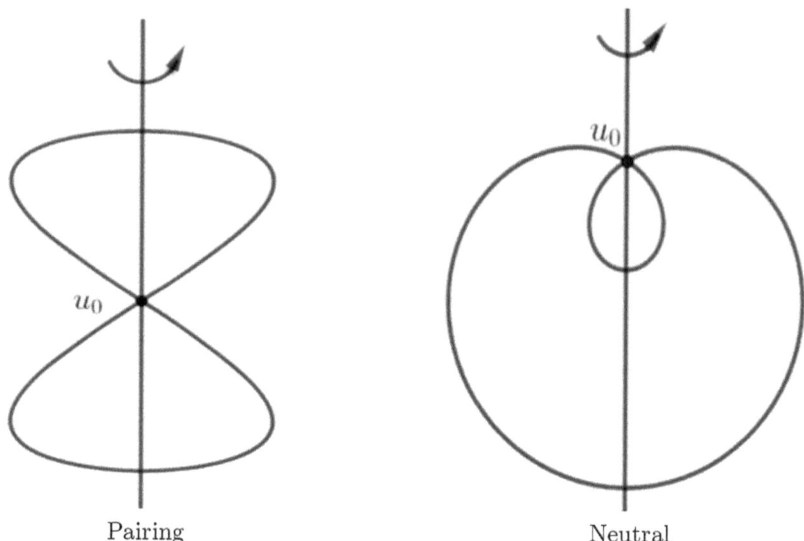



**Fig. 9.2** An example of each type of configuration at a saddle point (here, the curves are rotated around a vertical axis passing though the saddle)

where $r_h^-(x)$ (resp. $r_h^+(x)$) is the number of connected components of $\mathbb{S}^3 \backslash h_x^{-1}(\{0\})$ on which $h_x : \mathbb{S}^3 \to \mathbb{R}$ is negative (resp. positive). Of course, the index $r_h$ remains constant on each connected component of $\mathbb{R}^4 \backslash \mathcal{H}_h$. In particular, $r_h$ is equal to 0 on the unbounded component of $\mathbb{R}^4 \backslash \mathcal{H}_h$. It is worth noting that the value of $r_h(x)$ must obviously decrease by one unit as $x$ orthogonally crosses $\mathcal{H}_h$ at a simple elliptic point $x_0 = x_h(u_0)$ in the direction of $u_0 \in \mathbb{S}^3$. When the crossing occurs at a simple hyperbolic point $x_0 = x_h(u_0)$, it is necessary to distinguish two cases depending on whether the saddle point $u_0$ of $\mathcal{F}(\nabla h_{x_0})$ disconnects the connected component, say $\mathcal{L}_{u_0}$, of $(h_{x_0})^{-1}(\{0\})$ containing it or not. We can imagine an example of each of these two configurations by rotating the two figures shown in Fig. 9.2 around a vertical axis passing through the saddle point $u_0$ in $\mathbb{R}^3 \subset \mathbb{S}^3 = \mathbb{R}^3 \cup \{\infty\}$. If $\mathcal{L}_{u_0} \backslash \{u_0\}$ is disconnected, as in the first example, we can check that the value of $r_h(x)$ decreases by one unit as $x$ orthogonally crosses $\mathcal{H}_h$ at $x_0 = x_h(u_0)$ in the direction of $-u_0$. We say then that $x_0$ is an active hyperbolic point of $\mathcal{H}_h$, and that $u_0$ is an active saddle point of $\mathcal{F}(\nabla h_{x_0})$. If $\mathcal{L}_{u_0} \backslash \{u_0\}$ is connected, as in the second example where $\mathcal{L}_{u_0}$ is a torus pinched at $u_0$, we can check that the value of $r_h(x)$ does not change as $x$ orthogonally crosses $\mathcal{H}_h$ at $x_0 = x_h(u_0)$. We say then that $x_0$ is a neutral hyperbolic point of $\mathcal{H}_h$, and that $u_0$ is a neutral saddle point of $\mathcal{F}(\nabla h_{x_0})$. In short, the 'relative transverse orientation' of $\mathcal{H}_h$ (that is, the one that is associated with the $r_h$-index) is such that whenever $x_h(u)$ is a simple regular point of $\mathcal{H}_h$, then the normal line to $\mathcal{H}_h$ at $x_h(u)$ is oriented in the direction of $u$ (resp. $-u$) if $x_h(u)$ is elliptical (resp. hyperbolic and active), and is not oriented if $x_h(u)$ is a neutral hyperbolic point.

### 9.4.2 Effect of a Transverse Crossing of $\mathcal{F}_{\nabla h}$ and $i_{\mathcal{F}^2_{\nabla h}}$

Let $k \in [1, 3]$. As noticed above in Sect. 9.1.2, when $x \in \mathbb{R}^4$ transversally crosses the $k$th sheet $\mathcal{F}^k_{\nabla h}$ of $\mathcal{F}_{\nabla h}$ at a simple regular point of $\mathcal{F}^k_{\nabla h}$ in the direction of the transverse orientation, this crossing of $\mathcal{F}^k_{\nabla h}$ results in a two-units increase in the number of critical points of $h_x : \mathbb{S}^3 \to \mathbb{R}$, and the two critical points created have consecutive indices $k - 1$ and $k$. Of course, the same crossing but in the opposite direction results in the elimination of such a pair of critical points. Mathematically, such a transition can be regarded as a path in the vector space $C^\infty(\mathbb{S}^3; \mathbb{R})$ that crosses a 'codimension 1 non-Morse stratum' transversely at one point. We know from 'parametrized Morse theory' that: $(i)$ at the very moment of the transition, $h_x : \mathbb{S}^3 \to \mathbb{R}$ admits a so-called 'birth-death singularity'; $(ii)$ the transition itself can locally be described by a smooth family of functions $f_t : u \mapsto f_t(u)$ parametrized by $t$, $(t \in \mathbb{R})$, of the form

$$f_t(u_1, u_2, u_3) = u_1^3 - t u_1 + \varepsilon_2 u_2^2 + \varepsilon_3 u_3^2 + g(t),$$

where $\varepsilon_2, \varepsilon_3 \in \{\pm 1\}$, with respect to some local coordinates $(u_1, u_2, u_3)$; when $t < 0$, $f_t : u \mapsto f_t(u)$ is a Morse function without critical point; $f_0 : u \mapsto f_0(u)$ is a generalized Morse function with a birth-death singularity at $u = (0, 0, 0)$; when $t > 0$, $f_t : u \mapsto f_t(u)$ is a Morse function with two critical points of consecutive indices (these points were 'born' at $t = 0$).

When $x \in \mathbb{R}^4$ transversally crosses $\mathcal{F}_{\nabla h}$ at a simple regular point in the direction of the transverse orientation, we are thus in one of the following three cases:

| Case | Sheet crossed by $x$ | Pair of critical points that are born |
|---|:---:|---|
| $\varepsilon_2 = \varepsilon_3 = 1$ | $\mathcal{F}^1_{\nabla h}$ | 1 local minimal and 1 type 1 saddle |
| $\varepsilon_2 \neq \varepsilon_3$ | $\mathcal{F}^2_{\nabla h}$ | 1 type 1 saddle and 1 type 2 saddle |
| $\varepsilon_2 = \varepsilon_3 = -1$ | $\mathcal{F}^3_{\nabla h}$ | 1 local maxima and 1 type 2 saddle |

In the first (resp. third) case, a 'trivial bubble' (a 'trivial centre-saddle pairing' of $\mathcal{F}(\nabla h_x)$) was born as shown in Fig. 9.3a: the centre of this trivial centre-saddle pairing is of course located inside the bubble, and the corresponding saddle point is obviously active. In the second case, two saddle points of distinct types of $\mathcal{F}(\nabla h_x)$ were born. In the vicinity of each of them, the level set of the corresponding saddle point looks qualitatively like the piece of surface shown in Fig. 9.3b. Moreover, the two level sets corresponding to these two saddle points are in fact two pinched torus that are 'linked somehow like a Hopf link' [193]. In particular, these two saddle points are neutral.

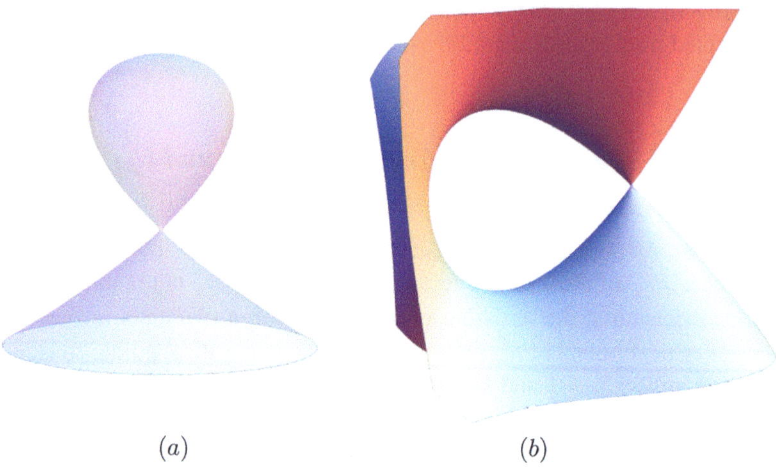

$$(a) \qquad\qquad\qquad (b)$$

**Fig. 9.3** Illustration of the three cases: first and third (a); second (b)

Let $x \in \mathbb{R}^4 \setminus \mathcal{F}_{\nabla h}$. For any regular value $d$ of $h_x : \mathbb{S}^3 \to \mathbb{R}$, denote by $g_{h_x}(d)$ the genus of the (not necessarily connected) surface $(h_x)^{-1}(\{d\})$. Now, for any critical value $c$ of $h_x$, put

$$g_{h_x}(c) := \min_{\varepsilon > 0} \left( \max \left( \left\{ g_{h_x}(d) \,|\, d \text{ regular value of } h_x \text{ in } ]c - \varepsilon, d + \varepsilon[ \right\} \right) \right),$$

and then

$$G_h(x) = \frac{1}{2} \sum_{c \in C_h(x)} g_{h_x}(c), \text{ where } C_h(x) = h_x \left( \left\{ u \in \mathbb{S}^3 \,|\, \nabla h_x(u) = 0 \right\} \right).$$

From the above study, when $x$ transversally crosses $\mathcal{F}_{\nabla h}$ at a simple regular point $x_0 \in \mathcal{F}_{\nabla h}$, the value of $G_h(x)$ changes if, and only if $x_0 \in \mathcal{F}^2_{\nabla h}$. In this case, the value of $G_h(x)$ increases (resp. decreases) by one unit if the crossing occurs in the direction of the transverse orientation of $\mathcal{F}_{\nabla h}$ (resp. in the opposite direction). Knowing that $G_h(x) = 0$ when $x$ is far from $\mathcal{H}_h$, we deduce the following result.

**Theorem 9.4.1** *Let $\mathcal{H}_h$ be a hedgehog of $\mathbb{R}^4$ such that $h \in C^\infty(\mathbb{S}^3; \mathbb{R})$, let $\mathcal{F}_{\nabla h}$ denote its focal surface, and let $\mathcal{F}^2_{\nabla h}$ be the second sheet of $\mathcal{F}_{\nabla h}$. We have: $\forall x \in \mathbb{R}^4 \setminus \mathcal{F}^2_{\nabla h}$,  $i_{\mathcal{F}^2_{\nabla h}}(x) = -G_h(x)$.*

As a corollary we obtain that: $\forall x \in \mathbb{R}^4 \setminus \mathcal{F}^2_{\nabla h}$,

$$Int\left( \mathcal{F}^2_{\nabla h} \right) = \left\{ x \in \mathbb{R}^4 \setminus \mathcal{F}^2_{\nabla h} \,|\, G_h(x) \neq 0 \right\};$$

and thus:

$$\left(x \in Int\left(\mathcal{F}^2_{\nabla h}\right)\right) \Longleftrightarrow \left(\begin{array}{c} h_x: \quad \mathbb{S}^3 \to \mathbb{R} \text{ admits a smooth} \\ \text{level surface with nonzero genus} \end{array}\right);$$

which is the statement of Theorem 9.2.2.

### 9.4.3   Critical Saddle Points

As in dimension 3, our idea will be to focus, for each $u \in \mathbb{S}^3$ such that $\mathcal{N}_{\nabla h}(u) \cap Sing(\mathcal{F}_{\nabla h}) = \varnothing$, on the values of $r$ such that $x = x_{h-r}(u)$ is a (type 1 or type 2) hyperbolic point of $\mathcal{H}_{h-r}$ (and thus, $u$ a saddle point of $\mathcal{F}(\nabla h_x)$) at which the relative transverse orientation of $\mathcal{H}_{h-r}$ switches (here, between the active and neutral modes): we will say then that $u$ is a critical (type 1 or type 2) saddle point of $\mathcal{F}(\nabla h_x)$, where $x = x_{h-r}(u)$.

**Definition 9.4.1**  Let $\mathcal{H}_h$ be a hedgehog of $\mathbb{R}^4$ such that $h \in C^\infty\left(\mathbb{S}^3; \mathbb{R}\right)$, and let $x = x_h(u)$ be a hyperbolic point of type $i$ of $\mathcal{H}_h$, ($i \in [1, 2]$). Then, any $v \in \mathbb{S}^3$ that is sufficiently close to $u$ on $\mathbb{S}^3$ is an $i$-saddle point of $\mathcal{F}\left(\nabla h_y\right)$, where $y = x_h(v)$. We say that such a $v$ disconnects $\left(h_y\right)^{-1}(\{0\})$ if $\left(h_y\right)^{-1}(\{0\})$ and $\left(h_y\right)^{-1}(\{0\})\backslash\{v\}$ does not have the same number of connected components. Now, the saddle point $u$ of $\mathcal{F}(\nabla h_x)$ is said to be critical if the disconnecting or nondisconnecting character of $v$ changes in every neighborhood of $u$ in the region of $(R_h)^{-1}(\mathbb{R}^*)$ containing it.

Return briefly to singularity and Morse theories viewpoints. The critical and singular sets, $\mathcal{C}_{\nabla h}$ and $\mathcal{S}_{\nabla h}$, can respectively be defined by

$$\begin{aligned} \mathcal{C}_{\nabla h} &:= \left\{(x, u) \in U\mathbb{R}^4 \,|\, \nabla(h_x)(u) = 0\right\} \\ &= \left\{(x, u) \in U\mathbb{R}^4 \,|\, x \in \mathcal{N}_{\nabla h}(u)\right\}, \end{aligned}$$

where $\mathcal{N}_{\nabla h}(u) := \{\nabla h(u)\} + \mathbb{R}u$ is the normal line to $\mathcal{H}_h$ at $x_h(u)$, and

$$\begin{aligned} \mathcal{S}_{\nabla h} &:= \left\{(x, u) \in \mathcal{C}_{\nabla h} \,\big|\, \exists k \in [1, 3], x = c^k_{\nabla h}(u)\right\} \\ &= \left\{(x, u) \in \mathcal{C}_{\nabla h} \,\big|\, \left(\det \circ \nabla^2\right)(h_x)(u) = 0\right\}. \end{aligned}$$

For each $i \in [1, 2]$, consider the following subsets of $\mathcal{C}_{\nabla h} \setminus \mathcal{S}_{\nabla h}$:

$$\Sigma_i(\nabla h) := \{(x, u) \in \mathcal{C}_{\nabla h} \setminus \mathcal{S}_{\nabla h} \,|\, u \text{ is an } i\text{-saddle point of } \mathcal{F}(\nabla h_x)\},$$

and

$$\Sigma^c_i(\nabla h) := \{(x, u) \in \Sigma_i(\nabla h) \,|\, u \text{ is a critical } i\text{-saddle point of } \mathcal{F}(\nabla h_x)\}.$$

For each $i \in [1, 2]$, $\Sigma_i (\nabla h) \setminus \Sigma_i^c (\nabla h)$ is the following union of disjoint open subsets: $\Sigma_i (\nabla h) \setminus \Sigma_i^c (\nabla h) = \Sigma_i^a (\nabla h) \bigsqcup \Sigma_i^n (\nabla h)$, where:

$$\Sigma_i^a (\nabla h) := \{(x, u) \in \Sigma_i (\nabla h) \mid u \text{ is an active } i\text{-saddle point of } \mathcal{F}(\nabla h_x)\}$$

and

$$\Sigma_i^n (\nabla h) := \{(x, u) \in \Sigma_i (\nabla h) \mid u \text{ is a neutral } i\text{-saddle point of } \mathcal{F}(\nabla h_x)\}.$$

More precisely, for each $i \in [1, 2]$, $\Sigma_i^c (\nabla h)$ separates $\Sigma_i^a (\nabla h)$ and $\Sigma_i^n (\nabla h)$ in $\Sigma_i (\nabla h)$. We are now prepared to prove our Theorem 9.2.3.

### 9.4.4   Proofs of Theorem 9.2.3 and Corollary 9.2.3

***Proof of Theorem 9.2.3*** Since $\mathcal{N}_{\nabla h} (u) \cap \mathcal{S}ing (\mathcal{F}_{\nabla h}) = \varnothing$, the 3 centers of principal curvature $c_{\nabla h}^1 (u)$, $c_{\nabla h}^2 (u)$ and $c_{\nabla h}^3 (u)$ are pairwise distinct. When $r \in \,\,]R_h^1 (u), R_h^2 (u)[$ tends to $R_h^1 (u)$, $u$ is an active 1-saddle point of $\mathcal{F}(\nabla h_x)$, where $x = x_{h-r} (u)$, whereas when $r \in \,\,]R_h^1 (u), R_h^2 (u)[$ tends to $R_h^2 (u)$, $u$ is a neutral 1-saddle point of $\mathcal{F}(\nabla h_x)$, where $x = x_{h-r} (u)$. More precisely, $(x_{h-r} (u), u) \in \Sigma_i^a (\nabla h)$ (resp. $(x_{h-r} (u), u) \in \Sigma_i^n (\nabla h)$) when $r \in \,\,]R_h^1 (u), R_h^2 (u)[$ tends to $R_h^1 (u)$ (resp. $R_h^2 (u)$). From this it follows that there exists some $r_1 \in \,\,]R_h^1 (u), R_h^2 (u)[$ such that $u$ is a critical 1-saddle point of $\mathcal{F}(\nabla h_{x_1})$, where $x_1 = x_{h-r_1} (u)$. Since $u$ is a critical 1-saddle point of $\mathcal{F}(\nabla h_{x_1})$, there exists some $v \in \mathbb{S}^3 \setminus \{u\}$ that is a critical point of $\mathcal{F}(\nabla h_{x_1})$ lying on the connected component, say $\mathcal{L}_u$, of the level set of $h_{x_1}$ that contains $u$. Such a $v$ is of course such that $x_1 = x_{h-r_1} (v)$. Since $\mathcal{N}_{\nabla h} (u) \cap \mathcal{S}ing (\mathcal{F}_{\nabla h}) = \varnothing$, if $v$ is a degenerate critical point of $(h - r_1)_{x_1}$, then $v$ is such that $R_{h-r_1} (v) = 0$ (since $v$ is degenerate), $\nabla R_{h-r_1} (v) \neq 0$ and $rk \left[ T_v x_{h-r_1} \right] = 2$ (otherwise $x_1$ would belong to $\mathcal{S}ing (\mathcal{F}_{\nabla h})$). It follows that under our assumptions, $\mathcal{H}_{h-r_1}$ has then a cusp singularity at $x_1 = x_{h-r_1} (v)$; that is, the image under $x_{h-r_1}$ of a neighborhood $\mathcal{V}$ of $v$ is diffeomorphic to a neighborhood of 0 in $\left\{ (x_1, x_2, x_3, x_4) \in \mathbb{R}^4 \mid x_2^2 = x_1^3 \right\}$. But in such a configuration, the relative transverse orientation of $x_{h-r_1} (v)$ is fully determined in the vicinity of $x_1$ by the index of the principal radius of curvature that vanishes at $v$: indeed, if $R_{h-r_1}^2 (v) = 0$, then the hyperbolic points of $x_{h-r_1} (\mathcal{V})$ are neutral in the vicinity of $x_1$, and if $R_{h-r_1}^1 (v) = 0$ (resp. $R_{h-r_1}^3 (v) = 0$) the relative transverse orientation of $x_{h-r_1} (\mathcal{V})$ at a regular point $x_{h-r_1} (w)$ that is close enough to $x_1 = x_{h-r_1} (v)$ is given by:

$$sgn \left[ R_{h-r_1} (w) \right] w \quad (\text{resp. } -sgn \left[ R_{h-r_1} (w) \right] w ).$$

Thus if $v$ is a degenerate critical point of $(h - r_1)_{x_1}$, then no switch of the relative transverse orientation (between the active and neutral modes) at $u$ can be due to the crossing of $x_{h-r_1}(\mathcal{V})$ with the image under $x_{h-r_1}$ of a neighborhood of $u$ in $\mathbb{S}^3$. Therefore, $v$ is a nondegenerate critical point of $(h - r_1)_{x_1}$ and thus a 1-saddle point of $\mathcal{F}(\nabla h_{x_1})$ by the way it is obtained. Thus $r_1 \in \,]R_h^1(u),\, R_h^2(u)[$ is such that $x_{h-r_1}(u) = x_h(u) - r_1 u$ is an (at least) type 1 double hyperbolic point of $\mathcal{H}_{h-r_1}$.

Finally, we can adapt the above arguments to prove that there exists some $r_2 \in \,]R_h^2(u),\, R_h^3(u)[$ such that $x_{h-r_2}(u) = x_h(u) - r_2 u$ is an (at least) type 2 double hyperbolic point of $\mathcal{H}_{h-r_2}$.  ∎

***Proof of Corollary 9.2.3*** Let $u \in \mathbb{S}^3$ such that $\mathcal{N}_{\nabla h}(u) \cap \mathcal{S}ing(\mathcal{F}_{\nabla h}) = \varnothing$. For such a point $u$, we know from our proof of Theorem 9.2.3 that there exists $(r_1, r_2) \in \mathbb{R}^2$ such that, for each $i \in [1, 2]$, $x_{h-r_i}(u) = x_h(u) - r_i u$ is an (at least) type $i$ double hyperbolic point of $\mathcal{H}_{h-r_i}$, and $u$ a critical saddle point of $\mathcal{F}(\nabla h_{x_i})$, where $x_i = x_{h-r_i}(u)$.

Assume first that $x_1 = x_h(u) - r_1 u$ is in $\mathbb{R}^4 \setminus \mathcal{F}_{\nabla h}$. Since $x_1$ is an (at least) double type 1 hyperbolic point, we then have $S_1(x_1) \geq 2$. Now (see "Index decomposition" in Sect. 9.1.2)

$$\sum_{k=1}^{2} i_{\mathcal{F}_{\nabla h}^k}(x_1) = (1 - C_m(x_1)) + (C_m(x_1) - S_1(x_1) - 1) = -S_1(x_1),$$

and consequently

$$i_{\mathcal{F}_{\nabla h}}(x_1) = \sum_{k=1}^{3} i_{\mathcal{F}_{\nabla h}^k}(x_1) = -S_1(x_1) + i_{\mathcal{F}_{\nabla h}^3}(x_1) \leq -S_1(x_1).$$

Therefore $i_{\mathcal{F}_{\nabla h}}(x_1) \leq -2$, and thus $N_{\nabla h}(x_1) = 2\left(1 - i_{\mathcal{F}_{\nabla h}}(x_1)\right) \geq 6$.

Now, if $x_1 \in \mathcal{F}_{\nabla h}$, we can proceed as in the proof of Corollary 9.2.1 to show that arbitrarily close to $x_1$, there exists some $y_1 \in \mathbb{R}^4 \setminus \mathcal{F}_{\nabla h}$ such that $S_1(y_1) \geq 2$, and thus $N_{\nabla h}(y_1) = 2\left(1 - i_{\mathcal{F}_{\nabla h}}(y_1)\right) \geq 6$. Therefore, in any case, $x_1 = x_h(u) - r_1 u$ is in the closure of $\left\{x \in \mathbb{R}^4 \setminus \mathcal{F}_{\nabla h} \,|\, N_{\nabla h}(x) \geq 6\right\}$.

But, in fact, we can be more precise. Since $u$ is a critical saddle point of $\mathcal{F}(\nabla h_{x_1})$, we know that arbitrarily close to $x_1$, there exists some $y_1 \in \mathbb{R}^4 \setminus \mathcal{F}_{\nabla h}$ that is a neutral hyperbolic point and thus a point of $Int\left(\mathcal{F}_{\nabla h}^2\right)$. Furthermore, since $u$ is a critical saddle point of $\mathcal{F}(\nabla h_{x_1})$, we have $C_m(x_1) \geq 2$, and we know [64] that arbitrarily close to $x_1$, there are points $y_1 \in \mathbb{R}^3 \setminus \mathcal{F}_{\nabla h}$, and that if $y_1 \in \mathbb{R}^3 \setminus \mathcal{F}_{\nabla h}$ is close enough to $x$, then $C_m(y_1) \geq 2$ and thus $y_1 \in Int\left(\mathcal{F}_{\nabla h}^1\right)$. Therefore, arbitrarily close to $x_1$, there exists some $y_1 \in \mathbb{R}^4 \setminus \mathcal{F}_{\nabla h}$ such that $y_1 \in Int\left(\mathcal{F}_{\nabla h}^1\right) \cap Int\left(\mathcal{F}_{\nabla h}^2\right)$.

Of course, we can prove in the same way that $x_2 = x_h(u) - r_2 u$ is also in the closure of $\left\{x \in \mathbb{R}^4 \setminus \mathcal{F}_{\nabla h} \,|\, N_{\nabla h}(x) \geq 6\right\}$, and more precisely in the closure of $Int\left(\mathcal{F}_{\nabla h}^2\right) \cap Int\left(\mathcal{F}_{\nabla h}^3\right)$.  ∎

## 9.5   Further Results and Remarks

### 9.5.1   *Minimal Spherical Shell of a Convex Body*

As we said in Sect. 9.2, E. Heil proved the concurrent normal conjecture in $\mathbb{R}^2$ (resp. $\mathbb{R}^3$) using of the existence of a minimal spherical shell for any convex body of $\mathbb{R}^2$(resp. $\mathbb{R}^3$), which had been established by T. Bonnesen [22] (resp. N. Kritikos [80]). In 1988, I. Bárány extended this existence to higher dimensions [12]. Given a convex body $K$ in $\mathbb{R}^{n+1}$, he defined

$$r\left(x\right) := \max\left(\{r \in \mathbb{R}_+ \,|B\left(x,r\right) \subseteq K\}\right),$$

and

$$R\left(x\right) := \min\left(\{R \in \mathbb{R}_+ \,|K \subseteq B\left(x,r\right)\}\right),$$

where $B\left(x,r\right)$ denotes the closed ball with center $x$ and radius $r$, and proved the following:

**Theorem 9.5.1 (I. Bárány [12])** *There exists a unique point $x_0$ in $K$ at which the function $R - r : K \to \mathbb{R}_+$, $x \mapsto R\left(x\right) - r\left(x\right)$ attains its minimum value.*

The set $C\left(x_0; r\left(x_0\right), R\left(x_0\right)\right) := \left\{x \in \mathbb{R}^{n+1}\,|r\left(x_0\right) \leq \|x - x_0\| \leq R\left(x_0\right)\right\}$ is called the **minimal spherical shell** of $K$. On each sphere bounding $C\left(x_0; r\left(x_0\right), R\left(x_0\right)\right)$ there are at least two points of $K$. More precisely, $C\left(x_0; r\left(x_0\right), R\left(x_0\right)\right)$ can be characterized as follows.

**Theorem 9.5.2 (I. Bárány [12])** *The point $x_0 \in K$ is the center of the minimal spherical shell of $K$ if, and only if, there exist $\left(u_1, \ldots, u_p\right)\left(\mathbb{S}^n\right)^p$ and $\left(v_1, \ldots, v_q\right) \in \left(\mathbb{S}^n\right)^q$, $\left(p, q \geq 1\right)$, such that:*

*(i)*

$$Conv\left(\{u_1, \ldots, u_p\}\right) \cap Conv\left(\{v_1, \ldots, v_q\}\right) \neq \varnothing,$$

*where $Conv$ denotes the convex hull;*

*(ii)*

$$x_0 + r\left(x_0\right) u_i \in \partial K \text{ and } x_0 + R\left(x_0\right) v_j \in \partial K \text{ for all } \left(i, j\right) \in [1, p] \times [1, q],$$

*where $\partial K$ denotes the boundary of $K$.*

### 9.5.2   *Extension to Hedgehogs*

Any hedgehog $\mathcal{H}_h$ of $\mathbb{R}^{n+1}$ that is such that $h \in C^2\left(\mathbb{S}^n; \mathbb{R}\right)$ can be regarded as a parallel hypersurface to the boundary of some convex body of class $C_+^2$, say $K$, in

$\mathbb{R}^{n+1}$ (see Remark 2 in Sect. 2.2.1): there exists $\rho \in \mathbb{R}$ such that $h = h_K - \rho$, where $h_K$ is the support function of $K$. We can thus define the ***minimal spherical shell of*** $\mathcal{H}_h$ to be the transversely oriented shell that is bounded by the transversely oriented spheres or point spheres

$$S\left(x_0; r\left(x_0\right) - \rho\right) \quad \text{and} \quad S\left(x_0; R\left(x_0\right) - \rho\right),$$

where $C\left(x_0; r\left(x_0\right), R\left(x_0\right)\right)$ is the minimal spherical shell of $K$, and where $S\left(a; r\right)$ denotes the sphere of radius $|r|$ centered at $a$ that is transversely oriented by its outward (resp. inward) pointing normals if $r > 0$ (resp. $r < 0$) holds, and the point sphere $\{a\}$ if $r = 0$ holds. This definition is of course independent from the choice of the convex body $K$ of class $C_+^2$ the boundary of which is parallel to $\mathcal{H}_h$.

**Remark** All these minimal spherical shells can be interpreted in terms of pedal hypersurfaces. Recall that the pedal hypersurface of the hedgehog $\mathcal{H}_h \subset \mathbb{R}^{n+1}$ with respect to a point $x \in \mathbb{R}^{n+1}$ is parametrized by

$$\mathbb{S}^n \to \mathbb{R}^{n+1}$$
$$u \longmapsto x + h_x\left(u\right) u,$$

where $h_x\left(u\right) = h\left(u\right) - \langle x, u \rangle$. In other words, to any $u \in \mathbb{S}^n$ is associated the foot of the perpendicular line from $x$ to the support hyperplane $\mathcal{H}_h$ at $x_h\left(u\right)$. If $C\left(x_0; r\left(x_0\right), R\left(x_0\right)\right)$ is the minimal spherical shell of some convex body $K$ with support function $h \in C^2\left(\mathbb{S}^n; \mathbb{R}\right)$ in $\mathbb{R}^{n+1}$, we can easily check that: $(i)$ the pedal hypersurface of the hedgehog $\mathcal{H}_{h-(r(x_0)+R(x_0))/2}$ with respect to $x_0$ is contained in the closed ball $B\left(x_0; \frac{R(x_0)-r(x_0)}{2}\right)$; $(ii)$ no other pedal hypersurface of $\mathcal{H}_h$ is contained in a smaller closed ball of $\mathbb{R}^{n+1}$; $(iii)$ Properties $(i)$ and $(ii)$ permit to characterize the minimal spherical shell of $K$. Moreover, this interpretation extends to hedgehogs.

Let us give some examples. Figure 9.4 shows: $(a)$ the ellipse with support function $\theta \in \mathbb{S}^1 = \mathbb{R}/2\pi\mathbb{Z} \mapsto h\left(\theta\right) := \sqrt{9\cos^2\theta + 25\sin^2\theta}$, the minimal spherical shell of which is $C\left(0_{\mathbb{R}^2}; 3, 5\right)$ where $0_{\mathbb{R}^2}$ is the origin of $\mathbb{R}^2$; $(b)$ the hedgehog with support function $h - \left(r\left(0_{\mathbb{R}^2}\right) + R\left(0_{\mathbb{R}^2}\right)\right)/2 = h - 4$, which is parallel to the ellipse; $(c)$ the pedal $\mathcal{P}_0\left(\mathcal{H}_{h-4}\right)$ of $\mathcal{H}_{h-4}$ with respect to the origin; $(d)$ all these curves together with the circle $C\left(0_{\mathbb{R}^2}; 1\right)$, which has center $0_{\mathbb{R}^2}$ and radius $\frac{1}{2}\left(5 - 3\right) = 1$. We can note that in this example, the minimal spherical shell $C\left(0_{\mathbb{R}^2}; -1, 1\right)$ of $\mathcal{H}_{h-4}$ is reduced to a single circle: it is in fact formed by two coincident circles that are equipped with opposite transverse orientations. Figure 9.5 shows the hedgehog with support function $h - \frac{7}{2}$, its pedal with respect to the origin, and both curves together with the minimal spherical shell (abbreviated into m.s.s) of $\mathcal{H}_{h-\frac{7}{2}}$.

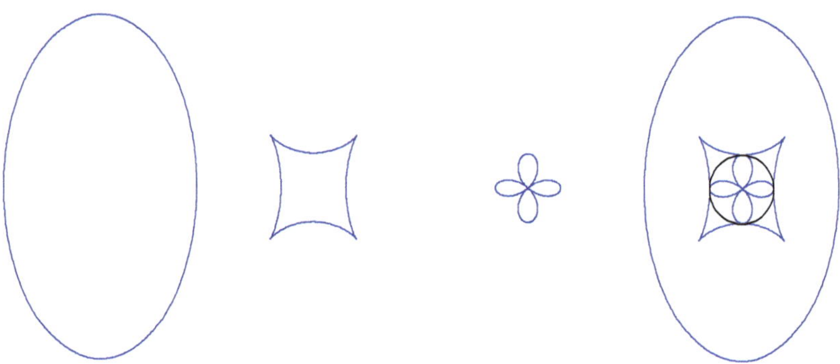

**Fig. 9.4**  (*a*) $\mathcal{H}_h$ (*b*) $\mathcal{H}_{h-4}$ (*c*) $\mathcal{P}_0(\mathcal{H}_{h-4})$; (*d*) These curves and $C(0_{\mathbb{R}^2}; 1)$

**Fig. 9.5**  $\mathcal{H}_{h-\frac{7}{2}}$, $\mathcal{P}_0\left(\mathcal{H}_{h-\frac{7}{2}}\right)$, these curves with the m.s.s. of $\mathcal{H}_{h-\frac{7}{2}}$

Bárány's characterization theorem of the minimal spherical shell can of course be adapted to hedgehogs:

**Theorem 9.5.3**  *Let $\mathcal{H}_h$ be a hedgehog of $\mathbb{R}^{n+1}$ such that $h \in C^2\left(\mathbb{S}^n; \mathbb{R}\right)$, and let $x \in \mathbb{R}^{n+1}$ and $(r, R) \in \mathbb{R}^2$ be such that $r \leq h_x \leq R$, where $h_x(u) = h(u) - \langle x, u \rangle$, $(u \in \mathbb{S}^n)$. Then $C(x; r, R)$ is the minimal spherical shell of $\mathcal{H}_h$ if, and only if, there exist $(u_1, \ldots, u_p)(\mathbb{S}^n)^p$ and $(v_1, \ldots, v_q) \in (\mathbb{S}^n)^q$, $(p, q \geq 1)$, such that:*

*(i)*

$$Conv\left(\{u_1, \ldots, u_p\}\right) \cap Conv\left(\{v_1, \ldots, v_q\}\right) \neq \varnothing,$$

*where $Conv$ denotes the convex hull;*

*(ii)*

$$h_x(u_i) = r, \ h_x(v_j) = R, \ and \ \nabla h_x(u_i) = \nabla h_x(v_j) = 0\,.$$

*for all* $(i, j) \in [1, p] \times [1, q]$.

Now, let us see how we can construct $h \in C^2\left(\mathbb{S}^3; \mathbb{R}\right)$ such that:

*(i)* $|h| \leq 1$;
*(ii)* the set $\left\{u \in \mathbb{S}^3 \,|\, h(u) = -1\right\}$ has exactly two distinct elements $u_1, u_2$, and the set $\left\{u \in \mathbb{S}^3 \,|\, h(u) = 1\right\}$ has exactly two distinct elements $v_1, v_2$;
*(iii)* $[u_1 u_2] \cap [v_1 v_2] \neq \varnothing$;
*(iv)* $h$ has exactly six critical points: 2 minima, 2 maxima and two saddle points.

Let $(e_1, e_2, e_3, e_4)$ be the canonical basis of $\mathbb{R}^4$. Put $u_1 = e_1, u_2 = -e_1, v_1 = -e_4, v_2 = e_4, w_1 = -e_2, w_2 = e_2$, and identify $\mathbb{S}^3 \setminus \{e_4\}$ to $\mathbb{R}^3$ through the stereographic projection from $e_4$ onto the hyperplane $Vect(e_1, e_2, e_3)$. The point $v_1 = -e_4$ is thus identified with the origin of $\mathbb{R}^3$. We then see that we can choose $h$ so that $h(u_1) = h(u_2) = -1, h(w_1) = -1/2, h(w_2) = 1/2$, and $h(v_1) = h(v_2) = -1$, such that the foliation of $\mathbb{R}^3$ generated by the level surfaces of $h$ is made by leaves diffeomorphic to $\mathbb{S}^2$, with the exception of the leaves shown in Fig. 9.6 (on which the dashed line is the circle $\mathbb{S}^2 \cap Vect(e_1, e_2)$). On the first picture of Fig. 9.6, the curve has to be rotated around the axis $\mathbb{R}e_2$.

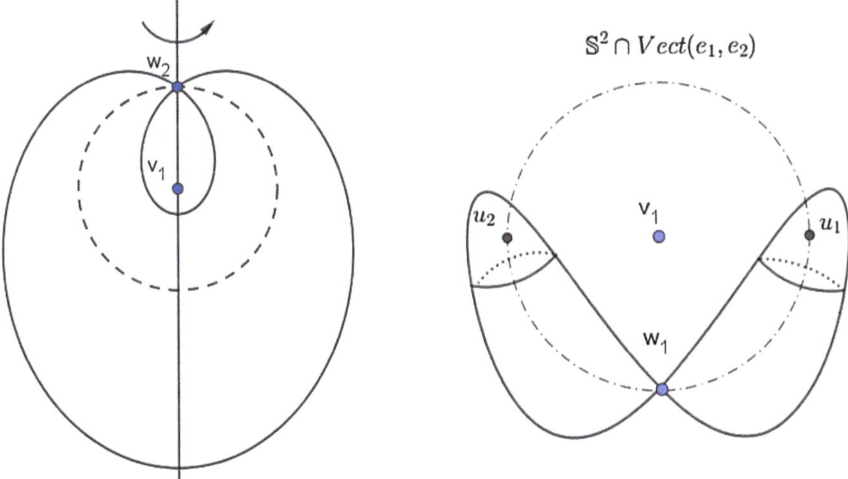


**Fig. 9.6**  Illustration for the proof of Theorem 9.5.4

For such a $h \in C^2\left(\mathbb{S}^3; \mathbb{R}\right)$, $C\left(0_{\mathbb{R}^4}; -1, 1\right)$ is the minimal spherical shell of $\mathcal{H}_h$, and there are only six normal lines to $\mathcal{H}_h$ passing through $0_{\mathbb{R}^4}$. Therefore:

**Theorem 9.5.4**  *It is not true that for any convex body $K$ of $\mathbb{R}^4$, there are at least 8 normal lines passing through the center of the minimal spherical shell of $K$.*

# Chapter 10
# Miscellaneous Questions Regarding Hedgehogs

## 10.1 Convolution of Hedgehogs

As observed by H. Görtler in [55] and [56], we can define the convolution product of two plane hedgehogs $\mathcal{H}_f$ and $\mathcal{H}_g$ in $\mathbb{R}^2$ as the plane hedgehog whose support function is given by

$$(f * g)(\theta) = \frac{1}{2\pi} \int_0^{2\pi} f(\theta - \alpha) g(\alpha) \, d\alpha,$$

for all $\theta \in \mathbb{S}^1$. We can then check at once that $\left(\mathcal{H}^2, +, ., *\right)$ is a commutative and associative algebra. H. Görtler also observed that the convolution product of two plane convex bodies remains a plane convex body. The interest of convolution of hedgehogs is that properties of one factor are often transmitted to the product. Of course, we think immediately of regularity properties but we also mentioned the following properties in [100]: to be centered (centrally symmetric with center at the origin), to be projective (i.e., to have an antisymmetric support function), to be of constant width. For convolution of Fuchsian hedgehogs, we refer the reader to Sect. 7.2.5.

A natural way of defining a (non-abelian) convolution product on the vector space $\mathcal{H}^{n+1}$ of arbitrary hedgehogs of $\mathbb{R}^{n+1}$ is to proceed as follows: (1) First, we identify $\mathbb{S}^n$ with the homogeneous space $G/H$, where $G$ is the group $SO(n+1)$ of rotations of $\mathbb{R}^{n+1}$ and $H$ the stabilizer subgroup of $G$ with respect to the north pole, say $v$, of $\mathbb{S}^n$ (that is, the subgroup $H$ of $G$ formed by the rotations $r \in G$ that leave $v$ fixed); any support function $h : \mathbb{S}^n \to \mathbb{R}$ can thus be regarded as a function $h : G \to \mathbb{R}$ such that $h(rs) = h(r)$ for all $(r, s) \in G \times H$; (2) Next, given any two arbitrary hedgehogs $\mathcal{H}_f$ and $\mathcal{H}_g$ of $\mathbb{R}^{n+1}$, we can define their convolution product $\mathcal{H}_f * \mathcal{H}_g$

© The Author(s), under exclusive license to Springer Nature Switzerland AG 2026
Y. Martinez-Maure, *Hedgehog Theory*, Lecture Notes in Mathematics 2381,
https://doi.org/10.1007/978-3-032-04298-9_10

as the hedgehog $\mathcal{H}_{f*g}$ with support function

$$(f * g)(r) = \int_G f\left(rt^{-1}\right) g(t)\, dm_G(t) \quad \text{for all } r \in G,$$

where $m_G$ is the normalized Haar measure on $G$. This construction of $\mathcal{H}_f * \mathcal{H}_g$ is essentially due to E. Grinberg and G. Zhang [58]. As expected, this convolution product behaves well with respect to expansions in series of spherical harmonics, and properties of one factor are often transmitted to the product (for instance, to be centered, projective, convex, of constant width, or a zonoid).

But of course, in the case of hedgehogs of $\mathbb{R}^4$ it is simpler to make use of quaternions and thus to define the convolution product $\mathcal{H}_f * \mathcal{H}_g$ of $\mathcal{H}_f$ and $\mathcal{H}_g$ in $\mathbb{R}^4$ to be the hedgehog $\mathcal{H}_{f*g}$ with support function

$$(f * g)(u) = \int_{\mathbb{S}^3} f(v\overline{u})\, g(v)\, d\sigma(v) \quad \text{for all } u \in \mathbb{S}^1_{\mathbb{H}} \cong \mathbb{S}^3,$$

where $\sigma$ is the spherical Lebesgue measure on $\mathbb{S}^3$.


## 10.2   Eversion of $\mathbb{S}^2$ Through Generic Paths of Hedgehogs

### 10.2.1   Introduction

In 1958, S. Smale published an abstract result of which an astonishing consequence was that any sphere of $\mathbb{R}^3$ can be everted (i.e. turned inside-out) by means of a regular homotopy [173]. Smale's proof made it possible to understand the theoretical possibility of carrying out such an eversion but did not make it possible to imagine an explicit eversion. In 1961, A. Shapiro was the first to devise a procedure to carry out an explicit eversion of the sphere. Shapiro's idea was to transform the sphere into a double cover of Boy's surface, which is an immersion of the real projective plane with a threefold axis of symmetry (see Sect. 2.2.2), and then to exchange the two sheets through the Boy's surface. Shapiro communicated this idea to B. Morin [48], but did not publish it. It was not until 1966 and an article by A. Phillips in the journal Scientific American to have the first illustrations of the eversion imagined by Shapiro [147].

In this section, which presents the results of one of our paper of 2015 [107], we are again interested in the eversion of the sphere in $\mathbb{R}^3$ but with different constraints on the transformations that are allowed. We will no longer require that the homotopy be regular, but on the other hand we will require that all the intermediate surfaces (possibly singular) form a generic path of hedgehogs, that is, of surfaces parametrized by their Gauss map.

This work is motivated by Problem 2.6.1 "*Does there exist a generic projective hedgehog without any swallowtail?*". Recall that generic singularities of $C^\infty$-

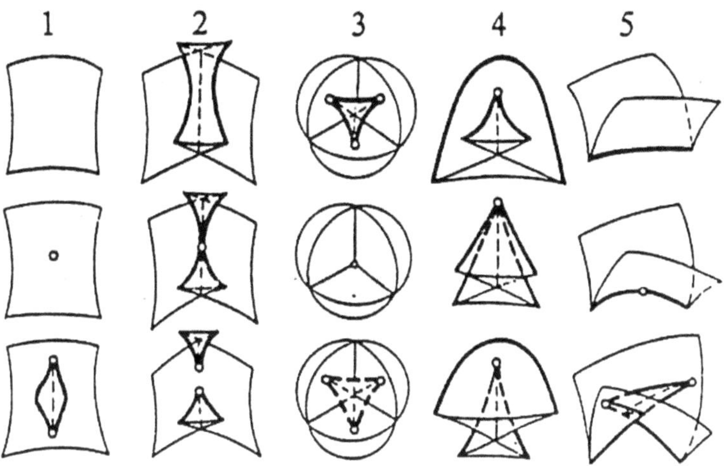

**Fig. 10.1** The 5 types of metamorphoses occurring in generic 1-parameter families of smooth hedgehogs (extracted from [6])

hedgehogs are cuspidal edges and swallowtails in $\mathbb{R}^3$ (see Sect. 2.6.3). We can distinguish two types of swallowtails, negative or positive, according to the sign of the Gaussian curvature on the tail (see Fig. 2.14b and c).

The question that will therefore be the subject of our study here will be to understand what are the generic singularities that are inevitable during an eversion of the sphere following a generic path of hedgehogs.

Let $\mathcal{H}^3$ denote in this subsection the linear space of $C^\infty$-hedgehogs, which can be regarded as $C^\infty\left(\mathbb{S}^2; \mathbb{R}\right)$ if we identify $C^\infty$-hedgehogs with their support functions. The study of the five types of wave front metamorphoses occurring in generic 1-parameter families of $C^\infty$-hedgehogs (see Fig. 10.1) allows us to affirm that if a generic path of $C^\infty$-hedgehogs $\gamma \; : \; [0, 1] \to \mathcal{H}^3, t \mapsto \mathcal{H}_{h_t}$ carries out an eversion of the sphere (or more generally of a closed convex surface), then there exist values of $t$ for which $\mathcal{H}_{h_t}$ has at least a pair of swallowtails. Moreover, it allows us to say that when $t$ continuously increases from 0 to 1, the first swallowtails that appears are necessarily negative ones. Now, nothing indicates a priori that such a generic path necessarily includes some hedgehog $\mathcal{H}_{h_t}$ having a pair of positive swallowtails. The main aim of this subsection is precisely to establish the existence of such a hedgehog $\mathcal{H}_{h_t}, \; (t \in [0, 1])$.

**Theorem 10.2.1** *If a generic path of $C^\infty$-hedgehogs $\gamma \; : \; [0, 1] \to \mathcal{H}^3, t \mapsto \mathcal{H}_{h_t}$ carries out an eversion of the sphere $\mathbb{S}^2$ (and thus $\mathcal{H}_{h_0}$ and $\mathcal{H}_{h_1}$ are such that $x_{h_0}\left(\mathbb{S}^2\right) = x_{h_1}\left(\mathbb{S}^2\right)$, but equipped with opposite usual transverse orientations), then there exists some $t \in [0, 1]$ such that $\mathcal{H}_{h_t}$ has at least a pair of positive swallowtails.*

For the sake of convenience we recall here the 5 types of metamorphoses occurring in generic 1-parameter families of $C^\infty$-hedgehogs (Fig. 10.1).

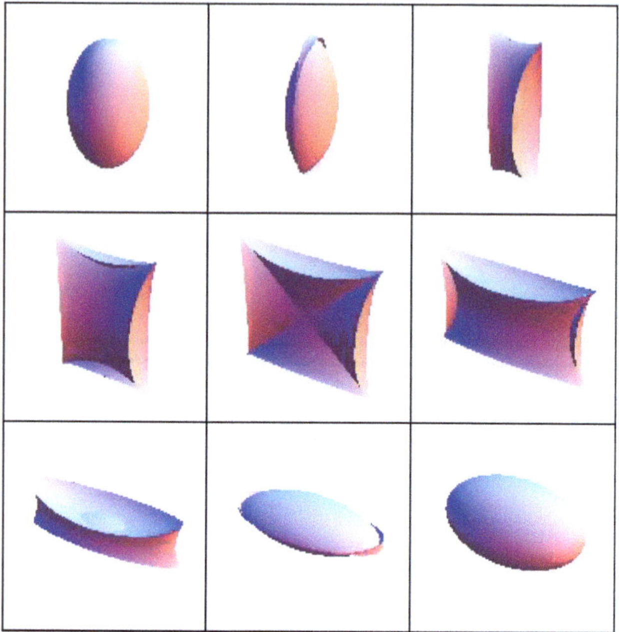

**Fig. 10.2**  A simple example of generic eversion

We will see below a simple example of a generic eversion (see Fig. 10.2). It is worth remembering here that any hedgehog $\mathcal{H}_h = x_h\left(\mathbb{S}^2\right) \in \mathcal{H}^3$ is equipped with a transverse orientation such that, at each regular point $x_h\left(u\right)$ of $\mathcal{H}_h$, the normal line is oriented by the vector $sgn\left[R_h\left(u\right)\right]u$, where $sgn\left(.\right)$ is the signum function and $R_h$ the curvature function of $\mathcal{H}_h$. For every $h \in C^\infty\left(\mathbb{S}^2; \mathbb{R}\right)$, the hedgehogs $\mathcal{H}_h$ and $\mathcal{H}_{\widetilde{h}}$, where $\widetilde{h}\left(u\right) = -h\left(-u\right)$ for all $u \in \mathbb{S}^2$, have the same geometrical realizations (that is, $x_h\left(\mathbb{S}^2\right) = x_{\widetilde{h}}\left(\mathbb{S}^2\right)$ but are equipped with opposite transverse orientations, since $x_{\widetilde{h}}\left(-u\right) = x_h\left(u\right)$ for all $u \in \mathbb{S}^2$. An eversion of a hedgehog $\mathcal{H}_h \in \mathcal{H}^3$ is thus the data of a path $\gamma : [0, 1] \to \mathcal{H}^3, t \mapsto \mathcal{H}_{h_t}$ such that $\mathcal{H}_{h_0} = \mathcal{H}_h$ and $\mathcal{H}_{h_1} = \mathcal{H}_{\widetilde{h}}$. For an eversion of the sphere $\mathbb{S}^2$, $h_0$ and $h_1$ will simply be the constant functions -1 and 1 on $\mathbb{S}^2$.

### *10.2.2  Proof of Theorem 10.2.1*

The proof of Theorem 10.2.1 will be based on the $j$-index introduced in Sect. 2.9 and on the associated transverse orientation, called absolute. Recall that, given a hedgehog $\mathcal{H}_h \in \mathcal{H}^3$, the $j_h$-index of a point $x \in \mathbb{R}^3 \backslash \mathcal{H}_h$ with respect to $\mathcal{H}_h$ is given by:

$$j_h\left(x\right) := 1 - c_h\left(x\right),$$

where $c_h(x)$ is the number of connected components of $(h_x)^{-1}(\{0\})$ on $\mathbb{S}^2$. We know that $j_h : x \mapsto j_h(x)$ remains constant on each connected component of $\mathbb{R}^3 \setminus \mathcal{H}_h$. The absolute transverse orientation of $\mathcal{H}_h$ is such that whenever $x_h(u)$ is a simple regular point of $\mathcal{H}_h$, then the normal line to $\mathcal{H}_h$ at $x_h(u)$, is oriented in the direction that $j_h$ decreases by one unit. On elliptic regions, the absolute transverse orientation is simply given by the direction of convexity. As we saw, a reversal of the absolute transverse orientation can only occur along certain self-intersection curves made of double hyperbolic points, which we call *inversion curves*. We introduced a function $\varepsilon_h : \mathbb{S}^2 \to \{-1, 0, 1\}$ whose sign indicates at regular points if the usual and the absolute transverse orientations coincide or not. If $x_h(u)$ is simple and regular, we defined $\varepsilon_h(u) \in \{-1, 1\}$ so that $v_h(u) := \varepsilon_h(u) R_h(u) u$ direct the normal line of $\mathcal{H}_h$ (equipped with its absolute transverse orientation) at $x_h(u)$. At such a point $u \in \mathbb{S}^2$, we define $\varepsilon_h(u) \in \{-1, 1\}$ in such a way that the unit vector

$$v_h(u) := \varepsilon_h(u)\, sgn\,[R_h(u)]\, u$$

direct the normal line of $\mathcal{H}_h$ at $x_h(u)$ when $\mathcal{H}_h$ is equipped with its absolute transverse orientation. Otherwise, we put $\varepsilon_h(u) = 0$. See in Sect. 2.9 the example of the Roman surface.

The notion of an inversion curve makes it possible to distinguish two types of positive swallowtails: those from which originates an inversion curve and the others. It is worth noting that the nature of a positive swallowtail indicates its position relative to the hyperbolic region that adjoins it. Let us cut out on a $C^\infty$-hedgehog $\mathcal{H}_h$ a neighborhood of a positive swallowtail such as the one represented in Fig. 2.14b. The curve of double points divides the portion of $\mathcal{H}_h$ considered into two parts: a part $A$ in the form of a tail and a part $B$ which consists only in hyperbolic points. Part $B$ has two sides: that of $A$ and the opposite side. If the $j_h$-index decreases when we cross $B$ outside of the curve of double points in the direction of $A$, it appears that the curve of double points is not an inversion curve. On the other hand, if $j_h(x)$ increases due to such a crossing, then it appears that the curve of double points is indeed an inversion curve. For these reasons, we will say that a positive swallowtail is *outgoing* if the curve of double points that comes from it is an inversion curve, and that it is *incoming* otherwise. In the example of the projective hedgehog version of the Roman surface, all the positive swallowtails are outgoing (see Fig. 2.7).

***Proof of Theorem 10.2.1*** Let $\gamma : [0, 1] \to \mathcal{H}^3, t \mapsto \mathcal{H}_{h_t}$ be a generic path of hedgehogs realizing an eversion of the sphere. Since $\varepsilon_{h_0}$ is constant equal to 1 and $\varepsilon_{h_1}$ constant equal to $-1$, there are necessarily values of $t \in \,]0, 1[$ for which the curves on which $\varepsilon_{h_t}$ changes sign disconnect $\mathbb{S}^2$. We know that inversion curves are curves made of double hyperbolic points. When $t$ increases continuously from 0 to 1, an inversion curve of $\mathcal{H}_{h_t}$ can only appear in two cases: either when two portions of hyperbolic regions come into contact; or when a pair of outgoing positive swallowtails are born. Now, if two portions of hyperbolic regions come into contact at the instant $t$, this can only be in the common image by $x_{h_t}$ of two antipodal points $-u$ and $u$ of $\mathbb{S}^2$: $x_{h_t}(-u) = x_{h_t}(u)$. In such a case, we have necessarily $\varepsilon_{h_t}(-v) =$

$-\varepsilon_{h_t}(v) \neq 0$ for every $v \neq u$ sufficiently close to $u$, which implies that there already existed an inversion curve disconnecting $\mathbb{S}^2$. Therefore, when $t$ increases continuously from 0 to 1, the first appearance of an inversion curve is necessarily concomitant with the appearance of a pair of outgoing positive swallowtails, which achieves the proof.                                                                  ∎

### *10.2.3   Complements*

Figure 10.2 describes the simplest example of a generic eversion that can be designed. This eversion is given by the path of hedgehogs $(h_t)_{t\in[0,1]}$ defined by:

$$h_t(x, y, z) := t\sqrt{x^2 + y^2 + 3z^2} - (1 - t)\sqrt{3x^2 + y^2 + z^2}.$$

This path of centered hedgehogs connects two ellipsoids equipped with opposite transverse orientations. The figure is read from left to right and from top to bottom. We start with the first ellipsoid. Two pairs of negative swallowtails born following type 1 metamorphoses (see Fig. 10.1). They then evolve and die following type 2 metamorphoses to arrive at the last hedgehog of the first line. This hedgehog has a hyperbolic ring-shaped region and two elliptical regions. Then two pairs of positive swallowtails born following type 1 metamorphoses and evolve until reaching the central hedgehog by a type 4 metamorphose. The second part of the eversion consists of passing from this central hedgehog to the second ellipsoid by performing operations opposite to those which we have just described.

   Finally, let us indicate how we can deduce from the projective hedgehog version of the Roman surface represented in Fig. 2.7 an example of a projective hedgehog having only generic singularities (i.e., cuspidal edges and swallowtails). The projective hedgehog version of the Roman surface admits a threefold axis of symmetry and three lines of self-intersections whose end points are three positive swallowtails and three singularities of type D4 (see the central figure of the type 4 metamorphose). Of course, if we count these numbers on the sphere $\mathbb{S}^2$ instead of counting them in the projective plane all these numbers must be doubled (since each point of the surface is obtained for two antipodal distinct points of $\mathbb{S}^2$). The idea is naturally to resort to a type 4 metamorphose while preserving the threefold axis of symmetry in order to obtain a projective hedgehog carrying exactly six swallowtails, all positive. This can be done easily and we can check that the hedgehog with support function

$$h(x, y, z) := x\left(x^2 - 3y^2\right) + 2z^3 + 2y\left(y^2 - 3x^2\right)z^2, \text{ where } (x, y, z) \in \mathbb{S}^2 \subset \mathbb{R}^3,$$

meets all our expectations. Figure 10.3 gives two views of this projective hedgehog $\mathcal{H}_h$, which has only generic singularities.

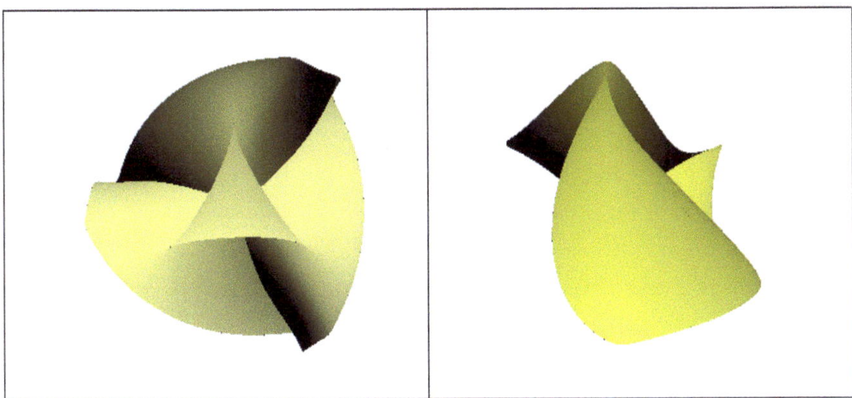

**Fig. 10.3** Two views of a generic projective hedgehog

## 10.3   Polytope in $\mathbb{R}^3$ with Prescribed Directions and Perimeters if the Facets

Hedgehogs are not only a generalization of convex bodies, but also a way of thinking about convex hypersurfaces in conjunction with their spherical images. The results of this subsection taken from [110][1] will illustrate this point.

In 1897, Hermann Minkowski studied the problem of prescribing the areas and outer unit normals of the facets of a 3-dimensional polytope. The existence and uniqueness theorem that he obtained is one the most fundamental results in the theory of polytopes. This subsection is devoted to the analogous problem of prescribing the perimeters and outer unit normals of the facets of a 3-dimensional polytope. Our main result (Theorem 10.3.4) gives a necessary and sufficient condition for the existence and uniqueness up to a translation of a 3-dimensional polytope $P$ in $\mathbb{R}^3$ having $N$ facets with given unit outward normals $n_1, \dots, n_N$ and corresponding facet perimeters $L_1, \dots, L_N$.

**Introduction to the Problem**

In this subsection, a polytope of $\mathbb{R}^3$ is the convex hull of finitely many points in $\mathbb{R}^3$. The classical Minkowski problem for polytopes in $\mathbb{R}^3$ concerns the following question:

> *Given a collection $n_1, \dots, n_N$ of $N$ pairwise distinct unit vectors in $\mathbb{R}^3$ and $F_1, \dots, F_N$ a collection of $N$ positive real numbers, is there a polytope $P$ in $\mathbb{R}^3$*

---

[1] This article was published in Discrete Math. 343, Y. Martinez-Maure, *Existence and uniqueness theorem for a 3-dimensional polytope in* $\mathbb{R}^3$ *with prescribed directions and perimeters of the facets*, 111770, 6 p. Copyright Elsevier 2020.

*having the $n_i$ as its facet unit outward normals and the $F_i$ as the corresponding facet areas $(1 \leq i \leq N)$, and, if so, is P unique up to translations?*

H. Minkowski proved the following uniqueness theorem (see [4, Theorem 9, p. 107]):

**Theorem 10.3.1 (H. Minkowski [125] and [127, pp. 103–121])** *A polytope in $\mathbb{R}^3$ is uniquely determined, up to translations, by the directions and the areas of its facets.*

A well-known necessary condition for the existence of a polytope having facet unit outward normals $n_1, \ldots, n_N$ and corresponding facet areas $F_1, \ldots, F_N$ is that:

$$\sum_{i=1}^{N} F_i n_i = 0.$$

An existence theorem of H. Minkowski ensures that this condition is both necessary and sufficient:

**Theorem 10.3.2 (H. Minkowski [125] and [127, pp. 103–121])** *Let $n_1, \ldots, n_N \in \mathbb{R}^3$ be N pairwise distinct unit vectors linearly spanning $\mathbb{R}^3$ and let $F_1, \ldots, F_N$ be N positive real numbers. There exists a polytope P in $\mathbb{R}^3$ having N facets with unit outward normals $n_1, \ldots, n_N$ and corresponding facet areas $F_1, \ldots, F_N$ if, and only if, we have*

$$\sum_{i=1}^{N} F_i n_i = 0.$$

Here, we have to mention that Theorem 10.3.2 is only the 3-dimensional version of the classical Minkowski existence and uniqueness theorem [167, p. 455], which is valid in $\mathbb{R}^d$ for all $d \geq 2$. The proof of the main result (Theorem 10.3.4) will make use of the 2-dimensional version, which is almost trivial:

**The Minkowski theorem for convex polygons in $\mathbb{R}^2$.**
*Let $n_1, \ldots, n_N \in \mathbb{R}^2$ be N pairwise distinct unit vectors linearly spanning $\mathbb{R}^2$ and let $l_1, \ldots, l_N$ be N positive real numbers. There exists a convex polygon P in $\mathbb{R}^2$ having N edges with unit outward normals $n_1, \ldots, n_N$ and corresponding edge lengths $l_1, \ldots, l_N$ if, and only if,*

$$\sum_{i=1}^{N} l_i n_i = 0.$$

This subsection is devoted to the analogue of the classical Minkowski problem obtained by replacing areas by perimeters. For this analogue, the following uniqueness result is known (see [4, p. 108]):

**Theorem 10.3.3** *A polytope in $\mathbb{R}^3$ is uniquely determined, up to translations, by the directions and the perimeters of its facets.*

Theorems 10.3.1 and 10.3.3 are similar uniqueness theorems which are both corollaries of a same general result by A.D. Alexandrov (see [4, Theorem 8, p. 107]). Thus, we are led to the natural question of the existence of an analogue to Theorem 10.3.2 for the existence of a polytope with prescribed directions and perimeters of the facets.

For convenience, we will restrict ourselves to 3-dimensional polytopes in $\mathbb{R}^3$. Recall that the dimension of a convex body in $\mathbb{R}^d$ is simply the dimension of its affine hull. Recall also that a facet of a 3-dimensional polytope $P$ is a (convex) polygonal face of $P$, and that its perimeter is defined to be the sum of the lengths of all its sides (edges).

**Difficulty of the Problem**

The above problem of prescribing the perimeters and outer unit normals of the facets of a 3-dimensional polytope has attracted the attention of geometers. Recently, a paper by V. Alexandrov highlighted its difficulty in explaining why a simple equation involving the prescribed perimeters cannot suffice to establish an analogue to Theorem 10.3.2 [3]. The main result of that paper reads as follows:

**Theorem   (V. Alexandrov [3])** *Let $n_1, \ldots, n_5$ in $\mathbb{R}^3$ be defined by the formulas*

$$n_1 := (0, 0, 1), \; n_2 := \left(\frac{1}{\sqrt{2}}, 0, \frac{1}{\sqrt{2}}\right), \; n_3 := \left(-\frac{1}{\sqrt{2}}, 0, \frac{1}{\sqrt{2}}\right),$$

$$n_4 := \left(0, \frac{1}{\sqrt{2}}, \frac{1}{\sqrt{2}}\right), \; n_5 := \left(0, -\frac{1}{\sqrt{2}}, \frac{1}{\sqrt{2}}\right).$$

*Let $\mathcal{L}(n_1, \ldots, n_5) \subset \mathbb{R}^5$ be the set of all points $(L_1, \ldots, L_5) \in \mathbb{R}^5$ with the following property: there exists a polytope $P \subset \mathbb{R}^3$ such that $n_1, \ldots, n_5$ (and no other vector) are the unit outward normals to the facets of $P$, and $L_k$ is the perimeter of the face with the outward normal $n_k$ for every $k \in \{1, \ldots, 5\}$. Then the set $\mathcal{L}(n_1, \ldots, n_5) \subset \mathbb{R}^5$ is not locally-analytic.*

This result was naturally interpreted by V. Alexandrov as a significant barrier to finding an existence theorem for a polytope with prescribed directions and perimeters of the facets. This was the major source of inspiration for the work presented in this subsection.

**Necessary Conditions for the Existence of a Solution**

Let $n_1, \ldots, n_N$ be a collection of $N$ pairwise distinct unit vectors linearly spanning $\mathbb{R}^3$ and let $L_1, \ldots, L_N$ be a collection of $N$ positive real numbers. The following set $\{(\mathbf{i}), (\mathbf{ii}), (\mathbf{iii}), (\mathbf{iv}), (\mathbf{v})\}$ of conditions is necessary for the existence of a 3-dimensional polytope $P$ in $\mathbb{R}^3$ having the $n_i$ as its facet outward unit normals and the $L_i$ as the corresponding facet perimeters.

(**i**) For each $i \in \{1, \dots, N\}$, there exists a decomposition of $L_i$ into a sum of $N$ non-negative real numbers,

$$L_i = \sum_{j=1}^{N} l_{ij},$$

in such a way that:

(**ii**) for all $(i, j) \in (\{1, \dots, N\})^2$, $l_{ij} = 0$ if $n_i$ and $n_j$ are colinear;

(**iii**) for all $(i, j) \in (\{1, \dots, N\})^2$, $l_{ji} = l_{ij}$.

In other words, conditions (**i**) through (**iii**) require the existence of a symmetric matrix with nonnegative entries $(l_{ij})_{1 \le i, j \le N}$, such that row $i$ sums to $L_i$, $(1 \le i \le N)$, and $l_{ij} = 0$ for colinear $n_i, n_j$, $(1 \le i, j \le N)$.

Indeed, if such a polytope $P$ exists, then denoting by $f_1, \dots, f_N$ the $N$ facets with respective unit outward normals $n_1, \dots, n_N$, the required relationships hold if we put:

$$l_{ij} := \begin{cases} 0 \text{ if } i = j \text{ or if } f_i \text{ and } f_j \text{ have no common edge} \\[2mm] \text{the length of the common edge otherwise.} \end{cases}$$

Our condition $(iv)$ is a consequence of the fact that the edge vectors of a facet (which are perpendicular to the unit normals of both incident facets), oriented in positive direction with respect to the unit normal of the facet, concatenate into a (simple) closed circuit:

$$(\textbf{iv}) \text{ For every } i \in \{1, \dots, N\}, \quad \sum_{j \in \{k \mid l_{ik} \neq 0\}} l_{ij} \left[ \frac{n_i \times n_j}{\sin(n_i, n_j)} \right] = 0,$$

where $\times$ denotes the cross product (here, we of course assume that $\mathbb{R}^3$ is oriented by its canonical basis), and $(n_i, n_j)$ denotes the length of the shortest arc of great circle joining $n_i$ to $n_j$ on the unit sphere $\mathbb{S}^2$ of $\mathbb{R}^3$ (recall that $n_i$ and $n_j$ are non-colinear by condition (**ii**) since $l_{ij} \neq 0$). Indeed, each facet $f_i$ of $P$ is a convex polygon the boundary of which is a closed polygonal line. Here, it is worth noting that, for all $j \in \{k \mid l_{ik} \neq 0\}$,

$$\overrightarrow{u_{i,j}} = (n_i \times n_j) / \sin(n_i, n_j)$$

is a unit vector that is such that the vector $\overrightarrow{v_{i,j}} := l_{ij}\overrightarrow{u_{i,j}}$ is of the form $\overrightarrow{MM'}$, where $M$ and $M'$ are two consecutive vertices of the oriented boundary of the face $f_i$ (see Fig. 10.4 where $v_1, \dots, v_m$ are the successive $\overrightarrow{v_{i,j}} := l_{ij}\overrightarrow{u_{i,j}}$, with $j \in \{k \mid l_{ik} \neq 0\}$).

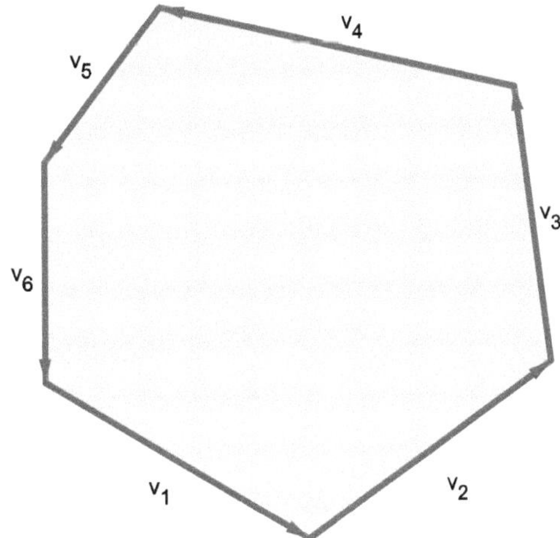

**Fig. 10.4** Illustration of the fact that $\displaystyle\sum_{j\in\{k\mid l_{ik}\neq 0\}} l_{ij}\,\overrightarrow{u_{ij}} = \sum_{l=1}^{m} v_l = 0$

Our last necessary condition $(v)$ will follow from Steinitz's theorem (e.g. see [197, Chapter 4, p. 103]), which characterizes in purely graph-theoretic terms those graphs that can be represented as the 1-skeleton of some 3-dimensional polytope:

**Theorem (Steinitz's Theorem)** *A graph can be represented as the 1-skeleton of some 3-dimensional polytope if, and only if, it is simple, planar, and 3-connected.*

For the convenience of the reader, we shall summarize some basic definitions and facts on graphs and 1-skeletons just before the proof of Theorem 10.3.4. From Steinitz's theorem, the following last condition is also necessary in our case:

$(v)$ The datum of the matrix $\left(l_{ij}\right)_{1\le i,j\le N}$ determines as follows a simple 3-connected planar graph $G$ drawn on the unit sphere $\mathbb{S}^2$ (so that no two of the edges intersect at a point other than a vertex): the vertices of $G$ are the unit vectors $n_1,\ldots,n_N$, and any pair of non-colinear vertices $\{n_i, n_j\}$ of $G$ is connected by an edge that is given by the shortest arc of great circle joining the two vertices on $\mathbb{S}^2$ if, and only if, $l_{ij} \neq 0$.

Note that the datum of $G$, drawn on $\mathbb{S}^2$, simply corresponds to that of the so-called **slope diagram representation** of the desired polytope $P$, say $SDR(P)$, in which facets, edges and vertices of $P$ are represented by points, spherical arcs and convex spherical polygons on $\mathbb{S}^2$. More precisely, in $SDR(P)$, each facet is represented by the end point of its outward unit normal vector, each edge is represented by the shortest arc of great circle joining the two points corresponding to the adjacent facets of the edge, and each vertex is represented by the convex spherical polygon bounded by the arcs corresponding to the edges of $P$ meeting at the vertex. In this subsection, we will also call $SDR(P)$ *the **spherical representation** of $P$.*

**The Main Result**

**Theorem 10.3.4** *Let $n_1, \ldots, n_N \in \mathbb{R}^3$ be $N$ distinct unit vectors linearly spanning $\mathbb{R}^3$ and let $L_1, \ldots, L_N$ be $N$ positive real numbers. There exists a 3-dimensional polytope $P$ in $\mathbb{R}^3$ having facet unit outward normals $n_1, \ldots, n_N$ and corresponding facet perimeters $L_1, \ldots, L_N$ if, and only if, the set of conditions $\{(\mathbf{i}), (\mathbf{ii}), (\mathbf{iii}), (\mathbf{iv}), (\mathbf{v})\}$ holds.*

**Important Remark** The set of conditions $\{(\mathbf{i}), (\mathbf{ii}), (\mathbf{iii}), (\mathbf{iv})\}$ is far from being sufficient to ensure that there exists a 3-dimensional polytope in $\mathbb{R}^3$ having facet unit outward normals $n_1, \ldots, n_N$ and corresponding facet perimeters $L_1, \ldots, L_N$. Indeed, many problems can arise if we drop condition $(\mathbf{v})$ from Theorem 10.3.4. If we retain only the first four conditions the data could correspond to a union of several polytopes. It is also possible for the data to be consistent with conditions $(\mathbf{i}) - (\mathbf{iv})$ while corresponding to non-convex polytopes, including such with non-convex facets, if one considers expanding the notion of unit outward normal. Of course, in our context, a reconstruction attempt as a convex polytope will then lead to problems. For instance, if we consider the unit vectors $n_1, \ldots, n_7 \in \mathbb{R}^3$ defined by $n_1 := (0, 0, -1)$, $n_2 := \left( \cos \frac{2\pi}{5}, \sin \frac{2\pi}{5}, 0 \right)$, $n_3 := \left( \cos \frac{6\pi}{5}, \sin \frac{6\pi}{5}, 0 \right)$, $n_4 := (1, 0, 0)$, $n_5 := \left( \cos \frac{4\pi}{5}, \sin \frac{4\pi}{5}, 0 \right)$, $n_6 := \left( \cos \frac{8\pi}{5}, \sin \frac{8\pi}{5}, 0 \right)$, $n_7 := (0, 0, 1)$, and the datum of the matrix

$$
\left( l_{ij} \right)_{1 \le i, j \le 7 :=}
\begin{pmatrix}
0 & 1 & 1 & 1 & 1 & 1 & 0 \\
1 & 0 & 1 & 0 & 0 & 1 & 1 \\
1 & 1 & 0 & 1 & 0 & 0 & 1 \\
1 & 0 & 1 & 0 & 1 & 0 & 1 \\
1 & 0 & 0 & 1 & 0 & 1 & 1 \\
1 & 1 & 0 & 0 & 1 & 0 & 1 \\
0 & 1 & 1 & 1 & 1 & 1 & 0
\end{pmatrix},
$$

for which the set of conditions $\{(\mathbf{i}), (\mathbf{ii}), (\mathbf{iii}), (\mathbf{iv})\}$ holds with $L_1 = L_7 = 5$ and $L_2 = L_3 = L_4 = L_5 = L_6 = 4$, we obtain a combinatorial prism. As per the given unit outward normals, the prism has parallel top and bottom pentagon facets, and five side facets with dihedral angle $\pi/5$ between adjacent sides. Connecting the side facets necessarily leads to self-intersections, thus there is no 3-dimensional polytope $P$ in $\mathbb{R}^3$ corresponding to these data. However, there exists a non-convex right prism with regular pentagram base (like the one shown in Fig. 10.5) that corresponds to these data with suitable adaptation of the notion of unit outward normal.

In other words, given $N$ distinct unit vectors $n_1, \ldots, n_N \in \mathbb{R}^3$ linearly spanning $\mathbb{R}^3$ and $N$ positive real numbers $L_1, \ldots, L_N$ such that the set of conditions $\{(\mathbf{i}), (\mathbf{ii}), (\mathbf{iii}), (\mathbf{iv})\}$ holds, we can of course associate a graph to the matrix $\left( l_{ij} \right)_{1 \le i, j \le N}$ but to be sure that this graph does correspond to the 1-skeleton of some 3-dimensional polytope, it is necessary and sufficient to assume that it satisfies the conditions of Steinitz's theorem. This is essentially what condition $(\mathbf{v})$ requires.

**Fig. 10.5** A pentagram

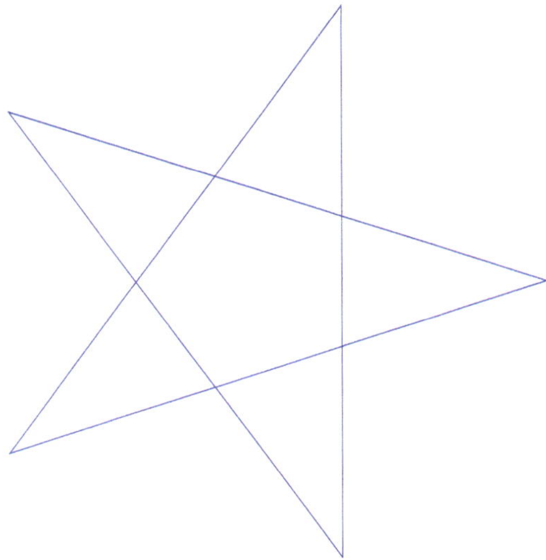

**Basic Definitions and Facts on Graphs and 1-Skeletons**

For the convenience of the reader, we summarize some basic definitions and facts on graphs and 1-skeletons:

- The 1-*skeleton* of a polytope $P$ is the graph whose vertices and edges are just the vertices and edges of $P$ with the same incidence relation.
- A graph is said to be *polyhedral* if it can be represented as the 1-skeleton of some 3-dimensional polytope.
- A graph is said to be *simple* if it contains neither multiple edges nor loops.
- A graph is said to be *planar* if it can be drawn in the plane so that no two of the edges intersect at a point other than a vertex.
- A graph is called 3-*connected* if it is connected, has at least 4 vertices, and remains connected whenever fewer than 3 vertices are removed.
- The dual of a polyhedral graph is also a polyhedral graph. More precisely, every polyhedral graph $G$ has a well-defined dual graph $G^*$ (independent of the plane embedding), corresponding to the 1-skeleton of the dual polytope.

***Proof of Theorem 10.3.4*** We have already seen that this set of conditions $\{(\mathbf{i}), (\mathbf{ii}), (\mathbf{iii}), (\mathbf{iv}), (\mathbf{v})\}$ is necessary for such a polytope $P$ to exist in $\mathbb{R}^3$.

Conversely assume that the set of conditions $\{(\mathbf{i}), (\mathbf{ii}), (\mathbf{iii}), (\mathbf{iv}), (\mathbf{v})\}$ holds. Recall that two polytopes $P$ and $P'$ are said to be ***combinatorially equivalent*** if there is a bijection between their faces that preserves the inclusion relation. It is well-known that the combinatorial structure of a 3-dimensional polytope $P$ is completely determined by its 1-skeleton [197, p. 105].

*(a) There exists a 3-dimensional polytope with the given combinatorial structure*
    By Steinitz's theorem, the simple 3-connected planar graph $G$ that is constructed on the sphere $\mathbb{S}^2$ in accordance with condition (**v**) and its geometric

dual graph $G^*$ (which is also simple, planar, and 3-connected) are polyhedral: they can be represented in $\mathbb{R}^3$ as the 1-skeletons of two dual 3-dimensional polytopes, say $Q$ and $Q^*$, respectively.

*(b)  The shape of one facet (resp. of one vertex-figure) can be chosen*

Moreover, by the following refinement by Barnette and Grünbaum, we can preassign the shape of a face of one of these two polytopes [15]:

**Theorem  (Barnette and Grünbaum)** *If one face of a 3-dimensional polytope $Q$ is an n-gon, then there exists a polytope $Q'$ combinatorially equivalent to $Q$, of which the corresponding face is any prescribed convex n-gon.*

Moreover, as noticed by Barnette and Grünbaum [15, p. 305]: "By an obvious application of duality, it follows from the theorem that the shape of one vertex-figure may be prescribed". In the part $(f)$ of the proof, we will make use of this dual form of the theorem to start our construction. Our desired polytope $P$ will be combinatorially equivalent to $Q^*$.

*(c)  Vocabulary convention*

In the remainder of the proof, the assembly of all the facets of a given 3-dimensional polytope that share a same vertex will be called a *corner* of the polytope. The *spherical representation* of such a corner, or of a 3-dimensional polytope, is defined as follows:

- A facet $f$ corresponding to a unit normal $n$ is represented on $\mathbb{S}^2$ by $n$;
- An edge is represented on $\mathbb{S}^2$ by the shortest arc of great circle joining the two points corresponding to the two adjacent facets of the edge;
- A vertex is represented on $\mathbb{S}^2$ by the convex spherical polygon that is bounded by the spherical arcs corresponding to the edges that are adjacent at the vertex.

*(d)  The polygons making up the desired facets are uniquely determined up to translations in space*

Our aim is to prove the existence of a 3-dimensional polytope $P$ that satisfies the set of conditions $\{(\mathbf{i}), (\mathbf{ii}), (\mathbf{iii}), (\mathbf{iv}), (\mathbf{v})\}$, and thus, the spherical representation of which is given by $G$.

Since condition $(\mathbf{iv})$ is satisfied, the Minkowski existence and uniqueness theorem for convex polygons ensures that, for any $i \in \{1, \ldots, N\}$, there exists in $n_i^{\perp}$ (the linear plane with unit normal $n_i$, endowed with the induced orientation) a positively oriented convex polygon $f_i$ whose edges $e_{ij}$ are directed by the unit vectors

$$\overrightarrow{u_{i,j}} = (n_i \times n_j) / \sin(n_i, n_j), \quad (j \in \{k \,|\, l_{ik} \neq 0\}),$$

and have corresponding lengths $l_{ij}$; and moreover, that this polygon $f_i$, of which the perimeter is $L_i$ by condition $(\mathbf{i})$, is unique up to translations in $n_i^{\perp}$. Note that in the above expression of $\overrightarrow{u_{i,j}}$, the vectors $n_i$ and $n_j$ are non-colinear by condition $(\mathbf{ii})$ since $l_{ij} \neq 0$.

Thus, *for any $i \in \{1, \ldots, N\}$, the desired oriented convex polygon $f_i$, with unit normal $n_i$, is well defined and unique up to translations in $\mathbb{R}^3$.*

*(e)  All the corners of the desired polytope are well defined and unique up to translations in space*

Now, let $\mathcal{P}_n$ be any positively oriented $n$-gon on $\mathbb{S}^2$ that is the oriented boundary of the closure of a connected component of the complementary of the graph $G$, which is drawn on $\mathbb{S}^2$ according to condition (**v**), ($n \geq 3$). Girard's theorem relates spherical angle excess and area of the spherical $n$-gon, which allows us to deduce that

$$\sum_{k=1}^{n} \alpha_k = (n - 2)\,\pi + Area\,(\mathcal{P}_n),$$

where $\alpha_1, \ldots, \alpha_n$ denote the interior angles of $\mathcal{P}_n$. Because of condition (**iv**), $\mathcal{P}_n$ has no reflex angle so that:

$$\sum_{k=1}^{n} \beta_k = 2\pi - Area\,(\mathcal{P}_n) < 2\pi,$$

where $\beta_1, \ldots, \beta_n$ denote the exterior angles of $\mathcal{P}_n$, (that is, $\beta_k := \pi - \alpha_k$ for all $k \in \{1, \ldots, n\}$). Now, for every for all $k \in \{1, \ldots, n\}$, the exterior angle $\beta_k$ at a vertex (say, $u_k$) of $\mathcal{P}_n$ can also be regarded as the interior angle of a convex polygon $f_k$ (which is the facet with unit outward normal $u_k$ of the desired polytope $P$) at the vertex of $f_k$ (say, $s_n$) that corresponds to $\mathcal{P}_n$. [This can be seen as follows. The two oriented sides of $\mathcal{P}_n$ that are adjacent at $u_k$ are part of two oriented great circles of $\mathbb{S}^2$, which are the oriented unit circles of two oriented linear planes of $\mathbb{R}^3$. For each of these two planes, consider the linear line that is orthogonal to the plane, and then the intersection point of this line with $\mathbb{S}^2$ that is on the same side of the plane as $\mathcal{P}_n$. By doing this, we obtain the end points of two unit vectors that are directing the two sides of $f_k$ adjacent at $s_n$ and that are pointing outward from these sides at $s_n$. Finally, we note that the geometric angle between these two unit vectors is nothing but the exterior angle $\beta_k$.].

The above inequality, which says that the sum $\sum_{k=1}^{n} \beta_k$ of the exterior angles of $\mathcal{P}_n$, is less than $2\pi$ can be regarded as a *nonnegative curvature condition* that is satisfied from our conditions. Therefore, taking into account condition (**iii**), the convex polygons $f_k$ that correspond to the unit vectors $u_k$ of which the end points are the vertices of $\mathcal{P}_n$ can be assembled (by gluing together their sides that correspond to a same edge of $\mathcal{P}_n$) to form a corner (of a 3-dimensional polytope), the spherical representation of which corresponds to $\mathcal{P}_n$ in $\mathbb{S}^2$. Here, the "convexity" at the corner is of course due to the nonnegative curvature condition.

Thus *all the corners of the desired polytope $P$ are well defined and unique up to translations in $\mathbb{R}^3$.*

*(f)* *They can be put together without contradiction*

Starting from any of these corners, we can construct by induction the desired polytope $P$, which is combinatorially equivalent to $Q^*$ and satisfies the set of conditions $\{(\mathbf{i}), (\mathbf{ii}), (\mathbf{iii}), (\mathbf{iv}), (\mathbf{v})\}$, by assembling at each step, an adjacent corner to the part of $P$, say $Part(P)$, that has already been constructed. Here, by "an adjacent corner" to $Part(P)$ we mean "a corner of $P$ that is not included in $Part(P)$ but that shares two facets with $Part(P)$". At each step, the spherical representation of the part of $P$ that is constructed is controlled by condition $(\mathbf{v})$, and the construction can continue until completion since we made sure that all the pieces had the required shape and dimensions.

It is worth noting that, as soon as the position of the first corner is fixed (with, for example, its vertex placed at the origin $O$), the position of any other vertex $S$ of $P$ is deduced from that of $O$ by a succession of translations from a vertex of $P$ to another: consider any succession of adjacent regions of $\mathbb{S}^2 \backslash G$ from that corresponding to $O$ to that corresponding to $S$ (two regions of $\mathbb{S}^2 \backslash G$ are said to be adjacent if their boundaries share an edge of $G$) and note that each crossing on $\mathbb{S}^2$ from one region $\mathcal{R}_i$ of $\mathbb{S}^2 \backslash G$ to an adjacent one $\mathcal{R}_j$ corresponds on $P$ to the translation from a vertex to another by a translation by a vector $\overrightarrow{v_{ij}}$ whose direction is determined by the arc of great circle $\gamma_{ij}$ separating the two regions on $\mathbb{S}^2$ ($\overrightarrow{v_{ij}}$ is orthogonal to it and oriented in the sense of the crossing) and whose norm $\left\| \overrightarrow{v_{ij}} \right\|$ is the length $l_{ij}$ corresponding to $\gamma_{ij}$. Of course, thanks to condition $(\mathbf{iv})$, the final position of the vertex (i.e. that of $S$) does not depend on the succession of adjacent regions that has been considered.

$\blacksquare$

**Remark** In higher dimensions, the problem should, of course, be much more difficult since there is no known analogue of ($k \geq 3$) Steinitz's theorem.

# Chapter 11
# List of Selected Problems

▶**Problem 2.6.1.** *Does there exist a generic projective hedgehog without any swallowtail in* $\mathbb{R}^3$? Such a hedgehog would be a model of the real projective plane parametrized by its Gauss map. See Sects. 2.2.2 and 2.6.3, and Sect. 10.2. Note that this problem as a discrete counterpart: see Sect. 2.6.3.

▶ *Is there a suitable notion of a* **Zindler surface in** $\mathbb{R}^3$? See Sect. 4.2.2.

▶**Subproblem of Problem 4.4.1.** *Does there exist a hyperbolic $C^k$-hedgehog in $\mathbb{R}^3$,* with $k \geq 3$? See Sect. 4.4.

▶*Existence of a* **Sturm-Hurwitz type theorem in higher dimensions?** *What could be the statement of a Sturm-Hurwitz type theorem in dimension 3 ?* See Sect. 4.6.

▶**The Minkowski problem for** $C^2$**-hedgehogs.**

($Q_1$) ***Existence of a $C^2$-solution:*** *What are necessary and sufficient conditions for a real continuous function $R$ on $\mathbb{S}^n$ to be the curvature function (that is, the inverse $\frac{1}{\kappa}$ of the Gauss curvature $\kappa$) of some $C^2$-hedgehog $\mathcal{H} = \mathcal{K}-\mathcal{L}$?*

($Q_2$) ***Uniqueness of*** **of a** *$C^2$-solution: If $R \in C(\mathbb{S}^n; R)$ is the curvature function of some hedgehog $\mathcal{H}$, what necessary and sufficient additional conditions must it satisfy in order that $\mathcal{H}$ be the unique hedgehog of which $R$ is the curvature function (up to an isometry of the space)?*

See Sects. 2.9 and 4.4, See Sect. 4.3.1, and Chap. 5. It would seem reasonable to start by considering the analytical case.

Note that the Minkowski problem for $C^2$-hedgehogs has a discrete counterpart. Finally, recall we solved the Christoffel-Minkowski problem for general planar hedgehogs in Sect. 4.7.2.

▶**Notion of complex curvature.** With the definitions given in Chap. 6, the complex circle $\mathcal{C}\left((a_0, a_1)\,;\, \rho\right)$ with equation $(X_1 - a_0)^2 + (X_2 - a_1)^2 = \rho^2$ in $\mathbb{C}^2$, where $\left(a_0, a_1, \rho\right) \in \mathbb{C}^2 \times \mathbb{C}^*$, can be locally regarded as a hedgehog with radius of curvature equal to $\rho$ (possibly after a suitable chart change on the Riemann

© The Author(s), under exclusive license to Springer Nature Switzerland AG 2026    349
Y. Martinez-Maure, *Hedgehog Theory*, Lecture Notes in Mathematics 2381,
https://doi.org/10.1007/978-3-032-04298-9_11

sphere). It would be interesting to better understand the meaning of this complex curvature.

▶**Concurrent normal conjecture.** Does every convex body in $n$-dimensional Euclidean space $\mathbb{R}^n$ has an interior point lying on normals through 2n distinct boundary points? See Sect. 9.2.

# References


1. Alexandrov, A.D.: On uniqueness theorem for closed surfaces (Russian). Doklady Akad. Nauk SSSR **22**, 99–102 (1939). English translation: Selected Works. Part 1: Selected Scientific Papers, pp. 149–153. Gordon and Breach Publishers, Amsterdam (1996)
2. Alexandrov, A.D.: Uniqueness theorems for surfaces in the large, I (Russian). Vestnik Leningrad. Univ. **11**, 5–17 (1956). English translation: Am. Math. Soc. Transl. **21**, 341–354 (1962)
3. Alexandrov, V.: Why there is no an existence theorem for a convex polytope with prescribed directions and perimeters of the faces? Abh. Math. Semin. Univ. Hamb. **88**, 247–254 (2018)
4. Alexandrov, A.D.: Convex Polyhedra. Springer Monographs in Mathematics, 539 pp. Springer, Berlin (2005)
5. Apéry, F.: Models of the real projective plane. Computer graphics of Steiner and Boy surfaces. Friedrich Vieweg & Sohn, Braunschweig/Wiesbaden (1987)
6. Arnol'd, V.I.: Critical points of smooth functions. In: Proc. int. Congr. Math. Vancouver 1974, vol. 1, pp. 19–39 (1975)
7. Arnol'd, V.I.: Méthodes mathématiques de la mécanique classique, Editions Mir. 470 pp. (1976). Translation into French of the Russian edition of 1974
8. Arnol'd, V.I.: Topological Invariants of Plane Curves and Caustics. AMS University Lecture Series, vol. 5, 60 pp. (1994)
9. Arnol'd, V.I.: Topological problems in wave propagation theory and topological economy principle in algebraic geometry. Proceedings in Honour to V. I. Arnold for his 60th birthday. AMS Fields Inst. Commun. **24**, 39–54 (1999)
10. Auerbach, H.: Sur un problème de M. Ulam concernant l'équilibre des corps flottants. Stud. Math. **7**, 121–142 (1938)
11. Banchoff, T., Gaffney, T., McCrory, C.: Cusps of Gauss Mappings. Research Notes in Math., vol. 55, 88 pp. (1981)
12. Bárány, I.: On the minimal ring containing the boundary of a convex body. Acta Sci. Math. **52**, 93–100 (1988)
13. Barbier, E.: Note sur le problème de l'aiguille et le jeu joint couvert. J. Math. Pures Appl. **5**, 273–286 (1860)
14. Bardet, M., Bayen, T.: On the degree of the polynomial defining a planar algebraic curves of constant width. arXiv 2013. https://arxiv.org/abs/1312.4358
15. Barnette, D., Grünbaum, B.: Preassigning the shape of a face. Pacific J. Math. **32**, 299–306 (1970)
16. Berger, M.: A Panoramic View of Riemannian Geometry, 824 pp. Springer, Berlin (2003)

17. Bieri, H.: Extremaleigenschaften rotationssymmetrischer Kegelstümpfe im gewöhnlichen Raum I; II. Elem. Math. **33** (1978). no. 1, pp. 7–14; no. 3, pp. 62–68

18. Blaschke, W.: Eine Frage über konvexe Körper. Jahresber. Deutsh. Math.-Verein. **25**, 121–125 (1916)

19. Blaschke, W.: Vorlesungen über Differentialgeometrie und geometrische Grundlagen von Einsteins Relativitätstheorie, vol. 3. Bearbeitet von G. Thomsen, J. Springer, Berlin (1929)

20. Blaschke, W., Rothe, H., Weitzenböck, R.: Aufgabe 552. Arch. Math. u. Phys. **27**, 82 (1917)

21. Bol, G.: Zur Theorie der konvexen Körper. Jahresber. Deutsche Math.-Ver. **49**, 113–123 (1939)

22. Bonnesen, T.: Über das isoperimetrische Defizit ebener Figuren. Math. Ann. **91**, 252–268 (1924)

23. Bonnesen, T., Fenchel, W.: Theory of Convex Bodies. BCS Associates, Moscow (1987). Translated from the German and edited by L. Boron, C. Christenson and B. Smith

24. Bracho, J., Montejano, L., Oliveros, D.: Carousels, Zindler curves and the floating body problem. Period. Math. Hung. **49**, 9–23 (2004)

25. Bricard, R.: Mémoire sur la théorie de l'octaèdre articulé. J. Math. Pures Appl. **3**, 113–150 (1897)

26. Calabi, E.: Book Review: The Minkowski multidimensional problem. Bull. Am. Math. Soc. (N.S.) **1**, 636–639 (1979)

27. Cannas da Silva, A.: Lectures on Symplectic Geometry. Lectures Notes in Mathematics, vol. 1764, 217 pp. Springer, Berlin (2001)

28. Cauchy, A.L.: Sur les polygones et polyèdres, second mémoire. J. Ecole Polytech. **19**, 87–98 (1813)

29. Cecil, T.E.: Lie Sphere Geometry. With Application to Submanifolds, 2nd edn. Universitext, 208 pp. Springer, New York (2008)

30. Chen, B.Y.: Pseudo-Riemannian Geometry, $\delta$-invariants and Applications. Word Scientific Publishing Co. Pte. Ltd., Hackensack (2011)

31. Cheng, S.Y., Yau, S.T.: On the regularity of the solution of the $n$-dimensional Minkowski problem. Commun. Pure Appl. Math. **29**, 495–516 (1976)

32. Cohn-Vossen, E.: Zwei Satze uber die Starrheit der Eilflchen. Nachr. Ges. Will. Gottingen, 125–134 (1927)

33. Connelly, R.: A counterexample to the rigidity conjecture for polyhedra. Inst. Haut. Etud. Sci., Publ. Math. **47**, 333–338 (1977)

34. Connelly, R., Sabitov, I., Walz, A.: The Bellows conjecture. Beitr. Algebra Geom. **38**, 1–10 (1997)

35. Curry, J., Ghrist, R., Robinson, M.: Euler calculus with applications to signals and sensing. In: Advances in Applied and Computational Topology. AMS short course on computational topology, New Orleans, 2011. Proceedings of Symposia in Applied Mathematics, vol. 70, pp. 75–145 (2012)

36. Dehn, M.: Über die Starrheit konvexer Polyeder. Math. Ann. **77**, 466–473 (1916)

37. Dumitrescu, A., Ebbers-Baumann, A., Grüne, A., Klein, R., Rote, G.: On the geometric dilation of closed curves, graphs, and point sets. Comput. Geom. **36**, 16–38 (2007)

38. Duval, J.: Around Brody lemma (2017). https://arxiv.org/pdf/1703.01850.pdf

39. Eggleston, H.G.: Convexity. Cambridge Tracts in Mathematics and Mathematical Physics, no. 47. Cambridge University Press, Cambridge (1958)

40. Epstein, C.L.: Envelopes of horospheres and Weingarten surfaces in hyperbolic 3-space. Preprint (1984)

41. Epstein, C.L.: The hyperbolic Gauss map and quasiconformal reflections. J. Reine Angew. Math. **372**, 96–135 (1986)

42. Epstein, C.L.: The asymptotic boundary of a surface imbedded in $\mathbb{H}^3$ with nonnegative curvature. Mich. Math. J. **34**, 227–239 (1987)

43. Falconer, K.: Fractal Geometry. Mathematical Foundations and Applications, 3rd edn., 368 pp. John Wiley & Sons, Hoboken (2014)

44. Fillastre, F.: Fuchsian convex bodies: basics of Brunn-Minkowski theory. Geom. Funct. Anal. **23**, 295–333 (2013)

45. Fillmore, J.P.: Barbier's theorem in the Lobachevski plane. Proc. Am. Math. Soc. **24**, 705–709 (1970)

46. Firey, W.: Christoffel's problem for general convex bodies. Mathematika **15**, 7–21 (1968)

47. Foote, R., Levi, M., Tabachnikov, S.: Tractrices, bicycle tire tracks, hatchet planimeters, and a 100-year-old conjecture. Am. Math. Mon. **120**, 199–216 (2013)

48. Francis, G., Morin, B.: Arnold Shapiro's eversion of the sphere. Math. Intell. **2**, 200–203 (1979)

49. Fujiwara, M.: On space curves of constant breadth. Tohoku Math. J. **5**, 180–184 (1914)

50. Fujiwara, M.: Über die Raumkurven konstenter Breite und ihre Beziehung mit der einseitigen Regelfläche. Tohoku Math. J. **8**, 1–10 (1915)

51. Fujiwara, M.: Über die Mittelkurve zweier geschlossenen konvexen Kurven in Bezug auf einen Punkt. Tôhoku Math. J. **10**, 99–103 (1916)

52. Gálvez, J.A., Mira, P.: Linearity of homogeneous solutions to degenerate elliptic equations in dimension three. J. Eur. Math. Soc. (2024). Published online first

53. Geppert, H.: Über den Brunn-Minkowskischen Satz. Math. Z. **42**, 238–254 (1937)

54. Gluck, H.: Manifolds with preassigned curvature - a survey. Bull. Am. Math. Soc. **81**, 313–329 (1975)

55. Görtler, H.: Erzeugung stützbarer Bereiche I. Deutsche Math. **2**, 454–456 (1937)

56. Görtler, H.: Erzeugung stützbarer Bereiche II. Deutsche Math. **3**, 189–200 (1937)

57. Grebennikov, A., Panina, G.: A note on the concurrent normal conjecture. Acta Math. Hungar. **167**, 529–532 (2022)

58. Grinberg, E., Zhang, G.; Convolutions, transforms, and convex bodies. Proc. Lond. Math. Soc. **78**, 73–115 (1999)

59. Groemer, H.: Minkowski addition and mixed volumes. Geom. Dedicata **6**, 141–163 (1977)

60. Grüne, A.: Geometric dilation and halving distance. PhD thesis, Rheinische Friedrich-Wilhelms-Universität Bonn (2008)

61. Guan, P., Wang, Z., Zhang, X.: A proof of Alexandrov's uniqueness theorem for convex surfaces in $\mathbb{R}^3$. Ann. Inst. H. Poincare Anal. Non Lineaire **33**, 329–336 (2016)

62. Hadwiger, H.: Altes und Neues über konvexe Körper. Birkhäuser, Basel (1955)

63. Han, Q., Nadirashvili, N., Yuan, Y.: Linearity of homogeneous order-one solutions to elliptic equations in dimension three. Commun. Pure Appl. Math. **56**, 425–432 (2003)

64. Heil, E.: Existenz eines 6-Normalenpunktes in einem konvexen Körper. Arch. Math. **32**, 412–416 (1979)

65. Heil, E.: Correction to 'Existenz eines 6-Normalenpunktes in einem konvexen Körper.' Arch. Math. **33**, 496 (1979/1980)

66. Heil, E.: Concurrent normals and critical points under weak smoothness assumptions. In: Discrete Geometry and Convexity, vol. 440, pp. 170–178. Ann. New York Acad. Sci., New York (1985)

67. Horváth, A.G.: Diameter, width and thickness in the hyperbolic plane. J. Geom. **112** (2021), Paper no. 47, 29 pp.

68. Hoschek, J.: Räumliche Zindlerkurven. Math. Balk. **4**, 253–260 (1974)

69. Hoschek, J.: Räumliche Zindlerkurven. Manuscr. Math. **13**, 175–185 (1974)

70. Ivaki, M.N.: The second mixed projection problem and the projection centroid conjectures. J. Funct. Anal. **272**, 5144–5161 (2017)

71. Ivaki, M.N.: A local uniqueness theorem for minimizers of Petty's conjectured projection inequality. Mathematika **64**, 1–19 (2018)

72. Izmestiev, I.: Infinitesimal rigidity of smooth convex surfaces through the second derivative of the Hilbert-Einstein functional. Diss. Math. **492** p. 58, (2013)

73. Izmestiev, I.: Infinitesimal rigidity of convex polyhedra through the second derivative of the Hilbert-Einstein functional. Can. J. Math. **66**, 785–825 (2014)

74. Izumiya, S., Romero Fuster, M.C., Ruas, M.A.S.: Differential Geometry From a Singularity Theory Viewpoint, 368 pp. World Scientific, Hackensack (2016)

75. Jerónimo-Castro, J., Jimenez-Lopez, F.G.: A characterization of the hyperbolic disc among constant width bodies. Bull. Kor. Math. Soc. **54**, 2053–2063 (2017)

76. Kallay, M.: Reconstruction of a plane convex body from the curvature of its boundary. Israel J. Math. **17**, 149–161 (1974)

77. Khovanskiĭ, A.G.: Newton polyhedra and genus of complete intersections. Funktsional. Anal. i Prilozhen. **12**, 51–61 (1978). Funct. Anal. Appl. **12**, 38–46 (1978)

78. Koutroufiotis, D.: On a conjectured characterization of the sphere. Math. Ann. **205**, 211–217 (1973)

79. Klee, V.: Can a plane convex body have two equichordal points ? Am. Math. Mon. **76**, 54–55 (1969)

80. Kritikos, N.: Über konvexe Flachen und einschließende Kugeln. Math. Ann. **96**, 583–586 (1927)

81. Langevin, R., Rosenberg, H.: A maximum principle at infinity for minimal surfaces and applications. Duke Math. J. **57**, 819–828 (1988)

82. Langevin, R., Levitt, G., Rosenberg, H.: Hérissons et multihérissons (enveloppes paramétrées par leur application de Gauss). In: Singularities, pp. 245–253. Banach Center Publ. 20, Warsaw, PWN (1988)

83. Leichtweiss, K.: Curves of constant width in the non-Euclidean geometry. Abh. Math. Semin. Univ. Hamb. **75**, 257–284 (2005)

84. Lin, C.S.: The local isometric embedding in $\mathbb{R}^3$ of 2-dimensional Riemannian manifolds with nonnegative curvature. J. Differ. Geom. **21**, 213–230 (1985)

85. Lindquist, N.: Support functions of central convex bodies. Portugaliae Math. **34**, 241–252 (1975)

86. Lonke, Y.: Derivatives of the $L^p$-cosine transform. Adv. Math. **176**, 175–186 (2003)

87. López, R.: Differential geometry of curves and surfaces in Lorentz-Minkowski space. Int. Electron. J. Geom. **7**, 44–107 (2014)

88. Lutwak, E.: Mixed projection inequalities. Trans. Am. Math. Soc. **287**, 91–106 (1985)

89. Mampel, K.L.: Über Zindlerkurven. J. Reine Angew. Math. **234**, 12–44 (1969)

90. Martinez-Maure, Y.: Feuilletages des Surfaces et Hérissons de $\mathbb{R}^3$, Thèse de $3^{ème}$ cycle. Université Paris 7 (1985)

91. Martinez-Maure, Y.: A note on the tennis ball theorem. Am. Math. Mon. **103**, 338–340 (1996)

92. Martinez-Maure, Y.: De nouvelles inégalités géométriques pour les hérissons. Arch. Math. **72**, 444–445 (1999)

93. Martinez-Maure, Y.: Indice d'un hérisson: étude et applications. Publ. Mat. **44**, 237–255 (2000)

94. Martinez-Maure, Y.: Hedgehogs and zonoids. Adv. Math. **158**, 1–17 (2001)

95. Martinez-Maure, Y.: Contre-exemple à une caractérisation conjecturée de la sphère. C. R. Acad. Sci. Paris, Sér. I **332**, 41–44 (2001)

96. Martinez-Maure, Y.: A fractal projective hedgehog. Demonstratio Math. **34**, 59–63 (2001)

97. Martinez-Maure, Y.: Théorie des hérissons et polytopes. C. R. Acad. Sci. Paris, Sér. I **336**, 241–244 (2003)

98. Martinez-Maure, Y.: Les multihérissons et le théorème de Sturm-Hurwitz. Arch. Math. **80**, 79–86 (2003)

99. Martinez-Maure, Y.: A Brunn-Minkowski theory for minimal surfaces. Illinois J. Math. **48**(2), 589–607 (2004)

100. Martinez-Maure, Y.: Geometric study of Minkowski differences of plane convex bodies. Can. J. Math. **58**, 600–624 (2006). https://doi.org/10.4153/CJM-2006-025-x

101. Martinez-Maure, Y.: Dérivation des surfaces convexes de $\mathbb{R}^3$ dans l'espace de Lorentz et étude de leurs focales. C. R. Acad. Sci. Paris, Sér. I **348**, 1307–1310 (2010)

102. Martinez-Maure, Y.: New notion of index for hedgehogs of $\mathbb{R}^3$ and applications. Eur. J. Combin. **31**, 1037–1049 (2010)

103. Martinez-Maure, Y.: Uniqueness results for the Minkowski problem extended to hedgehogs. Cent. Eur. J. Math. **10**, 440–450 (2012)

104. Martinez-Maure, Y.: Gauss rigidity and volume preservation under curvature-preserving deformations for hedgehogs. Results Math. **63**, 973–983 (2013)
105. Martinez-Maure, Y.: Hedgehog theory via Euler Calculus. Beitraege zur Algebra und Geometrie **56**, 397–421 (2015)
106. Martinez-Maure, Y.: Plane Lorentzian and Fuchsian hedgehogs. Can. Math. Bull. **58**, 561–574 (2015). https://doi.org/10.4153/CMB-2014-053-7
107. Martinez-Maure, Y.: Tout chemin générique de hérissons réalisant un retournement de la sphère dans $\mathbb{R}^3$ comprend un hérisson porteur de queues d'aronde positives. Publ. Math. **59**, 339–351 (2015)
108. Martinez-Maure, Y.: A stability estimate for the Aleksandrov-Fenchel inequality under regularity assumptions. Monatsh. Math. **182**, 65–76 (2017)
109. Martinez-Maure, Y.: New insights on marginally trapped surfaces: the hedgehog theory point of view. Adv. Appl. Math. **101**, 320–353 (2018). https://doi.org/10.1016/j.aam.2018.07.009
110. Martinez-Maure, Y.: Existence and uniqueness theorem for a 3-dimensional polytope in $\mathbb{R}^3$ with prescribed directions and perimeters of facets. Discrete Math. **343**, 111770, 6 pp. (2020). https://doi.org/10.1016/j.disc.2019.111770
111. Martinez-Maure, Y.: Real and complex hedgehogs, their symplectic area, curvature and evolutes. J. Symplectic Geom. **19**, 567–606 (2021). https://doi.org/10.4310/JSG.2021.v19.n3.a3
112. Martinez-Maure, Y.: Du vélo de Reuleaux aux courbes algébriques de largeur constante. Images des Mathématiques. Published on January 6 2022. https://images-archive.math.cnrs.fr/Du-velo-de-Reuleaux-aux-courbes-algebriques-de-largeur-constante.html
113. Martinez-Maure,Y.: Noncircular algebraic curves of constant width: an answer to Rabinowitz. Can. Math. Bull. **65**, 552–556 (2022). https://doi.org/10.4153/S0008439521000473
114. Martinez-Maure, Y.: On the concurrent normals conjecture for convex bodies. Mathematika **68**, 620–650 (2022). https://doi.org/10.1112/mtk.12128
115. Martinez-Maure, Y., Panina, G.: Singularities of virtual polytopes. J. Geom. **105**, 343–357 (2014)
116. Martinez-Maure, Y., Rochera, D.: Zindler-type hypersurfaces in $\mathbb{R}^4$. J. Geom. Phys. **182**, 104664 (2022)
117. Martinez-Maure, Y., Rochera, D.: Hedgehogs and their width in elliptic and hyperbolic planes. J. Geom. Anal. **35**, Paper no. 47, 35 pp. (2025). https://doi.org/10.1112/mtk.12128
118. Martini, H., Wu, S.: On Zindler curves in normed planes. Can. Math. Bull. **55**, 767–773 (2012)
119. Martini, H., Wu, S.: Classical curve theory in normed planes. Comput. Aided Geom. Des. **31**, 373–397 (2014)
120. Martini, H., Montejano, L., Oliveros, D.: Bodies of Constant Width. An Introduction to Convex Geometry with Applications. Birkhäuser, Cham (2019)
121. McDuff, D.: Symplectic structures—a new approach to geometry. Not. Am. Math. Soc. **45**, 952–960 (1998)
122. McDuff, D.: What is symplectic geometry?. In: European Women in Mathematics, pp. 33–53. World Sci. Publ., Hackensack (2010)
123. McDuff, D., Salomon, D.: $J$-holomorphic Curves and Symplectic Topology. Colloquium Publications, vol. 52. American Mathematical Society, Providence (2012)
124. McMullen, P.: The polytope algebra. Adv. Math. **78**, 76–130 (1989)
125. Minkowski, H.: Allgemeine Lehrsätze über die convexen Polyeder. Gött. Nachr., 198–219 (1897)
126. Minkowski, H.: Volumen und Oberfläche. Math. Ann. **57**, 447–495 (1903)
127. Minkowski, H.: Gesammelte Abhandlungen von Hermann Minkowski. Band I. Teubner, Leipzig (1911)
128. Monterde, J., Rochera, D.: Holditch's ellipse unveiled. Am. Math. Mon. **124**, 403–421 (2017)
129. Mooney, C.: Minimizers of convex functionals with small degeneracy set. Calc. Var. Partial Differ. Equ. **59**, Paper no. 74, 19 pp. (2020)
130. Münzner, H.F.: Über eine spezielle Klasse von Nabelpunkten und analoge Singularitäten in der zentroaffinen Flächentheorie. Comment. Math. Helv. **41**, 88–104 (1966–1967)

131. Münzner, H.F.: Über Flächen mit einer Weingartenschen Ungleichung. Math. Zeitschr. **97**, 123–139 (1967)

132. Myroshnychenko, S.: On a functional equation related to a pair of hedgehogs with congruent projections. J. Math. Anal. Appl. **445**, 1492–1504 (2017)

133. Nadirashvili, N., Tkachev, V., Vlädut, S.: Nonlinear Elliptic Equations and Nonassociative Algebras. Mathematical Surveys and Monographs, vol. 200. AMS, Providence (2014)

134. Nirenberg, L.: The Weyl and Minkowski problems in differential geometry in the large. Commun. Pure Appl. Math. **6**, 337–394 (1953)

135. Ohtsuka, H.: Dirichlet problems of Riemann surfaces and conformal mappings. Nagoya Math. J. **3**, 91–137 (1951)

136. Osserman, R.: Bonnesen-style isoperimetric inequalities. Am. Math. Mon. **86**, 1–29 (1979)

137. Osserman, R.: A Survey of Minimal Surfaces, 2nd edn. Dover Publ., New York (1986)

138. Ostrowski, A.: On the Morse-Kuiper theorem. Aequationes Math. **1**, 66–76 (1968)

139. Ovsienko, V., Tabachnikov, S.: Projective Differential Geometry Old and New. From the Schwarzian Derivative to the Cohomology of Diffeomorphism Groups. Cambridge Tracts in Mathematics, vol. 165. Cambridge University Press, Cambridge (2005)

140. Panina, G.: Rigidity and flexibility of virtual polytopes. Cent. Eur. J. Math. **1**, 157–168 (2003)

141. Panina, G.: New counterexamples to A. D. Alexandrov's hypothesis. Adv. Geom. **5**, 301–317 (2005)

142. Panina, G.: On hyperbolic virtual polytopes and hyperbolic fans. Cent. Eur. J. Math. **1**, 270–293 (2006)

143. Panraksa, C., Washington, L.C.: Real algebraic curves of constant width. Period. Math. Hung. **74**, 235–244 (2017)

144. Pardon, J.: Concurrent normals to convex bodies and spaces of Morse functions. Math. Ann. **352**, 55–71 (2012)

145. Petty, C.M., Crotty, J.M.: Characterization of spherical neighborhoods. Can. J. Math. **22**, 431–435 (1970)

146. Penrose, R.: Gravitational collapse and space-time singularities. Phys. Rev. Lett. **14**, 57–59 (1965)

147. Phillips, A.: Turning a sphere inside out. Sci. Am. **214**, 112–120 (1966)

148. Pogorelov, A.V.: The Minkowski Multidimensional Problem. John Wiley & Sons, Washington (1978). Russian original: 1975

149. Pogorelov, A.V.: Solution of one problem of A. D. Aleksandrov's problems. Dokl. Math. **57**, 398–399 (1998)

150. Pogorelov, A.V.: Uniqueness theorems for closed convex surfaces. Dokl. Math. **59**, 454–456 (1999)

151. Poston, T., Stewart, I.: Catastrophe Theory and Its Applications. Dover Publications, Inc., Mineola (1996). With an Appendix by D.R. Olsen, S.R. Carter, A. Rockwood, Reprint of the 1978 original

152. Pottmann, H.: Über Zindler-Kurven in euklidischen Räumen. Manuscr. Math. **60**, 217–220 (1988)

153. Proppe, H., Stancu, A., Stern, R.J.: On Holditch's theorem and Holditch curves. J. Convex Anal. **24**, 239–259 (2017)

154. Pukhlikov, A.V., Khovanskii, A.G.: Finitely additive measures of virtual polytopes. St. Petersbg. Math. J. **4**, 337–356 (1993)

155. Rabinowitz, S.: A polynomial curve of constant width. Missouri J. Math. Sci **9**, 23–27 (1997)

156. Rochera, D.: Algebraic equations for constant width curves and Zindler curves. J. Symb. Comput. **113**, 139–147 (2022)

157. Rochera, D.: Zindler curves in non-Euclidean geometry. J. Geom. Phys. **209**, Article ID 105402, 11 pp. (2025)

158. Rodriguez, L., Rosenberg, H.: Rigidity of certain polyhedra in $\mathbb{R}^3$. Comment. Math. Helv. **75**, 478–503 (2000)

159. Rosenberg, H., Toubiana, E.: Complete minimal surfaces and minimal herissons. J. Differ. Geom. **28**, 115–132 (1988)

160. Rychlik, M.: A complete solution to the equichordal point problem of Fujiwara, Blaschke, Rothe and R. Weitzenböck. Invent. Math. **129**, 141–212 (1997)
161. Sabitov, I.Kh.: The volume of a polyhedron as a function of length of its edges (Russian). Fundam. Prikl. Mat. **2**, 305–307 (1996)
162. Sangwine-Yager, J.R.: The missing boundary of the Blaschke Diagram. Am. Math. Mon. **96**, 233–237 (1989)
163. Santaló, L.A.: Convexidad en el plano hyperbolico. Univ. Nac. Tucumán Rev. Ser. A **19**, 173–183 (1969)
164. Schapira, P.: Operations on constructible functions. J. Pure Appl. Algebra **72**, 83–93 (1991)
165. Schneider, R.: Über eine Integralgleichung in der Theorie der konvexen Körper. Math. Nachr. **44**, 55–75 (1970)
166. Schneider, R.: Smooth approximation of convex bodies. Rend. Circ. Mat. Palermo. II. Ser. **33**, 436–440 (1984)
167. Schneider, R.: Convex Bodies: The Brunn-Minkowski Theory, 2nd expanded edn. Encyclopedia of Mathematics and its Applications. Cambridge University Press, Cambridge (2014)
168. Schneider, R.: The middle hedgehog of a planar convex body. Beitr. Algebra Geom. **58**, 235–245 (2017)
169. Schneider, R.: The typical irregularity of virtual convex bodies. J. Convex Anal. **25**, 103–118 (2018)
170. Schneider, R., Weil, W.: Zonoids and related topics. In: Convexity and Its Applications, pp. 296–317. Birkhäuser, Basel (1983)
171. Schoen, R.: Uniqueness, symmetry, and embeddedness of minimal surfaces. J. Differ. Geom. **18**, 791–809 (1983)
172. Šmakal, S.: Curves of constant breadth. Czechoslov. Math. J. 1 **23**(98), 86–94 (1973)
173. Smale, S.: A classification of immersions of the two-sphere. Trans. Am. Math. Soc. **90**, 281–290 (1959)
174. Small, A.: Linear structures on the collections of minimal surfaces in $\mathbb{R}^3$ and $\mathbb{R}^4$. Ann. Global Anal. Geom. **12**, 97–101 (1994)
175. Spivak, M.: A Comprehensive Introduction to Differential Geometry, vol. IV, 2nd edn. Publish Perish, Berkeley (1979)
176. Stadtmüller, C.: Metric contact manifolds and their Dirac operators. Diplomarbeit. Humboldt Universität zu Berlin, Institut für Mathematik (2011)
177. Stoker, J.J.: Differential Geometry. Reprint of the 1969 original. Wiley Classics Library. A Wiley-Interscience Publication. John Wiley & Sons, Inc., New York (1989)
178. Strubecker, K.: Differentialgeometrie des isotropen Raumes. V. Zur Theorie der Eilinien. Math. Z **51**, 525–573 (1949)
179. Sullivan, D.: Combinatorial invariants of analytic spaces. In: Proc. Liverpool Singularities I. Lecture Notes in Math., vol. 192, pp. 165–168 (1971)
180. Tabachnikov, S.: On the bicycle transformation and the filament equation: results and conjectures. J. Geom. Phys. **115**, 116–123 (2017)
181. Teissier, B.: Du théorème de l'index de Hodge aux inégalités isopérimétriques. C. R. Acad. Sci. Paris Sér. A-B, **288**, 287–289 (1979)
182. Thurston, W.P.: Three-Dimensional Geometry and Topology, vol. 1. Princeton Mathematical Series, vol. 35. Princeton University Press, Princeton (1997). Edited by Silvio Levy
183. Valentine, F.A.: Convex Sets. McGraw-Hill, New York (1964)
184. van den Dries, L.: Tame Topology and o-Minimal Structures. London Math. Soc. Lecture Notes Series, vol. 248. Cambridge University Press, Cambridge (1998)
185. Viro, O.: Some integral calculus based on Euler characteristic. In: Topology and Geometry. Rohlin Semin. 1984–1986. Lect. Notes Math., vol. 1346, pp. 127–138 (1988)
186. Wegner, B.: Globale Sätze über Raumkurven konstanter Breite. Math. Nachr. **53**(1–6), 337–344 (1972)

187. Wegner, B.: Verallgemeinerte Zindlerkurven in euklidischen Räumen. Math. Balk. **6**, 298–306 (1976). https://arxiv.org/abs/physics/0701241v3
188. Wegner, F.: Floating bodies of equilibrium in 2D, the tire track problem and electrons in a parabolic magnetic field (2007). https://arxiv.org/pdf/physics/0701241
189. Weil, W.: Blaschkes Problem der lokalen Charakterisierung von Zonoiden. Arch. Math. **29**, 655–659 (1977)
190. Wendl, C.: Lectures on Holomorphic Curves in Symplectic and Contact Geometry (2014). https://arxiv.org/pdf/1011.1690.pdf
191. Weyl, H.: Über die Starrheit der Eiflächen und konvexer Polyeder. Berl. Ber. (1917), 250–266
192. Wikipedia web page on the 'Reuleaux triangle'. https://en.wikipedia.org/wiki/Reuleaux_triangle
193. Wikipedia web page on the 'Hopf link'. https://en.m.wikipedia.org/wiki/Hopf_link
194. Wirsing, E.: Zur Analytizität von Doppelspeichkurven. Arch. Math. **9**, 300–307 (1958)
195. Wunderlich, W.: Algebraische Beispiele ebener und räumlicher Zindler-Kurven. Publ. Math. Debr. **24**, 289–297 (1977)
196. Zamfirescu, T.: Points on infinitely many normals to convex surfaces. J. Reine Angew. Math. **350**, 183–187 (1984)
197. Ziegler, G.M.: Lectures on Polytopes. Graduate Texts in Mathematics, vol. 152, 370 pp. Springer, Berlin (1995)
198. Zindler, K.: Über konvexe Gebilde II. Monatshefte Math. Phys. **31**, 25–56 (1921)
199. Zuily, C.: Existence locale de solutions $C^\infty$ pour des équations de Monge-Ampère changeant de type. Commun. Partial Differ. Equ. **14**, 691–697 (1989)

# Index

*LECTURE NOTES IN MATHEMATICS*     Springer

Editors in Chief: J.-M. Morel, B. Teissier;

**Editorial Policy**

1. Lecture Notes aim to report new developments in all areas of mathematics and their applications – quickly, informally and at a high level. Mathematical texts analysing new developments in modelling and numerical simulation are welcome.

   Manuscripts should be reasonably self-contained and rounded off. Thus they may, and often will, present not only results of the author but also related work by other people. They may be based on specialised lecture courses. Furthermore, the manuscripts should provide sufficient motivation, examples and applications. This clearly distinguishes Lecture Notes from journal articles or technical reports which normally are very concise. Articles intended for a journal but too long to be accepted by most journals, usually do not have this "lecture notes" character. For similar reasons it is unusual for doctoral theses to be accepted for the Lecture Notes series, though habilitation theses may be appropriate.

2. Besides monographs, multi-author manuscripts resulting from SUMMER SCHOOLS or similar INTENSIVE COURSES are welcome, provided their objective was held to present an active mathematical topic to an audience at the beginning or intermediate graduate level (a list of participants should be provided).

   The resulting manuscript should not be just a collection of course notes, but should require advance planning and coordination among the main lecturers. The subject matter should dictate the structure of the book. This structure should be motivated and explained in a scientific introduction, and the notation, references, index and formulation of results should be, if possible, unified by the editors. Each contribution should have an abstract and an introduction referring to the other contributions. In other words, more preparatory work must go into a multi-authored volume than simply assembling a disparate collection of papers, communicated at the event.

3. Manuscripts should be submitted either online at www.editorialmanager.com/lnm to Springer's mathematics editorial in Heidelberg, or electronically to one of the series editors. Authors should be aware that incomplete or insufficiently close-to-final manuscripts almost always result in longer refereeing times and nevertheless unclear referees' recommendations, making further refereeing of a final draft necessary. The strict minimum amount of material that will be considered should include a detailed outline describing the planned contents of each chapter, a bibliography and several sample chapters. Parallel submission of a manuscript to another publisher while under consideration for LNM is not acceptable and can lead to rejection.

4. In general, **monographs** will be sent out to at least 2 external referees for evaluation.

   A final decision to publish can be made only on the basis of the complete manuscript, however a refereeing process leading to a preliminary decision can be based on a pre-final or incomplete manuscript.

   Volume Editors of **multi-author works** are expected to arrange for the refereeing, to the usual scientific standards, of the individual contributions. If the resulting reports can be

forwarded to the LNM Editorial Board, this is very helpful. If no reports are forwarded
or if other questions remain unclear in respect of homogeneity etc, the series editors may
wish to consult external referees for an overall evaluation of the volume.

5. Manuscripts should in general be submitted in English. Final manuscripts should contain
   at least 100 pages of mathematical text and should always include

   - a table of contents;
   - an informative introduction, with adequate motivation and perhaps some historical
     remarks: it should be accessible to a reader not intimately familiar with the topic
     treated;
   - a subject index: as a rule this is genuinely helpful for the reader.
   - For evaluation purposes, manuscripts should be submitted as pdf files.

6. Careful preparation of the manuscripts will help keep production time short besides
   ensuring satisfactory appearance of the finished book in print and online. After
   acceptance of the manuscript authors will be asked to prepare the final LaTeX source
   files (see LaTeX templates online: https://www.springer.com/gb/authors-editors/book-
   authors-editors/manuscriptpreparation/5636) plus the corresponding pdf- or zipped ps-
   file. The LaTeX source files are essential for producing the full-text online version of
   the book, see http://link.springer.com/bookseries/304 for the existing online volumes
   of LNM). The technical production of a Lecture Notes volume takes approximately 12
   weeks. Additional instructions, if necessary, are available on request from lnm@springer.
   com.

7. Authors receive a total of 30 free copies of their volume and free access to their book on
   SpringerLink, but no royalties. They are entitled to a discount of 33.3 % on the price of
   Springer books purchased for their personal use, if ordering directly from Springer.

8. Commitment to publish is made by a *Publishing Agreement*; contributing authors of
   multiauthor books are requested to sign a *Consent to Publish form*. Springer-Verlag
   registers the copyright for each volume. Authors are free to reuse material contained in
   their LNM volumes in later publications: a brief written (or e-mail) request for formal
   permission is sufficient.

**Addresses:**
Professor Jean-Michel Morel, CMLA, École Normale Supérieure de Cachan, France
E-mail: moreljeanmichel@gmail.com

Professor Bernard Teissier, Equipe Géométrie et Dynamique,
Institut de Mathématiques de Jussieu – Paris Rive Gauche, Paris, France
E-mail: bernard.teissier@imj-prg.fr

Springer: Ute McCrory, Mathematics, Heidelberg, Germany,
E-mail: lnm@springer.com